ADAPTATION IN DYNAMICAL SYSTEMS

In the context of this book, adaptation is taken to mean a feature of a system aimed at achieving the best possible performance when mathematical models of the environment and the system itself are not fully available. This has applications ranging from theories of visual perception and the processing of information to the more technical problems of friction compensation and adaptive classification of signals in fixed-weight recurrent neural networks.

Largely devoted to the problems of adaptive regulation, tracking and identification, this book presents a unifying system-theoretic view on the problem of adaptation in dynamical systems. Special attention is given to systems with nonlinearly parametrized models of uncertainty. Concepts, methods, and algorithms given in the text can be successfully employed in wider areas of science and technology. The detailed examples and background information make this book suitable for a wide range of researchers and graduates in cybernetics, mathematical modeling, and neuroscience.

IVAN TYUKIN is an RCUK Academic Fellow in the Department of Mathematics, University of Leicester. His research and scientific interests cover many areas, including the analysis, modeling, and synthesis of systems with fragile, nonlinear, chaotic, and meta-stable dynamics.

ADAPTATION IN DYNAMICAL SYSTEMS

IVAN TYUKIN
University of Leicester and
Saint-Petersburg State Electrotechnical University

CAMBRIDGE
UNIVERSITY PRESS

CAMBRIDGE
UNIVERSITY PRESS

University Printing House, Cambridge CB2 8BS, United Kingdom

One Liberty Plaza, 20th Floor, New York, NY 10006, USA

477 Williamstown Road, Port Melbourne, VIC 3207, Australia

314-321, 3rd Floor, Plot 3, Splendor Forum, Jasola District Centre, New Delhi - 110025, India

103 Penang Road, #05-06/07, Visioncrest Commercial, Singapore 238467

Cambridge University Press is part of the University of Cambridge.

It furthers the University's mission by disseminating knowledge in the pursuit of education, learning and research at the highest international levels of excellence.

www.cambridge.org
Information on this title: www.cambridge.org/9780521198196

First published 2011

A catalogue record for this publication is available from the British Library

Library of Congress Cataloging in Publication data
Tyukin, Ivan.
Adaptation in Dynamical Systems / Ivan Tyukin.
p. cm
Includes bibliographical references and index.
ISBN 978-0-521-19819-6 (hardback)
1. Dynamics. 2. Control theory–Mathematical models.
3. Neurosciences–Mathematics. I. Title.
QA845.T94 2011
515′.39–dc22 2010051427

ISBN 978-0-521-19819-6 Hardback

Contents

Preface

Adaptation is amongst the most familiar and wide spread phenomena in nature. Since the early days of the nineteenth century it has puzzled researchers in broad areas of science. Since it had often been observed in responsive behaviors of biological systems, adaptation was initially understood as a regulatory mechanism that helps an animal to survive in a changing environment. Later the notion of adaptation was adopted in wider fields of science and engineering.

As a theoretical discipline it began to emerge as a branch of control theory during the first half of the twentieth century. Its beginning was marked by publications discussing basic principles of adaptation and its merits for engineering. Imprecise technology and mechanisms were, perhaps, amongst the strongest practical motivations for such a theory at that time. Various notions of adaptation were adopted by engineers and theoreticians in order to grasp, understand, and implement relevant features of this phenomenon in practice. The first applications of the new theory were simple schemes for extremal control of mechanical systems; these systems could be described by just a few linear ordinary differential equations. Since then adaptive controllers have evolved to encompass substantially more complex devices. The controlling devices themselves can now be viewed as nonlinear dynamical systems with specific input–output properties. Methods for the design and analysis of such systems are currently recognized by many in terms of the theory of adaptive control and systems identification.

Because the initial motivation to develop a theory of adaptation was driven mainly by the demands of mechanical engineering and the need for robust design of otherwise imprecise machines, the domain of application of the theories of adaptation was naturally restricted to the realm of artificial devices and engineering. The focus of the developing theory was restricted, in particular, to the problems of control of a relatively narrow class of well-studied and modeled mechanical systems, many of which were stable in the Lyapunov sense, for which the values of some parameters and variables are unknown and cannot be measured explicitly.

Yet, the potential role of the theory of adaptation was much wider and broader. It has become evident recently that there exists a demand for a systematic theory of adaptation outside of the domain of engineering.

Understanding basic mechanisms and principles of adaptation and regulation is recognized as relevant in physics, chemistry, biology, and brain sciences (Sontag 2004; Fradkov 2005). Because of the huge complexity of the phenomena studied in these domains, using the standard language of each particular science for systematic studies of the phenomenon of adaptation might not be adequate. Therefore in these areas system-theoretic views, irrespective of the particular subjects of study, have exceptional potential.

Apart from in the natural sciences, the needs for further development of the theory of adaptation are evident in handling complex artificial systems. This is especially true when changes in the working environment cannot be predicted a priori or there is a substantial degree of uncertainty about the system's internal state. Although there is a large literature on the theory of adaptive systems, both in the theoretical and in the applied domain, there are several issues preventing explicit application of classical recipes of adaptive control in these fields. These issues with classical schemes are

- the necessity to have a precise mathematical model of a controlled system,
- the requirement that models of uncertainties are linear or convex with respect to unknown or uncertain variables,
- the assumption that the target dynamics is stable in the Lyapunov sense,
- the assumption that a corresponding Lyapunov function for the target motions is available (Sastry and Bodson 1989; Narendra and Annaswamy 1989; Krstić *et al.* 1995; Ljung 1999; Eykhoff 1975; Bastin *et al.* 1992; Fradkov *et al.* 1999).

Every one of these requirements alone limits the role of the existing theory of adaptive systems in solving relevant problems in science. Altogether they constitute the "standard" approach which applies to several canonical cases, which are limited even within the realm of engineering.

The purpose of this work is to contribute towards extending the existing theory of adaptation and adaptive control beyond the scope of its usual applications in engineering to new and non-conventional areas, such as neuroscience and mathematical modeling of biological systems. It is hoped that this extension will create additional opportunities for control theorists to apply their expertise in novel and still developing fields of science; it will also help to expand the synthetic and analytical functions of systems and control theory into the natural sciences.

The focus of this book on the analysis of possible adaptation mechanisms in systems with nonlinear parametrization and unstable target dynamics was influenced by the author's work in the Laboratory for Perceptual Dynamics, RIKEN Brain Science Institute, Japan from November 2001 to March 2007. Neural systems of living organisms, and ultimately the human brain, were the source of inspiration. It became clear very quickly that the standard tools and methods in the arsenal of conventional adaptive control theory do not offer an acceptable explanation for the versatility and robustness of neural systems working in an uncertain environment. The aim therefore was to enhance the theory by making it suitable for the analysis and synthesis of adaptive schemes for nonlinear dynamical systems:

- with potentially Lyapunov-unstable and non-equilibrium target dynamics;
- when explicit definition of the target sets is not possible;
- using minimal, *qualitative*, macro-information about the system, and also allowing substantial uncertainty about the specific mathematical model of the system;
- allowing uncertainty models that are maximally adequate to describe the physical laws of processes and phenomena in the system.

The necessary ingredients of this extended theory of adaptation follow naturally from the logic of its development: from basic principles of the system's organization in the presence of uncertainties to specific laws of regulation. These ingredients include

(1) *methods for analysis* of basic input–output properties of the nonlinear systems; they should allow incomplete knowledge of equations describing the system dynamics; and they also should apply both to stable and to unstable systems;
(2) *principles and methods* of adaptation to disturbances that are unknown a priori and unavailable for measurement; the principles should rely exclusively on the fundamental physical properties of the systems considered; and the adaptation mechanisms should be able to realize these principles using adequate physical models of uncertainties and requiring a minimal amount of measurement information.

The following topics received particular attention: analysis of the completeness, realizability, and state boundedness of interconnections of uncertain dynamical systems; conditions ensuring convergence of the system's state to the target sets and their neighborhoods; designing laws of adaptive regulation and parameter estimation of nonlinearly parametrized models; characterizing the quality of the transient dynamics in systems with uncertainties; and parametric, signal/functional perturbations. In order to provide the reader with the necessary background and also to support our own argumentation a brief review of major classical concepts of adaptation is included.

The content of the book is based largely on the work I had the privilege to carry out together with my colleagues and co-authors.[1] The structure of the book can be summarized as follows. The text is organized into three large parts. The first part (Chapters 1–3) contains mainly introductory and preliminary results. Proofs of lemmas and theorems presented in this introductory part are kept within the main text.

In Chapter 1 we provide an informal discussion of the notion of adaptation followed by an overview of the range of specific problems considered in the text.

Chapter 2 contains background and preliminary results such as basic notions of stability, a very brief introduction to the method of Lyapunov functions, and a particularly important result on the exponential stability of the origin for a class of linear systems of ordinary equations with skew-symmetric matrices.

In Chapter 3 we review and analyze conventional approaches to the problem of adaptive control of nonlinear systems. We formulate the main theoretical and practical issues arising in these standard approaches (Fomin *et al.* 1981; Fradkov 1990; Narendra and Annaswamy 1989; Krstić *et al.* 1995) and their mathematical statements of the problem. These issues include the ambiguity of standard mathematical notions of an adaptive system, performance measures, limitations on defining the system's target sets,[2] restricted classes of the uncertainty models, and requirements for precise knowledge of the mathematical model of a system.

The second part, Chapters 4 and 5, presents the main theoretical results developed in the monograph. In order to preserve the integrity of the text, proofs of statements formulated in this part are given in appendices at the ends of these chapters.

In Chapter 4 we consider nonlinear systems defined in terms of their "input-to-output" and "input-to-state" characterizations given by mappings, or operators in functional, L_p, spaces. We introduce mathematical tools for the analysis of interconnections of dynamical systems with input–output (input–state) operators that are locally bounded in state and provide a formal statement of the problem for functional synthesis of an adaptive system. We demonstrate how this problem can be solved. The solution to the problem of functional synthesis of an adaptive system allows us to formulate various principles of its organization at the macroscopic level: the separation principle, the bottle-neck principle, and the emergence of weakly attracting sets in the interconnections of systems with contracting and wandering dynamics. The latter result is based on Tyukin *et al.* (2008a).

[1] This includes earlier texts such as Tyukin and Terekhov (2008).

[2] One of the most severe restrictions is the requirement for the target dynamics to be globally stable in the sense of Lyapunov. In addition, there is a necessity to specify target sets of the adaptive system a priori. The latter condition either requires prior identification of the system, which contradicts the very essence of adaptive behavior, or leads to enforcing motions that are not necessarily inherent and, hence, optimal to a physical system itself.

In Chapter 5 we utilize the principles derived in the previous chapter in order to provide an adequate statement of the problem of adaptive control and regulation of nonlinear dynamical systems. Its distinctive features are that the uncertainty models are allowed to be nonlinearly parametrized, mathematical models of the system need not be known precisely, the target dynamics is not restricted exclusively to globally Lyapunov-stable motions, and the target sets could be defined implicitly – as invariant sets of an auxiliary dynamical system. Generally, the problem is stated as that of *regulating the influence of uncertainties on the target dynamics to some functional space*. This allows one to refrain from explicit use of the method of Lyapunov functions and, hence, avoid its limitations.

We also consider several specific problems that have substantial theoretical and practical interest:

- adaptive regulation to invariant sets;
- adaptation in interconnected systems;
- state and parameter inference for systems with nonlinear parametrization of uncertainty.

In order to solve these problems two synthesis strategies were developed: the method of the *virtual adaptation algorithm* presented in Tyukin *et al.* (2007b) and the strategy based on purposeful introduction of unstable attracting sets into the system's state space (Tyukin *et al.* 2008a).

In the third part of the book (Chapters 6–8) we illustrate how the theory can be used to solve a number of practical problems of control, processing of information, and identification in mechanics, experimental biophysics, and computer and cognitive science. In particular, we consider the problem of adaptive classification in neural networks with fixed weights, the problem of identifying the dynamics of neuronal cells, and the problem of invariant recognition of spatially distributed information. We discuss why existing techniques cannot be successfully applied to solve these problems, or their application yields practically inefficient outcomes. The content of this part is based on Tyukin *et al.* (2008b), Tyukin *et al.* (2009), and Fairhurst *et al.* (2010).

This book would never have seen the light of day without the continuous support, help, and encouragement I received from many people with whom I have had the honor of working. I would like to express my deep gratitude to Professor V. A. Terekhov, my teacher, friend, and co-author, for his help, fruitful and motivating discussions of the philosophical foundations of the problem of adaptation, and unlimited patience. I am grateful to my colleagues and co-authors Cees van Leeuwen, Danil Prokhorov, Henk Nijmeijer, Erik Steur, David Fairhurst, Alexey Semyanov, and Inseon Song who contributed to the development of the ideas in the monograph. I am grateful to Dr Steven Holt and his colleagues for proof-reading and editing the monograph at the final stage of production. Finally, I am indebted to my dear wife Tanya, who contributed to the applied side of the project, assisted with the artwork, and also helped me enormously to summarize the results during the later stage of the production of the manuscript. My own personal role was limited to mere listening, interpretation, and writing. As is unfortunately the case in scientific endeavors, errors are inevitable companions. Even though I tried to avoid these unwelcome companions, my own journey is unlikely to be an exception, for which I fully accept sole responsibility. I would therefore be extremely grateful to readers, should they wish to help by contacting me when an error is found.

Notational conventions

Throughout the text the following notational conventions apply.

- Symbol $\mathbb{R}$ defines the field of real numbers and $\mathbb{R}_{\geq c} = \{x \in \mathbb{R} | x \geq c\}$; $\mathbb{N}$ defines the set of natural numbers; and $\mathbb{Z}$ denotes the set of whole numbers or integers.
- Symbol $\mathbb{R}^n$ stands for an n-dimensional linear space over the field of reals.
- $\mathcal{C}^k$ denotes the space of functions that are at least k times differentiable.
- Symbol $\mathcal{K}$ denotes the class of all strictly increasing functions $\kappa : \mathbb{R}_{\geq 0} \to \mathbb{R}_{\geq 0}$ such that $\kappa(0) = 0$; symbol $\mathcal{K}_\infty$ denotes the class of all functions $\kappa \in \mathcal{K}$ such that $\lim_{s\to\infty} \kappa(s) = \infty$.
- Let Ω be a set, then by $\mathcal{S}\{\Omega\}$ we denote the set of all subsets of Ω.
- $\|\mathbf{x}\|$ denotes the Euclidian norm of $\mathbf{x} \in \mathbb{R}^n$.
- The notation $|\cdot|$ stands for the absolute value of a scalar.
- The notation $\text{sign}(\cdot)$ denotes the signum function.
- By $L^n_p[t_0, T]$, where $t_0 \geq 0$, $T \geq t_0$, $p \geq 1$, we denote the space of all functions $\mathbf{f} : \mathbb{R}_{\geq 0} \to \mathbb{R}^n$ such that

$$\|\mathbf{f}\|_{p,[t_0,T]} = \left(\int_{t_0}^{T} \|\mathbf{f}(\tau)\|^p \, d\tau \right)^{1/p} < \infty.$$

- The notation $\|\mathbf{f}\|_{p,[t_0,T]}$ denotes the $L^n_p[t_0, T]$-norm of $\mathbf{f}(t)$.
- By $L^n_\infty[t_0, T]$, $t_0 \geq 0$, $T \geq t_0$, we denote the space of all functions $\mathbf{f} : \mathbb{R}_{\geq 0} \to \mathbb{R}^n$ such that

$$\|\mathbf{f}\|_{\infty,[t_0,T]} = \text{ess sup}\{\|\mathbf{f}(t)\|, t \in [t_0, T]\} < \infty,$$

  and $\|\mathbf{f}\|_{\infty,[t_0,T]}$ stands for the $L^n_\infty[t_0, T]$-norm of $\mathbf{f}(t)$.
- Let $\mathcal{A}$ be a set in $\mathbb{R}^n$, $\mathbf{x} \in \mathbb{R}^n$, and let $\|\cdot\|$ be the usual Euclidean norm in $\mathbb{R}^n$. By the symbol $\|\cdot\|_{\mathcal{A}}$ we denote the following induced norm:

$$\|\mathbf{x}\|_{\mathcal{A}} = \inf_{\mathbf{q}\in\mathcal{A}} \{\|\mathbf{x} - \mathbf{q}\|\}.$$

- Let $\Delta \in \mathbb{R}_{\geq 0}$, then the notation $\|\mathbf{x}\|_{\mathcal{A}_\Delta}$ stands for the following equality:

$$\|\mathbf{x}\|_{\mathcal{A}_\Delta} = \begin{cases} \|\mathbf{x}\|_{\mathcal{A}} - \Delta, & \|\mathbf{x}\|_{\mathcal{A}} > \Delta, \\ 0, & \|\mathbf{x}\|_{\mathcal{A}} \leq \Delta. \end{cases}$$

- The symbol $\|\cdot\|_{\mathcal{A}_\infty,[t_0,t]}$ is defined as follows:

$$\|\mathbf{x}(\tau)\|_{\mathcal{A}_\infty,[t_0,t]} = \sup_{\tau\in[t_0,t]} \|\mathbf{x}(\tau)\|_{\mathcal{A}}.$$

- Let $\mathbf{f} : \mathbb{R}^n \to \mathbb{R}^m$ be given. The function $\mathbf{f}(\mathbf{x}) : \mathbb{R}^n \to \mathbb{R}^m$ is said to be locally bounded if for any $\|\mathbf{x}\| < \delta$, $\delta \in \mathbb{R}_{>0}$ there exists a constant $D(\delta) > 0$ such that $\|\mathbf{f}(\mathbf{x})\| \leq D(\delta)$.
- Let Γ be an $n \times n$ square matrix, then $\Gamma > 0$ denotes a positive definite (symmetric) matrix. (Γ^{-1} is the inverse of Γ). By $\Gamma \geq 0$ we denote a positive semi-definite matrix.
- We reserve $\|\mathbf{x}\|_\Gamma^2$ to denote the quadratic form $\mathbf{x}^T\Gamma\mathbf{x}$, where $\mathbf{x} \in \mathbb{R}^n$ and $\mathbf{x}^T$ is the transpose of $\mathbf{x}$.
- Symbols $\lambda_{\min}(\Gamma)$ and $\lambda_{\max}(\Gamma)$ stand for the minimal and maximal eigenvalues of Γ, respectively.
- By the symbol I we denote the identity matrix.
- The solution of a system of differential equations $\dot{\mathbf{x}} = \mathbf{f}(\mathbf{x}, t, \boldsymbol{\theta}, \mathbf{u}(t))$, $\mathbf{u} : \mathbb{R}_{\geq 0} \to \mathbb{R}^m$, $\boldsymbol{\theta} \in \mathbb{R}^d$ passing through point $\mathbf{x}_0$ at $t = t_0$ will be denoted for $t \geq t_0$ as $\mathbf{x}(t, \mathbf{x}_0, t_0, \boldsymbol{\theta}, \mathbf{u})$, or simply as $\mathbf{x}(t)$ if it is clear from the context what the values of $\mathbf{x}_0$ and $\boldsymbol{\theta}$ are and how the function $\mathbf{u}(t)$ is defined.
- Let $\mathbf{u} : \mathbb{R}^n \times \mathbb{R}^d \times \mathbb{R}_{\geq 0} \to \mathbb{R}^m$ be a function of state $\mathbf{x}$, parameters $\hat{\boldsymbol{\theta}}$, and time t. Let in addition both $\mathbf{x}$ and $\hat{\boldsymbol{\theta}}$ be functions of t. Then, when the arguments of $\mathbf{u}$ are clearly defined by the context, we will simply write $\mathbf{u}(t)$ instead of $\mathbf{u}(\mathbf{x}(t), \hat{\boldsymbol{\theta}}(t), t)$.
- When dealing with vector fields and partial derivatives we will use the following extended notion of the Lie derivative of a function. Let it be the case that $\mathbf{x} \in \mathbb{R}^n$ and $\mathbf{x}$ can be partitioned as follows: $\mathbf{x} = \mathbf{x}_1 \oplus \mathbf{x}_2$, where $\mathbf{x}_1 \in \mathbb{R}^q$, $\mathbf{x}_1 = (x_{11}, \ldots, x_{1q})^T$, $\mathbf{x}_2 \in \mathbb{R}^p$, $\mathbf{x}_2 = (x_{21}, \ldots, x_{2p})^T$, $q + p = n$, and $\oplus$ denotes concatenation of two vectors. We define $\mathbf{f} : \mathbb{R}^n \times \mathbb{R}^d \times \mathbb{R} \to \mathbb{R}^n$ such that $\mathbf{f}(\mathbf{x}, \boldsymbol{\theta}, t) = \mathbf{f}_1(\mathbf{x}, \boldsymbol{\theta}, t) \oplus \mathbf{f}_2(\mathbf{x}, \boldsymbol{\theta}, t)$, where $\mathbf{f}_1 : \mathbb{R}^n \times \mathbb{R}^d \times \mathbb{R} \to \mathbb{R}^q$, $\mathbf{f}_1(\cdot) = (f_{11}(\cdot), \ldots, f_{1q}(\cdot))^T$, $\mathbf{f}_2 : \mathbb{R}^n \times \mathbb{R}^d \times \mathbb{R} \to \mathbb{R}^p$, and $\mathbf{f}_2(\cdot) = (f_{21}(\cdot), \ldots, f_{2p}(\cdot))^T$. Then $L_{\mathbf{f}_i(\mathbf{x},\boldsymbol{\theta},t)}\psi(\mathbf{x}, t)$, $i \in \{1, 2\}$, denotes the Lie derivative of the function $\psi(\mathbf{x}, t)$ with respect to the vector field $\mathbf{f}_i(\mathbf{x}, \boldsymbol{\theta}, t)$:

$$L_{\mathbf{f}_i(\mathbf{x},\boldsymbol{\theta},t)}\psi(\mathbf{x}, t) = \sum_{j=1}^{\dim \mathbf{x}_i} \frac{\partial \psi(\mathbf{x}, t)}{\partial x_{ij}} f_{ij}(\mathbf{x}, \boldsymbol{\theta}, t).$$

- Let $\mathbf{f}, \mathbf{g} : \mathbb{R}^n \to \mathbb{R}^n$ be differentiable vector fields. Then the symbol $[\mathbf{f}, \mathbf{g}]$ stands for the Lie bracket:

$$[\mathbf{f}, \mathbf{g}] = \frac{\partial \mathbf{f}}{\partial \mathbf{x}}\mathbf{g} - \frac{\partial \mathbf{g}}{\partial \mathbf{x}}\mathbf{f}.$$

  The adjoint representation of the Lie bracket is defined as

$$\mathrm{ad}^0_f \mathbf{g} = \mathbf{g}, \qquad \mathrm{ad}^k_{\mathbf{f}} \mathbf{g} = [\mathbf{f}, \mathrm{ad}^{k-1}_{\mathbf{f}} \mathbf{g}].$$

Part I

Introduction and preliminaries

1

Introduction

Consider a living organism or an artificial mechanism, which we shall refer to for the moment as a system, aiming to perform optimally in an uncertain environment. Despite the fact that the environment may be uncertain, we will suppose that we know the structure of the physical laws of the environment determining plausible motions of the system. Suppose that we even know what the system's action might be and assume that criteria of optimality according to which the system must determine its actions are available. Would we be able to decide a priori which particular action a system must execute or how it should adjust itself in order to maintain its behavior at the optimum?

Depending on the language describing the system's behavior, environment, and uncertainties a number of theoretical frameworks can be employed to find an answer to this non-trivial question. If the available information about the system is limited to a statistical description of the events and their likelihoods are known, then a good methodological candidate is the theory of statistical decision making. On the other hand, if the more sophisticated and involved apparatus of stochastic calculus is used to formalize the behavior of a system in an uncertain environment then a reasonable way to approach the analysis of such an object is to employ the theory of stochastic control and regulation. Despite these differences in how the behavior of a system may be described in various settings, there is a fundamental similarity in the corresponding theoretical frameworks. This similarity, if expressed informally, is that every framework should contain a description of the system's *actions*, *mechanisms for maintaining* and *adjusting* its behavior, and *criteria of optimality* or *goals*. These are in essence components of what we usually understand when calling a system adaptive.

In biology, according to the *Encyclopedia Britannica*, adaptation is described as a "process by which an animal or plant species becomes fitted to its environment; it is the result of natural selection's acting upon heritable variation. Even the simpler organisms must be adapted in a great variety of ways: in their structure, physiology, and genetics, in their locomotion or dispersal, in their means of defense and attack, in their reproduction and development, and in other respects." Actions, regulation and adjustments, criteria of optimality (fitness) are all present in this definition.

In systems theory there is less consensus on what the term "an adaptive system" describes. According to Evleigh (1967) a system is called adaptive if it "is a system which is provided with a means of continuously monitoring its own performance in relation to a given figure of merit or optimal condition and a means of modifying its own parameters by a closed-loop action so as to approach this optimum." Other definitions of an adaptive system have been provided by e.g. L. Zadeh, R. Bellman and R. Kalaba, J. G. Truxall, and V. A. Yakubovich, which we will consider in detail in Chapter 3. Yet they all share the very same ingredients such as actions, adjustments, and criteria of optimality. In this book we will also use the same general understanding of what an adaptive system means, though we will allow some technical deviations from these classical definitions.

Because the phenomenon of adaptation is generally understood as a special regulatory process in which a system maintains its performance at the optimum by adjusting itself and its actions, a natural language to analyze the phenomenon of adaptation is the language of systems and control theories. There are many inspiring and excellent monographs covering the general topic of adaptation. A non-exhaustive list of influential texts includes Tsypkin (1968), Tsypkin (1970), Narendra and Annaswamy (1989), Sastry and Bodson (1989), Krstić *et al.* (1995), and Fradkov *et al.* (1999). Hence it is reasonable to ask whether anything new can be added to this wealth of intellectual resources by one more text. As is often the case in science, novelty is a frequent consequence of a new formulation of a known problem, or it emerges as a result of answering new questions about familiar objects.

The purpose of this monograph is to contribute to the theory of adaptive systems by presenting a list of challenging questions and providing a unified theory that would allow one to find answers to these questions in a rigorous and systematic way. There are numerous examples illustrating the benefits of mathematical analysis of the phenomenon of adaptation: they range from solving the problems of crisis predictions (Gorban *et al.* 2010) to explaining plausible mechanisms of cell functioning in biology (Moreau and Sontag 2003), understanding the evolution of species (Gorban 2007), and motor learning (Smith *et al.* 2006). It is the author's hope that the methods developed here will also be useful for addressing open questions in science.

Below we present several examples of these questions emerging across the disciplines ranging from brain modeling to the issues of precise perturbation compensation in engineering and the problems of signal classification and pattern analysis in artificial intelligence. These examples are split into two major groups related to the problems of observation and regulation. For each of these groups we provide informal statements of the corresponding adaptation problems. These statements are not to be considered final and we will reshape them later on in the text. The function of these statements is to emphasize different facets of the problem of adaptation. There was no specific reason for choosing particular subject areas from which the examples are taken except probably the author's personal interests and bias.

1.1 Observation problems

The problem of state and parameter reconstruction of dynamical systems from the values of just few variables is a common task in the domain of mathematical modeling. Despite the fact that this problem received substantial attention in the past (see e.g. Bastin and Dochain (1990) and Ljung (1999)), there are gray spots in the literature for which finding a computationally plausible and theoretically rigorous solution remains a non-trivial task. The usual sources of difficulties are the presence of nonlinear parametrization, and the fact that we are not allowed to influence the system's behavior by varying its inputs in a reasonably broad class of functions.

There are numerous observation problems of this kind in physics. We start by presenting two examples from the domains of biophysics and neuroscience.

1.1.1 Example: quantitative modeling in biophysics and neuroscience

Let us consider the problem of simultaneous state and parameter reconstruction of models describing the dynamics of neural cells. Most of the available models of individual biological neurons are systems of ordinary differential equations describing the cell's response to stimulation; their parameters characterize variables such as time constants, conductances, and response thresholds, which are important for relating the model responses to the behavior of biological cells. Even the simplest models in this class, such as the Morris–Lecar model (Morris and Lecar 1981), are a great source of inspiration from the modeler's perspective (see Figure 1.1). This model is defined by the following system of equations:

$$\begin{aligned} \dot{V} &= \frac{1}{C}(-\bar{g}_{\mathrm{Ca}} m_{\infty}(V)(V - E_{\mathrm{Ca}}) - \bar{g}_{\mathrm{K}} w(V - E_{\mathrm{K}}) - \bar{g}_{\mathrm{L}}(V - E_{\mathrm{L}})) + I, \\ \dot{w} &= -\frac{1}{\tau(V)} w + \frac{w_{\infty}(V)}{\tau(V)}, \end{aligned} \tag{1.1}$$

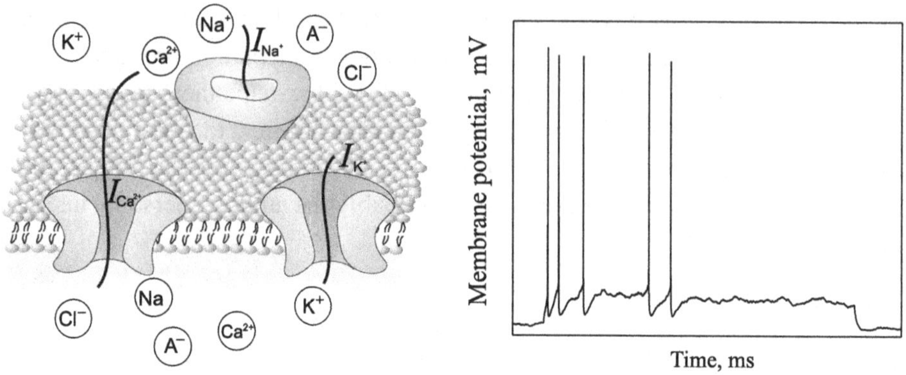

Physical variables

Measured variable

Figure 1.1 Incompleteness of information in quantitative modeling of a cell's behavior. The diagram on the left shows a basic phenomenological description of how currents propagate through a patch of the cell's membrane. There is a number of voltage-dependent channels, such as for Ca, Na, and K depicted in the figure. These channels pump ions through the membrane, and each of these channels has its own dynamics. The problem is that recording currents through every single channel in the membrane simultaneously is not always possible. Thus they must be estimated from available measurements, such as the membrane potentials depicted in the right diagram.

where

$$m_\infty(V) = 0.5\left(1 + \tanh\left(\frac{V - V_1}{V_2}\right)\right),$$

$$w_\infty(V) = 0.5\left(1 + \tanh\left(\frac{V - V_3}{V_4}\right)\right),$$

$$\tau(V) = T_0 \frac{1}{\cosh\left((V - V_3)/2V_4\right)}.$$

The variable V in (1.1) corresponds to the measured membrane potential, and I models an external stimulation current. The parameters $\bar{g}_{\mathrm{Ca}}$, $\bar{g}_{\mathrm{K}}$, and $\bar{g}_{\mathrm{L}}$ stand for the maximal conductances of the calcium, potassium, and leakage currents, respectively; C is the membrane capacitance; V_1, V_2, V_3, and V_4 are the parameters of the gating variables; T_0 is the parameter regulating the time scale of ionic currents; E_{Ca} and E_{K} are the Nernst potentials of the calcium and potassium currents; and E_{L} is the rest potential.

The total number of parameters in system (1.1) is 12, excluding the stimulation current I. Some of these parameters can be considered typical. For example the values of the Nernst potentials for calcium and potassium channels, E_{Ca} and E_{K}, are known and usually are set as $E_{\mathrm{Ca}} = 100\,\mathrm{mV}$ and $E_{\mathrm{K}} = -70\,\mathrm{mV}$ (Koch 2002).

The value of the rest potential, E_{L}, can be measured explicitly. The values of the parameters, $\bar{g}_{\mathrm{Ca}}$, $\bar{g}_{\mathrm{K}}$, $\bar{g}_{\mathrm{L}}$, and T_0, however, may vary substantially from one cell to another, and in general they are dependent on the conditions of the experiment. For example, the values of $\bar{g}_{\mathrm{Ca}}$, $\bar{g}_{\mathrm{K}}$, and $\bar{g}_{\mathrm{L}}$ depend on the density of ion channels in a patch of the membrane; and the value of T_0 is dependent on temperature. Hence, to be able to model the dynamics of individual cells, we have to recover these values from data.

Another example of the same nature is a model predicting the force generated by rat skeletal muscles during brief isometric contractions (Wexler *et al.* 1997). The model consists of three coupled nonlinear differential equations,

$$\begin{aligned}\dot{F} &= aT\left(1-\frac{F}{F_{\mathrm{m}}}\right)-\frac{F}{\tau_1+\tau_2 T/T_0},\\ \dot{T} &= k_1 T_0 C^2-(k_1C^2+k_2)T,\\ \dot{C} &= 2(k_1C^2+k_2)T-2k_1T_0C^2+kC_0-(k+k_0)C,\end{aligned} \tag{1.2}$$

where F is the force generated by the muscles, T is the concentration of Ca^{2+}–troponin complex, and C is the concentration of Ca^{2+} in the sarcoplasmic reticulum. The parameters τ_1, C_0, and k are fixed, while the parameters k_0, k_1, k_2, τ_2, F_{m}, a, and T_0 are free. The values of T and C are not available for direct observation, and the values of F over time can be measured. The question is whether it is possible to reconstruct the free parameters of the model together with the values of the concentrations T and C from the measurements of F. As in the previous example, we are dealing with an uncertain system in which the unknown parameters enter the right-hand side of the corresponding differential equations nonlinearly.

1.1.2 Example: adaptive classification in neural networks

The problem of estimating parameters of ordinary differential equations is not limited to the domain of modeling. It has an important relative in the field of artificial intelligence, namely the problem of adaptive classification of signals. An example of this problem is provided below.

Consider a set of signals defined as

$$\begin{aligned}&\mathcal{F}=\{f_i(\xi(t),\theta_i)\},\ i\in\{1,\ldots,N_f\},\\ &f_i:\mathbb{R}\times\mathbb{R}\to\mathbb{R},\ f_i(\cdot,\cdot)\in\mathcal{C}^0,\\ &\xi:\mathbb{R}_{\geq 0}\to\mathbb{R},\ \xi(\cdot)\in\mathcal{C}^1\cap L_\infty[0,\infty],\end{aligned} \tag{1.3}$$

where $\theta_i \in \Omega_\theta \subset \mathbb{R}$ are parameters of which the values are unknown a priori, $\Omega_\theta = [\theta_{\min}, \theta_{\max}]$ is a bounded interval, and $\xi(t)$ is a known and bounded function.

Signals $f_i(\xi(t), \theta_i)$ constitute the set of variables chosen to represent the state of an object.

Let $s \in \mathcal{F}$ be an element of class $\mathcal{F}$. The values of $s(t, \theta)$ are fed into the following system of differential equations:

$$\dot{x}_j = \sum_{m=1}^{N} c_{j,m}\sigma(\mathbf{w}_{j,m}^{\mathrm{T}}\mathbf{x} + w_{s,j,m}s(t) + w_{\xi,j,m}\xi + b_{j,m}),$$
$$j \in \{1, \ldots, N_x\},$$
$$\mathbf{x} = \mathrm{col}(x_1, \ldots, x_{N_x}),\ \mathbf{x}(t_0) = \mathbf{x}_0. \tag{1.4}$$

System (1.4) is often referred to as the recurrent neural network with standard multilayer perceptron structure. Here $c_{j,m}$, $\mathbf{w}_{j,m}$, $w_{s,j,m}$, $w_{\xi,j.m}$, and $b_{j,m}$ are parameters of which the values are fixed, and the function $\sigma : \mathbb{R} \to \mathbb{R}$ is sigmoidal:

$$\sigma(p) = \frac{1}{1 + e^{-p}}.$$

The problem of classification can now be stated as follows: is there a network of type (1.4) that is able to recover uncertain parameters i and θ_i from the input $s(t)$ (see Figure 1.2)? Informally, this means that there exist two sets of functions of the network state $\mathbf{x}$ and input $s(t)$:

$$\{h_{f,j}(\mathbf{x}(t), s(t))\},\ \{h_{\theta,j}(\mathbf{x}(t), s(t))\},$$
$$h_{f,j} : \mathbb{R}^{N_x} \times \mathbb{R} \to \mathbb{R},\ h_{\theta,j} : \mathbb{R}^{N_x} \times \mathbb{R} \to \mathbb{R},\ j \in \{1, \ldots, N_f\},$$

such that the values of i and θ_i can be inferred from $\{h_{f,j}(\mathbf{x}(t), s(t))\}$ and $\{h_{\theta,j}(\mathbf{x}(t), s(t))\}$, respectively, within a given finite interval of time.

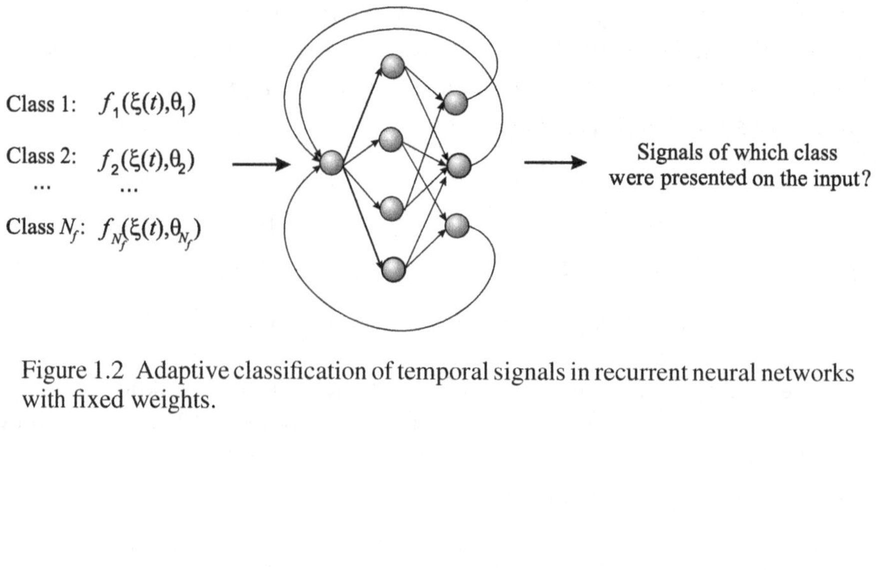

Figure 1.2 Adaptive classification of temporal signals in recurrent neural networks with fixed weights.

Networks (1.4) form an important class of computational structures of which the practical utility and capabilities are widely acknowledged in the literature (Haykin 1999). This class has been shown to be successful in dealing with a wide range of classification problems, including that of classifying signals from (1.3), provided that the values of θ_i in (1.3) are known. Empirical studies suggest that recurrent neural networks of this type are able to solve the classification problem (Feldkamp and Puskorius 1997; Prokhorov *et al.* 2002a) even if θ_i are unknown. The question, however, is how to show that this is indeed the case.

The problem of adaptive classification may look different from the previous examples in the domain of modeling. Indeed, here we have an existence question, whereas in the examples before we asked for a specific estimation algorithm. Despite these differences, there is substantial similarity in these problems. To be able to see this similarity, we would like to state the observation problem in a more general context below.

1.1.3 Preliminary statement of the problem

Let us generalize model (1.1) to the following class of dynamical systems:

$$\begin{aligned} \dot{\mathbf{x}} &= \mathbf{f}(\mathbf{x},\boldsymbol{\theta}) + \mathbf{g}(\mathbf{x},\boldsymbol{\theta})u(t), \quad \mathbf{x}(t_0) \in \Omega_x \subset \mathbb{R}^n, \\ y &= h(\mathbf{x}), \quad \mathbf{x} \in \mathbb{R}^n, \quad \boldsymbol{\theta} \in \Omega_\theta,\ \Omega_\theta \subset \mathbb{R}^d, \quad y \in \mathbb{R}, \end{aligned} \tag{1.5}$$

where $\mathbf{f},\mathbf{g} : \mathbb{R}^n \times \mathbb{R}^d \to \mathbb{R}^n$, $h : \mathbb{R}^n \to \mathbb{R}$, and $u : \mathbb{R} \to \mathbb{R}$ are continuous and differentiable functions. The variable x stands for the state vector, $u \in \mathcal{U} \subset \mathcal{C}^1[t_0,\infty)$ is the known input, $\boldsymbol{\theta}$ is the vector of unknown parameters, and y is the output of (1.5). Given that the right-hand side of (1.5) is differentiable, for any $\mathbf{x}' \in \Omega_x$, $u \in \mathcal{C}^1[t_0,\infty)$ there exists a time interval $\mathcal{T} = [t_0,t_1]$, $t_1 > t_0$ such that a solution $\mathbf{x}(t,\mathbf{x}')$ of (1.5) passing through $\mathbf{x}'$ at t_0 exists for all $t \in \mathcal{T}$. Hence, $y(t) = h(\mathbf{x}(t))$ is defined for all $t \in \mathcal{T}$. For the sake of convenience we will assume that the interval $\mathcal{T}$ of the solutions is large enough or even coincides with $[t_0,\infty)$ when necessary.

Taking these notations into account, we can now state the observation problem as follows: suppose that we are able to measure the values of $y(t)$ precisely; can the values of $\mathbf{x}'$ and the parameter vector $\boldsymbol{\theta}$ be recovered from the observations of $y(t)$, and, if so, how? In particular, we are interested in finding a computational algorithm

$$\dot{\boldsymbol{\xi}} = \mathbf{p}(\boldsymbol{\xi},t,u(t),y(t)), \qquad \boldsymbol{\xi}_0 = \boldsymbol{\xi}(t_0) \in \Omega_\xi, \tag{1.6}$$

such that for some known functions $\mathbf{h}_x(\boldsymbol{\xi})$ and $\mathbf{h}_\theta(\boldsymbol{\xi})$ and given number $\varepsilon > 0$ the following property holds:

$$\begin{aligned} \limsup_{t\to\infty} \|\mathbf{h}_x(\boldsymbol{\xi}(t,\boldsymbol{\xi}_0)) - \mathbf{x}(t)\| &\le \varepsilon, \\ \limsup_{t\to\infty} \|\mathbf{h}_\theta(\boldsymbol{\xi}(t,\boldsymbol{\xi}_0)) - \boldsymbol{\theta}\| &\le \varepsilon \ \forall\, \boldsymbol{\xi}_0 \in \Omega_\xi. \end{aligned} \tag{1.7}$$

In order to see how this statement is related to the adaptive classification problem in neural networks it is sufficient to notice that (1) the right-hand side of (1.4) can approximate an arbitrary continuous function in a bounded domain (hence it can model the right-hand side of (1.6)), and (2) the function s in (1.4) may be modeled as an output of system (1.5).

System (1.5) can be viewed as an external object or environment, and computational algorithm (1.6) and the functions $\mathbf{h}_x(\boldsymbol{\xi})$ and $\mathbf{h}_\theta(\boldsymbol{\xi})$ constitute the adapting system. The system responds to changes in the environment so that its performance (defined here by (1.7)) reaches an acceptable level and is maintained at this level indefinitely. If (1.5) were linearly parametrized, i.e. the functions $\mathbf{f}(\mathbf{x},\boldsymbol{\theta})$ and $\mathbf{g}(\mathbf{x},\boldsymbol{\theta})$ were linear in $\boldsymbol{\theta}$, then in order to answer this question we could employ the well-developed machinery of standard adaptive observers design (Marino and Tomei 1995b). Yet, as model (1.1) illustrates, the assumption of linear parametrization does not always hold. Hence alternative methods are needed.

This question (as well as other related issues of parameter estimation of nonlinear ordinary equations) is discussed in detail in Chapter 5. In addition to presenting sufficient conditions stipulating the mere existence of solutions to the observation problem, we provide specific computational algorithms (1.6) satisfying the required asymptotic properties (1.7). Special attention is paid to the analysis of the convergence rates of these algorithms. One may expect that the rates of convergence are likely to depend on the classes of nonlinearities in the models. This is indeed the case, as we illustrate in Chapter 5.

1.2 Regulation problems

Suppose now that we are not interested in reconstructing the values of the state and parameters of system (1.5). We do, however, require that the system's state is regulated to a given set in the system's state space for all $\boldsymbol{\theta} \in \Omega_\theta$. Consider for example the following system:

$$\begin{aligned} \dot{x}_1 &= x_2, \\ \dot{x}_2 &= -x_1 - x_2 + g(x_1, x_2, \boldsymbol{\theta}) + u, \end{aligned} \tag{1.8}$$

where x_1 and x_2 are the state variables, $\boldsymbol{\theta} \in \Omega_\theta$, $\Omega_\theta \subset \mathbb{R}^d$ is the vector of unknown parameters, $g : \mathbb{R} \times \mathbb{R} \times \mathbb{R}^d \to \mathbb{R}$ is a continuous function, and $u : \mathbb{R} \to \mathbb{R}$ is an input. Equations (1.8) describe a large class of mechanical and chemical systems. If we accept a simplified interpretation in which x_1 is the position of an object in space and x_2 is its velocity then $g(x_1, x_2, \boldsymbol{\theta})$ could stand for the friction terms (Canudas de Wit and Tsiotras 1999). If (1.8) is a model of a bio-reactor then x_1 and x_2 are the substrate concentrations and $g(x_1, x_2, \boldsymbol{\theta})$ could stand for the standard Michaelis–Menten nonlinearity (Bastin and Dochain 1990). In all these cases the function $g(x_1, x_2, \boldsymbol{\theta})$ is nonlinear in $\boldsymbol{\theta}$. The question is whether there is a function $u(x_1, x_2, \hat{\boldsymbol{\theta}})$ such that the solutions of (1.8) converge to the origin for all $\boldsymbol{\theta} \in \Omega_\theta$.

1.2.1 Example: non-dominating adaptive regulation

If no additional constraints are imposed then the above problem can be easily solved within the framework of dominating functions (Lin and Qian 2002b; Putov 1993). In this framework the original nonlinearly parametrized uncertainty $g(x_1, x_2, \boldsymbol{\theta})$ is replaced by a dominating linearly parametrized one $|g(x_1, x_2, \boldsymbol{\theta})| \leq \bar{g}(x_1, x_2)^{\mathrm{T}} \boldsymbol{\eta}$ and the problem is then solved using the standard method of Lyapunov functions (see Lin and Qian (2002b) for details). Although practical, this approach is not necessarily optimal for systems with limited resources. If the system is a living organism then using resources excessively may be an important limiting factor. The same argument applies for artificial yet autonomous systems. For these classes of systems a reasonable assumption is that the system is penalized for excessive use of domination terms in control.

One of the simplest examples of such non-dominating control schemes is the compensatory control $u = -g(x_1, x_2, \boldsymbol{\theta})$. If the value of $\boldsymbol{\theta}$ were known then this feedback would be able to steer the system to the origin. The problem, however, is that the values of $\boldsymbol{\theta}$ are unknown and the function $g(x_1, x_2, \boldsymbol{\theta})$ is nonlinearly parametrized. A possible strategy would be to make an initial guess at $\boldsymbol{\theta}$ and then adjust its value over time. The question, however, is how should one do this? This is a typical example of the non-dominating adaptation problem, of which a more formal statement is provided at the end of this section.

1.2.2 Example: adaptive tuning to bifurcations

In the previous case the set to which the system solutions are to converge was a priori known. There are systems for which information of this kind is not explicitly available. Their goal is not to reach a given state in the system's state space but rather to maintain adaptively a certain functional property of the system. An interesting example is the problem of adaptive self-tuning of a hearing nerve cell (Moreau and

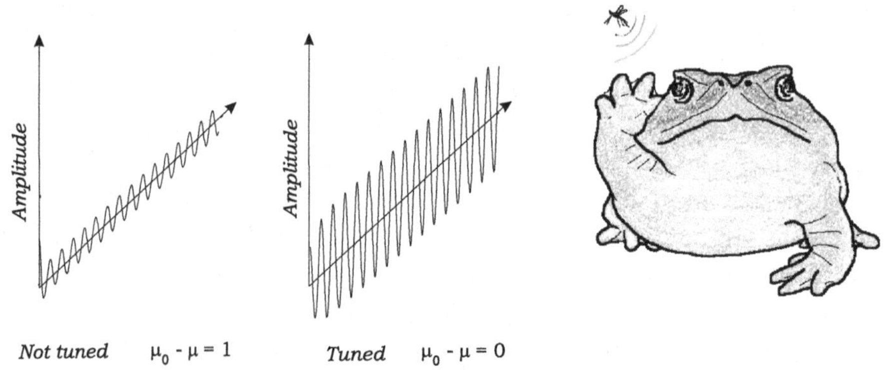

Figure 1.3 A diagram illustrating sensitivity control in the hearing cells via tuning to Andronov–Hopf bifurcation. The left panel shows response of the model, (1.9), $\lambda = 1, \omega = 1$, to a stimulus $u(t) = \sin(t)$ at $\mu_0 - \mu = 1$. The middle panel shows the response of the "tuned" model with $\mu_0 = \mu$ to the same stimulus. We see that the amplitude of oscillations in the tuned cell is many times larger than that in the untuned one.

Sontag 2003). The dynamics of the cell can be described by a nonlinear oscillator

$$\ddot{x} + (\mu_0 - \mu)\dot{x} + \lambda\dot{x}^3 + \omega^2 x = u(t), \ \lambda \in \mathbb{R}_{>0}, \tag{1.9}$$

where μ is the parameter to be adjusted and $u(t)$ is the input (stimulus). When the value of μ is set close to μ_0 the system's dynamics approaches supercritical Andronov–Hopf bifurcation. This leads to the possibility of substantial amplifications of signals in the specific frequency range (Camalet *et al.* 2000) (see Figure 1.3). The question, however, is what are the mechanisms ensuring that the system is always operating in close proximity to the bifurcation?

It has been shown in Moreau and Sontag (2003) that a simple adaptation procedure,

$$\dot{\mu} = -a \log \sqrt{x^2 + \dot{x}^2/\omega^2} - b, \ a, b \in \mathbb{R}_{>0}, \ a < b^2,$$

provides the required property. The value of this and similar results is difficult to overestimate, for they provide plausible adaptation models that can be searched for and validated in experiments. Moreover, the example motivates us to generalize this question even further and ask whether there exists a general recipe for deriving such feedbacks. This is the class of problems also known as adaptive tuning to bifurcations.

1.2.3 Example: adaptive regulation to invariant sets

The question of adaptive tuning to bifurcations is closely related to another interesting problem, that of adaptive regulation to invariant sets. The need to pose the problem of regulation as that of steering to a given invariant set emerges under those conditions, when the target set is not completely known. For example, we may know that the target set is necessarily an equilibrium or a periodic orbit, yet the precise location and shape of these sets might not be known. In Chapter 5 we present a set of results that allow us to solve both the problem of adaptive tuning to bifurcation and the problem of regulation to invariant sets in a unified manner using the method of a virtual algorithm of adaptation.

1.2.4 Preliminary statement of the problem

Let us now summarize the regulation problems considered above. Suppose that the system's motions are governed by the following set of equations:

$$\dot{\mathbf{x}} = \mathbf{f}_0(\mathbf{x}) + \mathbf{f}(\mathbf{x}, \boldsymbol{\theta}) + \mathbf{g}(\mathbf{x})u, \tag{1.10}$$

where $\mathbf{f}_0$, $\mathbf{f}$ and $\mathbf{g}$ are continuous functions and $\boldsymbol{\theta}$ is the vector of unknown parameters. The standard adaptive regulation question is that of whether there is a feedback

$$\begin{aligned} u &= u(\mathbf{x}, \hat{\boldsymbol{\theta}}), \\ \dot{\hat{\boldsymbol{\theta}}} &= A(\mathbf{x}) \end{aligned} \tag{1.11}$$

such that the system's state is stirred asymptotically to the origin. As our examples motivate, in addition to this standard requirement, we may wish to require that a functional of $u(\mathbf{x}, \boldsymbol{\theta})$ is optimized over time. A reasonable requirement could be that

$$\lim_{t\to\infty} \int_0^t (u(\mathbf{x}(\tau, \mathbf{x}_0), \boldsymbol{\theta}) - u(\mathbf{x}(\tau, \mathbf{x}_0), \hat{\boldsymbol{\theta}}(\tau)))^2 \, d\tau \to \min.$$

In the next chapters we shall see when, how, and in what sense such requirements may be satisfied.

Let us now suppose that system (1.10) undergoes a certain bifurcation at $\boldsymbol{\theta} = 0$ and its operating conditions require that this regime is maintained adaptively. Yet the values of $\boldsymbol{\theta}$ may change abruptly. In this case the adaptive regulation problem can be stated as that of looking for the functions (1.11) such that

$$\lim_{t\to\infty} \mathbf{f}(\mathbf{x}(t, \mathbf{x}_0), \boldsymbol{\theta}) + \mathbf{g}(\mathbf{x}(t, \mathbf{x}_0))u(\mathbf{x}(t, \mathbf{x}_0), \hat{\boldsymbol{\theta}}(t)) = 0.$$

The question, however, is how do we find such algorithms?

1.3 Summary

The examples of problems presented above are a sample of the sorts of challenges we wish to attack in this book. Although they originate in rather different fields, they are connected together by the need for theoretical assessment of how adaptation may be organized and analyzed in these systems. The systems considered in the examples contain nonlinearly parametrized uncertainties, and their target behavior need not necessarily be stable. Even the definition of the target sets is allowed to bear a degree of uncertainty. The main theoretical focus of this book is to provide a systematic extension of the existing theories of adaptation so that all these challenging problems, irrespective of their field of origin, can be addressed in a rigorous and unified manner.

The main strategy in our quest to create such an extension can be described as that of looking beyond the usual presumptions in the domain of analysis and synthesis of adaptive systems. In particular, we will concentrate on breaking through the following constraints (presumptions) which are often implicitly or explicitly imposed in the standard statements of the problem of adaptation:

(1) a practically successful adaptive system must be stable in the sense of Lyapunov;
(2) the analysis and synthesis methods should operate exclusively and at all times with those variables of which the values are available for direct observations;
(3) a successful system should be able to maintain its optimal performance over infinitely long intervals of time.

In order to be able to avoid these constraints when their presence in the problem is not at all necessary, we shall present a systematic view on the problem of adaptation starting from the very basic principles of a system's organization and passing on to the laws implementing these principles in particular settings. As a result of this hierarchical approach, a likely object of our analysis would be a system that is adapting, albeit not necessarily being globally stable in the sense of Lyapunov.

A possible way to develop a feeling for why some constraints are important whereas others can be removed from the problem is to look at the problem retrospectively (Lakatos 1976). In the next two chapters we review the most influential concepts of adaptation in the literature of systems and control theories and justify the research program that guided the development of our own contribution.

2

Preliminaries

Determining asymptotic properties of dynamical systems, including the formulation of a qualitative picture of the system's trajectories over large intervals of time, is one of the central questions of modern theory for adaptive systems. This is not surprising, for the very reason for adaptation is the lack of available measurement information. If such information is not available a priori, and carrying out numerical or physical experiments is not a feasible option, assessment of the qualitative properties of the system's behavior is often the only way to characterize the system. What are these qualitative properties? Informally, from these properties we should be able to tell, for example, how a system might respond to external perturbations, or how the system's variables behave over long intervals of time. Formally, we may wish to know whether the system is stable in some sense, whether its trajectories are bounded, and to what sets these trajectories will be confined with time.

In this chapter we shall provide a brief summary and necessary background about these qualitative properties of dynamical systems. We do not wish, however, to present an exhaustive review of all concepts. There are many excellent texts devoted to detailed analysis of every single issue mentioned above. Here we will rather review these concepts with a level of detail and generality just sufficient for developing a qualitative understanding of the problem of adaptation and the basics of methods of adaptive regulation. Let us start with the simplest and at the same time the most difficult notion for analysis: the notion of an *attracting set.*

2.1 Attracting sets and attractors

In order to introduce the notion of an attracting set it is often useful to think of a system as a family of parametrized maps $\mathbf{x} : \mathbb{R} \times \mathbb{R}^n \rightarrow \mathbb{R}^n$. In the modeling language this will restrict our attention to the following models: $\mathbf{x}(t, \mathbf{x}_0)$, or *flows*, where t stands for the time instance and $\mathbf{x}_0$ is the value of the system's state at

$t = 0$. Usually an additional semi-group property is imposed on $\mathbf{x}(t, \mathbf{x}_0)$:

$$\mathbf{x}(t', \mathbf{x}(t'', \mathbf{x}_0)) = \mathbf{x}(t' + t'', \mathbf{x}_0).$$

Although this assumption is not entirely necessary for producing the definition, we shall keep this possibility in mind, for it provides us with a link to physical reality. We would also like to notice that models $\mathbf{x}(t, \mathbf{x}_0)$ do not yet have any explicit reference to any inputs or other factors acting on a real system externally. These factors are all hidden in the model. The main reason for this is that we want to keep the notation compact. However, should such a necessity arise, one can easily modify the definitions below so that all external variables are made explicitly visible.

Before we proceed with a formal definition explaining what we will understand under the term *attracting set* we will need to introduce one additional notion. This is the notion of an *invariant set* with respect to a given flow $\mathbf{x}(t, \mathbf{x}_0)$.

Definition 2.1.1 A set $\mathcal{A} \subset \mathbb{R}^n$ is called invariant with respect to the flow $\mathbf{x}(t, \mathbf{x}_0)$ iff for all $\mathbf{x}_0 \in \mathcal{A}$, $t \in \mathbb{R}$ the following property holds:

$$\mathbf{x}(t, \mathbf{x}_0) \in \mathcal{A}.$$

It is sometimes useful to distinguish between forward-invariant and backward-invariant sets, of which the definitions are provided below.

Definition 2.1.2 A set $\mathcal{A} \subset \mathbb{R}^n$ is called forward-invariant with respect to the flow $\mathbf{x}(t, \mathbf{x}_0)$ iff for all $\mathbf{x}_0 \in \mathcal{A}$, $t \in \mathbb{R}_{\geq 0}$ we have that $\mathbf{x}(t, \mathbf{x}_0) \in \mathcal{A}$. The set is backward-invariant iff $\mathbf{x}(t, \mathbf{x}_0) \in \mathcal{A}$ for all $\mathbf{x}_0 \in \mathcal{A}$, $t \in \mathbb{R}_{\leq 0}$.

Simple examples of invariant sets are equilibria, limit cycles, or just orbits of autonomous systems in the state space (see Figure 2.1). Whether a given set is invariant or not is an important item of information for the analysis of uncertain systems. Indeed, if we know that $\mathcal{A}$ is invariant then all trajectories passing through at least one point of $\mathcal{A}$ will necessarily remain there for all t. Despite its clear benefits for analysis, the notion of invariance of a set with respect to $\mathbf{x}(t, \mathbf{x}_0)$ is not a very instrumental property from the viewpoint of regulation. Suppose that we know that $\mathcal{A} \subset \mathbb{R}^n$ is forward-invariant with respect to $\mathbf{x}(t, \mathbf{x}_0)$, and let us suppose that the system's state passes through a point that does not belong to $\mathcal{A}$. Let us finally assume that for some reason we wish to know whether the system's state will reach $\mathcal{A}$ or its arbitrarily small vicinity in finite time. For example, if $\mathbf{x}(t, \mathbf{x}_0)$ models trajectories of an organism in space and $\mathcal{A}$ is the set of locations of food then a question of vital importance for this organism is whether it should employ its resources to initiate movements towards the set $\mathcal{A}$ or whether it should wait for some time until external forces such as the flow of water or wind eventually bring it to $\mathcal{A}$ in a reasonable amount of time. Answering this question might not be a feasible

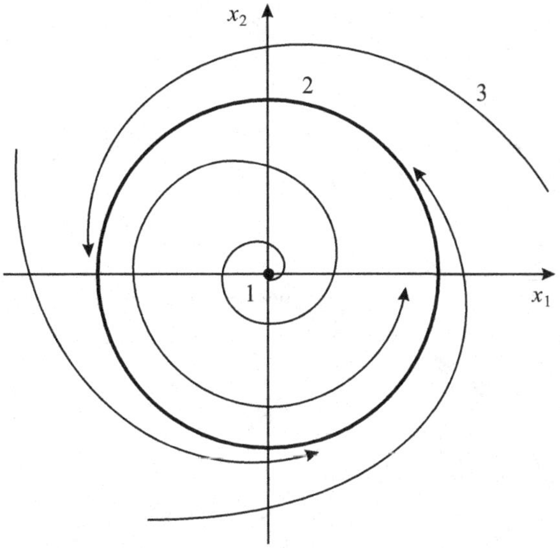

Figure 2.1 Examples of invariant sets of autonomous dynamical systems: 1, equilibrium; 2, a limit cycle; and 3, an orbit.

exercise in the absence of additional information about the system. We may still view this property as preferable and desirable. An invariant set possessing such a property is often referred to as *attracting*. Formally the notion of an attracting set is provided in the next definition

Definition 2.1.3 A closed[1] invariant set $\mathcal{A} \subset \mathbb{R}^n$ is called attracting iff

(1) there is a neighborhood $U(\mathcal{A})$ of $\mathcal{A}$ such that

$$\mathbf{x}(t, \mathbf{x}_0) \in U(\mathcal{A}) \; \forall \, \mathbf{x}_0 \in U(\mathcal{A}), t \in \mathbb{R}_{\geq 0}; \tag{2.1}$$

(2) the following limiting property holds

$$\lim_{t \to \infty} \|\mathbf{x}(t, \mathbf{x}_0)\|_{\mathcal{A}} = 0 \; \forall \, \mathbf{x}_0 \in U(\mathcal{A}). \tag{2.2}$$

According to Definition 2.1.3 a closed invariant set $\mathcal{A}$ is attracting if there is a forward-invariant neighborhood $U(\mathcal{A})$ such that all trajectories starting in $U(\mathcal{A})$ converge to $\mathcal{A}$ asymptotically. At first glance the definition is rather general and clear. Although this is indeed the case, there are situations in which a generalization of this notion may be required. Let us consider an example.

[1] Let us remind the reader that a set $\mathcal{A}$ is closed iff it contains all of its limit points. For example, if $\mathcal{A}$ is closed and $a_i \in \mathcal{A}$ $i = 1, \ldots, \infty$ is a sequence then $\lim_{i \to \infty} a_i$ (if exists) should also belong to $\mathcal{A}$. If $\mathcal{A}$ is an interval then it is closed iff $\mathcal{A}$ contains its boundaries. A point in $\mathbb{R}^n$ is obviously a closed set. In addition to these simple examples there are more exotic instances of closed sets such as the Cantor set (also known as "Cantor dust").

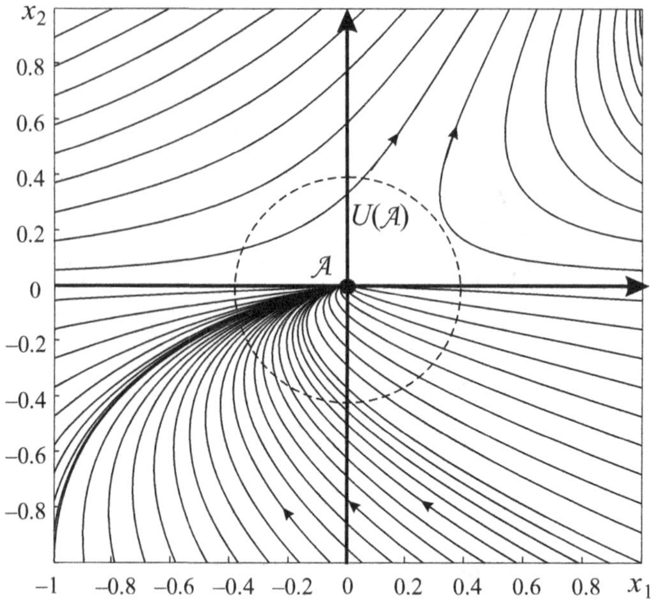

Figure 2.2 The phase portrait of system (2.3).

Example 2.1.1 Suppose that the system dynamics is governed, up to a coordinate transformation, by the following set of ordinary differential equations:

$$\begin{aligned} \dot{x}_1 &= -x_1 + x_2, \\ \dot{x}_2 &= |x_2|. \end{aligned} \tag{2.3}$$

The solution of the second equation in system (2.3) is a non-decreasing function of t for all initial conditions. Furthermore, for all $x_2(0) \leq 0$ we have $\lim_{t\to\infty} x_2(t, x_2(0)) = 0$; and $\lim_{t\to\infty} x_2(t, x_2(0)) = \infty$ for all $x_2(0) > 0$. From this simple analysis we can conclude that solutions of the system will necessarily approach the origin asymptotically for all $x_2(0) \leq 0$, and will move away from the equilibrium for arbitrarily large distances if $x_2(0) > 0$. This is illustrated in Figure 2.2 depicting the phase portrait of system (2.3). This figure demonstrates that for any neighborhood $U(\mathcal{A})$ of the origin $\mathcal{A}$ there are points $\mathbf{x}' \in U(\mathcal{A})$ such that solutions $\mathbf{x}(t, \mathbf{x}')$ escape the neighborhood $U(\mathcal{A})$ and never come back. Hence, according to Definition 2.1.3, $\mathcal{A}$ cannot be called an attracting set. On the other hand, there are points $\mathbf{x}'' \in U(\mathcal{A})$ such that $\lim_{t\to\infty} \mathbf{x}(t, \mathbf{x}'') = 0$. If $U(\mathcal{A})$ is an open circle, then the number of such points is as large as the number of points corresponding to the solutions escaping $U(\mathcal{A})$. Thus the set $\mathcal{A}$ bears an overall signature of attractivity.

Contradiction of the type we discussed in this example was noticed and analyzed by many authors, e.g. in Gorban and Cheresiz (1981).[2] This led to the emergence of the new notion of a *weakly attracting set*, which was formally defined by J. Milnor in his seminal work (Milnor 1985):

Definition 2.1.4 A set $\mathcal{A}$ is a weakly attracting, or Milnor attracting, set iff

(1) it is closed, invariant, and
(2) for some set $\mathcal{V}$ (not necessarily a neighborhood of $\mathcal{A}$) with *strictly positive measure* and for all $\mathbf{x}_0 \in \mathcal{V}$ the following limiting relation holds:

$$\lim_{t\to\infty} \mathbf{x}(t,\mathbf{x}_0) = \mathcal{A} \; \forall \; \mathbf{x}_0 \in \mathcal{V}(\mathcal{A}). \tag{2.4}$$

The key difference of the notion of a weakly attracting set from that provided in Definition 2.1.3 is that the domain of attraction $\mathcal{V}$ is not necessarily a neighborhood of $\mathcal{A}$. Despite the fact that this difference may look small and insignificant at first glance, it becomes very instrumental for successful statement and solution of particular problems of adaptation. In Chapter 5 we will present a large class of problems for which solutions might not even exist if the standard definition of attracting sets were exclusively used in the formal statement of the problem. Although we are not yet ready to provide these examples now, we still would like to point out the existence of these two rather different views on what an attracting set may mean.[3]

So far we have defined the notions of invariance and attractivity of a set with respect to a flow. In the context of adaptation, invariance and attractivity are often desirable asymptotic characterizations of the preferred domain to which the state of an adapting system must be able to move. The question, however, is whether these properties characterize the preferred state with minimal ambiguity. To some degree, thanks to the requirement of invariance in the definitions, this issue is already taken into account. In order to illustrate this point let us suppose that the invariance property in Definitions 2.1.3 and 2.1.4 is replaced with forward-invariance.

Consider system (2.3) from Example 2.1.1. If we replace the invariance requirement with forward-invariance in Definition 2.1.4 then the equilibrium of this system will still be weakly attracting. One can easily see that in this case the equilibrium will not be the only attracting set in the state space. In fact, if we were to replace invariance with mere forward-invariance, the bottom half of every disk centered at the point $(0,0)$ would be a weakly attracting set too. Indeed, all sets defined in this

[2] See also Gorban (2004) for a more recent and extended review.

[3] We would like to note that Definitions 2.1.3 and 2.1.4 do not exhaust all of the possibilities for defining attracting sets of dynamical systems. There are many other alternatives, such as in Bhatia and Szego (1970). Our choice of particular notions is motivated mostly by the scope of the problems we will consider in this book.

way are forward-invariant according to Definition 2.1.2, and for every such set there exists a set $\mathcal{V}(\mathcal{O})$ (e.g. pick $\mathcal{V}(\mathcal{O}) = \{(x_1, x_2)|x_2 \leq 0\}$) satisfying condition (2.4). Thus the number of weakly attracting sets in system (2.3) would be infinite and not even countable. Hence an object specified in terms of mere forward-invariance and attraction can in principle bear a substantial degree of ambiguity.

In order to disambiguate the asymptotic behavior of dynamical systems even further, the attractivity property of a set is often considered, together with its minimality. Informally the minimality property can be viewed as a requirement that an attracting set $\mathcal{A}$ should not contain any other attracting sets strictly smaller than $\mathcal{A}$. Formally this can be stated as the requirement that for every $\mathbf{x}_0 \in \mathcal{A}$ the trajectory $\mathbf{x}(t, \mathbf{x}_0)$ is dense in $\mathcal{A}$. Attracting sets having this latter property are often referred to as *attractors*. Similarly to attracting sets, there are standard and weak attractors, and we shall be able to see the advantage of both notions in the next chapters.

So far we have provided formal definitions for invariance, attracting sets, and attractors. It is natural now to ask how we can tell whether a set is invariant, attracting, or is an attractor for a given dynamical system. In other words, in addition to the definitions we need to have instrumental criteria for establishing at least the existence of the sets with the aforementioned properties. The role of these criteria in the domain of analysis and synthesis of adaptive systems is that these criteria will provide specific *target constraints* an adapting system should implement in order to be able to fulfill its goals.

In the literature on adaptive control there are many criteria of this kind. Here we consider only those criteria that are necessary in order to understand state-of-the-art statements of the problem of adaptation which we discuss in Chapter 3. These are Barbalat's lemma, stability, persistency of excitation of a vector-function, and one special class of dynamical systems of which the asymptotic behavior can be easily analyzed analytically. Let us start with the simplest of them – Barbalat's lemma.

2.2 Barbalat's lemma

An inherent feature of many adaptive systems is that they operate in conditions under which information about the environment and their own dynamics is lacking. A simple example is that of an organism that may be able to measure its relative position in space with a certain tolerance but is not able to measure its velocity. Yet, it needs to detect conditions under which the velocity is converging to zero asymptotically. More generally, let $h : \mathbb{R} \rightarrow \mathbb{R}$ be a function of which the value is physically relevant, but we do not know this function precisely. Suppose that we know some integral characterization of the function, such as the upper and lower bounds of its integral over a family of intervals. What can we say about the asymptotic properties of the function? Is there a limit of $h(t)$ at $t \rightarrow \infty$, and, if so, what is its value? The answer to this question is partially provided by Barbalat's lemma.

In order to state the lemma let us recall the property of uniform continuity of a function of a real variable.

Definition 2.2.1 A function $h : \mathbb{R} \to \mathbb{R}$ is called uniformly continuous iff for every $\varepsilon > 0, \varepsilon \in \mathbb{R}$ there exists $\delta > 0, \delta \in \mathbb{R}$ such that for all $t, \tau \in \mathbb{R}$ the following inequality holds:

$$|t - \tau| < \delta \Rightarrow |h(t) - h(\tau)| < \varepsilon. \tag{2.5}$$

The lemma now can be formulated as follows.

Lemma 2.1 *Let $h : \mathbb{R} \to \mathbb{R}$ be a uniformly continuous function and suppose that the following limit exists:*

$$\lim_{t\to\infty} \int_{t_0}^{t} h(\tau)d\tau = a, \; t_0 \in \mathbb{R}, \; a \in \mathbb{R}. \tag{2.6}$$

Then

$$\lim_{t\to\infty} h(t) = 0. \tag{2.7}$$

Proof of Lemma 2.1. Suppose that (2.7) does not hold. This implies that there exists a diverging sequence of t_n, $n = 1, \ldots, \infty$ such that

$$|h(t_n)| > \varepsilon, \; \varepsilon \in \mathbb{R}, \; \varepsilon > 0.$$

Because the function $h(t)$ is uniformly continuous we have that

$$\forall\, \varepsilon_1 > 0, \varepsilon_1 \in \mathbb{R} \; \exists\, \delta_1 > 0, \; \delta_1 \in \mathbb{R} : \; |t - t_n| < \delta_1 \Rightarrow |h(t) - h(t_n)| < \varepsilon_1.$$

Let $\varepsilon_1 = \varepsilon/2$, then

$$|h(t)| + |h(t_n) - h(t)| \geq |h(t_n)| \Rightarrow |h(t)| \geq |h(t_n)| - |h(t) - h(t_n)| \geq \varepsilon/2 \tag{2.8}$$

$\forall\, t \in [t_n, t_n + \delta_1]$. Consider now

$$\left| \int_{t_0}^{t_n+\delta_1} h(\tau)d\tau - \int_{t_0}^{t_n} h(\tau)d\tau \right| = \left| \int_{t_n}^{t_n+\delta_1} h(\tau)d\tau \right|. \tag{2.9}$$

Given that (2.6) holds, we can conclude that there exists a number n' such that

$$\left| \int_{t_n}^{t_n+\delta_1} h(\tau)d\tau \right| \leq \varepsilon_2, \; \varepsilon_2 > 0, \; \varepsilon_2 \in \mathbb{R} \; \forall\, n \geq n', \tag{2.10}$$

where ε_2 is an arbitrarily small number. On the other hand, on applying the mean-value theorem to the right-hand side of (2.9) and using (2.8) we obtain that the estimate

$$\left|\int_{t_n}^{t_n+\delta_1} h(\tau)d\tau\right| = \left|\delta_1 h(t')\right|, \; t' \in [t_n, t_n+\delta] \; \Rightarrow$$

$$\left|\int_{t_n}^{t_n+\delta_1} h(\tau)d\tau\right| \geq \delta_1\varepsilon/2 \tag{2.11}$$

must hold for all $n = 1, \ldots, \infty$. Thus, taking (2.11) and (2.10) into account, we can conclude that

$$\varepsilon_2 > \left|\int_{t_n}^{t_n+\delta_1} h(\tau)d\tau\right| \geq \delta_1\varepsilon/2, \; \forall\, n \geq n'.$$

Given that the value of ε_2 can be chosen arbitrarily small and that $\delta_1\varepsilon/2 > 0$, we obtain a contradiction. Hence the assumption that $h(t)$ does not converge to zero is not true. □

An instrumental function of Lemma 2.1 in the domain of synthesis and analysis of adaptive systems is that it constitutes a simple convergence criterion. If we know that the state vector $\mathbf{x}$ of a system satisfies the integral inequality

$$\int_{t_0}^{t} \|\mathbf{x}(\tau, \mathbf{x}_0)\|^2 \, d\tau < B, \;\; B \in \mathbb{R}_{\geq 0}, \; \forall\, t \geq t_0,$$

and the derivative of $\mathbf{x}(t, \mathbf{x}_0)$ with respect to t is bounded, we can conclude that $\mathbf{x}(t, \mathbf{x}_0) \to 0$ at $t \to \infty$. In other words, the system's state will have to approach the origin asymptotically. Although simple, this argument is a common component of convergence proofs in the domain of adaptive regulation.

Despite their simplicity and practical utility, the analysis arguments based exclusively on Lemma 2.1 have obvious limitations. This is because the lemma does not characterize the transient properties of the converging functions. For example, we may be interested in knowing how fast a function approaches its limit values, or how large the excursions of the state vector in the system's state space may become before it will settle in close proximity to the origin. The answers to these important questions cannot be derived explicitly from Lemma 2.1. The lemma does not guarantee that the convergence is going to be fast or slow, or that the state does not deviate much from the origin over time. In order to be able to produce these more delicate predictions, additional characterizations of the system's flow rather than simply uniform continuity are needed. One such characterization is the notion of *stability*.

2.3 Basic notions of stability

Let us ask ourselves what we usually mean by referring to some system or process as being stable. Intuitively and in vague everyday language we link stability with the property of a system that a given variable or perhaps a set of variables will not change much in a certain sense in response to perturbations of some kind. In order to state the very same definition formally, one needs to clarify what these variables and perturbations are and what this phrase "will not change much" means. Fortunately, all necessary clarifications usually follow explicitly from the nature of the problem and our own understanding of the goals of the analysis. However, depending on the problem, these specific clarifications vary from one case to another. This gives rise to a rich family of stability definitions. Here we will consider only those few which from the author's point of view are immediately relevant for the analysis of classical mathematical statements of the problem of adaptive regulation provided. These are the notions of *Lyapunov stability*, *Poincaré stability*, and *Poisson stability*. Other basic stability notions, such as *input-to-state* and *input-to-output stability*, which will be instrumental for the further development of the problem of adaptation, are introduced in Chapter 4.

Definition 2.3.1 Let $\mathbf{x}(t, \mathbf{x}_0) : \mathbb{R} \times \mathbb{R}^n \to \mathbb{R}^n$ be a solution of a dynamical system defined for all $t \geq t_0$, $t_0, t \in \mathbb{R}$ and passing through $\mathbf{x}_0 \in \mathbb{R}^n$ at $t = t_0$. Solution $\mathbf{x}(t, \mathbf{x}_0)$ is globally stable in the sense of Lyapunov iff for every $\varepsilon > 0$, $\varepsilon \in \mathbb{R}$ there exists $\delta > 0$, $\delta \in \mathbb{R}$, such that the following holds:[4]

$$\|\mathbf{x}_0 - \mathbf{x}'_0\| \leq \delta \Rightarrow \|\mathbf{x}(t, \mathbf{x}_0) - \mathbf{x}(t, \mathbf{x}'_0)\| \leq \varepsilon \; \forall\, t \geq t_0. \tag{2.12}$$

Alternatively,

$$\|\mathbf{x}_0 - \mathbf{x}'_0\| \leq \delta \Rightarrow \left\|\mathbf{x}(t, \mathbf{x}_0) - \mathbf{x}(t, \mathbf{x}'_0)\right\|_{\infty,[t_0,\infty]} \leq \varepsilon. \tag{2.13}$$

If property (2.12) holds only in a neighborhood of $\mathbf{x}(t, \mathbf{x}_0)$ then the stability is local. The property of Lyapunov stability of a solution has a very simple interpretation. Let us view the flow $\mathbf{x}(t, \mathbf{x}_0)$ as a mapping from the space $\mathbb{R}^n$ of initial conditions $\mathbf{x}_0$ into the space of trajectories $\mathbf{x}(t, \mathbf{x}_0)$, and let the space of trajectories be endowed with the standard uniform norm $\|\cdot\|_{\infty,[t_0,\infty]}$. Then stability of a solution in the sense of Lyapunov is analogous to the usual notion of continuity of the mapping $\mathbf{x} : \mathbb{R}^n \to L^n_\infty[t_0, \infty]$. This is precisely what expression (2.13) in Definition 2.3.1 states. In other words, small variations of initial conditions lead to small variations of $\mathbf{x}(t, \mathbf{x}_0)$ over all $t \geq t_0$. If $\mathbf{x}(t, \mathbf{x}_0)$ is stable in the sense of Lyapunov then we can make sure that the value of an observed trajectory $\mathbf{x}(t, \mathbf{x}'_0)$

[4] Here and in other definitions of stability, when this applies, we assume that $\mathbf{x}(t, \mathbf{x}'_0)$ is also defined for all $t \geq t_0$.

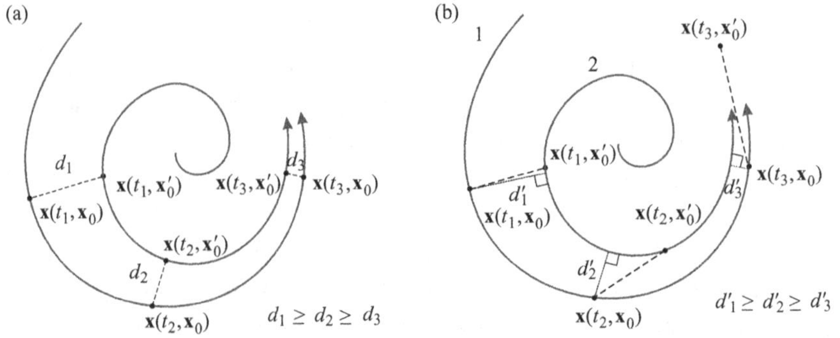

Figure 2.3 Diagrams illustrating the notions of Lyapunov stability of a solution (a) ($d_i = \|\mathbf{x}(t_i,\mathbf{x}_0) - \mathbf{x}(t_i,\mathbf{x}_0')\|$) and Poincaré stability of an orbit (b) ($d_i' = \|\mathbf{x}(t_i,\mathbf{x}_0')\|_{\mathcal{A}}$, $\mathcal{A} = \{\mathbf{p} \in \mathbb{R}^n | \mathbf{p} = \mathbf{x}(t,\mathbf{x}_0), t \in \mathbb{R}\}$).

at any t would not be far from the value of $\mathbf{x}(t,\mathbf{x}_0)$ at the same t, provided that the perturbation in $\mathbf{x}_0$ is sufficiently small (i.e. $\mathbf{x}_0'$ is sufficiently close to $\mathbf{x}_0$).

In some cases knowing that the deviations are guaranteed to be small, provided that the perturbations in initial conditions are small, might not be enough. For example, asymptotic convergence of a perturbed solution to its unperturbed version may be required. In this case we will use the notion of asymptotic Lyapunov stability of solutions

Definition 2.3.2 A solution $\mathbf{x}(t,\mathbf{x}_0)$ is (globally) asymptotically stable in the sense of Lyapunov iff it is (globally) stable in the sense of Definition 2.3.1 and

$$\lim_{t\to\infty} \mathbf{x}(t,\mathbf{x}_0') - \mathbf{x}(t,\mathbf{x}_0) = 0. \tag{2.14}$$

The *isochronous* property of Lyapunov stability of a solution is illustrated in Figure 2.3(a). In order to tell whether $\mathbf{x}(t,\mathbf{x}_0)$ is stable we have to compare the values of $\mathbf{x}(t,\mathbf{x}_0)$ and $\mathbf{x}(t,\mathbf{x}_0')$ at the same values of t.

Clearly, Lyapunov stability does not exhaust the whole spectrum of plausible asymptotic descriptions of solutions of a dynamical system with respect to each other. Consider an example. Let $\mathbf{x}(t,\mathbf{x}_0')$ and $\mathbf{x}(t,\mathbf{x}_0)$ be two solutions of the same system, and $\mathbf{x}_0' \neq \mathbf{x}_0$. Then a possible characterization of their relative position in the state space could be

$$\begin{aligned} \rho(t,\mathbf{x}(t,\mathbf{x}_0'),\mathbf{x}(t,\mathbf{x}_0)) &= \|\mathbf{x}(t,\mathbf{x}_0')\|_{\mathcal{A}}, \\ \mathcal{A} &= \{\mathbf{p} \in \mathbb{R}^n | \mathbf{p} = \mathbf{x}(t,\mathbf{x}_0),\ t \in \mathbb{R}\}. \end{aligned} \tag{2.15}$$

In (2.15) solution $\mathbf{x}(t,\mathbf{x}_0)$ is viewed as an invariant set of the system (see the comment after Definition 2.1.1 and also Figure 2.1); the closeness of the solutions to each other at the given time instant t is determined as the distance from the point $\mathbf{x}(t, \mathbf{x}'_0)$ to the curve $\mathcal{A}$. On defining the closeness of solutions or trajectories as (2.15), we arrive at the notion of *stability in the sense of Poincaré*.

Definition 2.3.3 Let $\mathbf{x}(t, \mathbf{x}_0) : \mathbb{R} \times \mathbb{R}^n \to \mathbb{R}^n$ be a solution of the system defined for all $t \geq t_0$, with $t_0, t \in \mathbb{R}$, and passing through a point $\mathbf{x}_0 \in \mathbb{R}^n$ at $t = t_0$. Let $\mathcal{A}$ denote an invariant set induced by $\mathbf{x}(t, \mathbf{x}_0)$:

$$\mathcal{A} = \{\mathbf{p} \in \mathbb{R}^n | \mathbf{p} = \mathbf{x}(t, \mathbf{x}_0),\ t \geq t_0,\ t \in \mathbb{R}\}.$$

Solution $\mathbf{x}(t, \mathbf{x}_0)$ is stable in the sense of Poincaré iff for every $\varepsilon > 0$, $\varepsilon \in \mathbb{R}$ there exists $\delta > 0$, $\delta \in \mathbb{R}$ such that

$$\|\mathbf{x}'_0\|_{\mathcal{A}} \leq \delta \Rightarrow \|\mathbf{x}(t, \mathbf{x}'_0)\|_{\mathcal{A}} \leq \varepsilon \ \forall\, t \geq t_0. \tag{2.16}$$

The main difference of the stability notions in the senses of Lyapunov and Poincaré are illustrated in Figure 2.3(b). We can see that an unstable solution in the sense of Lyapunov can in principle be stable in the sense of Poincaré. In this respect Poincaré stability is a weaker requirement.

The difference between these notions can be further illustrated with a simple thought-experiment. Let us imagine a car moving along a road on a flat surface with constant velocity. Suppose that the driver is to follow a point on a path specified by a curve within the boundaries of the road. The point moves with the same velocity as the car. Let us denote the trajectory of the car by $\mathbf{x}(t, \mathbf{x}_0)$, and the trajectory the driver should follow by $\mathbf{x}(t, \mathbf{x}'_0)$. The difference $\|\mathbf{x}_0 - \mathbf{x}'_0\|$ stands for the initial distance of the car from the curve, and the driver aims to minimize the value of $\|\mathbf{x}(t, \mathbf{x}'_0) - \mathbf{x}(t, \mathbf{x}_0)\|$. When the path is an infinitely long straight line this difference would not exceed the value of $\|\mathbf{x}_0 - \mathbf{x}'_0\|$. Thus the motion would be stable in the sense of Lyapunov. Let us imagine now that the path $\mathbf{x}(t, \mathbf{x}'_0)$ is not a straight line but a curved one, for example, a circle. Elementary physics tells us that when the curvature of $\mathbf{x}(t, \mathbf{x}'_0)$ exceeds a certain critical value the friction forces would not be able to support the motion of the car along the path $\mathbf{x}(t, \mathbf{x}'_0)$. Hence eventually, even if $\|\mathbf{x}_0 - \mathbf{x}'_0\| = 0$, the car's trajectory $\mathbf{x}(t, \mathbf{x}_0)$ would deviate from $\mathbf{x}(t, \mathbf{x}'_0)$. Therefore this motion of $\mathbf{x}(t, \mathbf{x}_0)$ with respect to $\mathbf{x}(t, \mathbf{x}'_0)$ cannot be defined as stable in the sense of Lyapunov. Moreover, if it were stable for all non-zero velocities and all circle curvatures then such a motion would contradict the laws of physics.

Does this imply that stable and at the same time physics-consistent motions are not achievable in this example? Apparently not, provided that we allow the driver to change the velocity of the car. Although in this case we may not be able to ensure that the motions are stable in the sense of Lyapunov, we will be able to invent a driving strategy that makes these motions stable in the sense of Definition 2.3.3.

Indeed, steering the car towards the path (now it is viewed as a set $\mathcal{A}$) and then driving along the path with sufficiently slow velocity would be a plausible solution.

This example, although informal and simple, allows us to draw rather general conclusions. In the first case, when the velocities are fixed and equal, we considered a tracking problem in which the system (comprised of the driver and the car) is to follow trajectories $\mathbf{x}(t, \mathbf{x}_0')$ generated by a reference model. In the second case we considered a path-following problem. Tracking a reference trajectory is shown to be a stricter goal than simply traveling along a path. Similarly, stability of solutions in the sense of Definition 2.3.1 is a stricter requirement than stability in the sense of Definition 2.3.3. In some problems achieving the latter is a more realistic goal than achieving the former. Taking advantage of the possibility of using various stability notions allows us to formulate (or in some cases imagine) the system's goals which are most adequate to the constraints inherent to the system. This in turn enables us to avoid unnecessary complications from the beginning and thus allows us to concentrate on the very essence of the problem.

Let us proceed with the analysis of stability notions considered so far. The set $\mathcal{A}$ in the definition of Poincaré stability is determined by some trajectory of the same system (see Figure 2.3(b)). It is obvious that the set $\mathcal{A}$ thus defined cannot be arbitrary. Further generalization of this notion leads us to the notion of stability of a (positively) invariant set $\mathcal{A}$ in the sense of Lyapunov (Zubov 1964).

Definition 2.3.4 Let $\mathbf{x}(t, \mathbf{x}_0) : \mathbb{R} \times \mathbb{R}^n \to \mathbb{R}^n$ be a solution of a dynamical system defined for all $t \geq t_0$, $t_0, t \in \mathbb{R}$ and passing through $\mathbf{x}_0 \in \mathbb{R}^n$ at $t = t_0$; suppose that $\mathcal{A} \subset \mathbb{R}^n$ is a closed (forward) invariant set. The set $\mathcal{A}$ is stable in the sense of Lyapunov iff for every $\varepsilon > 0$, $\varepsilon \in \mathbb{R}$ there exists $\delta > 0$, $\delta \in \mathbb{R}$ such that

$$\|\mathbf{x}_0\|_{\mathcal{A}} \leq \delta \Rightarrow \|\mathbf{x}(t, \mathbf{x}_0)\|_{\mathcal{A}} \leq \varepsilon \; \forall \, t \geq t_0. \tag{2.17}$$

Alternatively,

$$\|\mathbf{x}_0\|_{\mathcal{A}} \leq \delta \Rightarrow \|\mathbf{x}(t, \mathbf{x}_0)\|_{\mathcal{A}_{\infty},[t_0,\infty]} \leq \varepsilon. \tag{2.18}$$

The simplest example of Lyapunov stability of invariant sets is the Lyapunov stability of equilibria.[5] In general Definition 2.3.4 allows us to define the stability of forward-invariant domains. The latter property is useful for those problems in which the precise location of the target set is unknown but information about the domain to which it belongs is available. Similarly to the case of stability of solutions, one can also define the (global) *asymptotic* stability of sets in the sense of Lyapunov. In order to do so, we require that in addition to (2.17) and (2.18) the following

[5] Notice that in this case all three of the versions of stability considered are equivalent.

property holds:

$$\lim_{t\to\infty} \|\mathbf{x}(t,\mathbf{x}_0)\|_{\mathcal{A}} = 0. \tag{2.19}$$

All stability notions considered so far relate the behavior of the system's solutions to a set or another trajectory over infinitely long and *connected* intervals of time. There are systems, however, for which the solutions do not stay near a given set indefinitely. Solutions of these systems may eventually escape any small neighborhood of the set. However, they always return to the same neighborhood after some time. The key property here is the recurrence of motion, and stability of such recurrence is formally specified by the notion of *Poisson stability*.

Definition 2.3.5 Let $\mathbf{x}(t,\mathbf{x}_0) : \mathbb{R} \times \mathbb{R}^n \to \mathbb{R}^n$ be a solution defined for all $t \geq t_0$, with $t_0, t \in \mathbb{R}$, and passing through $\mathbf{x}_0 \in \mathbb{R}^n$ at $t = t_0$. Point $\mathbf{x}_0$ is called stable in the sense of Poisson iff for all $\varepsilon > 0$, $\delta > 0$, $\varepsilon, \delta \in \mathbb{R}$ and any $t' \geq t_0$ there exists $t'' > t' + \delta$ such that

$$\|\mathbf{x}_0 - \mathbf{x}(t'',\mathbf{x}_0)\| \leq \varepsilon. \tag{2.20}$$

Poisson stability of a point implies that, should the system's trajectory pass through a point $\mathbf{x}_0$ once, it will visit an arbitrarily small neighborhood of $\mathbf{x}_0$ infinitely many times. Notice that, despite the fact that we refer to the point $\mathbf{x}_0$ as stable, the system's trajectories associated with this point are allowed to generate arbitrarily large (but finite) excursions in the state space. This property is illustrated in Figure 2.4.

One can clearly see that stability of a point in the sense of Poisson is a much weaker requirement than that of stability in the sense of Lyapunov. Generalization of the former, when property (2.20) holds for every point in a set leads to the notion of Poisson stability of a set. In Chapter 5 we will demonstrate how this property can

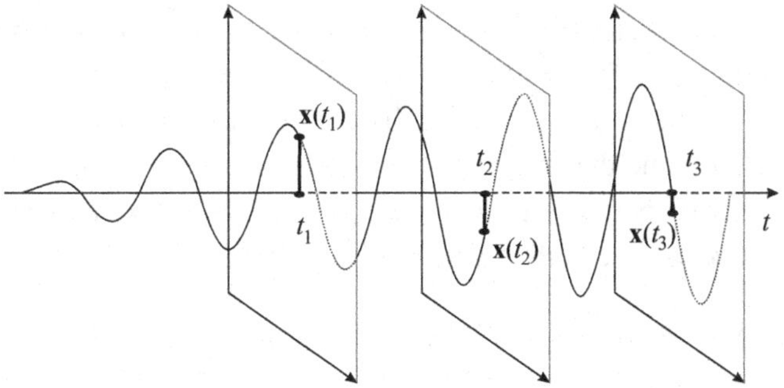

Figure 2.4 Stability in the sense of Poisson: despite the fact that the distances from $\mathbf{x}(t_i), i = 1, 2, \ldots$ to $\mathbf{x}_0$ (depicted as black solid bold lines) do not grow with time, the maximal deviation of the solution from $\mathbf{x}_0$ grows.

be used to solve a class of adaptive regulation problems for systems with nonlinear parametrization.

So far we have reviewed a number of stability notions determining various degrees of "smallness" of the system's response to perturbations. Even though we did not provide a detailed comparison of these notions in every respect, we illustrated the fact that the difference in how the "smallness" is defined may be an important factor both limiting and enabling solutions to specific problems of regulation. In the moving-car example considered earlier, however, we did not use any formal criteria for specifying the desired asymptotic behavior of the system. Instead we used our common-sense intuition and basic knowledge of physics. In order to be able to solve a wider range of problems such formal criteria and methods for assessing asymptotic properties of the system's solutions are needed. One such criterion has already been discussed (see Lemma 2.1). This criterion, although useful for establishing facts of asymptotic convergence of the solutions to zero, does not tell us enough about other asymptotic properties of the system, such as stability. In the next section we will present a brief review of one of the most powerful and instrumental techniques for deriving such stability criteria – *the method of Lyapunov functions*. Our introduction of the method is kept here at the very elementary level, which is just sufficient for one to understand how the method is used in the domain of adaptive regulation. Those interested in developing a more detailed familiarity with the method are referred to the excellent texts by Zubov (1964), Bhatia and Szego (1970), and Lyapunov (1892).

2.4 The method of Lyapunov functions

Let us suppose that the system's behavior is described by the following equation:

$$\dot{\mathbf{x}} = \mathbf{f}(\mathbf{x}, \boldsymbol{\theta}, \mathbf{u}, t), \quad \mathbf{f} : \mathbb{R}^n \times \mathbb{R}^d \times \mathbb{R}^m \times \mathbb{R} \to \mathbb{R}^n, \ \mathbf{f} \in \mathcal{C}^0, \tag{2.21}$$

where $\mathbf{x}$ is the state vector, $\boldsymbol{\theta}$ is the vector of parameters of which the value is unknown, and $\mathbf{u}$ stands for the vector of inputs. We will suppose, if not stated otherwise, that the inputs $\mathbf{u}$ are modeled by continuous functions $\mathbf{u} : \mathbb{R} \to \mathbb{R}^m$. In addition, we will assume that the right-hand side of (2.21) is *locally Lipschitz*, that is, for some given and bounded domains Ω_x, Ω_θ, Ω_u there exist constants D_x, D_θ, D_u such that $\forall\, \mathbf{x}, \mathbf{x}' \in \Omega_x$, $\boldsymbol{\theta}, \boldsymbol{\theta}' \in \Omega_\theta$, $\mathbf{u}, \mathbf{u}' \in \Omega_u$:

$$\|\mathbf{f}(\mathbf{x}, \boldsymbol{\theta}, \mathbf{u}, t) - \mathbf{f}(\mathbf{x}', \boldsymbol{\theta}', \mathbf{u}', t)\| \leq D_x \|\mathbf{x} - \mathbf{x}'\| + D_\theta \|\boldsymbol{\theta} - \boldsymbol{\theta}'\| + D_u \|\mathbf{u} - \mathbf{u}'\|. \tag{2.22}$$

What can we say about the global asymptotic properties of (2.21) from knowledge of some local properties of the system, for example property (2.22)? It is well known that continuity of the right-hand side of (2.21) guarantees local existence of the system's solutions, and property (2.22) ensures that solutions of (2.21) are uniquely defined locally (Arnold 1990). In order to provide further global characterizations of the system's behavior, additional information about the right-hand side of (2.21) is required. In the analysis of stability, defining such local information involves the notion of a positive definite function.

Definition 2.4.1 A function $V : \mathbb{R}^n \to \mathbb{R}$ is called positive definite iff $V(\mathbf{x}) \geq 0$ for all $\mathbf{x} \in \mathbb{R}^n$.

In the class of positive definite functions we will consider only those functions which satisfy the following additional constraint:

$$\rho_1(\|\mathbf{x}\|) \leq V(\mathbf{x}) \leq \rho_2(\|\mathbf{x}\|), \quad \rho_1(\cdot),\ \rho_2(\cdot) \in \mathcal{K}_\infty. \tag{2.23}$$

Constraint (2.23) enables us to use the functions $V(\mathbf{x})$ as the estimates of distance from a given point $\mathbf{x}$ to the origin. Indeed, one can easily see that if the function $V(\mathbf{x}(t, \mathbf{x}_0))$ does not grow with time then the corresponding solution $\mathbf{x}(t, \mathbf{x}_0)$ of (2.21) remains bounded in forward time. This and other properties can be deduced from a more general statement such as the *Lyapunov stability theorem*. Theorem 2.1 below is a special case of this theorem.

Theorem 2.1 *Let $\mathbf{x} = 0$ be an equilibrium of system (2.21), and there exists a positive definite and differentiable function $V(\mathbf{x})$ satisfying (2.23). Let us suppose that for all $\mathbf{x}$ the following property holds:*

$$\dot{V} \leq 0. \tag{2.24}$$

Then the equilibrium $\mathbf{x} = 0$ is (globally) stable in the sense of Lyapunov.

In addition, if there exists a positive definite function

$$W(\mathbf{x}) : \mathbb{R}^n \to \mathbb{R},\ \alpha_1(\|\mathbf{x}\|) \leq W(\mathbf{x}),\ \alpha_1(\cdot) \in \mathcal{K}, \tag{2.25}$$

such that

$$\dot{V} \leq -W(\mathbf{x}(t, \mathbf{x}_0)), \tag{2.26}$$

then the equilibrium $\mathbf{x} = 0$ is (globally) asymptotically stable in the sense of Lyapunov.

Proof of Theorem 2.1. Given that the right-hand side of (2.21) is locally Lipschitz in $\mathbf{x}$ and continuous in t, we can conclude that for every $\mathbf{x}_0 \in \mathbb{R}^n$ there exists an interval $[t_0, T]$, $T > t_0$, such that solution $\mathbf{x}(t, \mathbf{x}_0)$ of the system is defined for all $t \in [t_0, T]$. Furthermore, condition (2.24) guarantees that the solution $\mathbf{x}(t, \mathbf{x}_0)$ is defined for all $t \geq t_0$.

Indeed, let $[t_0, T]$ be the maximal interval of existence of the system's solution, and let T be finite. Consider the difference $V(\mathbf{x}(t,\mathbf{x}_0)) - V(\mathbf{x}(t_0,\mathbf{x}_0))$:

$$V(\mathbf{x}(t,\mathbf{x}_0)) - V(\mathbf{x}(t_0,\mathbf{x}_0)) = \int_{t_0}^{t} \frac{\partial V}{\partial \mathbf{x}} \mathbf{f}(\mathbf{x}(\tau,\mathbf{x}_0), \boldsymbol{\theta}, \mathbf{u}(\tau), \tau) d\tau.$$

On taking (2.24) into account, we obtain

$$V(\mathbf{x}(t,\mathbf{x}_0)) - V(\mathbf{x}(t_0,\mathbf{x}_0)) = \int_{t_0}^{t} \dot{V}(\tau) d\tau \leq 0.$$

Hence

$$V(\mathbf{x}(t,\mathbf{x}_0)) \leq V(\mathbf{x}_0) \; \forall \, t \in [t_0, T].$$

Moreover, in accordance with (2.23) the following holds:

$$\rho_1(\|\mathbf{x}(t,\mathbf{x}_0)\|) \leq \rho_2(\|\mathbf{x}_0\|) \; \forall \, t \in [t_0, T] \Rightarrow$$
$$\|\mathbf{x}(t,\mathbf{x}_0)\| \leq \rho_1^{-1}(\rho_2(\|\mathbf{x}_0\|)) \; \forall \, t \in [t_0, T],$$

where $\rho_1^{-1}(\rho_1(s)) = s \; \forall \, s \geq 0$. Consider the domain $\mathcal{D} = \{(t,\mathbf{x}),\ t \in [t_0, T],\ \mathbf{x} \in \mathbb{R}^n \,|\, \|\mathbf{x}\| \leq 2\rho_1^{-1}(\rho_2(\|\mathbf{x}_0\|))\}$; $\mathcal{D}$ is compact, and hence $\mathbf{x}(t,\mathbf{x}_0)$, $t \in [t_0, T]$ can be continued until the boundary of $\mathcal{D}$ (the right-hand side is Lipschitz in $\mathbf{x}$ and $\mathbf{u}$, and continuous in t). Because $\mathbf{x}(t,\mathbf{x}_0)$ cannot reach the boundary $\mathbf{x} = 2\rho_1^{-1}(\rho_2(\|\mathbf{x}_0\|))$ it must necessarily cross the boundary $t = T$. Given that the right-hand side of (2.21) is locally Lipschitz and continuous in t, the interval of existence of the solution $\mathbf{x}(t,\mathbf{x}_0)$ of (2.21) can be extended by a finite increment Δ. This, however, is in contradiction with the fact that T is finite. Hence we can conclude that $\mathbf{x}(t,\mathbf{x}_0)$ is defined for all $t \geq t_0$, and that it is bounded.

Notice that the composite $\rho_1^{-1}(\rho_2(s))$ is a non-decreasing function of s, and $\rho_1^{-1}(\rho_2(s)) \in \mathcal{K}_\infty$. Thus, denoting $\varepsilon(\delta) = \rho_1^{-1}(\rho_2(\delta))$, we arrive at

$$\|\mathbf{x}_0\| \leq \delta \Rightarrow \|\mathbf{x}(t,\mathbf{x}_0)\| \leq \rho_1^{-1}(\rho_2(\delta)) = \varepsilon(\delta).$$

The function $\varepsilon(\delta) \in \mathcal{K}_\infty$, hence its range coincides with $\mathbb{R}_{\geq 0}$. Therefore we can conclude now that for every $\tilde{\varepsilon} > 0$ there exists $\delta = \varepsilon^{-1}(\tilde{\varepsilon}) > 0$ such that

$$\|\mathbf{x}_0\| \leq \delta \Rightarrow \|\mathbf{x}(t,\mathbf{x}_0)\| \leq \varepsilon(\delta) = \varepsilon(\varepsilon^{-1}(\tilde{\varepsilon})) = \tilde{\varepsilon}.$$

In other words, according to Definition 2.3.4, the origin is globally stable in the sense of Lyapunov.

To prove the second part of the theorem we will follow the argument presented in Khalil (2002). Notice that inequality (2.26) automatically implies

$$\dot{V} \leq -\alpha_1(\|\mathbf{x}\|). \tag{2.27}$$

Therefore

$$V(\mathbf{x}(t,\mathbf{x}_0)) - V(\mathbf{x}_0) \le \int_{t_0}^{t} -\alpha_1(\|\mathbf{x}(\tau,\mathbf{x}_0)\|)d\tau \ \forall\, t \ge t_0, \tag{2.28}$$

and hence

$$\lim_{t\to\infty} \int_{t_0}^{t} \alpha_1(\|\mathbf{x}(\tau,\mathbf{x}_0)\|)d\tau \le V(\mathbf{x}_0) < \infty. \tag{2.29}$$

Moreover, $V(\mathbf{x}(t,\mathbf{x}_0))$ is a monotone function of t, and it is bounded from below because it is positive definite. Thus there exists $a \in \mathbb{R}_{\ge 0}$ such that $\lim_{t\to\infty} V(\mathbf{x}(t,\mathbf{x}_0)) = a$. Let us now show that $a = 0$. Suppose that $a > 0$. Inequality (2.23) implies that $\|\mathbf{x}\| \ge \rho_2^{-1}(V(\mathbf{x}))$, thus $\dot{V} \le -\alpha_1(\|\mathbf{x}\|) \le -\alpha_1(\rho_2(V(\mathbf{x}(t,\mathbf{x}_0)))) \le -\alpha_1(\rho_2^{-1}(a))$. This leads to the conclusion that the function $V(\mathbf{x}(t,\mathbf{x}_0)) \le V(\mathbf{x}(t_0,\mathbf{x}_0)) - (t-t_0)\alpha_1(\rho_2^{-1}(a))$ becomes negative in finite time. The latter, however, is not possible because $V(\mathbf{x})$ is assumed to be positive definite.

If the function $\mathbf{u} : \mathbb{R} \to \mathbb{R}^m$ on the right-hand side of (2.21) is bounded, the function $\mathbf{f}(\cdot)$ is bounded with respect to t, and the function $\alpha_1(\cdot)$ is differentiable, then the proof of the second part of the theorem can be easily completed by using Lemma 2.1. Indeed, in this case differentiability of $\alpha_1(\cdot)$, boundedness of $\mathbf{x}$, $\mathbf{u}$, and $\boldsymbol{\theta}$, and boundedness of the right-hand side of (2.21) with respect to t imply that the function $\alpha_1(\|\mathbf{x}(t,\mathbf{x}_0)\|)$ is uniformly continuous in t. According to Lemma 2.1, inequality (2.29) assures that

$$\lim_{t\to\infty} \alpha_1(\|\mathbf{x}(t,\mathbf{x}_0)\|) = 0,$$

and strict monotonicity of the function $\alpha_1(\cdot)$ implies that $\|\mathbf{x}(t,\mathbf{x}_0)\| \to 0$ as $t \to \infty$. □

The main benefit of Theorem 2.1 is that it allows us to reduce the analysis of asymptotic properties of the system's solutions to an easier problem of checking the algebraic inequalities (2.24) and (2.26). These inequalities can serve as target constraints determining the desired behavior of an adaptive system. Notice that these constraints do not require precise knowledge of the unknown parameters $\boldsymbol{\theta}$. This property allows us to consider the theorem (and many other similar statements) as a suitable tool for solving a range of synthesis problems in the domain of adaptive regulation.

In the proof of the second part of the theorem, regarding asymptotic stability, we considered a specific case illustrating how Lemma 2.1 can be used to show that $\mathbf{x}$ approaches the origin asymptotically. The main reason for using this particular technique is that the use in tandem of a stability proof ensuring boundedness of the system's solutions followed by the analysis of estimates (2.27)–(2.29) lies at the core of many stability proofs in the literature on adaptive control and regulation. This motivated us to use the very same idea in the proof of Theorem 2.1. We too will regularly use this tandem in the next chapters.

Despite its simplicity and generality, the method of Lyapunov functions has an obvious disadvantage. In order to be able to use the method one needs to find a function $V(\mathbf{x})$ satisfying properties (2.24) and (2.26). Finding such a function is a non-trivial operation. Yet, there are large classes of systems for which the corresponding Lyapunov functions are already known. One of these classes of systems is considered in the next section.

2.5 Linear skew-symmetric systems with time-varying coefficients

Let the system's dynamics be given by the following system of ordinary differential equations:

$$\begin{aligned} \dot{\mathbf{x}}_1 &= A\mathbf{x}_1 + B\phi^{\mathrm{T}}(t)\mathbf{x}_2, \\ \dot{\mathbf{x}}_2 &= -\phi(t)C\mathbf{x}_1, \end{aligned} \tag{2.30}$$

where $\mathbf{x}_1 \in \mathbb{R}^q, \mathbf{x}_2 \in \mathbb{R}^p, \phi(t) : \mathbb{R} \to \mathbb{R}^{p\times m}$ is a continuous function of t, and A, B, and C are $q \times q$, $q \times m$, and $m \times q$ matrices, respectively. We have already mentioned that a number of problems in the domain of adaptive regulation can be reduced to the analysis of (2.30) (see e.g. Narendra and Annaswamy (1989)). Therefore understanding the basic asymptotic properties of system (2.30) is desirable.

Let us first investigate the stability of the zero equilibrium of system (2.30). Suppose that there exists a positive definite and symmetric matrix $P = P^{\mathrm{T}}$:

$$\mathbf{x}^{\mathrm{T}}P\mathbf{x} > 0 \ \forall\, \mathbf{x} \neq 0 \tag{2.31}$$

such that

$$\begin{aligned} PA^{\mathrm{T}} + AP = -Q,\ \ Q = Q^{\mathrm{T}},\ \mathbf{x}^{\mathrm{T}}Q\mathbf{x} > 0\,\forall\, \mathbf{x} \neq 0, \\ PB = C^{\mathrm{T}}. \end{aligned} \tag{2.32}$$

Since $P = P^{\mathrm{T}}$ and $Q = Q^{\mathrm{T}}$ are positive definite, the eigenvalues of P and Q are real and positive, and moreover the following property holds:

$$\begin{aligned} \lambda_{\min}(P)\|\mathbf{x}\|^2 \leq \mathbf{x}^{\mathrm{T}}P\mathbf{x} \leq \lambda_{\max}(P)\|\mathbf{x}\|^2, \\ \lambda_{\min}(Q)\|\mathbf{x}\|^2 \leq \mathbf{x}^{\mathrm{T}}Q\mathbf{x} \leq \lambda_{\max}(Q)\|\mathbf{x}\|^2. \end{aligned} \tag{2.33}$$

Indeed, let λ be an eigenvalue of P (possibly complex), $\mathbf{x}_\lambda$ be its corresponding eigenvector, and λ^* and $\mathbf{x}_\lambda^*$ be the complex conjugates of λ and $\mathbf{x}_\lambda$ respectively.

Then, according to (2.31), the following holds:

$$0 < \mathbf{x}_\lambda^* P \mathbf{x}_\lambda = \lambda \|\mathbf{x}_\lambda\|^2 = \mathbf{x}_\lambda^{\mathrm{T}} P \mathbf{x}_\lambda^{*\mathrm{T}} = \lambda^* \|\mathbf{x}_\lambda\|^2.$$

Therefore, $\lambda \in \mathbb{R}$ and $\lambda > 0$. In order to see that (2.33) holds too, we notice that $P = P^{\mathrm{T}}$ is Hermitian. Hence there is a non-singular orthonormal $q \times q$ matrix T, $T^{\mathrm{T}}T = I$, consisting of the eigenvectors of P (see e.g. Bellman (1970) and Lancaster and Tismenetsky (1985)) such that $T^{\mathrm{T}}PT$ is a diagonal matrix with the eigenvalues of P placed on its main diagonal. Finally, let $\mathbf{x} = T\boldsymbol{\xi}$, and notice that $\|\mathbf{x}\|^2 = \boldsymbol{\xi}^{\mathrm{T}}T^{\mathrm{T}}T\boldsymbol{\xi} = \|\boldsymbol{\xi}\|^2$. Thus

$$\begin{aligned} \mathbf{x}^{\mathrm{T}} P \mathbf{x} &\leq \lambda_{\max}(P)\|\boldsymbol{\xi}\|^2 = \lambda_{\max}(P)\|\mathbf{x}\|^2, \\ \mathbf{x}^{\mathrm{T}} P \mathbf{x} &\geq \lambda_{\min}(P)\|\boldsymbol{\xi}\|^2 = \lambda_{\min}(P)\|\mathbf{x}\|^2. \end{aligned} \tag{2.34}$$

Let us proceed with the stability analysis of the zero equilibrium of (2.30). We will do so using the method of Lyapunov functions. According to the method, stability of the equilibrium is guaranteed if we find a function $V(\cdot)$ satisfying conditions (2.24) and (2.26). We don't know this function yet, hence a plausible option is to select a *candidate function* (Lyapunov candidate function) that satisfies the constraint of positive definiteness. After completing this step we can continue with checking whether the second condition, (2.26), holds too.

Let us pick the following Lyapunov candidate for system (2.30):

$$V(\mathbf{x}) = \mathbf{x}_1^{\mathrm{T}} P \mathbf{x}_1 + \mathbf{x}_2^{\mathrm{T}} \mathbf{x}_2. \tag{2.35}$$

Function $V(\mathbf{x})$ defined as in (2.35) satisfies condition (2.23). Hence Lyuapunov stability of the origin will follow if we show that

$$\dot{V} \leq 0.$$

For this purpose we consider

$$\begin{aligned} \dot{V} &= \mathbf{x}_1^{\mathrm{T}} P(A\mathbf{x}_1 + B\phi(t)^{\mathrm{T}}\mathbf{x}_2) + (A\mathbf{x}_1 + B\phi(t)^{\mathrm{T}}\mathbf{x}_2)^{\mathrm{T}} P\mathbf{x}_1 - 2\mathbf{x}_2^{\mathrm{T}}\phi(t)C\mathbf{x}_1 \\ &= \mathbf{x}_1^{\mathrm{T}}(PA + A^{\mathrm{T}}P)\mathbf{x}_1 + 2\mathbf{x}_1^{\mathrm{T}} PB\phi(t)^{\mathrm{T}}\mathbf{x}_2 - 2\mathbf{x}_2^{\mathrm{T}}\phi(t)C\mathbf{x}_1. \end{aligned} \tag{2.36}$$

Taking (2.32) into account, we obtain that

$$\dot{V} \leq -\mathbf{x}_1 Q\mathbf{x}_1 + 2\mathbf{x}_1^{\mathrm{T}} PB\phi(t)^{\mathrm{T}}\mathbf{x}_2 - 2\mathbf{x}_2^{\mathrm{T}}\phi(t)C\mathbf{x}_1 \leq -\mathbf{x}_1^{\mathrm{T}} Q\mathbf{x}_1 \leq 0. \tag{2.37}$$

The latter inequality, as follows from Theorem 2.1, guarantees that the zero equilibrium of (2.30) is stable. Formally this statement is summarized in the following lemma.

Lemma 2.2 *Consider system (2.30) and suppose that there exists a positive definite and symmetric matrix $P = P^{\mathrm{T}}$ satisfying the conditions (2.32). Then the zero equilibrium of (2.30) is globally stable in the sense of Lyapunov.*

If, in addition, the function $\phi(t)$ on the right-hand side of (2.30) is bounded uniformly in t,

$$\exists\, M \in \mathbb{R} : \ \|\phi(t)\| \leq M \ \forall\, t \in \mathbb{R}, \tag{2.38}$$

then

$$\lim_{t\to\infty} \mathbf{x}_1(t) = 0. \tag{2.39}$$

Proof of Lemma 2.2 The first part of the lemma is already proven. In order to see that the second part of the lemma holds too, one can employ the estimate

$$\lambda_{\min}(Q)\|\mathbf{x}_1\|^2 \leq \mathbf{x}_1^{\mathrm{T}} Q \mathbf{x}_1 \leq \lambda_{\max}(Q)\|\mathbf{x}_1\|^2$$

and apply Lemma 2.1 to

$$V(\mathbf{x}(t)) - V(\mathbf{x}(t_0)) \leq -\int_{t_0}^{t} \mathbf{x}_1(\tau)^{\mathrm{T}} Q \mathbf{x}(\tau) d\tau \leq -\int_{t_0}^{t} \lambda_{\min}(Q)\|\mathbf{x}_1(\tau)\|^2\, d\tau \Rightarrow$$

$$\int_{t_0}^{t} \lambda_{\min}(Q)\|\mathbf{x}_1(\tau)\|^2\, d\tau \leq V(\mathbf{x}(t_0)) \ \forall\, t \geq t_0. \tag{2.40}$$

□

Lemma 2.2 and its proof constitute a simple illustration of how the method of Lyapunov functions can be used in the analysis of stability of equilibria in system (2.30). Even though the proof is rather elementary, it offers a useful and instrumental interpretation in the context of adaptive regulation. Let us suppose, for example, that $\phi(t)$ is a disturbance acting on the system's dynamics. Property (2.39) can be viewed as the desired behavior of the system, and $\mathbf{x}_2$ are the system's internal variables, of which the function is to minimize the influence of the disturbance on the desired behavior. The first line of condition (2.32) serves as an existence hypothesis stipulating the possibility that the desired behavior (2.39) is realizable in the absence of perturbations.

The argument in the proof of Lemma 2.2 can be straightforwardly generalized to the case of nonlinear systems. We consider such cases in detail in Chapter 3. We will see that a sequence of steps very similar to that described by (2.36)–(2.40) is a common ingredient of many stability proofs in the domain of nonlinear adaptive regulation. In this respect the equations in (2.30) constitute a good prototype for the systematic study of such systems.

So far we have demonstrated how the method of Lyapunov functions together with Lemma 2.1 can be employed to derive conditions ensuring that a part of the system's state vector, $\mathbf{x}_1(t)$, converges to zero asymptotically. What can we say about the rest of the system's variables, namely about $\mathbf{x}_2(t)$? First of all, by applying Lemma 2.1 to (2.38) and (2.39), we can conclude that

$$\lim_{t\to\infty} \dot{\mathbf{x}}_1(t) = 0. \tag{2.41}$$

Hence

$$\lim_{t\to\infty} A\mathbf{x}_1(t) + B\phi(t)^{\mathrm{T}}\mathbf{x}_2(t) = \lim_{t\to\infty} B\phi(t)^{\mathrm{T}}\mathbf{x}_2(t) = 0. \tag{2.42}$$

If $\dot{\mathbf{x}}_2 = 0$ then property (2.42) implies that the vector $\mathbf{x}_2(t)$ is orthogonal to all of the rows of the matrix $\alpha(t) = B\phi(t)^{\mathrm{T}}$. A vector that is orthogonal to all of the basis vectors of a vector space must necessarily be zero. Hence, if $\alpha(t)$ spans $\mathbb{R}^{p\times q}$, then (2.42) implies that $\mathbf{x}_2 = 0$. In our case $\dot{\mathbf{x}}_2 \neq 0$, but it is nevertheless asymptotically vanishing. Therefore, it is intuitively clear that if $\alpha(t)$ has non-zero projections onto all matrices (vectors) in $\mathbb{R}^{p\times q}$, and $\alpha(t)$ moves sufficiently fast (e.g. so that its velocity is non-vanishing), then condition (2.42) could imply that $\mathbf{x}_2(t) \to 0$ as $t \to \infty$. These properties of $\alpha(t)$ having non-zero projections to every vector in $\mathbb{R}^{q\times p}$ together with the requirement that its velocity is non-vanishing are captured by the notion of persistency of excitation.

Definition 2.5.1 A function $\alpha : \mathbb{R} \to \mathbb{R}^{p\times q}$ is called persistently exciting iff there exist $T, \delta, \Delta \in \mathbb{R}_{>0}$, such that for all $t \in \mathbb{R}$ and every $\boldsymbol{\theta} \in \mathbb{R}^p$ the following inequality holds:

$$\delta\|\boldsymbol{\theta}\|^2 \leq \boldsymbol{\theta}^{\mathrm{T}}\left(\int_t^{t+T} \alpha(\tau)\alpha(\tau)^{\mathrm{T}}\, d\tau\right)\boldsymbol{\theta} \leq \Delta\|\boldsymbol{\theta}\|^2. \tag{2.43}$$

Persistency of excitation of a function admits a simple geometric interpretation. On applying the mean-value theorem to (2.43) we obtain that

$$\boldsymbol{\theta}^{\mathrm{T}}\left(\int_t^{t+T} \alpha(\tau)\alpha(\tau)^{\mathrm{T}}\, d\tau\right)\boldsymbol{\theta} = T(\boldsymbol{\theta}^{\mathrm{T}}\alpha(\tau))(\alpha(\tau)^{\mathrm{T}}\boldsymbol{\theta}),\ \tau \in [t, t+T].$$

Therefore, noticing that $\boldsymbol{\theta}^{\mathrm{T}}\alpha(\tau)^{\mathrm{T}}\alpha(\tau)\boldsymbol{\theta} = \|\alpha(\tau)\boldsymbol{\theta}\|^2$ and taking (2.43) into account we can conclude that there exist $T, \delta, \Delta \in \mathbb{R}_{>0}$:

$$\forall\, t \in \mathbb{R}\ \exists\, \tau \in [t, t+T]:\ \frac{\delta}{T}\|\boldsymbol{\theta}\|^2 \leq \|\alpha(\tau)\boldsymbol{\theta}\|^2 \leq \frac{\Delta}{T}\|\boldsymbol{\theta}\|^2. \tag{2.44}$$

If $\alpha(t)$ is a vector-function, i.e. $q = 1$, then

$$\alpha(\tau)\boldsymbol{\theta} = \|\alpha(\tau)\|\|\boldsymbol{\theta}\|\cos(\beta(\tau)),$$

where $\beta(\tau)$ is the angle between the vectors $\alpha(\tau)$ and $\boldsymbol{\theta}$, $\tau \in [t, t+T]$. Thus (2.44) in this case is equivalent to

$$\frac{\delta}{T} \leq \|\alpha(\tau)\|^2\cos^2(\beta(\tau)) \leq \frac{\Delta}{T}.$$

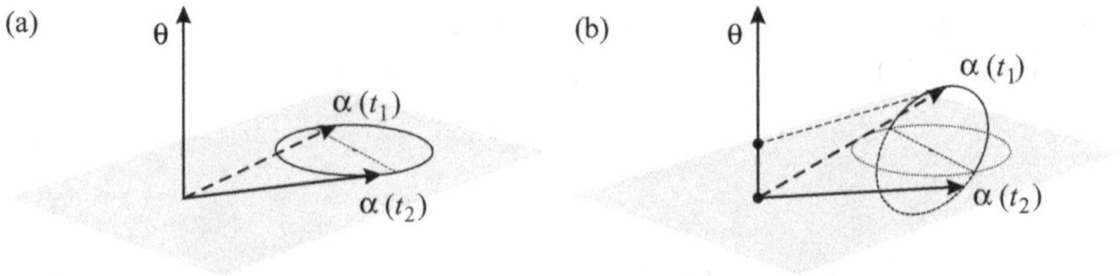

Figure 2.5 Persistency of excitation of the function $\alpha(t)$. In (a) $\alpha(t)$ is not persistently exciting. The vector $\alpha(t)$ stays in Ω for all $t \in \mathbb{R}$, where Ω is the plane normal to $\boldsymbol{\theta}$. In (b) $\alpha(t)$ is persistently exciting. The vector $\alpha(t)$ does not stay in Ω for all t. In particular, there are time instants t_1 and t_2 at which its projection on $\boldsymbol{\theta}$ is not zero.

This means that for all t there exists a time instant $\tau \in [t, t+T]$ such that the angle $\beta(\tau)$ deviates from $\pm\pi/2 + 2\pi k$, $k \in \mathbb{Z}$. If the length of $\|\alpha(\tau)\|$ is bounded from above by M_α then

$$|\cos(\beta(\tau))| \geq \frac{1}{M_\alpha}\sqrt{\frac{\delta}{T}} > 0,$$

and hence the vector $\alpha(\tau)$ must have a non-zero projection on $\boldsymbol{\theta}$. This is illustrated in Figure 2.5.

Let us show that persistency of excitation of the function $B\phi(t)^{\mathrm{T}}$ in (2.30) together with condition (2.39) ensures that

$$\lim_{t\to\infty} \mathbf{x}_2(t) = 0.$$

The result is formulated in Lemma 2.3

Lemma 2.3 *Consider the system*

$$\dot{\mathbf{x}}_2 = -\phi(t)C\mathbf{x}_1(t),$$

where $\mathbf{x}_1(t)$ is a bounded and continuous function satisfying conditions (2.39) and (2.41). Suppose that

$$\lim_{t\to\infty} B\phi(t)^{\mathrm{T}}\mathbf{x}_2(t) = 0,$$

where $\phi(t)$ is a bounded continuous function, and $B\phi(t)^{\mathrm{T}}$ is persistently exciting. Then

$$\lim_{t\to\infty} \mathbf{x}_2(t) = 0.$$

Proof of Lemma 2.3 Consider the interval $[t_0, \infty)$ as a union of the intervals $[t_i, t_{i+1}]$, $t_{i+1} = t_i + T$, $i = 0, 1, \ldots$ Given that $\mathbf{x}_1(t) \to 0$ as $t \to \infty$ and that the function $\phi(t)$ is bounded, there should exist a function $\delta_2(t) : \mathbb{R} \to \mathbb{R}^p$ such that

$$\mathbf{x}_2(\tau) = \mathbf{x}_2(t_i) + \delta_2(\tau), \ \forall\, \tau \in [t_i, t_{i+1}], \ \lim_{t\to\infty} \delta_2(t) = 0. \tag{2.45}$$

Let us denote $B\phi(t) = \alpha(t)$. The function $\alpha(t)$ is persistently exciting, hence there exist $\tau_i \in [t_i, t_i + T]$ such that

$$\|\alpha(\tau_i)\mathbf{x}_2(t_i)\| \geq \frac{\delta}{T}\|\mathbf{x}_2(t_i)\|. \tag{2.46}$$

On substituting (2.45) into (2.46) we obtain

$$\frac{\delta}{T}\|\mathbf{x}_2(t_i)\| \leq \|\alpha(\tau_i)(\mathbf{x}_2(\tau_i) - \delta_2(\tau_i))\| \leq \|\alpha(\tau_i)\delta_2(\tau_i)\| + \|\alpha(\tau_i)\mathbf{x}_2(\tau_i)\|,$$

and finally

$$\lim_{i\to\infty} \frac{\delta}{T}\|\mathbf{x}_2(t_i)\| \leq \lim_{i\to\infty} (\|\alpha(\tau_i)\delta_2(\tau_i)\| + \|\alpha(\tau_i)\mathbf{x}_2(\tau_i)\|) = 0.$$

Therefore, $\mathbf{x}_2(t_i) \to 0$ as $i \to \infty$. This, together with (2.45), implies that $\mathbf{x}_2(t) \to 0$ as $t \to \infty$. □

A remarkable property of system (2.30) subjected to conditions (2.32) is that persistency of excitation of $\phi(t)$ not only assures that both $\mathbf{x}_1(t)$ and $\mathbf{x}_2(t)$ converge to zero asymptotically but also guarantees that the rate of this convergence is exponential. Numerous versions of this result can be found in the literature on adaptive control (see e.g. Morgan and Narendra (1992), Sastry and Bodson (1989), and Narendra and Annaswamy (1989)). We will reproduce it here, providing, in addition, the rate of convergence expressed in terms of A, B, C, and $\phi(t)$. The result is generally based on the argument presented in Loria and Panteley (2002).

Theorem 2.2 *Consider system (2.30), and suppose that conditions (2.32) hold. Moreover, let the function $B\phi(t)^{\mathrm{T}}$ in (2.30) be persistently exciting:*

$$\exists\, \delta,\, T : \int_t^{t+T} \phi(\tau)B^{\mathrm{T}}B\phi(\tau)d\tau \geq \delta\, \forall\, t, \tag{2.47}$$

and[6]

$$\max\{\|\phi(t)\|, \|\dot{\phi}(t)\|\} \leq B_1.$$

Let $\Phi(t, t_0)$, $\Phi(t_0, t_0) = I$ be the fundamental system of solutions of (2.30), and let $\mathbf{p}$ be a vector from $\mathbb{R}^{p+q}$. Then

$$\|\Phi(t_2, t_1)\mathbf{p}\| \leq e^{-\rho(t_2 - t_1)}\|\mathbf{p}\|D_2, \ \forall\, t_2 \geq t_1 \geq t_0,$$

[6] In what follows, if not stated otherwise, the symbol $\|\cdot\|$ applied to a $q \times p$ matrix A will stand for the induced norm $\|A\| = \sup_{\mathbf{x}\in\mathbb{R}^p/\{0\}} \|A\mathbf{x}\|/\|\mathbf{x}\|$. It is clear from this definition that $\|A\mathbf{x}\| \leq \|A\|\|\mathbf{x}\|$. The value of $\|A\|$ is $\|A\| = \lambda_{\max}(A^{\mathrm{T}}A)^{1/2}$ (see e.g. Bernstein (2005)). In systems with uncertainties, finding the eigenvalues of $A^{\mathrm{T}}A$ can be a challenging task. In these cases the following estimate of $\|A\|$ is useful: $\|A\| \leq \sqrt{qp}\max_{i,j}|a_{i,j}|$.

where the parameters ρ and D_2 do not depend on t_0 and $\mathbf{p}$, *and can be expressed explicitly as functions of B_1, the constants δ and T in (2.47), and matrices A, B, C, P, and Q.*

Proof of Theorem 2.2 Let us start with the following lemma

Lemma 2.4 *Let* $\mathbf{x}(t) : \mathbb{R} \to \mathbb{R}^n$ *be a function satisfying*

$$\max\{\|\mathbf{x}\|_{2,[t,\infty]}, \|\mathbf{x}\|_{\infty,[t,\infty]}\} \leq c\|\mathbf{x}(t)\|, \; \forall \, t \geq t_0. \tag{2.48}$$

Then

$$\|\mathbf{x}(t)\| \leq c e^{1/2} e^{-\frac{t-t_1}{2c^2}} \|\mathbf{x}(t_1)\|, \; \forall \, t \geq t_1 \geq t_0. \tag{2.49}$$

Proof of Lemma 2.4. Notice that (2.48) implies

$$\int_t^\infty \|\mathbf{x}(\tau)\|^2 \, d\tau \leq c^2 \|\mathbf{x}(t)\|^2, \; \|\mathbf{x}\|^2_{\infty,[t,\infty]} \leq c^2 \|\mathbf{x}(t)\|^2.$$

Let $v(t) = \int_t^\infty \|\mathbf{x}(\tau)\|^2 \, d\tau$, then $\dot{v} = -\|\mathbf{x}(t)\|^2 \leq -(1/c^2) \int_t^\infty \|\mathbf{x}(\tau)\|^2 \, d\tau = -1/(c^2 v)$. Invoking the comparison lemma from Khalil (2002) results in

$$\int_{t_2}^\infty \|\mathbf{x}(\tau)\|^2 \, d\tau \leq e^{-\frac{t_2-t_1}{c^2}} \int_{t_1}^\infty \|\mathbf{x}(\tau)\|^2 \, d\tau, \; t_2 \geq t_1 \geq t_0.$$

Notice that

$$\begin{aligned} T\|\mathbf{x}(t_2+T)\|^2 &\leq T\|\mathbf{x}\|^2_{\infty[t_2+T,\infty]} \leq \int_{t_2}^{t_2+T} \|\mathbf{x}\|^2_{\infty,[\tau,\infty]} \, d\tau \\ &\leq \int_{t_2}^\infty c^2 \|\mathbf{x}(\tau)\|^2 \, d\tau \leq c^2 e^{-\frac{t_2-t_1}{c^2}} c^2 \|\mathbf{x}(t_1)\|^2. \end{aligned}$$

On letting $T = c^2$ we obtain that

$$\|\mathbf{x}(t_2+T)\| \leq c e^{\frac{T}{2c^2}} e^{-\frac{t_2+T-t_1}{2c^2}} \|\mathbf{x}(t_1)\| = c e^{\frac{1}{2}} e^{-\frac{t_2+T-t_1}{2c^2}} \|\mathbf{x}(t_1)\|.$$

Denoting $\tau = t_2 + T$, we get

$$\|\mathbf{x}(\tau)\| \leq c e^{\frac{1}{2}} e^{-\frac{\tau-t_1}{2c^2}} \|\mathbf{x}(t_1)\| \; \forall \, \tau \geq t_1 + T.$$

The lemma will be proven if we show that the same estimate holds for $t_0 \leq t_1 \leq \tau \leq t_1 + T$. Condition (2.48) implies that $\|\mathbf{x}(\tau)\| \leq c\|\mathbf{x}(t_1)\|$ for all $\tau \geq t_1$. On the other hand,

$$1 = e^{\frac{1}{2}} e^{-\frac{T}{2c^2}} \leq e^{\frac{1}{2}} e^{-\frac{\tau-t_1}{2c^2}} \leq e^{\frac{1}{2}} \; \forall \, t_1 \leq \tau \leq t_1 + T.$$

Hence $\|\mathbf{x}(\tau)\| \leq c\|\mathbf{x}(t_1)\| \leq c e^{\frac{1}{2}} e^{-\frac{\tau-t_1}{2c^2}} \|\mathbf{x}(t_1)\|$ for $t_1 \leq \tau \leq t_1 + T$ as well. □

Let us turn to the proof of the theorem. Consider $\mathbf{x} = \mathrm{col}(\mathbf{x}_1, \mathbf{x}_2)$. If we show that there exists $c \in \mathbb{R}_{>0}$ such that (2.48) holds then according to Lemma 2.4, we can conclude that

$$\|\mathbf{x}(t_2)\| \le c e^{\frac{1}{2}} e^{-\frac{1}{2c^2}(t_2 - t_1)} \|\mathbf{x}(t_1)\|, \ \forall\, t_2 \ge t_1 \ge t_0.$$

The result would then follow immediately if we let $\mathbf{x}(t_1) = \mathbf{p}$ and substitute $\mathbf{x}(t_2) = \Phi(t_2, t_1)\mathbf{p}$ into the inequality above. Let us now find a constant c such that (2.48) holds for system (2.30).

Consider the positive definite function $V(\mathbf{x}) = \mathbf{x}_1^{\mathrm{T}} P \mathbf{x}_1 + \|\mathbf{x}_2\|^2$, where P is a symmetric positive definite matrix satisfying (2.32). According to (2.35)–(2.37), we can conclude that

$$\min\{\lambda_{\min}(P), 1\}\|\mathbf{x}\|^2 \le V(\mathbf{x}) \le \max\{\lambda_{\max}(P), 1\}\|\mathbf{x}\|^2,$$

and that $\dot{V} \le -\mathbf{x}_1^{\mathrm{T}} Q \mathbf{x}_1 \le -\lambda_{\min}(Q)\|\mathbf{x}_1\|^2$. Thus

$$\begin{aligned} \|\mathbf{x}_1\|_{2,[t,\infty]} &\le \frac{\max\{\lambda_{\max}(P), 1\}^{1/2}}{\lambda_{\min}(Q)^{1/2}} \|\mathbf{x}(t)\| = c_1 \|\mathbf{x}(t)\|, \\ \|\mathbf{x}\|_{\infty,[t,\infty]} &\le \frac{\max\{\lambda_{\max}(P), 1\}^{1/2}}{\min\{\lambda_{\min}(P), 1\}^{1/2}} \|\mathbf{x}(t)\| = c_2 \|\mathbf{x}(t)\|, \ \forall\, t \ge t_0. \end{aligned} \tag{2.50}$$

Let us now estimate $\|\mathbf{x}_2\|_{2,[t,\infty]}$. In order to do so we introduce a new variable $\mathbf{z} = \mathbf{x}_2 - \phi(t) B^{\mathrm{T}} \mathbf{x}_1$ and consider its derivative w.r.t. t:

$$\begin{aligned} \dot{\mathbf{z}} = &-\phi(t) B^{\mathrm{T}} B \phi(t)^{\mathrm{T}} \mathbf{z} - [\phi(t) B^{\mathrm{T}} A + (\phi(t) B^{\mathrm{T}} B \phi(t)^{\mathrm{T}}) \phi(t) B^{\mathrm{T}} \\ &+ \phi(t) C + \dot{\phi}(t) B^{\mathrm{T}}] \mathbf{x}_1. \end{aligned} \tag{2.51}$$

To proceed further we will need the following lemma.

Lemma 2.5 *Consider the system*

$$\dot{\mathbf{z}} = -\Gamma \alpha(t) \alpha(t)^{\mathrm{T}} \mathbf{z}, \ \mathbf{z} \in \mathbb{R}^n, \alpha : \mathbb{R} \to \mathbb{R}^{n \times m}, \tag{2.52}$$

where α is persistently exciting (i.e. property (2.43) holds), $\|\alpha(t)\| \le M$, and $\Gamma = \Gamma^{\mathrm{T}}$ is a positive definite matrix. Let $\mathbf{z}(t, t_0)$ be a solution of (2.52) passing through $\mathbf{z}_0$ at $t = t_0$. Then there exist $\lambda, D \in \mathbb{R}_{>0}$ such that

$$\|\mathbf{z}(t, t_0)\| \le D e^{-\lambda(t - t_0)} \|\mathbf{z}_0\|, \ t \ge t_0,$$

where

$$D = \left(\frac{\lambda_{\max}(\Gamma)}{\lambda_{\min}(\Gamma)}\right)^{\frac{1}{2}} e^{\lambda T}, \ \lambda = \frac{\delta \lambda_{\min}(\Gamma)}{T(1 + \lambda_{\max}(\Gamma) M^2 T)^2}$$

independently of $\mathbf{z}_0$, t, and t_0.

Proof of Lemma 2.5. Consider the following positive definite function:

$$V(\mathbf{z}) = \|\mathbf{z}\|^2_{\Gamma^{-1}},\ \lambda_{\min}(\Gamma^{-1})\|\mathbf{z}\|^2 \leq V(\mathbf{z}) \leq \lambda_{\max}(\Gamma^{-1})\|\mathbf{z}\|^2.$$

Its derivative is $\dot{V} = -2\|\mathbf{z}^{\mathrm{T}}\alpha(t)\|^2$, and hence

$$V(\mathbf{z}(t_0+T)) - V(\mathbf{z}(t_0)) \leq -2\int_{t_0}^{t_0+T} \|\alpha(\tau)^{\mathrm{T}}\mathbf{z}(\tau)\|^2\, d\tau.$$

Given that $\mathbf{z}(\tau) = \mathbf{z}(t_0) - \Gamma\int_{t_0}^{\tau}\alpha(s)\alpha(s)^{\mathrm{T}}\mathbf{z}(s)ds$ and $(a-b)^2 \geq a^2\beta/(1+\beta) - \beta b^2$, $\beta \in \mathbb{R}_{>0}$, we have that

$$\|\alpha(\tau)^{\mathrm{T}}\mathbf{z}(\tau)\|^2 \geq \|\alpha(\tau)^{\mathrm{T}}\mathbf{z}(t_0)\|^2\frac{\beta}{1+\beta} - \beta\left(\int_{t_0}^{\tau}\alpha(\tau)^{\mathrm{T}}\Gamma\alpha(s)\alpha(s)^{\mathrm{T}}\mathbf{z}(s)ds\right)^2.$$

Thus, taking into account that $\alpha(t)$ is persistently exciting we can derive the following estimate:

$$\begin{aligned}
&2\int_{t_0}^{t_0+T}\|\alpha(\tau)^{\mathrm{T}}\mathbf{z}(\tau)\|^2\, d\tau \\
&\geq \frac{2\beta\delta}{1+\beta}\|\mathbf{z}(t_0)\|^2 - 2\int_{t_0}^{t_0+T}\int_{t_0}^{t_0+T}(\alpha(\tau)^{\mathrm{T}}\Gamma\alpha(s))^2\, ds \\
&\quad\times\int_{t_0}^{t_0+T}\|\alpha(s)^{\mathrm{T}}\mathbf{z}(s)\|^2\, ds\, d\tau \\
&\geq \frac{\beta}{1+\beta}\frac{2\delta}{\lambda_{\max}(\Gamma^{-1})}V(\mathbf{z}(t_0)) - \beta\lambda_{\max}(\Gamma)^2M^4T^2(V(\mathbf{z}(t_0)) - V(\mathbf{z}(t_0+T))).
\end{aligned}$$

Hence

$$\begin{aligned}
V(\mathbf{z}(t_0+T)) - V(\mathbf{z}(t_0)) &\leq -\frac{\beta}{1+\beta}\frac{2\delta}{\lambda_{\max}(\Gamma^{-1})}V(\mathbf{z}(t_0)) \\
&\quad + \beta\lambda_{\max}(\Gamma)^2M^4T^2(V(\mathbf{z}(t_0)) - V(\mathbf{z}(t_0+T))) \Rightarrow \\
V(\mathbf{z}(t_0+T)) &\leq \rho V(\mathbf{z}(t_0)),
\end{aligned}$$

where

$$\rho = \left(1 - \frac{2\delta}{\lambda_{\max}(\Gamma^{-1})}\frac{\beta}{(1+\beta)(1+\beta\lambda^2_{\max}(\Gamma)M^4T^2)}\right).$$

The value of ρ is minimized at $\beta = 1/(\lambda_{\max}(\Gamma)M^2T)$, in which case[7]

$$V(t_0+T) \leq \rho V(t_0),\ \rho = \left(1 - \frac{2\delta}{\lambda_{\max}(\Gamma^{-1})}\frac{1}{(1+\lambda_{\max}(\Gamma)M^2T)^2}\right).$$

[7] Notice that condition (2.43) implies that $0 < \rho < 1$, and hence $\ln(\rho)$ is defined.

Notice that

$$-\frac{\ln(1-\sigma)}{T} \geq \frac{\sigma}{T}, \ \sigma \in (0,1).$$

Therefore

$$V(\mathbf{z}(t_0+T)) \leq e^{\frac{\ln(\rho)}{T}T} V(\mathbf{z}(t_0)) \leq e^{-2\lambda T} V(\mathbf{z}(t_0)),$$

$$\lambda = \frac{\delta}{\lambda_{\max}(\Gamma^{-1})} \frac{1}{T(1+\lambda_{\max}(\Gamma)M^2T)^2}.$$

Since any $\Delta t \geq 0$ can be expressed as $\Delta t = nT + t'$, $t' \in [0, T)$, we have that

$$V(\mathbf{z}(t_0+\Delta t)) \leq \frac{e^{-2\lambda(nT+t')}}{e^{-2\lambda t'}} V(\mathbf{z}(t_0)) \leq \frac{e^{-2\lambda \Delta t}}{e^{-2\lambda T}} V(\mathbf{z}(t_0)).$$

Finally

$$\|\mathbf{z}(t_0+\Delta t)\| \leq De^{-\lambda \Delta t}\|\mathbf{z}(t_0)\|, \ D = \left(\frac{\lambda_{\max}(\Gamma^{-1})}{\lambda_{\min}(\Gamma^{-1})}\right)^{1/2} e^{\lambda T}.$$

The desired inequality now follows from obvious identities: $\lambda_{\min}(\Gamma^{-1}) = 1/\lambda_{\max}(\Gamma)$ and $\lambda_{\max}(\Gamma^{-1}) = 1/\lambda_{\min}(\Gamma)$. □

Let $\Phi_1(t, t_0)$, $\Phi_1(t_0, t_0) = I$ be the fundamental system of solutions of

$$\dot{\mathbf{z}} = -\phi(t)B^{\mathrm{T}}B\phi(t)^{\mathrm{T}}\mathbf{z}.$$

According to Lemma 2.5, we have that $\|\Phi_1(t_2, t_1)\mathbf{z}(t_1)\| \leq c_3 e^{-\tau(t_2-t_1)}\|\mathbf{z}(t_1)\|$, where[8]

$$c_3 = e^{\tau T}, \ \tau = \frac{\delta}{T(1+T^2\lambda_{\max}(B^{\mathrm{T}}B)B_1^2)^2}.$$

Given that any solution of (2.51) can be expressed as

$$\mathbf{z}(t) = \Phi_1(t,t_1)\mathbf{z}(t_1) + \int_{t_1}^{t} \Phi_1(t,s)\chi(s)\mathbf{x}_1(s)ds, \ t \geq t_1 \geq t_0,$$

where $\chi(s) = -(\phi(s)B^{\mathrm{T}}A + (\phi(s)B^{\mathrm{T}}B\phi(s)^{\mathrm{T}})\phi(s)B^{\mathrm{T}} + \phi(s)C + \dot{\phi}(s)B^{\mathrm{T}})$, and that

$$\|\chi(s)\| \leq B_1(\|A\|\|B\| + \|C\| + \|B\|) + B_1^3\|B\|^3 = c_4,$$

we can immediately derive that

$$\int_{t_1}^{t} \|\mathbf{z}(s)\|^2 ds \leq \|\mathbf{z}(t_1)\|^2 \frac{c_3^2}{2\tau} + c_4^2 c_3^2 \int_{t_1}^{t} \left(\int_{t_1}^{s} e^{-\tau(s-\sigma)}\|\mathbf{x}_1(\sigma)\|d\sigma\right)^2 ds.$$

[8] Here we use the fact that $\|B\phi(t)^{\mathrm{T}}\| \leq \|B\|\|\phi(t)^{\mathrm{T}}\|$, and that $\lambda_{\max}(G^{\mathrm{T}}G) = \lambda_{\max}(GG^{\mathrm{T}})$ for any $G \in \mathbb{R}^{p\times m}$ (see Bernstein (2005), Proposition 4.4.9, for more details). Hence we have that $\|\phi(t)^{\mathrm{T}}\| = \|\phi(t)\|$, $\|B^{\mathrm{T}}\| = \|B\|$, and $\|B\phi(t)^{\mathrm{T}}\| \leq \lambda_{\max}(B^{\mathrm{T}}B)^{1/2}B_1$.

According to Khalil (2002), page 200, the double integral above can be estimated as

$$\begin{aligned}
&\int_{t_1}^{t}\left(\int_{t_1}^{s} e^{-\tau(s-\sigma)}\|\mathbf{x}_1(\sigma)\|d\sigma\right)^2 ds \\
&\quad = \int_{t_1}^{t}\left(\int_{t_1}^{s} e^{-\tau(s-\sigma)/2}e^{-\tau(s-\sigma)/2}\|\mathbf{x}_1(\sigma)\|d\sigma\right)^2 ds \\
&\quad \le \int_{t_1}^{t}\int_{t_1}^{s} e^{-\tau(s-\sigma)}d\sigma \int_{t_1}^{s} e^{-\tau(s-\sigma)}\|\mathbf{x}_1(\sigma)\|^2\, d\sigma\, ds \\
&\quad \le \frac{1}{\tau}\int_{t_1}^{t}\int_{t_1}^{s} e^{-\tau(s-\sigma)}\|\mathbf{x}_1(\sigma)\|^2\, d\sigma\, ds = \frac{1}{\tau}\int_{t_1}^{t}\int_{\sigma}^{t} \|\mathbf{x}_1(\sigma)\|^2 e^{-\tau(s-\sigma)}\, ds\, d\sigma \\
&\quad = \frac{1}{\tau}\int_{t_1}^{t} \|\mathbf{x}_1(\sigma)\|^2 \int_{\sigma}^{t} e^{-\tau(s-\sigma)}\, ds\, d\sigma \le \frac{1}{\tau^2}\|\mathbf{x}_1\|^2_{2,[t_1,t]}.
\end{aligned}$$

Therefore

$$\|\mathbf{z}\|^2_{2,[t_1,t]} \le \frac{c_3^2}{2\tau}\|\mathbf{z}(t_1)\|^2 + \frac{c_4^2 c_3^2}{\tau^2}\|\mathbf{x}_1\|^2_{2,[t_1,t]}.$$

Given that $\mathbf{z} = \mathbf{x}_2 - \phi(t)B^{\mathrm{T}}\mathbf{x}_1$, the following estimate holds: $\|\mathbf{z}\|_{2,[t_1,t]} \ge \|\mathbf{x}_2\|_{2,[t_1,t]} - \|\phi(\cdot)B^{\mathrm{T}}\mathbf{x}_1\|_{2,[t_1,t]}$. Thus

$$\|\mathbf{x}_2\|_{2,[t_1,t]} \le \|\mathbf{z}\|_{2,[t_1,t]} + B_1\|B\|\,\|\mathbf{x}_1\|_{2,[t_1,t]},$$

and hence

$$\|\mathbf{x}_2\|_{2,[t_1,t]} \le \left[\frac{c_3}{\sqrt{2\tau}}\|\mathbf{z}(t_1)\| + \frac{c_4 c_3}{\tau}\|\mathbf{x}_1\|_{2,[t_1,t]}\right] + B_1\|B\|\,\|\mathbf{x}_1\|_{2,[t_1,t]}.$$

Given that

$$\|\mathbf{z}(t_1)\| \le (B_1\|B\|\,\|\mathbf{x}_1(t_1)\| + \|\mathbf{x}_2(t_1)\|) \le (1 + B_1\|B\|)\|\mathbf{x}(t_1)\|,$$

we obtain

$$\begin{aligned}
\|\mathbf{x}_2\|_{2,[t_1,t]} &\le \frac{c_3}{\sqrt{2\tau}}(1 + B_1\|B\|)\|\mathbf{x}(t_1)\| + \left[\frac{c_4 c_3}{\tau} + B_1\|B\|\right]\|\mathbf{x}_1\|_{2,[t_1,t]} \\
&= c_5\|\mathbf{x}(t_1)\| + c_6\|\mathbf{x}_1\|_{2,[t_1,t]}.
\end{aligned}$$

Thus, invoking (2.50), we can derive that

$$\|\mathbf{x}\|_{2,[t_1,t]} \le (c_5 + c_1 c_6)\|\mathbf{x}(t_1)\|.$$

Hence

$$\max\{\|\mathbf{x}\|_{2,[t_1,t]}, \|\mathbf{x}\|_{\infty,[t_1,t]}\} \leq c_7\|\mathbf{x}(t_1)\|, \; c_7 = \max\{(c_5 + c_1c_6), c_2\},$$

and

$$\|\Phi(t_0,t_1)\mathbf{p}\| \leq D_2 e^{-\rho(t-t_0)}\|\mathbf{p}\|, \; D_2 = c_7 e^{1/2}, \; \rho = 1/(2c_7^2).$$

Let us now summarize the main results of this chapter. In this chapter we introduced basic notions of asymptotic characterization of dynamical systems such as invariant sets, attracting sets, and attractors. We introduced various definitions of stability and discussed tools of stability-analysis such as the method of Lyapunov functions and Lemma 2.1. These tools are illustrated with the stability analysis problem for system (2.30). Our choice of the system is motivated primarily by the fact that equations of this type emerge in many problems of adaptive regulation. Even though stability analysis of the zero equilibrium in this system turns out to be a simple exercise, we will come back to this example many times. We have not yet explained, however, why other concepts and notions introduced in this chapter, such as Poisson and Poincaré stability, are no less important. The main reason for postponing more detailed motivation of these matters is that we wanted to jump to the analysis of modern methods of stable adaptive regulation as quickly as possible. This is presented in the next chapter. The analysis would allow us to formulate basic theoretical limitations of these classical methods. In this way the necessity to employ wider stability and convergence concepts emerges naturally as a consequence of such limitations.

3

The problem of adaptation in dynamical systems

In this chapter we provide an overview of mathematical formulations of the problem of adaptation in dynamical systems. Adaptation is considered here as a special regulatory process that emerges as a response of a physical system to changes in the environment. In order to understand the notion of adaptation the terms "regulation," "response," and "environment" need to be given some physical sense and precise mathematical definitions. Instead of inventing a new language of our own, we adopt these terms from the language of mathematical control theory. It is clear that this adoption of terms will create a certain bias in our approach. On the other hand, it will allow us to operate with rigorously defined and established objects that have already passed the test of time.

We start with a retrospective overview and analysis of main ideas in the existing literature on adaptation in the domain of control. These ideas gave rise to distinct mathematical statements of the problem of adaptation. We will discuss their strengths and limitations with respect to the demands of new real-world applications in biology, neuroscience, and engineering. As a result of the analysis, we will formulate a list of features that a modern theory of adaptation must inherit, and propose a program that will allow us to contribute to the development of such a theory in the following chapters.

3.1 Logical principles of adaptation

The theory of adaptive regulation, as a set of notions, methods, and tools for systematic study of systems adjusting their properties in response to changes in the environment, has a long history of development. As with many other theories, mathematical formalizations are often preceded by the development of a general common-sense understanding of the phenomenon. This common-sense understanding is then supplied with a range of models that are believed to be adequate to the phenomena being studied at the time. These models, together with empirical insights, are then expressed in precise mathematical language. Axioms, assumptions, and constraints are clearly an inherent part of such a description. Since basic models, axioms, and assumptions are influenced by paradigms and views prevailing at the time of their development, so are the resulting theories. Thus, in order to understand the boundaries and scope of existing theories of adaptation it is only natural to view existing classical results through the prism of their historical necessity. Here we do not wish to discuss these classical statements in great detail. We will, however, review core *logical principles* of the early theories of adaptation. These principles, though old, constitute the basis of ideas inherent to many recent results and statements.

3.1.1 Searching-based adaptation and extremal systems

The first theories of adaptive regulation emerged at the end of the 1930s. These results were initially motivated by the necessity to develop systems for automatic optimization of performance for industrial machines and combustion engines. Typically, a criterion determining the performance of a machine was defined by means of a function of measured physical variables. The main operational principle of such automatic optimization systems can be described as searching for the extremum of this function followed by regulatory actions aimed at keeping the system's performance at the optimum.

Examples of these early automatic extremum-seeking or peak-holding procedures in the context of adaptation were described by Y. S. Khlebtsevich and V. V. Kazakevich (Khlebtsevich 1965; Kazakevich 1946). In the 1950s extremum-seeking systems and the associated set of theoretical problems received substantial attention in the literature (Draper and Lee 1951, 1960; Tsien 1954; Ivakhnenko 1962; Kazakevich 1958; Aseltine *et al.* 1958; Feldbaum 1959; Morosanov 1957; Ostrovskii 1957; Pervozvanskii 1960). One of the first systematic studies of these systems was presented in the *Principles of Optimalizing Control Systems and an Application to the Internal Combustion Engine* (1951) by C. S. Draper and Y. T. Li. A large bibliography on adaptive systems can be found in Aseltine *et al.* (1958). In most of these early works the input–output behavior of the systems was described by static maps. In the 1990s these models were extended to the class of Wiener–Hammerstein systems explicitly incorporating some linear dynamics (see e.g. Wittenmark and Urquhart (1995)).

Despite the apparent simplicity of the models, the problem of ensuring the overall Lyapunov stability of the extremum-seeking adapting systems remained open for nearly 40 years. One of the first theoretically justified schemes of extremal controllers for a broad class of *dynamical systems* appeared in Krstić and Wang

(2000). The main idea is to use searching signals modeled by an external harmonic perturbation of small amplitude. The latter serves as a generator of certain exploratory motions of the system in the vicinity of its current state. Applications of adaptive systems based on extremum-searching are reviewed in Sternby (1980) and in Astrom and Wittenmark (1961). These applications range from control systems for combustion engines, turbines, and windmills to solar panels. Potential applications include automotive systems such as brake controls (Drakunov *et al.* 1995) and soft landing of valves (Peterson and Stefanopoulou 2004), and bioreactors (Guay *et al.* (2004). Yet, despite all these success stories, adaptation mechanisms based on the extremum-seeking procedures still have not gained broad popularity.

There are several reasons for the lack of wider popularity of these methods. Some of them are provided below. First of all, we would like to mention the necessity of exploring motions in these systems. Such motions, if present in the normal operating regime of a system, inevitably will interfere with the system's own goals. Second, the amount of time needed to locate the extremum is often a non-negligible quantity. Third, unmodeled dynamics, delays, and external perturbations combined with the necessity to follow pre-specified exploratory motions may give rise to the emergence of undesired oscillations in the behavior of the resulting adapting systems. Finally, the common assumption that the input–output behavior of the system is described by a unimodal function prevents straightforward application of the existing methods to systems with multiple extrema.

3.1.2 Principles of adaptation beyond searching-based optimization strategies

An alternative and an extension to the trial-and-error adaptation employed in systems with extremum-seeking regulatory mechanisms is the idea of direct, searchless optimization proposed in the 1960s. In the context of adaptive systems this alternative corresponds to a family of regulatory mechanisms that allow one to steer the system's state to the optimal values without resorting to exploration. Therefore, this class of systems is sometimes referred to as "analytical" or "searchless" self-tuning systems (Krasovskii 1963; Solodovnikov 1965; Kazakov and Evlanov 1965). The advantage of the latter concept is that it is no longer limited to extremum-seeking mechanisms and, hence, allows one to cover a wider range of problems within a unified theoretical framework. Yet, this generality comes at a price. The cost of such extension was that the definition of adaptation needed to be revised. Particular notions of adaptation inevitably affect the range and content of basic operational principles determining the resulting behavior of an adapting system or process. In order to be able to understand the development of these principles, let us review how the notion of adaptation evolved over time in the literature.

Adaptive systems. Point of view 1. According to Bellman and Kalaba (1960) an adapting, self-organizing, and self-regulating system is one that gradually improves its performance through observation of its inputs in some stochastic environment. More precisely, in the summary to Bellman and Kalaba (1960), Bellman and Kalaba write that "... in many engineering, economic, biological and statistical control processes, a decision-making device is called upon to perform under various conditions of uncertainty regarding underlying physical processes. These conditions range from complete knowledge to total ignorance. As the process unfolds, additional information may become available to the controlling element, which then has the possibility of learning to improve its performance based upon experience; i.e., the controlling element may adapt itself to its environment." Following this intuition R. Bellman and other authors defined adaptation (and adaptive control) as an iterative control strategy involving Bayesian estimation of uncertain random variables (e.g. parameters of the process models, statistical characteristics of input signals, and perturbations) combined with the ideas of optimality (Bellman and Kalaba 1960; Bellman 1961; Bellman and Kalaba 1965; Fu 1969; Kwakernaak 1969).

For example, in Kwakernaak (1969) adaptation necessarily includes (a) identification or estimation of a-posteriori statistics followed (b) by stochastic control. In particular, if the uncertainties are limited to parameters of the model and the values of this parameter are distributed in accordance with some probabilistic law then (b) can be viewed as a standard Bayesian optimization task.

Adaptive systems. Point of view 2. A number of authors attempted to define adaptation by considering specific ways in which adaptive behavior can be realized in a given system. For example, in Li and van der Valde (1961) the authors suggest the following hierarchy of regulatory actions, depending on their complexity, ensuring a certain degree of adaptivity:

(a) high-gain time-invariant compensatory feedbacks;
(b) modification (or switching) of the system's parameters according to a given program or a set of rules;
(c) modification of the system's parameters determined by a given performance criterion;
(d) modification of the systems's structure in order to optimize a given performance criterion.

This hierarchy offers a characterization of adaptive systems with respect to the preferred regulatory action. This characterization allows one to organize adaptive systems into four classes (a)–(d).

Class (a) consists of those systems in which an appropriately selected time-invariant control ensures satisfactory performance under various conditions. A simple example of such a system can be described as follows. Suppose that the system's dynamics is given by the equation

$$\dot{x} = d(t, x) + u, \tag{3.1}$$

where u is a regulatory or control input, and $d : \mathbb{R} \times \mathbb{R} \to \mathbb{R}$ is a continuous and bounded function, i.e. $\|d(t, x)\| \leq B$, $B \in \mathbb{R}_{>0}$. The function $d(\cdot, \cdot)$ is unknown, and the purpose of the system is to keep its state, x, in the ε-neighborhood of the origin. If the function $d(\cdot, \cdot)$ were known then the feedback

$$u = -kx - d(t, x), \ k \in \mathbb{R}_{>0} \tag{3.2}$$

would be the desired solution. This feedback would "compensate" for the influence of $d(\cdot, \cdot)$ on the system's dynamics and make the origin asymptotically stable. The problem, however, is that the function $d(\cdot, \cdot)$ is not known, and hence compensatory feedback (3.2) cannot be used. A reasonable strategy for the system would be to use a feedback compensating for the influence of all possible perturbations $d(\cdot, \cdot)$ and ensuring that the system's solutions satisfy

$$\limsup_{t \to \infty} |x(t, x_0)| \leq \varepsilon.$$

An example of such a feedback is

$$u = -\frac{B}{\varepsilon} x. \tag{3.3}$$

An obvious advantage of this control function is that it does not require any additional information about the function $d(\cdot, \cdot)$ apart from knowledge of the bounds for $\|d(t, x)\|$. The disadvantage is that the gain B/ε may become very large, and hence control (3.2) is not suitable when minimal regulatory actions are preferable.

The adaptation mechanisms in systems of class (b) are more sophisticated than mere high-gain domination of the uncertainties described above. Examples of these systems include cyclic switching, and hysteresis-based adaptation mechanisms (Middleton *et al.* 1988; Morse *et al.* 1988), and perhaps the wealth of adaptation schemes in the domain of supervisory control based on pre-routing (Morse 1995). In these works the adaptive controller consisted of an indexed family of elementary control functions. Each control function corresponds to a certain state of the regulated process, and it is chosen such that the process's dynamics is satisfactory with this controller. Adaptation in this system is realized via switching from one elementary control function to another according to a specific program.

Class (c) contains the vast majority of algorithms of parameter adaptation reported in the literature. These algorithms include gradient-based algorithms for parameter tuning and many other schemes described in the classical monographs by Narendra and Annaswamy (1989), Krstić *et al.* (1995), and Fomin *et al.* (1981). In order to illustrate these schemes, consider the system

$$\dot{x} = \theta d(t,x) + u. \tag{3.4}$$

This system is a modification of (3.1) in which the function $d(\cdot,\cdot)$ is multiplied by $\theta \in \mathbb{R}$. Let us suppose that the function $d(\cdot,\cdot)$ is known, but the value of θ is not available. The objective is to find a control input u that will steer the system's state to the origin. This objective can be attained if we choose the following control:

$$u = -kx - \hat{\theta} d(t,x), \; k \in \mathbb{R}_{>0},$$

where the parameter $\hat{\theta}$ is updated according to the following rule:

$$\dot{\hat{\theta}} = x d(t,x). \tag{3.5}$$

Observe that

$$V(x,\hat{\theta}) = \frac{1}{2}x^2 + \frac{1}{2}(\theta - \hat{\theta})^2$$

is a Lyapunov function for (3.4) and (3.5), and that

$$\dot{V} \leq -kx^2.$$

Hence, solutions of (3.4) and (3.5) exist and are bounded. Furthermore, by invoking Lemma 2.1 we can conclude that $\lim_{t\to\infty} x(t,x_0) = 0$.

Class (d) in the classification above includes regulatory mechanisms based on the automatic compensation of parameter uncertainties by a limit cycle in the control loop. Also systems with variable structure (Emelyanov 1967) and sliding mode-based control (Utkin 1992) can be related to this class. The utility of this class of regulatory mechanisms is that they do not generally require detailed knowledge of the system's dynamics, as is often the case for the mechanisms of class (c).

The notions of adaptation provided in variants 1 and 2 were criticized on account of the fact that they allow a substantial degree of flexibility, in interpretation. As a reaction to this flexibility, a new interpretation emerged.

Adaptive systems. Point of view 3. In Gibson (1961), Mishkin and Braun (1961), and Truxhall (1965) the authors criticize flexible application of the term "self-tuning" to "adaptive" systems classified as of class (a) or (d) in the classification above. The critique can be generally expressed as Truxhall's statement that adaptation and self-tuning are mere interpretations or points of view of an observer studying the behavior of a system (Truxhall 1965). This claim is a natural reaction to a number of attempts to classify any system with uncertainties as adaptive (Tou 1964). A system that does possess a certain degree of adaptation is likely to operate according to the following sequence of steps:

(a) determine relevant behavioral variables (characteristics) and criteria of optimal performance;
(b) compare current values or characteristics of these variables with the ones corresponding to optimal behavior;
(c) adjust itself in such a way that the variables previously selected attain optimal values.

Steps (b) and (c) are to be realized automatically by the system, whereas step (a) is the responsibility of the designer. The main logical stages inherent to developing any adaptive process according to Truxhall are *optimization*, *identification*, and *stability analysis*.

On analyzing classical points of view on the notions and principles of adaptation, we can conclude that, despite the fact that these views may differ substantially in terms of the definitions and interpretations, there are certain invariants that may allow us to produce a rather meaningful, albeit informal, characterization of an adaptive system. One of these is that adaptation is commonly recognized as a purposeful regulatory process in a system operating in an uncertain environment. This regulation is often related to the adjustments of the system's structure or tuning of its parameters constrained by a set of problem-specific optimality conditions. Almost all of the concepts considered share the same postulate that effective learning or adaptation should be based upon some prior information about the system and/or environment. Further specification of the concept requires a more detailed and precise description of the environment, regulated processes, and system. These are considered in the next section.

3.2 Formal definitions of adaptation and mathematical statements of the problem of adaptation

Formal statements of the problem of adaptation and adaptive regulation received great attention in a number of fundamental works (Tsien 1954; Ivakhnenko 1962; Feldbaum 1959, 1965; Li and van der Valde 1961; Margolis and Leondes 1961; Mishkin and Braun 1961; Florentin 1962; Zadeh 1963; Lee 1964; Yakubovich 1968; Krasovsky *et al.* 1977; Saridis 1977; Sragovich 1981; Fomin *et al.* 1981; Widrow *et al.* 1976; Tsypkin 1968, 1970). Despite the fact that these statements may vary in terms of the degree of mathematical rigor and their particular language, there are useful invariants in these statements that will help us to understand the evolution of the problem of adaptation over time. In what follows we review those statements which are not linked explicitly with any particular mechanism of adaptation and thus bear a substantial degree of generality.

3.2.1 The formulation of the problem of adaptation by R. Bellman and R. Kalaba

In the fundamental work by Bellman and Kalaba (1965) adaptation and adaptive (control) systems are discussed within the following framework of assumptions.[1] An adaptive control process is described by

$$p_1 = T(p, q, r),$$

where p is the system's state variable, q is the decision variable, r is a random variable with a fixed, but unknown, distribution function, and $T(\cdot, \cdot, \cdot)$ is a transformation describing the dynamics of the state over time. Let $dG(r)$ be an a priori estimate for the unknown distribution function of r. The physical state of the system is p, but the state of the control process is the couple (p, G).

The aim is to use a control policy maximizing the expected value of the "criterion function"

$$R_N = g(p, q_1, r_1) + g(p_1, q_2, r_2) + \cdots + g(p_{N-1}, q_N, r_N).$$

The control policy here is the sequence of values of the decision variable: $q_1, q_2, \ldots, q_N$.

The set of solutions to this problem is constrained by the following additional assumptions.

(1) We can observe the state of the system at each stage.
(2) At each stage, we regard the a-priori estimate as the actual distribution function. Expected values are obtained on this basis.
(3) We have a systematic procedure for modifying the a-priori distribution function as the process unfolds; this procedure itself may be an adaptive one.

As a result of the choice of q_1, we then have the transformations

$$\begin{aligned} p_1 &= T(p, q_1, r_1), \\ dG_1(r) &= S(r; p, q_1, r_1, G). \end{aligned} \tag{3.6}$$

In words, the new estimate of the distribution function depends upon the old distribution function, the value of r_1, if observed, the original state p, the new state p_1, and the decision q_1.

Let $f_N(p, G)$ be the expected value of R_N obtained using an optimal policy, starting in the stage (p, G). Then the principle of optimality yields the functional

[1] We present the description of an adaptive control process as it is stated in the original work by Bellman and Kalaba (1965), pages 102–103.

equation

$$f_N(p, G) = \max_{q_1} \left[\int [g(p, q_1, r_1) + f_{N-1}(T(p, q_1, r_1), G_1)] dG(r_1) \right],$$

where G_1 is as in (3.6), for $N \geq 2$ and for $N = 1$ we have

$$f_1(p, G) = \max_{q_1} \left[\int g(p, q_1, r_1) dG(r_1) \right].$$

In the conclusion to this statement the authors acknowledge that it is not too difficult to write down these relations, but it is quite difficult to obtain either analytical or numerical results from these equations.

3.2.2 Adaptivity according to L. Zadeh

An alternative approach for developing a constructive and precise notion of adaptivity is suggested in Zadeh (1963). According to L. Zadeh, finding a satisfactory definition of adaptation is a difficult problem. This is because of "... the lack of clear differentiation between the external manifestations of adaptive behavior on the one hand, and the internal mechanism by which it is achieved on the other. To subsume both under a single concise definition has proved to be an elusive objective, since it is very difficult – perhaps impossible – to find a way of characterizing in concrete terms the large variety of ways in which adaptive behavior can be realized."

In order to define external manifestations of adaptivity, L. Zadeh employs the following language. Consider a system $\mathcal{A}$ endowed with a family of functions $\mathcal{S}_\gamma$:

$$\mathcal{S}_\gamma = \{u | u : \mathbb{R}_{>0} \rightarrow \mathbb{R}^m\}.$$

The family $\mathcal{S}_\gamma$ defines all admissible inputs to the system. It is assumed that some components of $u(t)$ represent external inputs applied to $\mathcal{A}$ by a controller, whereas others model the influence of disturbances. In L. Zadeh's terminology the family $\mathcal{S}_\gamma$ constitutes a source, possibly, but not necessarily, with a probability measure defined on this family in such a way as to make S_γ a stochastic process. It is supposed that there is a specified family of sources $\{\mathcal{S}_\gamma\}$ parametrized by $\gamma \in \mathcal{G}$, where $\mathcal{G}$ is a set. Thus we have a system $\mathcal{A}$ for which the inputs are some functions $u(t)$ from $\mathcal{S}_\gamma$. The latter is a member of a family of sources $\{\mathcal{S}_\gamma\}$.

In order to define the performance of $\mathcal{A}$ an additional performance function P is introduced. A possible way to do this is as follows. Let $\mathcal{A}(\mathcal{S}_\gamma)$ denote a set of the system's responses to inputs $\mathcal{S}_\gamma$. For consistency we suppose that the set of responses is a family of functions parametrized by γ. Thus for every class of input signals $\mathcal{S}_\gamma$ there would be a corresponding set of responses $\mathcal{R}_\gamma$. Hence, one can now define P as a function from $\mathcal{G}$ into $\mathbb{R}^d$, and the value of $P(\gamma)$ quantitatively measures the system's performance.

Suppose now that $W \subset \mathbb{R}^d$ is a set specifying the *acceptable* performance of the system. In other words, we say that the performance of $\mathcal{A}$ is acceptable iff $P(\gamma) \in W$.

Definition 3.2.1 A system $\mathcal{A}$ is adaptive with respect to $\{\mathcal{S}_\gamma\}$ and W if it performs acceptably well, i.e. $P(\gamma) \in W$, with every source in the family $\{\mathcal{S}_\gamma\}$, $\gamma \in \mathcal{G}$. More compactly, $\mathcal{A}$ is adaptive with respect to $\mathcal{G}$ and W if it maps $\mathcal{G}$ into W.

According to L. Zadeh's comment regarding this definition, every system is adaptive with respect to some $\mathcal{G}$ and W. Thus what matters is not whether system $\mathcal{A}$ is adaptive, but with respect to which sets $\mathcal{G}$ and W its adaptive behavior is evident.

3.2.3 The problem of adaptation according to V. A. Yakubovich

Perhaps the most general and rigorous definitions of the term "adaptation" and subsequently formal statements of the problem of adaptive regulation are those provided in Yakubovich (1968), Saridis (1977), Sragovich (1981), Fomin *et al.* (1981), and Fradkov (1990). These definitions gave rise to a set of successful concepts of adaptation that resulted in the mathematical theory of adaptive control presented in the monographs by Fomin *et al.* (1981) and Fradkov (1990). Adaptation, as a regulatory process, is not restricted to a finite interval of time but can continue indefinitely. The goals of adaptation are some asymptotic regimes of the system's dynamics. These goals are thus not explicitly dependent on transients. For the sake of illustration let us consider a simplified statement of the problem of adaptation from Fomin *et al.* (1981).

Suppose that the system's dynamics is governed by the following system of ordinary differential equations:

$$\begin{aligned} \dot{\mathbf{x}} &= \mathbf{f}(\mathbf{x}, \boldsymbol{\theta}, t, \mathbf{u}), \\ \mathbf{y} &= \mathbf{h}(\mathbf{x}, \boldsymbol{\theta}, t, \mathbf{u}), \end{aligned} \tag{3.7}$$

where $\boldsymbol{\theta}$ is the vector of unknown parameters, $\Omega_\theta \subset \mathbb{R}^d$, and $\mathbf{u} \in \mathbb{R}^m$, $\mathbf{f} : \mathbb{R}^n \times \Omega_\theta \times \mathbb{R}_{\geq 0} \times \mathbb{R}^m \to \mathbb{R}^n$, $\mathbf{h} : \mathbb{R}^n \times \Omega_\theta \times \mathbb{R}_{\geq 0} \times \mathbb{R}^m \to \mathbb{R}^k$ are some sufficiently smooth functions. Without loss of generality, and if the nature of the problem requires it, one can replace the parameter $\boldsymbol{\theta}$ on the right-hand side of (3.7) with a continuous function of t: $\boldsymbol{\theta} : \mathbb{R}_{\geq 0} \to \Omega_\theta$, $\boldsymbol{\theta}(t) \in \Theta$. In this case the set Θ becomes a set of functions of a given class. The broadness of this class determines the *class of adaptivity* of the system.

The goal of adaptation (or adaptive regulation) is defined by means of a certain goal functional Q:

$$Q[\mathbf{x}(\tau), \mathbf{u}(\tau), 0 \leq \tau \leq t] \leq \Delta \; \forall t \geq t^*,$$

where $\Delta \geq 0$ is a threshold of acceptable performance. The functional $Q[\cdot, \cdot, \cdot]$ relates the system's state and corresponding control signals $\mathbf{u} \in \mathbb{R}^m$ to a value from $\mathbb{R}$. The latter is an indicator of the system's performance.

The problem of adaptive regulation is stated as that of finding a control law ensuring that the goal requirement is fulfilled. The values of the control input at time t should be explicitly computable from the available measurement information about the system's behavior in the past. More precisely, the following class of control functions is considered admissible:

$$\mathbf{u} = U_t[\mathbf{y}(\tau), \mathbf{u}(\tau), \boldsymbol{\beta}(\tau), 0 \leq \tau \leq t],$$
$$\boldsymbol{\beta} = B_t[\mathbf{y}(\tau), \mathbf{u}(\tau), \boldsymbol{\beta}(\tau), 0 \leq \tau \leq t],$$

where $\boldsymbol{\beta} \in \mathbb{R}^p$ is the vector of adjustable parameters of the control law $\mathbf{u}$. Once the functional $U_t[\cdot, \cdot, \cdot, \cdot]$ has been chosen, ensuring that the goal functional is satisfied and that the system's state is bounded depends solely upon a suitable choice of the adjustment law for $\boldsymbol{\beta}$. Overall the choice of $U_t[\cdot, \cdot, \cdot, \cdot]$ and $B_t[\cdot, \cdot, \cdot, \cdot]$ should guarantee that the goal requirement is satisfied for all $\boldsymbol{\theta} \in \Omega_\theta$ (or $\boldsymbol{\theta}(t) \in \Theta$) and initial conditions $\mathbf{x}(0)$ and $\boldsymbol{\beta}(0)$.

Before we conclude this section, let us briefly comment on the evolution of the problem of adaptation in the context of dynamical systems and regulation. Originally the notions of adaptation and mathematical formulations of the problem of adaptive regulation were confined to the framework of stochastic processes and iterative optimization. This is not surprising, for the uncertainties were understood as stochastic processes. This inevitably led to the perception that the most suitable tools for developing the notion of adaptation and general principles of adaptivity are the theories and language of stochastic analysis. Furthermore, adaptation itself was limited to a descriptive characterization of how an abstract dynamical system responds to a certain class of inputs. The lack of agreement on how a particular adaptive behavior is to be realized resulted in the vagueness of the notion of adaptation which led to L. Zadeh's and J. Truxhall's critique that a safer and more meaningful position would be to define adaptation in terms of its external manifestations.

This was the case before the emergence of parametric formulations of the notions and problems of adaptation. Despite the fact that these formulations did not inherit the same level of generality as the non-parametric ones, they enabled a more precise and specific description of what an adaptive system is. According to these formulations, adaptation is a specific process of tuning of the parameters of a controller.

This process must ensure a satisfactory performance of the system in the presence of uncertainties. In this text we accept this particular view on the problem of adaptation. We restrict our attention to deterministic systems and aim to provide a set of methods applicable to solve the observation and regulation problems considered in Chapter 1. In order to understand what sort of methods we need, let us first review existing methods in the domain of adaptive control and regulation. We do not wish to provide an exhaustive analysis of all of the methods available in the literature. Our goal is more modest. We will focus on a few basic ideas forming the core of many existing methods for solving the problem of adaptive regulation in various parametric formulations.

3.3 Adaptive control for nonlinear deterministic dynamical systems

Suppose that the class of dynamical systems for which an adaptive regulatory (or control) mechanism is to be derived is specified as follows:

$$\dot{\mathbf{x}} = \mathbf{f}(\mathbf{x}, \boldsymbol{\theta}, t, \mathbf{u}), \tag{3.8}$$

where $\mathbf{f} : \mathbb{R}^n \times \mathbb{R}^d \times \mathbb{R} \times \mathbb{R}^m \to \mathbb{R}^n$ is a locally Lipschitz function with respect to $\mathbf{x}$, $\boldsymbol{\theta}$, and $\mathbf{u}$, uniformly in t, $\mathbf{x} \in \mathbb{R}^n$ is the state vector, $\boldsymbol{\theta} \in \Omega_\theta$, where $\Omega_\theta \subset \mathbb{R}^d$ is the vector of parameters, and $\mathbf{u}$ is the control input. The values of $\boldsymbol{\theta}$ are supposed to be unknown a priori.

The main objective is to find a control input such that the state $\mathbf{x}$ is steered into a small neighborhood of a given set $\Omega_x \subset \mathbb{R}^n$, or tracks asymptotically some trajectory $\mathbf{r}(t) : \mathbb{R} \to \mathbb{R}^n$ for all $\boldsymbol{\theta} \in \Omega_\theta \subset \mathbb{R}^d$. The desired control input $\mathbf{u}$ should comply with the following requirements:

(1) $\mathbf{u}$ is not a function of the time-derivatives of $\mathbf{x}$;
(2) $\mathbf{u}$ does not depend explicitly on $\boldsymbol{\theta}$.

The class of possible control inputs will be limited to

$$\mathbf{u} : \mathbb{R}^n \times \mathbb{R}^p \times \mathbb{R} \to \mathbb{R}^m, \ \mathbf{u} = \mathbf{u}(\mathbf{x}, \hat{\boldsymbol{\theta}}, t), \tag{3.9}$$

where $\hat{\boldsymbol{\theta}}$ is generated by

$$\begin{aligned} \dot{\boldsymbol{\xi}} &= \mathbf{f}_\xi(\boldsymbol{\xi}, \mathbf{x}, t, \mathbf{u}), \ \mathbf{f}_\xi : \mathbb{R}^\xi \times \mathbb{R}^n \times \mathbb{R} \times \mathbb{R}^m \to \mathbb{R}^\xi, \\ \hat{\boldsymbol{\theta}} &= \mathbf{h}_\xi(\boldsymbol{\xi}, \mathbf{x}, t), \ \mathbf{h}_\xi : \mathbb{R}^\xi \times \mathbb{R}^n \times \mathbb{R} \to \mathbb{R}^p. \end{aligned} \tag{3.10}$$

The right-hand side of (3.10) is supposed to be locally Lipschitz; $\boldsymbol{\xi}$ is the vector of the internal state of the controller, and $\hat{\boldsymbol{\theta}}$ is the vector of adjustable parameters.

3.3.1 Velocity gradient

The method of velocity gradient was proposed in Fradkov (1979) and further developed in Fomin *et al.* (1981), Fradkov (1990), and Fradkov *et al.* (1999). The main idea of the method is two-fold: (1) the process of adaptation is viewed as an optimization procedure; and (2) this procedure is to minimize the instantaneous velocity of a goal functional $Q(\mathbf{x},t) : \mathbb{R}^n \times \mathbb{R}_{\geq 0} \rightarrow \mathbb{R}$. Let us now consider this method in more detail.

Suppose that the system's dynamics is given by (3.8), and that the objective is to find the function $\mathbf{u}$ and corresponding auxiliary system (3.10), such that for all $\boldsymbol{\theta} \in \Omega_\theta$ the following property holds:

$$\lim_{t\rightarrow\infty} Q(\mathbf{x}(t,\mathbf{x}_0),t) = 0. \tag{3.11}$$

The function $Q(\mathbf{x},t)$ in (3.11) is supposed to be a positive definite function with respect to $\mathbf{x}$ uniformly in t. In particular, we suppose that

$$\exists\, \rho_1(\cdot), \rho_2(\cdot) \in \mathcal{K}_\infty : \ \rho_1(\|\mathbf{x}\|) \leq Q(\mathbf{x},t) \leq \rho_2(\|\mathbf{x}\|)\ \forall\, t \in \mathbb{R}. \tag{3.12}$$

The method requires that the following two assumptions hold for (3.8) and $Q(\mathbf{x}(t,\mathbf{x}_0))$.

Assumption 3.1 *There exists and is known a function* $\mathbf{u} : \mathbb{R}^n \times \mathbb{R}^p \times \mathbb{R} \rightarrow \mathbb{R}^m$, *such that for every* $\boldsymbol{\theta} \in \Omega_\theta$ *on the right-hand side of (3.8) there is a vector* $\hat{\boldsymbol{\theta}}^*$ *ensuring that the zero equilibrium of the combined system (3.8) with control* $\mathbf{u}(\mathbf{x}, \hat{\boldsymbol{\theta}}^*, t)$ *is globally asymptotically stable in the sense of Lyapunov. Furthermore, the following holds:*

$$\frac{\partial Q(\mathbf{x},t)}{\partial \mathbf{x}} \mathbf{f}(\mathbf{x},\boldsymbol{\theta},t,\mathbf{u}(\mathbf{x},\hat{\boldsymbol{\theta}}^*, t)) + \frac{\partial Q(\mathbf{x},t)}{\partial t} \leq -\alpha_1(Q),\ \alpha_1 \in \mathcal{K}. \tag{3.13}$$

Assumption 3.1 simply states that for every $\boldsymbol{\theta} \in \Omega_\theta$ there exists a feedback $\mathbf{u}(\mathbf{x}, \hat{\boldsymbol{\theta}}, t)$ globally (asymptotically) stabilizing the origin. Furthermore, the function $Q(\mathbf{x},t)$ specifying the goal of adaptation in (3.12) is the corresponding Lyapunov function.

Assumption 3.2 *There exists and is known a function satisfying Assumption 3.1, and furthermore for all* $\boldsymbol{\theta}, \hat{\boldsymbol{\theta}}, \hat{\boldsymbol{\theta}}' \in \Omega_\theta$, $\mathbf{x} \in \mathbb{R}^n$, *and* $t \in \mathbb{R}$ *the following inequality holds:*

$$\begin{aligned}(\hat{\boldsymbol{\theta}}' - \hat{\boldsymbol{\theta}})^{\mathrm{T}} &\left(\frac{\partial}{\partial \hat{\boldsymbol{\theta}}}\left[\frac{\partial Q}{\partial \mathbf{x}} \mathbf{f}(\mathbf{x},\boldsymbol{\theta},t,\mathbf{u}(\mathbf{x},\hat{\boldsymbol{\theta}},t))\right]\right)^{\mathrm{T}} \\ &\geq \frac{\partial Q(\mathbf{x},t)}{\partial \mathbf{x}} \mathbf{f}(\mathbf{x},\boldsymbol{\theta},t,\mathbf{u}(\mathbf{x},\hat{\boldsymbol{\theta}},t)) - \frac{\partial Q(\mathbf{x},t)}{\partial \mathbf{x}} \mathbf{f}(\mathbf{x},\boldsymbol{\theta},t,\mathbf{u}(\mathbf{x},\hat{\boldsymbol{\theta}}', t)).\end{aligned} \tag{3.14}$$

Assumption 3.2 is a sort of convexity constraint. It requires that the partial derivative

$$\frac{\partial Q}{\partial \mathbf{x}}\mathbf{f}(\mathbf{x},\boldsymbol{\theta},t,\mathbf{u}(\mathbf{x},\hat{\boldsymbol{\theta}},t)) \tag{3.15}$$

be convex with respect to $\hat{\boldsymbol{\theta}}$. Notice that Assumption 3.2 is a stronger requirement than simply an extra convexity constraint imposed on the function $\mathbf{u}(\mathbf{x},\hat{\boldsymbol{\theta}},t)$. Indeed, even for functions $\mathbf{u}(\mathbf{x},\hat{\boldsymbol{\theta}},t)$ that are convex with respect to $\hat{\boldsymbol{\theta}}$ the actual sign of inequality (3.14) depends on $\partial Q(\mathbf{x},t)/\partial \mathbf{x}$.

Let us suppose that Assumptions 3.1 and 3.2 hold, and consider the evolution of $Q(\mathbf{x},t)$ over time. This evolution satisfies the following differential equation:

$$\dot{Q} = \frac{\partial Q}{\partial \mathbf{x}}\mathbf{f}(\mathbf{x},\boldsymbol{\theta},t,\mathbf{u}(\mathbf{x},\hat{\boldsymbol{\theta}},t)) + \frac{\partial Q(\mathbf{x},t)}{\partial t}.$$

By invoking Theorem 2.1 we can conclude that condition (3.11) is automatically satisfied if the inequality

$$\dot{Q} \leq -\alpha_1(Q)$$

holds for all $\mathbf{x}$, t, and $\boldsymbol{\theta}$. Thus affecting the system dynamics in such a way that the derivative $\dot{Q}$ is decreasing with time may be a plausible control-and-adaptation strategy. In the framework of parametric adaptation we can influence the system's dynamics by adjusting parameters of the controller. Following this logic, we can propose a plausible adaptation algorithm, in which parameters of the controller are adjusted proportionally to the gradient of (3.15) or $\dot{Q}$ with respect to $\hat{\boldsymbol{\theta}}$. In particular, we set

$$\dot{\hat{\boldsymbol{\theta}}} = -\Gamma\left(\frac{\partial}{\partial \hat{\boldsymbol{\theta}}}\left[\frac{\partial Q}{\partial \mathbf{x}}\mathbf{f}(\mathbf{x},\boldsymbol{\theta},t,\mathbf{u}(\mathbf{x},\hat{\boldsymbol{\theta}},t))\right]\right)^{\mathrm{T}}, \quad \Gamma > 0. \tag{3.16}$$

The quantity $\dot{Q}$ can be viewed as a sort of "velocity" of $Q(\mathbf{x},t)$; hence the above proposal is referred to as the method of velocity gradient.

Clearly the right-hand side of (3.16) may depend on $\boldsymbol{\theta}$ explicitly. Hence, in order to be able to implement this algorithm without violating the original statement that the values of $\boldsymbol{\theta}$ are unknown, we shall impose an additional technical assumption. This assumption is that the partial derivative

$$\frac{\partial}{\partial \hat{\boldsymbol{\theta}}}\left[\frac{\partial Q}{\partial \mathbf{x}}\mathbf{f}(\mathbf{x},\boldsymbol{\theta},t,\mathbf{u}(\mathbf{x},\hat{\boldsymbol{\theta}},t))\right]$$

does not depend on $\boldsymbol{\theta}$ explicitly.

Asymptotical properties of the velocity-gradient adaptation scheme are formulated in the next theorem (see also Fomin *et al.* (1981)).

Theorem 3.1 *Consider system (3.8) and suppose that there is a function $Q(\mathbf{x}, t)$ satisfying condition (3.12). Furthermore, let us suppose that there exists a function $u(\mathbf{x}, t, \hat{\boldsymbol{\theta}})$ satisfying Assumptions 3.1 and 3.2.*

Then solutions of the closed-loop system with control $\mathbf{u}(\mathbf{x}, \hat{\boldsymbol{\theta}}, t)$ and adaptation algorithm (3.16) exist for all $t \geq t_0$ and are bounded for every $\boldsymbol{\theta} \in \Omega_\theta$ and all initial conditions $\mathbf{x}(t_0) = \mathbf{x}_0 \in \mathbb{R}^n$. Moreover, property (3.11) holds.

Proof of Theorem 3.1. The proof of the theorem can be easily completed following a logic similar to that we used when we analyzed system (2.30). We will start with demonstrating the existence and boundedness of the system's solutions by invoking the method of Lyapunov functions. Then we will apply Lemma 2.1 to show that (3.11) holds.

Consider the following function:

$$V(\mathbf{x}, \hat{\boldsymbol{\theta}}, \boldsymbol{\theta}^*, t) = Q(\mathbf{x}, t) + \frac{1}{2}\|\hat{\boldsymbol{\theta}} - \hat{\boldsymbol{\theta}}^*\|^2_{\Gamma^{-1}},$$

where $\|\hat{\boldsymbol{\theta}} - \hat{\boldsymbol{\theta}}^*\|^2_{\Gamma^{-1}}$ stands for $(\hat{\boldsymbol{\theta}} - \hat{\boldsymbol{\theta}}^*)^{\mathrm{T}}\Gamma^{-1}(\hat{\boldsymbol{\theta}} - \hat{\boldsymbol{\theta}}^*)$. The function V is positive definite in $\mathbb{R}^{n+p}$ and hence satisfies condition (2.33). Taking (3.16) into account, we obtain

$$\dot{V} = \dot{Q} - (\hat{\boldsymbol{\theta}} - \hat{\boldsymbol{\theta}}^*)^{\mathrm{T}}\Gamma^{-1}\Gamma\left(\frac{\partial}{\partial\hat{\boldsymbol{\theta}}}\left[\frac{\partial Q}{\partial \mathbf{x}}\mathbf{f}(\mathbf{x}, \boldsymbol{\theta}, t\mathbf{u}(\mathbf{x}, \hat{\boldsymbol{\theta}}, t))\right]\right)^{\mathrm{T}}.$$

Property (3.14) implies that

$$\begin{aligned}\dot{V} &= \dot{Q} - (\hat{\boldsymbol{\theta}} - \hat{\boldsymbol{\theta}}^*)^{\mathrm{T}}\left(\frac{\partial}{\partial\hat{\boldsymbol{\theta}}}\left[\frac{\partial Q}{\partial \mathbf{x}}\mathbf{f}(\mathbf{x}, \boldsymbol{\theta}, t, \mathbf{u}(\mathbf{x}, \hat{\boldsymbol{\theta}}, t))\right]\right)^{\mathrm{T}} \\ &\leq \dot{Q} - \frac{\partial Q(\mathbf{x}, t)}{\partial \mathbf{x}}[\mathbf{f}(\mathbf{x}, \boldsymbol{\theta}, t, \mathbf{u}(\mathbf{x}, \hat{\boldsymbol{\theta}}, t)) - \mathbf{f}(\mathbf{x}, \boldsymbol{\theta}, t, \mathbf{u}(\mathbf{x}, \hat{\boldsymbol{\theta}}^*, t))].\end{aligned}$$

Thus, taking into account the equality

$$\begin{aligned}&\frac{\partial Q(\mathbf{x}, t)}{\partial \mathbf{x}}[\mathbf{f}(\mathbf{x}, \boldsymbol{\theta}, t, \mathbf{u}(\mathbf{x}, \hat{\boldsymbol{\theta}}, t)) - \mathbf{f}(\mathbf{x}, \boldsymbol{\theta}, t, \mathbf{u}(\mathbf{x}, \hat{\boldsymbol{\theta}}^*, t))] \\ &\quad = \frac{\partial Q(\mathbf{x}, t)}{\partial \mathbf{x}}[\mathbf{f}(\mathbf{x}, \boldsymbol{\theta}, t, \mathbf{u}(\mathbf{x}, \hat{\boldsymbol{\theta}}, t)) - \mathbf{f}(\mathbf{x}, \boldsymbol{\theta}, t, \mathbf{u}(\mathbf{x}, \hat{\boldsymbol{\theta}}^*, t))] \\ &\qquad + \frac{\partial Q(\mathbf{x}, t)}{\partial t} - \frac{\partial Q(\mathbf{x}, t)}{\partial t} \\ &\quad = \dot{Q} - \frac{\partial Q(\mathbf{x}, t)}{\partial \mathbf{x}}\mathbf{f}(\mathbf{x}, \boldsymbol{\theta}, t, \mathbf{u}(\mathbf{x}, \hat{\boldsymbol{\theta}}^*, t)) - \frac{\partial Q(\mathbf{x}, t)}{\partial t}\end{aligned}$$

and invoking (3.13), we conclude that

$$\dot{V} \leq \frac{\partial Q(\mathbf{x}, t)}{\partial \mathbf{x}}\mathbf{f}(\mathbf{x}, \boldsymbol{\theta}, t, \mathbf{u}(\mathbf{x}, \hat{\boldsymbol{\theta}}^*, t)) + \frac{\partial Q(\mathbf{x}, t)}{\partial t} \leq -\alpha_1(Q) \leq 0. \tag{3.17}$$

This demonstrates that the point $(\mathbf{x}, \hat{\boldsymbol{\theta}}) = (0, \hat{\boldsymbol{\theta}}^*)$ in the extended state space of the system (3.8) and (3.16) is globally stable in the sense of Lyapunov. Thus the existence and boundedness of the solutions follow.

Notice that (3.17) implies that the following integral inequality holds:

$$\int_{t_0}^{t} \alpha_1(Q(\mathbf{x}(\tau), \tau))d\tau \leq V(\mathbf{x}(t), \hat{\boldsymbol{\theta}}(t_0), \hat{\boldsymbol{\theta}}^*, t_0) \leq \infty \; \forall \, t_0, \, t \in \mathbb{R}.$$

Then, according to Barbalat's lemma (Lemma 2.1),

$$\lim_{t\to\infty} \alpha_1(Q(\mathbf{x}(t), t)) = 0 \Rightarrow \lim_{t\to\infty} Q(\mathbf{x}(t), t) = 0.$$

□

Remark 3.1 The concept of velocity gradient can, in principle, be applied to a wide range of systems (3.8) with linear and nonlinear parametrization of the right-hand side with respect to the vector of unknown parameters $\boldsymbol{\theta}$. One needs to ensure, however, that the convexity requirement, as specified by (3.2), holds. This requirement can be viewed as a constraint on the class of admissible regulatory inputs $\mathbf{u}(\mathbf{x}, \hat{\boldsymbol{\theta}}, t)$ and functionals $Q(\mathbf{x}, t)$. Notice that this constraint does not explicitly limit the class of parametrizations of $\mathbf{f}(\mathbf{x}, \boldsymbol{\theta}, t, \mathbf{u})$. It does, however, impose certain restrictions on the choice of $\mathbf{u}(\mathbf{x}, \hat{\boldsymbol{\theta}}, t)$. Let us consider an example. Suppose that the system's dynamics is described by the equation

$$\dot{x} = \cos(\theta_1 x + \theta_2)x^2 + u, \tag{3.18}$$

where θ_1 and θ_2 are some unknown real numbers, and the goal is to steer the system's state to the origin by selecting an appropriate input u. Clearly, system (3.18) is nonlinearly parametrized with respect to θ_1 and θ_2. Let

$$u = -\hat{\theta}x - x^3. \tag{3.19}$$

This choice of the control input satisfies Assumptions 3.1 and 3.2. Indeed, Assumption 3.2 holds when the function u is linear in $\hat{\theta}$. In order to see that Assumption 3.1 holds too, one may pick $Q(x, t) = \frac{1}{2}x^2$ and derive

$$\dot{Q} = -x^2(\hat{\theta} - \cos(\theta_1 x + \theta_2)x + x^2).$$

Given that the inequality $\hat{\theta}^* > 1/4 + \rho$ ensures that the estimate

$$\hat{\theta}^* - \cos(\theta_1 x + \theta_2)x + x^2 \geq \rho, \; \rho > 0, \; \rho \in \mathbb{R}$$

holds for all x, it also implies that

$$\dot{Q} \leq -2\rho Q.$$

The latter inequality guarantees that Assumption 3.1 holds. Thus, one can apply the method of velocity gradient to derive a feedback ensuring that the state of system (3.18) is adaptively regulated to the origin. Further generalizations of this approach can be found in a number of works, e.g. Putov (1993) and Lin and Qian (2002b).

Remark 3.2 An interesting special case to which the theorem might apply is the class of systems that are affine in control with linear parametrization of admissible control functions:

$$\dot{\mathbf{x}} = \mathbf{f}(\mathbf{x}, \boldsymbol{\theta}, t) + \mathbf{g}(\mathbf{x}, t)\mathbf{u},$$
$$\mathbf{u} = \Phi(\mathbf{x}, t)\hat{\boldsymbol{\theta}}, \quad \Phi : \mathbb{R}^n \times \mathbb{R} \to \mathbb{R}^p.$$

In this case algorithm (3.16) reduces to

$$\dot{\hat{\boldsymbol{\theta}}} = -\Gamma \left(\frac{\partial Q(\mathbf{x}, t)}{\partial \mathbf{x}} \mathbf{g}(\mathbf{x}, t)\Phi(\mathbf{x}, t) \right)^{\mathrm{T}}. \tag{3.20}$$

Denoting

$$\left(\frac{\partial Q(\mathbf{x}, t)}{\partial \mathbf{x}} \mathbf{g}(\mathbf{x}, t)\Phi(\mathbf{x}, t) \right)^{\mathrm{T}} = \phi(t), \; x_1 = Q(\mathbf{x}, t), \; \mathbf{x}_2 = \hat{\boldsymbol{\theta}} - \boldsymbol{\theta},$$

and assuming that (3.13) can be written as an equality,

$$\frac{\partial Q(\mathbf{x}, t)}{\partial \mathbf{x}} [\mathbf{f}(\mathbf{x}, \theta, t) + \mathbf{g}(\mathbf{x}, t)\Phi(\mathbf{x}, t)\hat{\boldsymbol{\theta}}^*)] + \frac{\partial Q(\mathbf{x}, t)}{\partial t} = -\alpha Q, \tag{3.21}$$

where $\alpha \in \mathbb{R}_{>0}$, we arrive at the following description of the closed-loop system:

$$\frac{d}{dt}\begin{pmatrix} x_1 \\ \mathbf{x}_2 \end{pmatrix} = \begin{pmatrix} -\alpha & \phi(t)^{\mathrm{T}} \\ -\Gamma\phi(t) & 0 \end{pmatrix} \begin{pmatrix} x_1 \\ \mathbf{x}_2 \end{pmatrix}. \tag{3.22}$$

In Chapter 2 (Lemmas 2.2 and 2.3 and Theorem 2.2) we formulated some of the asymptotic properties for such systems.

As follows from the remarks above, successful application of the method of velocity gradient depends on the choice of functions $Q(\mathbf{x}, t)$ and $\mathbf{u}(\mathbf{x}, \hat{\boldsymbol{\theta}}^*, t)$. The former serves as a goal functional and at the same time doubles as the Lyapunov function linked to the class of admissible control inputs $\mathbf{u}(\mathbf{x}, \hat{\boldsymbol{\theta}}^*, t)$. Finding these functions for general nonlinear systems is outside the method's scope. These functions are assumed to be given. There are, however, classes of systems for which this assumption may be lifted. These systems and corresponding methods are presented in the next section.

3.3.2 Adaptive integrator back-stepping

Below we present the method of adaptive integrator back-stepping as it is described in the original work by Krstić *et al.* (1992). Consider the class of systems with lower-triangular structure:

$$\dot{x}_i = x_{i+1} + \boldsymbol{\theta}^{\mathrm{T}}\boldsymbol{\phi}_i(x_1, \ldots, x_i), \tag{3.23}$$

$$\dot{x}_n = \phi_0(\mathbf{x}) + \boldsymbol{\theta}^{\mathrm{T}}\boldsymbol{\phi}_n(\mathbf{x}) + \beta_0(\mathbf{x})u, \tag{3.24}$$

where $\boldsymbol{\theta} \in \mathbb{R}^p$ is the vector of unknown parameters, and $\phi_0(\cdot)$, $\boldsymbol{\phi}_i(\cdot)$, and $\beta_0(\cdot)$ are sufficiently smooth functions. We suppose that the function $\beta_0(\mathbf{x})$ is separated away from zero for all $\mathbf{x} \in \mathbb{R}^n$:

$$\exists\, \delta \in \mathbb{R}_{>0} : |\beta_0(\mathbf{x})| \geq \delta \; \forall\, \mathbf{x} \in \mathbb{R}^n.$$

Let our objective be to find a control input steering the variable x_1 to $x_1 = x^{\mathrm{e}} = 0$ and simultaneously stabilizing the system in some sense. In order to find such a control input Krstić *et al.* (1992) developed the following iterative procedure.

Step 1. Let us denote $z_1 = x_1$ and $z_2 = x_2 - \alpha_1$ and rewrite the first equation of the original system in the above notation $\dot{x}_1 = x_2 + \boldsymbol{\theta}^{\mathrm{T}}\boldsymbol{\phi}_1(x_1)$:

$$\dot{z}_1 = z_2 + \alpha_1 + \boldsymbol{\theta}^{\mathrm{T}}\boldsymbol{\phi}_1(x_1). \tag{3.25}$$

The variable α_1 is considered as a *virtual* control. Now consider

$$V_1(z_1, \hat{\boldsymbol{\theta}}) = \frac{1}{2}z_1^2 + \|\boldsymbol{\theta} - \hat{\boldsymbol{\theta}}\|^2_{\Gamma^{-1}},\; \Gamma > 0,$$

where $\hat{\boldsymbol{\theta}} : \mathbb{R}_{\geq 0} \to \mathbb{R}^p$ is a differentiable function yet to be defined. Differentiating $V_1(z_1, \hat{\boldsymbol{\theta}})$ and taking (3.25) into account, we obtain

$$\begin{aligned}\dot{V}_1 &= z_1(z_2 + \alpha_1 + \boldsymbol{\theta}^{\mathrm{T}}\boldsymbol{\phi}_1(x_1)) + (\hat{\boldsymbol{\theta}} - \boldsymbol{\theta})\Gamma^{-1}\dot{\hat{\boldsymbol{\theta}}} \\ &= z_1(z_2 + \alpha_1 + \hat{\boldsymbol{\theta}}^{\mathrm{T}}\boldsymbol{\phi}_1(x_1)) + (\hat{\boldsymbol{\theta}} - \boldsymbol{\theta})\Gamma^{-1}(\dot{\hat{\boldsymbol{\theta}}} - \Gamma z_1\boldsymbol{\phi}_1(x_1)).\end{aligned} \tag{3.26}$$

If the system (3.23) and (3.24) consisted of just one equation and x_2 were the actual control input u, i.e. $x_2 = \alpha_1 = u$, then, on setting $\boldsymbol{\tau}_1(x_1) = \dot{\hat{\boldsymbol{\theta}}}_1$,

$$\boldsymbol{\tau}_1 = \Gamma z_1(x_1)\boldsymbol{\phi}_1(x_1), \tag{3.27}$$

$$\alpha_1(x_1, \hat{\boldsymbol{\theta}}) = -c_1 z_1 - \hat{\boldsymbol{\theta}}^{\mathrm{T}}\boldsymbol{\phi}_1(x_1),\; c_1 \in \mathbb{R}_{>0}, \tag{3.28}$$

would ensure that $\dot{V}_1 = -c_1 z_1^2 = -c_1 x_1^2 \leq 0$. Hence this choice would guarantee asymptotic convergence of the system's state to the set $x_1 = 0$. The function

$\alpha_1(x_1, \hat{\boldsymbol{\theta}})$, therefore, can be called the first stabilizing function, and the function $\boldsymbol{\tau}_1(x_1)$ is the corresponding first tuning function.

Since x_2 is not the actual control input, $z_2 \neq 0$, and hence our previous choice $u = \alpha_1(x_1, \hat{\boldsymbol{\theta}})$, $\dot{\hat{\boldsymbol{\theta}}} = \boldsymbol{\tau}_1(x_1)$ is not justified. For the moment, let us simply rewrite $\dot{z}_1$ and $\dot{V}_1$ using the notations above:

$$\dot{V}_1 = -c_1 z_1^2 + z_1 z_2 + (\hat{\boldsymbol{\theta}} - \boldsymbol{\theta})\Gamma^{-1}(\dot{\hat{\boldsymbol{\theta}}} - \boldsymbol{\tau}_1(x_1)), \tag{3.29}$$

$$\dot{z}_1 = -c_1 z_1 + z_2 + (\boldsymbol{\theta} - \hat{\boldsymbol{\theta}})^{\mathrm{T}} \boldsymbol{\phi}_1(x_1). \tag{3.30}$$

Step 2. Let us introduce the new variable $z_3 = x_3 - \alpha_2$ and rewrite the second equality in (3.23) and (3.24), namely $\dot{x}_2 = x_3 + \boldsymbol{\theta}^{\mathrm{T}} \boldsymbol{\phi}_2(x_1, x_2)$, as follows:

$$\dot{z}_2 = z_3 + \alpha_2 + \boldsymbol{\theta}^{\mathrm{T}} \boldsymbol{\phi}_2(x_1, x_2) - \frac{\partial \alpha_1(x_1, \hat{\boldsymbol{\theta}})}{\partial x_1}(x_2 + \boldsymbol{\theta}^{\mathrm{T}} \boldsymbol{\phi}_1(x_1)) - \frac{\partial \alpha_1(x_1, \hat{\boldsymbol{\theta}})}{\partial \hat{\boldsymbol{\theta}}} \dot{\hat{\boldsymbol{\theta}}}. \tag{3.31}$$

Suppose that we can use α_2 as a control input, and consider the function $V_2 = V_1 + \frac{1}{2} z_2^2$. According to (3.30) and (3.31) the time-derivative of V_2 is

$$\begin{aligned} \dot{V}_2 = -c_1 z_1^2 + z_2 \bigg[z_1 + z_3 + \alpha_2 - \frac{\partial \alpha_1}{\partial x_1} x_2 - \frac{\partial \alpha_1}{\partial \hat{\boldsymbol{\theta}}} \dot{\hat{\boldsymbol{\theta}}} \\ + \hat{\boldsymbol{\theta}}^{\mathrm{T}} \left(\boldsymbol{\phi}_2(x_1, x_2) - \frac{\partial \alpha_1}{\partial x_1} \boldsymbol{\phi}_1(x_1) \right) \bigg] \\ + (\hat{\boldsymbol{\theta}} - \boldsymbol{\theta})^{\mathrm{T}} \Gamma^{-1} \left[\dot{\hat{\boldsymbol{\theta}}} - \Gamma \left(z_1 \boldsymbol{\phi}_1(x_1) + z_2 \left(\boldsymbol{\phi}_2(x_1, x_2) - \frac{\partial \alpha_1}{\partial x_1} \boldsymbol{\phi}_1(x_1) \right) \right) \right]. \end{aligned} \tag{3.32}$$

If x_3 were the actual control u then the value of z_3 in (3.31) and (3.32) would be zero. Thus, by compensating for the term $\hat{\boldsymbol{\theta}} - \boldsymbol{\theta}$ in (3.32) by setting $\dot{\hat{\boldsymbol{\theta}}} = \boldsymbol{\tau}_2(x_1, x_2, \hat{\boldsymbol{\theta}})$, where

$$\begin{aligned} \boldsymbol{\tau}_2(x_1, x_2, \hat{\boldsymbol{\theta}}) &= \Gamma \bigg[z_1(x_1) \boldsymbol{\phi}_1(x_1) \\ &\quad + z_2(x_1, x_2, \hat{\boldsymbol{\theta}}) \left(\boldsymbol{\phi}_2(x_1, x_2) - \frac{\partial \alpha_1(x_1, \hat{\boldsymbol{\theta}})}{\partial x_1} \boldsymbol{\phi}_1(x_1) \right) \bigg] \\ &= \boldsymbol{\tau}_1(x_1) + \Gamma z_2 \left(\boldsymbol{\phi}_2(x_1, x_2) - \frac{\partial \alpha_1(x_1, \hat{\boldsymbol{\theta}})}{\partial x_1} \boldsymbol{\phi}_1(x_1) \right), \end{aligned} \tag{3.33}$$

and assigning

$$\alpha_2(x_1,x_2,\hat{\boldsymbol{\theta}}) = -z_1(x_1) - c_2 z_2(x_1,x_2,\hat{\boldsymbol{\theta}}) + \frac{\partial \alpha_1(x_1,\hat{\boldsymbol{\theta}})}{\partial x_1} x_2 \\ + \frac{\partial \alpha_1(x_1,\hat{\boldsymbol{\theta}})}{\partial \hat{\boldsymbol{\theta}}} \boldsymbol{\tau}_2(x_1,x_2,\hat{\boldsymbol{\theta}}) \\ - \hat{\boldsymbol{\theta}}^{\mathrm{T}} \left(\boldsymbol{\phi}_2(x_1,x_2) - \frac{\partial \alpha_1(x_1,\hat{\boldsymbol{\theta}})}{\partial x_1} \boldsymbol{\phi}_1(x_1) \right) \tag{3.34}$$

we would ensure that

$$\dot{V}_2 = -c_1 z_1^2 - c_2 z_2^2.$$

However, the variable x_3 is not the actual control input u, and therefore $z_3 \neq 0$. Yet, taking the notation (3.34) into account, $\dot{V}_2$ can still be expressed in the following simplified form:

$$\dot{V}_2 = -c_1 z_1^2 - c_2 z_2^2 + z_2 z_3 + \left[z_2 \frac{\partial \alpha_1}{\partial \hat{\boldsymbol{\theta}}} + (\boldsymbol{\theta} - \hat{\boldsymbol{\theta}})^{\mathrm{T}} \Gamma^{-1} \right] (\boldsymbol{\tau}_2 - \dot{\hat{\boldsymbol{\theta}}}), \tag{3.35}$$

and (3.31) may be rewritten as

$$\dot{z}_2 = -z_1 - c_2 z_2 + z_3 + (\boldsymbol{\theta} - \hat{\boldsymbol{\theta}})^{\mathrm{T}} \left(\boldsymbol{\phi}_2 - \frac{\partial \alpha_1}{\partial x_1} \boldsymbol{\phi}_1 \right) + \frac{\partial \alpha_1}{\partial \hat{\boldsymbol{\theta}}} (\boldsymbol{\tau}_2 - \dot{\hat{\boldsymbol{\theta}}}). \tag{3.36}$$

Step 3. Let us introduce the new variable $z_4 = x_4 - \alpha_3$ and, taking this denotation into account, rewrite the equality $\dot{x}_3 = x_4 + \boldsymbol{\theta}^{\mathrm{T}} \boldsymbol{\phi}_3(x_1,x_2,x_3)$ as follows:

$$\dot{z}_3 = z_4 + \alpha_3 + \boldsymbol{\theta}^{\mathrm{T}} \boldsymbol{\phi}_3 - \frac{\partial \alpha_2}{\partial x_1} (x_2 + \boldsymbol{\theta}^{\mathrm{T}} \boldsymbol{\phi}_1) - \frac{\partial \alpha_2}{\partial x_2} (x_3 + \boldsymbol{\theta}^{\mathrm{T}} \boldsymbol{\phi}_2) - \frac{\partial \alpha_2}{\partial \hat{\boldsymbol{\theta}}} \dot{\hat{\boldsymbol{\theta}}}. \tag{3.37}$$

Using the same logic as in Steps 1 and 2, we suppose that α_3 is a control input, and introduce the following Lyapunov candidate function for the extended system: $V_3 = V_2 + \frac{1}{2} z_3^2$ (the state vector of this system is $(z_1, z_2, z_3, \hat{\boldsymbol{\theta}})$). Consider $\dot{V}_3$:

$$\dot{V}_3 = -c_1 z_1^2 - c_2 z_2^2 + z_2 \frac{\partial \alpha_1}{\partial \hat{\boldsymbol{\theta}}} (\boldsymbol{\tau}_2 - \dot{\hat{\boldsymbol{\theta}}}) \\ + z_3 \left[z_2 + z_4 + \alpha_3 - \frac{\partial \alpha_2}{\partial x_1} x_2 - \frac{\partial \alpha_2}{\partial x_2} x_3 - \frac{\partial \alpha_2}{\partial \hat{\boldsymbol{\theta}}} \dot{\hat{\boldsymbol{\theta}}} \right. \\ \left. + \hat{\boldsymbol{\theta}}^{\mathrm{T}} \left(\boldsymbol{\phi}_3 - \frac{\partial \alpha_2}{\partial x_1} \boldsymbol{\phi}_1 - \frac{\partial \alpha_2}{\partial x_2} \boldsymbol{\phi}_2 \right) \right]$$

$$+ (\hat{\boldsymbol{\theta}} - \boldsymbol{\theta})^{\mathrm{T}}\Gamma^{-1}\left[\dot{\hat{\boldsymbol{\theta}}} - \Gamma\left(z_1\boldsymbol{\phi}_1 + z_2\left(\boldsymbol{\phi}_2 - \frac{\partial\alpha_1}{\partial x_1}\boldsymbol{\phi}_1\right)\right.\right.$$
$$\left.\left. + z_3\left(\boldsymbol{\phi}_3 - \frac{\partial\alpha_2}{\partial x_1}\boldsymbol{\phi}_1 - \frac{\partial\alpha_2}{\partial x_2}\boldsymbol{\phi}_2\right)\right)\right] \tag{3.38}$$

and compensate for the term $\hat{\boldsymbol{\theta}} - \boldsymbol{\theta}$ in (3.38) by setting $\dot{\hat{\boldsymbol{\theta}}} = \boldsymbol{\tau}_3$:

$$\boldsymbol{\tau}_3 = \Gamma\left[z_1\boldsymbol{\phi}_1 + z_2\left(\boldsymbol{\phi}_2 - \frac{\partial\alpha_1}{\partial x_1}\boldsymbol{\phi}_1\right) + z_3\left(\boldsymbol{\phi}_3 - \frac{\partial\alpha_2}{\partial x_1}\boldsymbol{\phi}_1 - \frac{\partial\alpha_2}{\partial x_2}\boldsymbol{\phi}_2\right)\right]$$
$$= \boldsymbol{\tau}_2 + \Gamma z_3\left(\boldsymbol{\phi}_3 - \frac{\partial\alpha_2}{\partial x_1}\boldsymbol{\phi}_1 - \frac{\partial\alpha_2}{\partial x_2}\right). \tag{3.39}$$

Given that

$$\dot{\hat{\boldsymbol{\theta}}} - \boldsymbol{\tau}_2 = \dot{\hat{\boldsymbol{\theta}}} - \boldsymbol{\tau}_3 + \boldsymbol{\tau}_3 - \boldsymbol{\tau}_2 = \dot{\hat{\boldsymbol{\theta}}} - \boldsymbol{\tau}_3 + \Gamma z_3\left(\boldsymbol{\phi}_3 - \frac{\partial\alpha_2}{\partial x_1}\boldsymbol{\phi}_1 - \frac{\partial\alpha_2}{\partial x_2}\boldsymbol{\phi}_2\right), \tag{3.40}$$

we can rewrite (3.38) as follows:

$$\dot{V}_3 = -c_1z_1^2 - c_2z_2^2 + z_2\frac{\partial\alpha_1}{\partial\hat{\boldsymbol{\theta}}}\left(\boldsymbol{\tau}_3 - \dot{\hat{\boldsymbol{\theta}}}\right)$$
$$+ z_3\left[z_2 + z_4 + \alpha_3 - \frac{\partial\alpha_2}{\partial x_1}x_2 - \frac{\partial\alpha_2}{\partial x_2}x_3 - \frac{\partial\alpha_2}{\partial\hat{\boldsymbol{\theta}}}\dot{\hat{\boldsymbol{\theta}}}\right.$$
$$\left. + \left(\hat{\boldsymbol{\theta}}^{\mathrm{T}} - z_2\frac{\partial\alpha_1}{\partial\hat{\boldsymbol{\theta}}}\Gamma\right)\left(\boldsymbol{\phi}_3 - \frac{\partial\alpha_2}{\partial x_1}\boldsymbol{\phi}_1 - \frac{\partial\alpha_2}{\partial x_2}\boldsymbol{\phi}_2\right)\right]$$
$$+ (\hat{\boldsymbol{\theta}} - \boldsymbol{\theta})^{\mathrm{T}}\Gamma^{-1}(\dot{\hat{\boldsymbol{\theta}}} - \boldsymbol{\tau}_3). \tag{3.41}$$

If x_4 (or equivalently α_3) were the actual control then $z_4 = 0$ and the choice

$$\alpha_3 = -z_2 - c_3z_3 + \frac{\partial\alpha_2}{\partial x_1}x_2 + \frac{\partial\alpha_2}{\partial x_2}x_3 + \frac{\partial\alpha_2}{\partial\hat{\boldsymbol{\theta}}}\boldsymbol{\tau}_3$$
$$+ \left(z_2\frac{\partial\alpha_1}{\partial\hat{\boldsymbol{\theta}}}\Gamma - \hat{\boldsymbol{\theta}}^{\mathrm{T}}\right)\left(\boldsymbol{\phi}_3 - \frac{\partial\alpha_2}{\partial x_1}\boldsymbol{\phi}_1 - \frac{\partial\alpha_2}{\partial x_2}\boldsymbol{\phi}_2\right) \tag{3.42}$$

would ensure that

$$\dot{V}_3 = -c_1z_1^2 - c_2z_2^2 - c_3z_3^2.$$

Since this is not the case, we have that

$$\dot{V}_3 = -c_1z_1^2 - c_2z_2^2 - c_3z_3^2 + z_3z_4$$
$$+ \left[z_2\frac{\partial\alpha_1}{\partial\hat{\boldsymbol{\theta}}} + z_3\frac{\partial\alpha_2}{\partial\hat{\boldsymbol{\theta}}} + (\boldsymbol{\theta} - \hat{\boldsymbol{\theta}})^{\mathrm{T}}\Gamma^{-1}\right](\boldsymbol{\tau}_3 - \dot{\hat{\boldsymbol{\theta}}}), \tag{3.43}$$

and, taking (3.37) and (3.42) into account, the right-hand side of $\dot{z}_3$ can be determined as

$$\begin{aligned}\dot{z}_3 = & -z_2 - c_3 z_3 + z_4 + (\boldsymbol{\theta} - \hat{\boldsymbol{\theta}})^{\mathrm{T}} \left(\boldsymbol{\phi}_3 - \frac{\partial \alpha_2}{\partial x_1} \boldsymbol{\phi}_1 - \frac{\partial \alpha_2}{\partial x_2} \boldsymbol{\phi}_2 \right) \\ & + \frac{\partial \alpha_2}{\partial \hat{\boldsymbol{\theta}}} (\boldsymbol{\tau}_3 - \dot{\hat{\boldsymbol{\theta}}}) + z_2 \frac{\partial \alpha_1}{\partial \hat{\boldsymbol{\theta}}} \Gamma \left(\boldsymbol{\phi}_3 - \frac{\partial \alpha_2}{\partial x_1} \boldsymbol{\phi}_1 - \frac{\partial \alpha_2}{\partial x_2} \boldsymbol{\phi}_2 \right). \end{aligned} \tag{3.44}$$

Step i. Consider $z_{i+1} = x_{i+1} - \alpha_i$ and rewrite the equation $\dot{x}_i = x_{i+1} + \boldsymbol{\theta}^{\mathrm{T}} \boldsymbol{\phi}_i(x_1, \ldots, x_i)$ accordingly:

$$\dot{z}_i = z_{i+1} + \alpha_i + \boldsymbol{\theta}^{\mathrm{T}} \boldsymbol{\phi}_i - \sum_{k=1}^{i-1} \frac{\partial \alpha_{i-1}}{\partial x_k} (x_{k+1} + \boldsymbol{\theta}^{\mathrm{T}} \boldsymbol{\phi}_k) - \frac{\partial \alpha_{i-1}}{\partial \hat{\boldsymbol{\theta}}} \dot{\hat{\boldsymbol{\theta}}}. \tag{3.45}$$

Let α_i be a virtual control of the system in $(z_1, \ldots, z_i)$ coordinates. Looking for stability conditions for the extended system with state $(z_1, \ldots, z_i) \oplus \hat{\boldsymbol{\theta}}$, we introduce the Lyapunov candidate function $V_i = V_{i-1} + \frac{1}{2} z_i^2$. Its time-derivative is

$$\begin{aligned}\dot{V}_i = & -\sum_{k=1}^{i-1} c_k z_k^2 + \left(\sum_{k=1}^{i-2} z_{k+1} \frac{\partial \alpha_k}{\partial \hat{\boldsymbol{\theta}}} \right) (\boldsymbol{\tau}_{i-1} - \dot{\hat{\boldsymbol{\theta}}}) \\ & + z_i \Bigg[z_{i-1} + z_{i+1} + \alpha_i - \sum_{k=1}^{i-1} \frac{\partial \alpha_{i-1}}{\partial x_k} x_{k+1} - \frac{\partial \alpha_{i-1}}{\partial \hat{\boldsymbol{\theta}}} \dot{\hat{\boldsymbol{\theta}}} \\ & \qquad + \hat{\boldsymbol{\theta}}^{\mathrm{T}} \left(\boldsymbol{\phi}_i - \sum_{k=1}^{i-1} \frac{\partial \alpha_{i-1}}{\partial x_k} \boldsymbol{\phi}_k \right) \Bigg] \\ & + (\hat{\boldsymbol{\theta}} - \boldsymbol{\theta})^{\mathrm{T}} \Gamma^{-1} \left[\dot{\hat{\boldsymbol{\theta}}} - \Gamma \sum_{l=1}^{i} z_l \left(\boldsymbol{\phi}_l - \sum_{k=1}^{l-1} \frac{\partial \alpha_{l-1}}{\partial x_k} \boldsymbol{\phi}_k \right) \right]. \end{aligned} \tag{3.46}$$

On choosing the function $\boldsymbol{\tau}_i$ in $\dot{\hat{\boldsymbol{\theta}}} = \boldsymbol{\tau}_i$ as

$$\boldsymbol{\tau}_i = \Gamma \sum_{l=1}^{i} z_l \left(\boldsymbol{\phi}_l - \sum_{k=1}^{l-1} \frac{\partial \alpha_{l-1}}{\partial x_k} \boldsymbol{\phi}_k \right) = \boldsymbol{\tau}_i + \Gamma z_i \left(\boldsymbol{\phi}_i - \sum_{k=1}^{i-1} \frac{\partial \alpha_{i-1}}{\partial x_k} \boldsymbol{\phi}_k \right) \tag{3.47}$$

we can annihilate the influence of $\hat{\boldsymbol{\theta}} - \boldsymbol{\theta}$ in (3.46). Furthermore, given that

$$\dot{\hat{\boldsymbol{\theta}}} - \boldsymbol{\tau}_{i-1} = \dot{\hat{\boldsymbol{\theta}}} - \boldsymbol{\tau}_i + \boldsymbol{\tau}_i - \boldsymbol{\tau}_{i-1} = \dot{\hat{\boldsymbol{\theta}}} - \boldsymbol{\tau}_i + \Gamma z_i \left(\boldsymbol{\phi}_i - \sum_{k=1}^{i-1} \frac{\partial \alpha_{i-1}}{\partial x_k} \boldsymbol{\phi}_k \right), \tag{3.48}$$

we can rewrite $\dot{V}_i$ as

$$\dot{V}_i = -\sum_{k=1}^{i-1} c_k z_k^2 + \sum_{k=1}^{i-2} z_{k+1} \frac{\partial \alpha_k}{\partial \hat{\boldsymbol{\theta}}} (\boldsymbol{\tau}_i - \dot{\hat{\boldsymbol{\theta}}})$$

$$+ z_i \left[z_{i-1} + z_{i+1} + \alpha_i - \sum_{k=1}^{i-1} \frac{\partial \alpha_{i-1}}{\partial x_k} x_{k+1} - \frac{\partial \alpha_{i-1}}{\partial \hat{\boldsymbol{\theta}}} \dot{\hat{\boldsymbol{\theta}}} + \left(\hat{\boldsymbol{\theta}}^{\mathrm{T}} - \sum_{k=1}^{i-2} z_{k+1} \frac{\partial \alpha_k}{\partial \hat{\boldsymbol{\theta}}} \Gamma \right) \right.$$

$$\left. \times \left(\boldsymbol{\phi}_i - \sum_{k=1}^{i-1} \frac{\partial \alpha_{i-1}}{\partial x_k} \boldsymbol{\phi}_k \right) \right] + (\hat{\boldsymbol{\theta}} - \boldsymbol{\theta})^{\mathrm{T}} \Gamma^{-1} (\dot{\hat{\boldsymbol{\theta}}} - \boldsymbol{\tau}_i). \tag{3.49}$$

If x_{i+1} were the actual control input, i.e. $\alpha_i = u = x_{i+1}$, then consequently setting

$$\alpha_i = -z_{i-1} - c_i z_i + \sum_{k=1}^{i-1} \frac{\partial \alpha_{i-1}}{\partial x_k} x_{k+1} + \frac{\partial \alpha_{i-1}}{\partial \hat{\boldsymbol{\theta}}} \boldsymbol{\tau}_i$$

$$+ \left[\sum_{k=1}^{i-2} z_{k+1} \frac{\partial \alpha_k}{\partial \hat{\boldsymbol{\theta}}} \Gamma - \hat{\boldsymbol{\theta}}^{\mathrm{T}} \right] \left(\boldsymbol{\phi}_i - \sum_{k=1}^{i-1} \frac{\partial \alpha_{i-1}}{\partial x_k} \boldsymbol{\phi}_k \right) \tag{3.50}$$

would ensure that

$$\dot{V}_i = -\sum_{k=1}^{i} c_k z_k^2.$$

Since this is not the case we have

$$\dot{V}_i = -\sum_{k=1}^{i} c_i z_i^2 + z_i z_{i+1} + \left[\sum_{k=1}^{i-1} \frac{\partial \alpha_k}{\partial \hat{\boldsymbol{\theta}}} + (\boldsymbol{\theta} - \hat{\boldsymbol{\theta}})^{\mathrm{T}} \Gamma^{-1} \right] (\boldsymbol{\tau}_i - \dot{\hat{\boldsymbol{\theta}}}), \tag{3.51}$$

and z_i satisfy the following equations:

$$\dot{z}_i = -z_{i-1} - c_i z_i + z_{i+1} + (\boldsymbol{\theta} - \hat{\boldsymbol{\theta}})^{\mathrm{T}} \left(\boldsymbol{\phi}_i - \sum_{k=1}^{i-1} \frac{\partial \alpha_{i-1}}{\partial x_k} \boldsymbol{\phi}_k \right)$$

$$+ \frac{\partial \alpha_i}{\partial \hat{\boldsymbol{\theta}}} (\boldsymbol{\tau}_i - \dot{\hat{\boldsymbol{\theta}}}) + \left(\sum_{k=1}^{i-2} z_{k+1} \frac{\partial \alpha_k}{\partial \hat{\boldsymbol{\theta}}} \Gamma \right) \left(\boldsymbol{\phi}_i - \sum_{k=1}^{i-1} \frac{\partial \alpha_{i-1}}{\partial x_k} \boldsymbol{\phi}_k \right). \tag{3.52}$$

Step n. Let $z_n = x_n - \alpha_{n-1}$. Observe that the variable α_{n-1} is already defined at the previous, $n-1$, step of the procedure. Thus, taking $\dot{x}_n = \phi_0(\mathbf{x}) + \boldsymbol{\theta}^{\mathrm{T}} \boldsymbol{\phi}_n(\mathbf{x}) + \beta_0(\mathbf{x}) u$, (3.51), and (3.52) into account, we can express $\dot{z}_n$ as

$$\dot{z}_n = \phi_0 + \boldsymbol{\theta}^{\mathrm{T}} \boldsymbol{\phi}_n + \beta_0 u - \sum_{k=1}^{n-1} \frac{\partial \alpha_{n-1}}{\partial x_k} (x_{k+1} + \boldsymbol{\theta}^{\mathrm{T}} \boldsymbol{\phi}_k) - \frac{\partial \alpha_{n-1}}{\partial \hat{\boldsymbol{\theta}}} \dot{\hat{\boldsymbol{\theta}}}. \tag{3.53}$$

Let us now introduce the following function $V_n = V_{n-1}+\frac{1}{2}z_n^2$. Clearly, the function V_n is positive definite with respect to z_i and $(\hat{\boldsymbol{\theta}} - \boldsymbol{\theta})$, and its derivative satisfies

$$\dot{V}_n = -\sum_{k=1}^{n-1} c_k z_k^2 + \left(\sum_{k=1}^{n-2} z_{k+1}\frac{\partial \alpha_k}{\partial \hat{\boldsymbol{\theta}}}\right)(\boldsymbol{\tau}_{n-1} - \dot{\hat{\boldsymbol{\theta}}})$$
$$+ z_n\left[z_{n-1} + \beta_0 u + \phi_0 - \sum_{k=1}^{n-1}\frac{\partial \alpha_{n-1}}{\partial x_k}x_{k+1} - \frac{\partial \alpha_{n-1}}{\partial \hat{\boldsymbol{\theta}}}\dot{\hat{\boldsymbol{\theta}}}\right.$$
$$\left. + \hat{\boldsymbol{\theta}}^{\mathrm{T}}\left(\boldsymbol{\phi}_n - \sum_{k=1}^{n-1}\frac{\partial \alpha_{n-1}}{\partial x_k}\boldsymbol{\phi}_k\right)\right] + (\hat{\boldsymbol{\theta}} - \boldsymbol{\theta})^T\Gamma^{-1}\left[\dot{\hat{\boldsymbol{\theta}}} - \Gamma\sum_{l=1}^{n} z_l\left(\boldsymbol{\phi}_l - \sum_{k=1}^{l-1}\boldsymbol{\phi}_k\right)\right]. \tag{3.54}$$

In order to compensate for the term $(\hat{\boldsymbol{\theta}} - \boldsymbol{\theta})$ in (3.54) we set

$$\dot{\hat{\boldsymbol{\theta}}} = \boldsymbol{\tau}_n = \Gamma\sum_{l=1}^{n} z_l\left(\boldsymbol{\phi}_l - \sum_{k=1}^{l-1}\frac{\partial \alpha_{l-1}}{\partial x_k}\boldsymbol{\phi}_k\right)$$
$$= \boldsymbol{\tau}_{n-1} + \Gamma z_n\left(\boldsymbol{\phi}_n - \sum_{k=1}^{n-1}\frac{\partial \alpha_{n-1}}{\partial x_k}\boldsymbol{\phi}_k\right). \tag{3.55}$$

Taking (3.54) and

$$\hat{\boldsymbol{\theta}} - \boldsymbol{\tau}_{n-1} = \boldsymbol{\tau}_n - \boldsymbol{\tau}_{n-1} = \Gamma z_n\left(\boldsymbol{\phi}_n - \sum_{k=1}^{n-1}\frac{\partial \alpha_{n-1}}{\partial x_k}\boldsymbol{\phi}_k\right) \tag{3.56}$$

into account, we can conclude that

$$\dot{V}_n = \sum_{k=1}^{n-1} c_k z_k^2 + z_n\left[z_{n-1} + \beta_0 u + \phi_0 - \sum_{k=1}^{n-1}\frac{\partial \alpha_{n-1}}{\partial x_k}x_{k+1} - \frac{\partial \alpha_{n-1}}{\partial \hat{\boldsymbol{\theta}}}\dot{\hat{\boldsymbol{\theta}}}\right.$$
$$\left. + \left(\hat{\boldsymbol{\theta}}^{\mathrm{T}} - \sum_{k=1}^{n-2} z_{k+1}\frac{\partial \alpha_k}{\partial \hat{\boldsymbol{\theta}}}\Gamma\right)\left(\boldsymbol{\phi}_n - \sum_{k=1}^{n-1}\frac{\partial \alpha_{n-1}}{\partial x_k}\boldsymbol{\phi}_k\right)\right]. \tag{3.57}$$

Thus choosing u as

$$u = \frac{1}{\beta_0}\left[-z_{n-1} - c_n z_n - \phi_0 + \sum_{k=1}^{n-1}\frac{\partial \alpha_{n-1}}{\partial x_k}x_{k+1} + \frac{\partial \alpha_{n-1}}{\partial \hat{\boldsymbol{\theta}}}\boldsymbol{\tau}_n\right.$$
$$\left. + \left(\sum_{k=1}^{n-2} z_{k+1}\frac{\partial \alpha_k}{\partial \hat{\boldsymbol{\theta}}}\Gamma - \hat{\boldsymbol{\theta}}^{\mathrm{T}}\right)\left(\boldsymbol{\phi}_n - \sum_{k=1}^{n-1}\frac{\partial \alpha_{n-1}}{\partial x_k}\boldsymbol{\phi}_k\right)\right] \tag{3.58}$$

will ensure that the derivative $\dot{V}_n$ is non-positive:

$$\dot{V}_n = -\sum_{k=1}^{n} c_k z_k^2,$$

and consequently that z_i and $\hat{\boldsymbol{\theta}}$ are bounded along the solutions of (3.23) and (3.24). Given that

$$\begin{aligned}
&x_1 = z_1,\ x_2 = z_2 + \alpha_1(z_1, \hat{\boldsymbol{\theta}}),\\
&x_3 = z_3 + \alpha_2(x_1, x_2, \hat{\boldsymbol{\theta}}) = z_3 + \alpha_2(z_1, z_2 + \alpha_1(z_1, \hat{\boldsymbol{\theta}}), \hat{\boldsymbol{\theta}}), \quad \ldots
\end{aligned}$$

and that the functions $\alpha_i(x_1, \ldots, x_i, \hat{\boldsymbol{\theta}})$ are smooth, we can conclude that $\mathbf{x}(t, \mathbf{x}_0) = (x_1(t, \mathbf{x}_0), \ldots, x_n(t, \mathbf{x}_0))$ is bounded and defined for all $t \geq t_0$. Moreover, according to Lemma 2.1,

$$\lim_{t \to \infty} z_i(t) = 0. \tag{3.59}$$

In order to see that $\mathbf{x}(t, \mathbf{x}_0)$ converges to a limit at $t \to \infty$, let us rewrite (3.23) and (3.24) in the z_i-coordinates. On denoting

$$\mathbf{w}_i(x_1, \ldots, x_i, \hat{\boldsymbol{\theta}}) = \boldsymbol{\phi}_i(x_1, \ldots, x_i) - \sum_{k=1}^{i-1} \frac{\partial \alpha_{i-1}}{\partial x_k} \boldsymbol{\phi}_k(x_1, \ldots, x_k) \tag{3.60}$$

we obtain that

$$\begin{aligned}
\dot{z}_1 &= -c_1 z_1 + z_2 + (\boldsymbol{\theta} - \hat{\boldsymbol{\theta}})^{\mathrm{T}} \mathbf{w}_1(x_1, \hat{\boldsymbol{\theta}}),\\
\dot{z}_2 &= -z_1 - c_2 z_2 + z_3 + (\boldsymbol{\theta} - \hat{\boldsymbol{\theta}})^{\mathrm{T}} \mathbf{w}_2(x_1, x_2, \hat{\boldsymbol{\theta}})\\
&\quad - \sum_{k=3}^{n} \frac{\partial \alpha_1}{\partial \hat{\boldsymbol{\theta}}} \Gamma z_k \mathbf{w}_k(x_1, \ldots, x_k, \hat{\boldsymbol{\theta}}),\\
\dot{z}_3 &= -z_2 - c_3 z_3 + z_4 + (\boldsymbol{\theta} - \hat{\boldsymbol{\theta}})^{\mathrm{T}} \mathbf{w}_3(x_1, x_2, x_3, \hat{\boldsymbol{\theta}})\\
&\quad - \sum_{k=4}^{n} \frac{\partial \alpha_2}{\partial \hat{\boldsymbol{\theta}}} \Gamma z_k \mathbf{w}_k(x_1, \ldots, x_k, \hat{\boldsymbol{\theta}}) + z_2 \frac{\partial \alpha_1}{\partial \hat{\boldsymbol{\theta}}} \Gamma \mathbf{w}_3(x_1, x_2, x_3, \hat{\boldsymbol{\theta}}),\\
&\ \vdots
\end{aligned}$$

$$
\begin{aligned}
z_i &= -z_{i-1} - c_i z_i + z_{i+1} + (\boldsymbol{\theta} - \hat{\boldsymbol{\theta}})^{\mathrm{T}} \mathbf{w}(x_1, \ldots, x_i, \hat{\boldsymbol{\theta}}) \\
&\quad - \sum_{k=i+1}^{n} \frac{\partial \alpha_{i-1}}{\partial \hat{\boldsymbol{\theta}}} \Gamma z_k \mathbf{w}_k(x_1, \ldots, x_k, \hat{\boldsymbol{\theta}}) \\
&\quad + \sum_{k=1}^{i-2} z_{k+1} \frac{\partial \alpha_k}{\partial \hat{\boldsymbol{\theta}}} \Gamma \mathbf{w}_i(x_1, \ldots, x_i, \hat{\boldsymbol{\theta}}), \\
&\vdots \\
\dot{z}_n &= -z_{n-1} - c_n z_n + (\boldsymbol{\theta} - \hat{\boldsymbol{\theta}})^{\mathrm{T}} \mathbf{w}_n(x_1, \ldots, x_n, \hat{\boldsymbol{\theta}}) \\
&\quad + \sum_{k=1}^{n-2} z_{k+1} \frac{\partial \alpha_k}{\partial \hat{\boldsymbol{\theta}}} \Gamma \mathbf{w}_n(x_1, \ldots, x_n, \hat{\boldsymbol{\theta}}), \\
\dot{\hat{\boldsymbol{\theta}}} &= \Gamma \sum_{i=1}^{n} z_i \mathbf{w}_l(x_1, \ldots, x_i, \hat{\boldsymbol{\theta}}).
\end{aligned}
\tag{3.61}
$$

The right-hand side of (3.61) is a smooth vector field. Hence, $\ddot{z}_i$ are bounded due to the fact that the variables $\mathbf{x}$, z_i, and $\hat{\boldsymbol{\theta}}$ are bounded. According to Lemma 2.1, (3.59) implies that

$$\lim_{t \to \infty} \dot{z}(t) = 0.$$

Hence, in accordance with (3.61), the following limits exist:

$$\lim_{t \to \infty} (\boldsymbol{\theta} - \hat{\boldsymbol{\theta}}(t)) \mathbf{w}_i(x_1(t), \ldots, x_i(t), \hat{\boldsymbol{\theta}}(t)) = 0. \tag{3.62}$$

Let us demonstrate that $x_i(t)$ also converge to the corresponding limits as $t \to \infty$. Indeed, since $x_1 = z_1$, the equation $\lim_{t \to \infty} z_1(t) = 0$ implies that $\lim_{t \to \infty} x_1(t) = 0$. Notice that $\mathbf{w}_1(x_1, \hat{\boldsymbol{\theta}}) = \boldsymbol{\phi}_1(x_1)$ and

$$
\begin{aligned}
&\lim_{t \to \infty} (\boldsymbol{\theta} - \hat{\boldsymbol{\theta}}(t))^{\mathrm{T}} \mathbf{w}_1(x_1(t), \hat{\boldsymbol{\theta}}(t)) = \lim_{t \to \infty} (\boldsymbol{\theta} - \hat{\boldsymbol{\theta}}(t))^{\mathrm{T}} \boldsymbol{\phi}_1(x_1(t)) = 0 \Rightarrow \\
&\lim_{t \to \infty} \boldsymbol{\theta}^{\mathrm{T}} \boldsymbol{\phi}_1(x_1(t)) = \lim_{t \to \infty} \hat{\boldsymbol{\theta}}^{\mathrm{T}} \boldsymbol{\phi}_1(x_1(t)) = \boldsymbol{\theta}^{\mathrm{T}} \boldsymbol{\phi}_1(0);
\end{aligned}
$$

hence,

$$
\begin{aligned}
\lim_{t \to \infty} x_2(t) &= \lim_{t \to \infty} z_2(t) + \alpha_1(z_1(t), \hat{\boldsymbol{\theta}}(t)) \\
&= \lim_{t \to \infty} z_2(t) - c_1 z_1(t) - \hat{\boldsymbol{\theta}}(t)^{\mathrm{T}} \boldsymbol{\phi}_1(x_1(t)) = -\boldsymbol{\theta}^{\mathrm{T}} \boldsymbol{\phi}_1(0).
\end{aligned}
$$

Thus, using (3.62) and (3.61), we arrive at

$$\lim_{t\to\infty}(\boldsymbol{\theta}-\hat{\boldsymbol{\theta}}(t))^{\mathrm{T}}\mathbf{w}_2(x_1,x_2,\hat{\boldsymbol{\theta}}(t)) = \lim_{t\to\infty}(\boldsymbol{\theta}-\hat{\boldsymbol{\theta}}(t))^{\mathrm{T}}\boldsymbol{\phi}_2(x_1(t),x_2(t)) = 0 \Rightarrow$$

$$\lim_{t\to\infty}\hat{\boldsymbol{\theta}}^{\mathrm{T}}(t)\boldsymbol{\phi}_2(x_1(t),x_2(t)) = \boldsymbol{\theta}^{\mathrm{T}}\boldsymbol{\phi}_2(0,x_2^{\mathrm{e}}).$$

According to (3.50) the variables α_i satisfy

$$\alpha_i = -z_{i-1} - c_i z_i + \sum_{k=1}^{i-1}\frac{\partial\alpha_{i-1}}{\partial x_k}\left(x_{k+1}+\hat{\boldsymbol{\theta}}^{\mathrm{T}}\boldsymbol{\phi}_k\right) + \frac{\partial\alpha_{i-1}}{\partial\hat{\boldsymbol{\theta}}}\boldsymbol{\tau}_i$$
$$+\left[\sum_{k=1}^{i-2} z_{k+1}\frac{\partial\alpha_k}{\partial\hat{\boldsymbol{\theta}}}\Gamma - \hat{\boldsymbol{\theta}}^{\mathrm{T}}\right]\boldsymbol{\phi}_i;$$

hence,

$$\lim_{t\to\infty}\alpha_2(t) = \lim_{t\to\infty} = -\boldsymbol{\theta}^{\mathrm{T}}\boldsymbol{\phi}_2(0,x_2^{\mathrm{e}}),$$

and $\lim_{t\to\infty} x_3(t) = \lim_{t\to\infty} z_3(t)+\alpha_2(t) = -\boldsymbol{\theta}^{\mathrm{T}}\boldsymbol{\phi}_2(0,x_2^{\mathrm{e}})$. On repeating these steps for $i = 3,\ldots,n$ we obtain

$$\lim_{t\to\infty}\hat{\boldsymbol{\theta}}^{\mathrm{T}}(t)\boldsymbol{\phi}_i(x_1(t),\ldots,x_i(t)) = \boldsymbol{\theta}^{\mathrm{T}}\boldsymbol{\phi}_i(0,x_2^{\mathrm{e}},\ldots,x_i^{\mathrm{e}}),$$
$$\lim_{t\to\infty} x_i(t) = -\boldsymbol{\theta}^{\mathrm{T}}\boldsymbol{\phi}_{i-1}(0,x_2^{\mathrm{e}},\ldots,x_{i-1}^{\mathrm{e}}).$$

Summarizing the results, we can now state the following theorem (Krstić *et al.* 1992).

Theorem 3.2 *Let the system (3.23) and (3.24) be given, with u defined as in (3.58). Then the equilibrium* $\mathbf{x}^{\mathrm{e}}\oplus\boldsymbol{\theta}$, $\mathbf{x}^{\mathrm{e}} = (x_1^{\mathrm{e}},\ldots,x_n^{\mathrm{e}})$ *of the extended system is globally stable in the sense of Lyapunov. Furthermore,*

$$\lim_{t\to\infty}\mathbf{x}(t) = \mathbf{x}^{\mathrm{e}}.$$

The method of adaptive integrator back-stepping resolves the issue of searching for the goal functional $Q(\mathbf{x},t)$ required in the method of velocity gradient. Although the method applies to a narrower class of systems, it is fully constructive and hence constitutes a convenient solution to the problem of adaptive regulation. One can also view this method as a synthesis technology.

Before we proceed further, let us comment that both of the methods considered so far assume (or impose, should we consider dominance-based generalizations) linear or convex parametrization of the corresponding control input. This inevitably motivates the question about the possibility of ensuring a certain degree of adaptivity in systems with nonlinearly parametrized controls. In the next sections we review the ideas offering answers to this question.

3.3.3 Minimax and domination-based algorithms of adaptive regulation

A technique aimed at addressing the issue of nonlinear parametrization was proposed in Loh *et al.* (1999). The authors consider systems in the so-called error-model form:

$$\dot{e}_{\mathrm{c}} = -k_1 e_{\mathrm{c}} + k_2 \left[\boldsymbol{\varphi}_\ell^{\mathrm{T}}(\boldsymbol{\alpha}_\ell - \hat{\boldsymbol{\alpha}}_\ell) + \sum_{i=1}^{m} \Big(f_i(\boldsymbol{\phi}_i, \theta_i) - f_i(\boldsymbol{\phi}_i, \hat{\theta}_i) \Big) - u_{\mathrm{a}}(t) \right], \tag{3.63}$$

where e_{c} stands for the error (or state); $\boldsymbol{\alpha}_\ell \in \mathbb{R}^\ell$ and $\theta_i \in \Theta_i \subset \mathbb{R}$ are parameters of which the values are unknown, and $\hat{\boldsymbol{\alpha}}_\ell \in \mathbb{R}^\ell$, $\hat{\theta}_i$ are their estimates. The domains Θ_i are supposed to be closed intervals in $\mathbb{R}$ with known boundaries, $k_1 \in \mathbb{R}_{>0}$ and $k_2 \in \mathbb{R}$ are known parameters of the model, and $\boldsymbol{\phi}_i : \mathbb{R}_{\geq 0} \to \mathbb{R}^p$, $\boldsymbol{\varphi}_\ell : \mathbb{R}_{\geq 0} \to \mathbb{R}^\ell$, and $f_i : \mathbb{R}^p \times \mathbb{R} \to \mathbb{R}$ are some known continuous functions. The variable $u_{\mathrm{a}}(t)$ is the auxiliary control input.

The objective is to find an auxiliary input $u_{\mathrm{a}}(t)$ and an adaptation algorithm for $\hat{\boldsymbol{\alpha}}_\ell$ and $\hat{\theta}_i$ such that the values of e_{c} and $\hat{\boldsymbol{\alpha}}_\ell$ and $\hat{\theta}_i$ are bounded.

Let us introduce the new variable e_ϵ, $\epsilon \in \mathbb{R}_{>0}$,

$$e_\epsilon = e_{\mathrm{c}} - \epsilon S\left(\frac{e_{\mathrm{c}}}{\epsilon}\right), \quad S\left(\frac{e_{\mathrm{c}}}{\epsilon}\right) = \begin{cases} 1, & e_c/\epsilon \geq 1, \\ e_c/\epsilon, & |e_c/\epsilon| < 1, \\ -1, & e_c/\epsilon \leq -1, \end{cases} \tag{3.64}$$

and choose the class of admissible adaptation algorithms and auxiliary control inputs as

$$u_{\mathrm{a}} = \mathrm{sign}(k_2) S\left(\frac{e_{\mathrm{c}}}{\epsilon}\right) \sum_{i=1}^{m} a_i^*(\hat{\theta}_i, t), \tag{3.65}$$

$$\dot{\hat{\boldsymbol{\alpha}}}_\ell = \mathrm{sign}(k_2) \Gamma_\alpha e_\epsilon \boldsymbol{\varphi}_\ell, \; \Gamma_\alpha > 0, \tag{3.66}$$

$$\dot{\hat{\theta}}_i = \mathrm{sign}(k_2) \gamma_{\theta_i} e_\epsilon \omega_i^*(\hat{\theta}_i, t), \; \gamma_{\theta_i} > 0. \tag{3.67}$$

The functions $a_i^* : \mathbb{R} \times \mathbb{R} \to \mathbb{R}$ and $\omega_i^* : \mathbb{R} \times \mathbb{R} \to \mathbb{R}$ in (3.65)–(3.67) must satisfy the following equations:

$$a_i^*(\hat{\theta}_i, t) = \min_{\omega_i \in \mathbb{R}} \max_{\theta_i \in \Theta_i} \mathrm{sign}(e_\epsilon k_2) \left[f_i(\boldsymbol{\phi}_i(t), \theta_i) - f_i(\boldsymbol{\phi}_i(t), \hat{\theta}_i) + (\hat{\theta}_i - \theta_i)\omega_i \right]; \tag{3.68}$$

$$\omega_i^*(\hat{\theta}_i, t) = \arg \min_{\omega_i \in \mathbb{R}} \max_{\theta_i \in \Theta_i} \mathrm{sign}(e_\epsilon k_2) \left[f_i(\boldsymbol{\phi}_i(t), \theta_i) - f_i(\boldsymbol{\phi}_i(t), \hat{\theta}_i) + (\hat{\theta}_i - \theta_i)\omega_i \right]. \tag{3.69}$$

The asymptotic properties of the system (3.63)–(3.69) are formulated in the next theorem (Loh *et al.* 1999).

Theorem 3.3 *Consider the system (3.63), where the variables* $\hat{\theta}_i$ *and* $\hat{\boldsymbol{\alpha}}_\ell$, $u_{\mathrm{a}}(t)$ *satisfy (3.64)–(3.69).*

Then solutions of the system (3.63)–(3.69) are bounded for all initial conditions $\hat{\boldsymbol{\alpha}}_\ell(t_0) \in \mathbb{R}^\ell$, $e_{\mathrm{c}}(t_0) \in \mathbb{R}$. *If, in addition, the functions* $\boldsymbol{\varphi}_\ell(\cdot)$ *and* $\boldsymbol{\phi}_i(\cdot)$, $i = 1, \ldots, m$ *are globally bounded then*

$$\lim_{t\to\infty} e_\epsilon(t) = e_{\mathrm{c}}(t) - \epsilon S\left(\frac{e_{\mathrm{c}}(t)}{\epsilon}\right) = 0. \tag{3.70}$$

Proof of Theorem 3.3. Consider the function

$$V = \frac{1}{2}\left(e_\epsilon^2 + |k_2|(\hat{\boldsymbol{\alpha}}_\ell - \boldsymbol{\alpha}_\ell)^{\mathrm{T}}\Gamma_{\alpha_\ell}^{-1}(\hat{\boldsymbol{\alpha}}_\ell - \boldsymbol{\alpha}_\ell) + |k_2|\sum_{i=1}^{m}\gamma_{\theta_i}^{-1}(\hat{\theta}_i - \theta_i)^2\right)$$

and compute its time-derivative:

$$\dot{V} = \begin{cases} 0, \ |e_c| \le \epsilon, \\ -k_1 e_c e_\epsilon + k_2 e_\epsilon \sum_{i=1}^{m}\Big[f_i(\boldsymbol{\phi}_i(t), \theta_i) - f_i(\boldsymbol{\phi}_i(t), \hat{\theta}_i) \\ + (\hat{\theta}_i - \theta_i)\omega_i^* - a_i^* \operatorname{sign}(k_2) S(e_{\mathrm{c}}/\epsilon)\Big], \ |e_{\mathrm{c}}| > \epsilon. \end{cases} \tag{3.71}$$

Taking the definition of e_ϵ into account (see (3.65)), we can conclude that the second equation in (3.71) implies that

$$\dot{V} \le |k_2|\left(-\frac{k_1}{|k_2|}e_\epsilon^2 + e_\epsilon \sum_{i=1}^{m}\left[\operatorname{sign}(k_2)\left(f_i(\boldsymbol{\phi}_i(t), \theta_i) - f_i(\boldsymbol{\phi}_i(t), \hat{\theta}_i)\right)\right.\right.$$
$$\left.\left. + (\hat{\theta}_i - \theta_i)\omega_i^* - a_i^* S\left(\frac{e_c}{\epsilon}\right)\right]\right). \tag{3.72}$$

Now consider two cases, (1) $e_\epsilon \ge 0$ and (2) $e_\epsilon < 0$, and show that the inequality $\dot{V} \le 0$ holds in both cases.

Case 1. Let $e_\epsilon \ge 0$ and $|e_{\mathrm{c}}| > \epsilon$. Given that $S(e_{\mathrm{c}}/\epsilon) = \operatorname{sign}(e_{\mathrm{c}})$ for $|e_{\mathrm{c}}| > \epsilon$, we first notice that the inequality $\dot{V} \le 0$ would be satisfied if we could show that

$$a_i^* \ge \operatorname{sign}(k_2)\left[f_i(\boldsymbol{\phi}_i(t), \theta_i) - f_i(\boldsymbol{\phi}_i(t), \hat{\theta}_i) + (\hat{\theta}_i - \theta_i)\omega_i^*\right] \tag{3.73}$$

holds for all $\theta_i \in \Theta_i$. Clearly (3.73) holds if for some a-priori-chosen ω_i^* the following condition is satisfied:

$$a_i^*(\hat{\theta}_i, t) = \max_{\theta_i \in \Theta_i} \operatorname{sign}(k_2)\left[f_i(\boldsymbol{\phi}_i(t), \theta_i) - f_i(\boldsymbol{\phi}_i(t), \hat{\theta}_i) + (\hat{\theta}_i - \theta_i)\omega_i^*\right].$$

Observe that the functions a_i^* serve as auxiliary dominating terms in (3.65). Thus choosing the functions ω_i^* such that the values of $a_i^*(\hat{\theta}_i, t)$ are minimized leads to

$$\begin{aligned} a_i^*(\hat{\theta}_i, t) &= \min_{\omega_i \in \mathbb{R}} \max_{\theta_i \in \Theta_i} \operatorname{sign}(k_2) \left[f_i(\boldsymbol{\phi}_i(t), \theta_i) - f_i(\boldsymbol{\phi}_i(t), \hat{\theta}_i) + (\hat{\theta}_i - \theta_i)\omega_i \right], \\ \omega_i^* &= \arg\min_{\omega_i \in \mathbb{R}} \max_{\theta_i \in \Theta_i} \operatorname{sign}(k_2) \left[f_i(\boldsymbol{\phi}_i(t), \theta_i) - f_i(\boldsymbol{\phi}_i(t), \hat{\theta}_i) + (\hat{\theta}_i - \theta_i)\omega_i \right]. \end{aligned} \tag{3.74}$$

Case 2. Let $e_c < 0$, then the required inequality $\dot{V} \leq 0$ is satisfied if

$$a_i^*(\hat{\theta}_i, t) = \max_{\theta_i \in \Theta_i} \operatorname{sign}(k_2) \left[f_i(\boldsymbol{\phi}_i(t), \hat{\theta}_i) - f_i(\boldsymbol{\phi}_i(t), \theta_i) + (\theta_i - \hat{\theta}_i)\omega_i^* \right]$$

for all $\theta_i \in \Theta_i$. Imposing the minimality constraints leads to the following:

$$\begin{aligned} a_i^*(\hat{\theta}_i, t) &= \min_{\omega_i \in \mathbb{R}} \max_{\theta_i \in \Theta_i} \operatorname{sign}(k_2) \left[f_i(\boldsymbol{\phi}_i(t), \hat{\theta}_i) - f_i(\boldsymbol{\phi}_i(t), \theta_i) + (\theta_i - \hat{\theta}_i)\omega_i \right], \\ \omega_i^*(\hat{\theta}_i, t) &= \arg\min_{\omega_i \in \mathbb{R}} \max_{\theta_i \in \Theta_i} \operatorname{sign}(k_2) \left[f_i(\boldsymbol{\phi}_i(t), \hat{\theta}_i) - f_i(\boldsymbol{\phi}_i(t), \theta_i) + (\theta_i - \hat{\theta}_i)\omega_i \right]. \end{aligned} \tag{3.75}$$

Finally, notice that (3.68) and (3.69) specify the same functions as the combination of conditions (3.74) for $e_\epsilon \geq 0$ and (3.75) for $e_\epsilon < 0$. Therefore, we can conclude that $\dot{V} \leq 0$ for all $t \geq t_0$, provided that the theorem conditions hold. Boundedness of the function V implies that solutions of the system are bounded.

In order to see that (3.70) holds, notice that

$$\dot{V} \leq -k_1 e_\epsilon^2.$$

Hence, boundedness of the functions $\boldsymbol{\varphi}_\ell$ and $\boldsymbol{\phi}_i$ implies that $\dot{e}_c$ is bounded. Thus $\dot{e}_\epsilon$ is bounded, and e_ϵ is uniformly continuous for all $t \geq t_0$. In this case property (3.70) follows immediately from Barbalat's lemma (Lemma 2.1). *The theorem is proven.* □

In addition to the minimax method there are other alternatives for dealing with nonlinear parametrization of uncertainties. One of the ideas can be briefly presented as follows. Instead of presenting a nonlinearly parametrized compensator in the controller (which leads to error models (3.63) that are nonlinearly parametrized w.r.t. $\hat{\theta}_i$) one can search for an admissible controller that is nonlinear in $\mathbf{x}$ but linear in $\hat{\theta}_i$ (or, generally, a controller satisfying Assumptions 3.2 and 3.1). The literature about various formulations of this idea is large (see e.g. Lin and Qian (2002a,b), Ding (2001), Boskovic (1995), and Putov (1993)). These methods are similar to the minimax-based techniques in that the control functions on the right-hand side are used to dominate (majorize) the nonlinearity rather than to compensate for it precisely. In addition, we wish to mention that Karsenti *et al.* (1996) and Astolfi and Ortega (2003) offer alternative solutions to the problem of adaptive regulation in the presence of nonlinear parametrization. These solutions bear an overall similarity to the adaptation algorithms in finite form which we present in Chapter 5. Therefore we do not discuss them in detail here.

3.4 Applicability issues of conventional methods of adaptive control and regulation

The synthesis and analysis methods considered so far obviously do not exhaust the vast variety of intricate techniques available in the literature on adaptive control and regulation to date. Variations of these general ideas are many; they are detailed in a number of monographs (see e.g. Fradkov *et al.* (1999)), and reviewing all of these with the same level of detail would divert the discussion away from the main topic of this book. Yet an overview of basic statements of the problem of adaptation and, consequently, methods of adaptive control and regulation, from Bellman's classical works (Bellman 1961; Bellman and Kalaba 1965) to Krstić *et al.* (1995), Ioannou and Sun (1996), Fradkov *et al.* (1999), Kolesnikov (2000), and Astolfi and Ortega (2003), allows us to formulate a number of fundamental problems and challenges in the area. In what follows we specify four general issues that motivated the development of results presented in this book.

The first class of problems is related to the very notion of *adaptation in regulated dynamical systems.* Truxhall's general thesis that adaptation is a mere viewpoint of a designer is too ambiguous and not very constructive to serve as a formal definition of an *adaptive* system. On the other hand, more formal and classical statements of the problem of adaptation, such as those provided by L. Zadeh (Zadeh 1963) and V. A. Yakubovich (Yakubovich 1968; Fomin *et al.* 1981), despite their having guided the development of the theory over many decades, do not go far beyond the usual presumption about what an adaptive system is. In particular, within this framework adaptive regulation (and, hence, adaptation) is associated with a special behavior of a set of the system's variables (outputs) of which the values behave in a certain sense predictably in the presence of perturbations and unknowns regarding the part of the system to be regulated. This definition of adaptivity leads to a number of issues, of which some examples are listed in the preface to the special issue of *System & Control Letters* devoted to the problems of adaptive control (Polderman and Pait 2003). Polderman and Pait stress that adaptivity on the one hand and the overall reduction of uncertainty about the regulated system as a result of adaptation on the other are not sufficiently well connected in most of the conventional statements of the problem of adaptation. Thus, usual interpretations of adaptive regulation without any additional references to, for example, optimality blur the boundaries between the general problem of regulation and adaptive regulation. As an illustration, consider the following statement of the output-regulation problem (Byrnes and Isidori 2003). The mathematical model of the system to be regulated is

$$\begin{aligned}\dot{\mathbf{x}} &= \mathbf{f}(\mathbf{x}, \omega, \mathbf{u}),\\ \dot{\omega} &= S(\omega),\\ \mathbf{y} &= \mathbf{k}(\mathbf{x}, \omega),\\ \mathbf{e} &= \mathbf{h}(\mathbf{x}, \omega),\end{aligned} \tag{3.76}$$

where $\mathbf{y} \in \mathbb{R}^p$ is the vector of measured variables, $\mathbf{e} \in \mathbb{R}^q$ is the vector of regulated variables, $\mathbf{u} \in \mathbb{R}^m$ is the control input, and $\omega(t, \omega_0, t_0)$ is the perturbation function defined as a solution of the following initial-value problem

$$\dot{\omega} = S(\omega), \;\; \omega(t_0) = \omega_0.$$

The functions $\mathbf{f}, \mathbf{h}, \mathbf{k}$, and S are supposed to be C^k-smooth. The goal is to ensure that

$$\lim_{t\to\infty} \mathbf{e}(t) = 0 \tag{3.77}$$

holds for all $\mathbf{x}(t_0) \in \mathbb{R}^n$ and $\omega(t_0) \in \Omega_\omega \subset \mathbb{R}^{\mathrm{d}}$. The problem constituted by (3.76) and (3.77) is stated as a typical regulation problem. Despite, however, the fact that the term adaptation is not explicitly mentioned in Byrnes and Isidori (2003), this statement of the problem is very similar to that of Yakubovich (1968) and Fomin *et al.* (1981). This suggests, if we leave some mathematical details aside, that the latter statement in principle is no more general than the former. It is, therefore, clear that adaptive regulation must be considered a more specific problem. One such specific distinction could be, for example, the addition of a formal requirement for an increase of the system's performance with time.

The second class of problems is the general issue of the *goals of adaptation*. Standard statements of the adaptive-regulation problem require that the adaptation goal be defined as

$$Q[\mathbf{x}(\tau), \mathbf{u}(\tau), 0 \leq \tau \leq t] \leq \Delta, \; \forall\, t > t^*,$$

where $Q(\cdot)$ is a known sign-definite function(al). In many cases, e.g. in Narendra and Annaswamy (1989), Ioannou and Sun (1996), Fradkov (1990), and Fradkov *et al.* (1999), applicability of the results additionally requires that the functions $Q(\cdot)$ satisfy the following list of properties:

- $Q(\mathbf{x}, t) : \mathbb{R}^n \times \mathbb{R}_{\geq 0} \to \mathbb{R}_{\geq 0}$, $Q \in C^1$;
- $\lim_{\|\mathbf{x}\|\to\infty} Q(\mathbf{x}, t) = \infty, \forall\, t \in \mathbb{R}_{\geq 0}$;
- $Q(0, t) = 0$ (if regulation to the origin is required).

These properties, coupled with the standard Assumption 3.1 for the velocity-gradient method (Fradkov 1979, 1990), or with a similar condition for other gradient-based methods (Narendra and Annaswamy 1989), automatically imply that Q is a corresponding Lyapunov function for the system in which the parameters are "frozen" and are set to the optimal values. Thus availability of a Lyapunov function for the unperturbed system (no parametric perturbations) is necessary for many standard adaptive-regulation schemes. Existing improvements (Panteley *et al.* 2002) allow one to replace the requirement of availability of a Lyapunov function with the knowledge that such a function exists. However, even if the mere existence of the Lyapunov function is required, such a requirement implies that the system's target motions have to be globally stable in the sense of Lyapunov. This obviously restricts the range of admissible motions in such adaptive systems a priori. Hence dealing with unstable, multi-stable, or even locally stable target dynamics is problematic within this classical framework.

On the other hand, there exist systems in which multi-stability is natural. These systems include lasers (Brambilla *et al.* 1991; Chizhevsky 2000, 2001; Chizhevsky and Corbalan 2002), chemical reactors, metabolic networks, and biological and social systems (Bak and Pakzusci 1995; Sneppen *et al.* 1995; Ito *et al.* 2003; Hindmarsh and Rose 1984; Ditchburn and Ginzburg 1952; Martinez-Conde *et al.* 2004; Solé *et al.* 1999; Bak and Sneppen 1993; Makarova 2002; Malinetskii 2006). The class of problems in which only locally stable motions are generally plausible includes the interesting problem of bifurcation control (Krener *et al.* 2004) in the presence of uncertainties. Unstable and chaotic target dynamics also occur in large-scale models of the human brain and information processing (van Leeuwen *et al.* 1997; Kaneko and Tsuda 2000; Kaneko 1990, 1994). There are examples of dynamical systems in which unstable attractors not only exist and are plausible but also prevail and have a certain behavioral function (Timme *et al.* 2002).

All these systems and processes do not admit a single globally asymptotically stable equilibrium, and, hence, finding the corresponding goal function $Q(\cdot)$ is hardly possible without invoking additional regularization procedures (e.g. extending the system's state space, introducing a reference model, etc.). The sole purpose of such procedures is to transform the problem into the usual framework for which a Lyapunov function may be found. The questions, however, are is this regularization always possible, is it actually needed, and is it natural? Or, perhaps, would enabling unstable target dynamics and developing appropriate analysis and synthesis techniques be more beneficial? This motivates the development of novel methods of adaptive regulation that are suitable for systems with unstable target dynamics.

The third general class of problems is the *problem of performance* of adaptive systems. Often the performance of an adaptive system is considered acceptable if, in addition to some asymptotic properties of the system, the overall Lyapunov stability is guaranteed (Fomin *et al.* 1981; Fradkov 1990; Fradkov *et al.* 1999; Sastry 1999; Sastry and Bodson 1989; Narendra and Annaswamy 1989; Krstić *et al.* 1995) in an extended state space. Let $\mathcal{Z}$ be such extended state space including the state vector of the system, the vector of unknown parameters, and the vector of other internal variables of the system. Let $\mathbf{z}_0$ be an element of $\mathcal{Z}$, let $\Omega_z \subset \mathcal{Z}$ be the target set, and denote solutions of the extended system passing through $\mathbf{z}_0$ at $t = t_0$ by $\mathbf{z}(t, \mathbf{z}_0)$. Stability of Ω_z in the sense of Lyapunov implies that for any $\varepsilon > 0$ there exists $\delta > 0$ such that $\|\mathbf{z}_0\|_{\Omega_z} < \delta \Rightarrow \|\mathbf{z}(t, z_0)\|_{\Omega_z} < \varepsilon$ for all $t \geq t_0$. If for a given small ε the value of δ could be made reasonably large then the performance of such a system would be very satisfactory.

In practice, however, this is rarely the case, at least in the domain of adaptive control. What we are likely to observe is a more conservative situation in which the value of δ is small when the value of ε is small. In order to illustrate this point, consider the case in which $\Omega_z = \mathbf{z}^*$ is a point and $V(\mathbf{z}, \mathbf{z}^*)$ is a Lyapunov function satisfying the usual conditions $\gamma(\|\mathbf{z}-\mathbf{z}^*\|) \leq V(\mathbf{z}, \mathbf{z}^*) \leq \alpha(\|\mathbf{z}-\mathbf{z}^*\|), \alpha, \gamma \in \mathcal{K}_\infty$. Thus $\dot{V} \leq 0$ implies that $V(\mathbf{z}(t, \mathbf{z}_0), \mathbf{z}^*) \leq V(\mathbf{z}_0, \mathbf{z}^*) \leq \alpha(\|\mathbf{z}_0 - \mathbf{z}^*\|)$. On the other hand, $\gamma(\|\mathbf{z}(t, \mathbf{z}_0) - \mathbf{z}^*\|) \leq V(\mathbf{z}(t, \mathbf{z}_0), \mathbf{z}^*)$, and hence $\|\mathbf{z}(t, \mathbf{z}_0) - \mathbf{z}^*\| \leq \gamma^{-1}(\alpha(\|\mathbf{z}_0 - \mathbf{z}^*\|))$. If we now demand that $\|\mathbf{z}(t, \mathbf{z}_0) - \mathbf{z}^*\| < \varepsilon$, this would be automatically satisfied if $\mathbf{z}_0$ satisfied the inequality $\gamma^{-1}(\alpha(\|\mathbf{z}_0 - \mathbf{z}^*\|)) < \varepsilon$. This is equivalent to saying that $\|\mathbf{z}_0 - \mathbf{z}^*\| \leq \alpha^{-1}(\gamma(\varepsilon)) = \delta(\varepsilon)$. Notice that this usual procedure results in an estimate of δ of which the value is a monotone function of ε, and hence the smaller ε the smaller δ.

While this analysis is acceptable in many applications, it has clear limitations in the domain of adaptive systems. This is because substantial changes in the system's operational conditions ($\mathbf{z}_0$ becomes distant from $\mathbf{z}^*$) constitute the main reason for adaptation. Thus performance measures concerning small deviations due to small perturbations might not make a lot of sense in problems of this kind.[2] In order to find a remedy to this issue, a stronger requirement of asymptotic and even exponential Lyapunov stability is sometimes imposed (Sastry and Bodson 1989). The ability to satisfy these new requirements often comes at a cost: the need to ensure that certain variables of a system satisfy a sort of persistency-of-excitation condition (Lee and Narendra 1988; Morgan and Narendra 1992) (see Definition 2.5.1). Given that the actual trajectories of these variables will generally depend on a number of uncertain quantities, checking whether this condition holds constitutes a separate issue.

[2] This view that adaptation (as a re-tuning) becomes advantageous (and, hence, makes sense) over other approaches to regulation, when the degree of uncertainty is relatively high, is illustrated in French (2002).

Alternative characterizations of the performance of adaptive systems include upper bounds of L_2- and L_∞-norms for the state, parameter estimates, control inputs, and error signals (Ioannou and Sun 1996; Krstić *et al.* 1995; French *et al.* 2000; French 2002). An interesting performance measure involving an adaptation time is introduced and studied in Timofeev (1988). In all these cases, however, the class of systems is limited to a specific system of ordinary differential equations, or the analysis is restricted to one or just a few algorithms of adaptation. In addition to these results there is a number of working heuristics aimed at improving the performance of adaptive systems, see e.g. Narendra and Balakrishnan (1994, 1997) or Krstić *et al.* (1992, 1994), and Krstić and Kokotović (1993). In Krstić *et al.* (1994), for example, the authors make a very interesting observation that a special class of nonlinear adaptation schemes constructed for linear models outperforms standard adaptation schemes for the same model class. This suggests that the standard schemes are indeed too conservative and that there is room for performance improvement. The problem, however, is that finding a systematic procedure for generating such algorithms is a non-trivial issue.

The fourth class of problems is the problem of the adequacy of mathematical models of reality used for the creation and analysis of adaptive regulatory mechanisms. As we mentioned earlier, the standard models considered in the domain of adaptive control are ordinary differential equations, in which the uncertainties are linearly parametrized functions:

$$\mathbf{f}(\mathbf{x}, \boldsymbol{\theta}) = \sum_{i=1}^{m} \boldsymbol{\phi}_i(\mathbf{x})\theta_i, \ \boldsymbol{\phi}_i : \mathbb{R}^n \to \mathbb{R}^n, \ \boldsymbol{\theta} = (\theta_1, \theta_2, \ldots, \theta_m)^{\mathrm{T}}.$$

A positive aspect of these models is the availability of a wide spectrum of analysis tools and adaption algorithms aimed at dealing with this class of system (Sastry and Bodson 1989; Sastry 1999; Fomin *et al.* 1981; Narendra and Annaswamy 1989; Ioannou and Sun 1996; Fradkov *et al.* 1999). To a first approximation such modeling is valid and is sometimes very instrumental. In particular, linearity with respect to $\boldsymbol{\theta}$ allows one to extend the applicability of standard adaptation algorithms in which knowledge of $\mathbf{x}(t)$ is needed to those schemes in which only a function $\mathbf{x}(t)$ (measured outputs) is available for direct observation (Kreisselmeier 1977; Krstić and Kokotović 1996; Marino and Tomei 1993; Marino 1990; Nikiforov 1998). On the other hand, models of a large class of physical processes are genuinely nonlinearly parametrized. This is the case e.g. for models of biological and chemical reactors (Boskovic 1995; Stigter and Keesman 2004), models of friction in mechanical and biomedical systems (Armstrong-Helouvry 1991, 1993; Canudas de Wit and Tsiotras 1999; Pacejka and Bakker 1993; ÓLeary *et al.* 2003), induction motors and magnet suspensions (Costic *et al.* 2000; Ghosh *et al.* 2000), electromechanical valves (Peterson and Stefanopoulou 2004), and models of engines and power plants (Bachmayer *et al.* 2000; Gelfi *et al.* 2003).

Generally adaptive control methods for these models invoke domination or majorization techniques (Loh *et al.* 1999; Annaswamy *et al.* 1998; Kojic and Annaswamy 2002; Ding 2001; Lin and Qian 2002a,b; Boskovic 1995). This leads to excessive overcompensation of "unwanted" nonlinearities, which in turn negatively affects the performance of the whole system in the long term. Non-dominating adaptation algorithms are either local (Karsenti *et al.* 1996) or not practical enough (Pomet 1992; Ilchman 1997; Martensson 1985; Martensson and Polderman 1993).[3]

3.5 Summary

Analysis of the evolution of the problem of adaptive regulation and control allows us to formulate several key limiting characterizations of standard and commonly used mathematical statements of the problem (for deterministic systems). Relaxation of one or all of these limiting factors is a feature that we consider highly desirable.

In the rest of the book we will develop a set of results that will allow us to construct and analyze adaptation mechanisms with these desired features. We will depart from an understanding that the target dynamics of a system is not restricted to stable motions in the sense of Lyapunov. Regarding the information we will require about the systems that are adaptively controlled, we will try to keep these requirements at the minimal levels that are adequate to the needs of analysis. In order to develop general principles of adaptation and adaptive regulation we will need to

(1) use a language that allows us to describe dynamical systems without the need to know the corresponding differential equations in detail (or without even assuming that the model is specified by ordinary differential equations);

(2) introduce and develop analysis methods for studying asymptotic properties of such objects; no stability assumptions about asymptotic regimes must be used unless stated overwise.

These results should allow us to formulate general principles for the functioning and organization of an adaptive system in the presence of uncertainties. The principles are expected to take the shape of certain constraints specifying general input–output and input–state properties of such systems. These principles of the general functioning and organization of an adaptive system will then be used to formulate specific goals of adaptation. Since no stability requirements would be introduced at the stage of development of these principles, it is natural to expect

[3] These algorithms constitute a mathematical proof that an adaptation law exists but cannot be considered as practical solutions to the problem (Martensson 1985; Martensson and Polderman 1993).

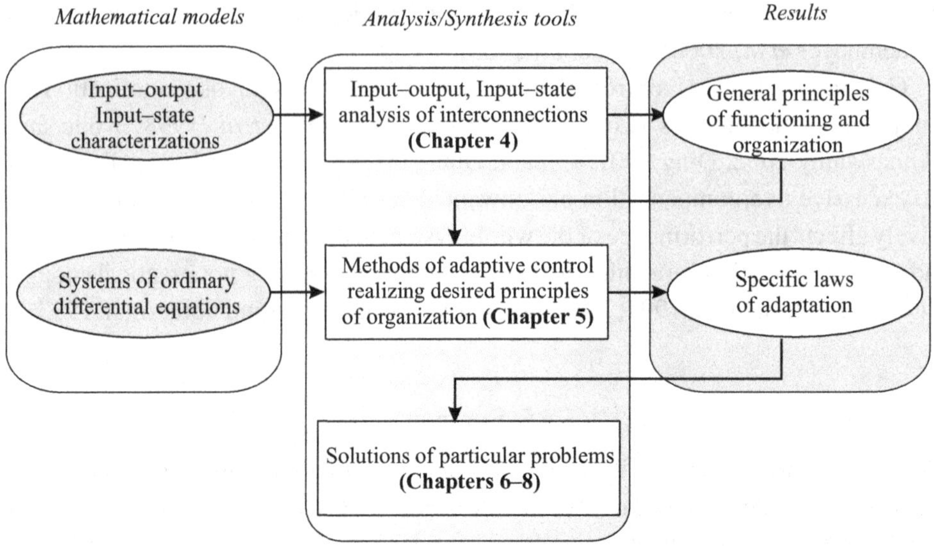

Figure 3.1 A hierarchical outline of methods and results presented in the text.

that no stability constraints would be needed in order to ensure that these goals are realized by a specific adaptation mechanism.

A diagram illustrating this strategy for creating a set of tools and methods suitable to solve the five issues mentioned above is shown in Figure 3.1. This diagram shows the general strategy the author personally pursued. Naturally, it reflects the structure of the book.

Part II

Theory

4

Input–output analysis of uncertain dynamical systems

In this chapter we provide analysis tools for dynamical systems described as input–output and input–state mappings (or simply *operators*) in the corresponding spaces. Such a description is advantageous and natural when mathematical models of the systems are vaguely known and uncertain. We will see that the basic properties of these input–output and input–state mappings (such as boundedness and continuity) constitute important information for our understanding of the various ways in which an adaptation can be organized in these systems.

In particular, we will see that some basic stability notions (Lyapunov stability of invariant sets (LaSalle and Lefschetz 1961), stability of solutions in the sense of Lyapunov, and input-to-state, input-to-output, output-to-state, and input–output stability (Zames 1966)) are equivalent to continuity of a certain mapping characterizing the dynamics of the system (Theorems 4.1 and 4.3).

As we have said earlier, real physical systems, however, are not always stable and hence their input–output and input–state characterizations are not always continuous. Moreover, the target dynamics of these systems should not necessarily admit continuous input–output or input–state description. Indeed, continuity of a mapping S at a given point u_0 in essence reflects the fact that the value of the mapping $S(u)$, can be made arbitrarily close to $S(u_0)$, provided that u remains sufficiently close to u_0. In reality this is a quite idealistic picture, and in most cases such infinitesimal closeness is not needed. What is needed, however, is that the system's behavior (the quantity $S(u)$) does not change much if u remains close to u_0. However subtle the difference between these wordings might appear, the objects of study are quite different mathematically. In the former case they are continuous mappings, whereas in the latter case they are not. Therefore the results available for systems for which the input–output or input–state mappings are continuous cannot be extended automatically to the realm of systems for which these mappings are discontinuous. Thus results that would allow us to analyze the asymptotic properties of such systems and their interconnections are needed.

In this chapter we will study the asymptotic properties of interconnections of systems whose behavior can be described by locally bounded mappings and that are not necessarily continuous. We will provide conditions specifying the existence, completeness, and boundedness of variables in such interconnections (Theorems 4.4 and 4.5), and estimate the locations of limit sets in these systems (Corollary 4.1). These results about the general properties of interconnections of systems with locally bounded input–output mappings will be used to state the problem of functional synthesis of a general adaptive regulator.

Two general solutions of the problem will be provided. One solution is called here the *bottle-neck* or *jamming* synthesis principle. It will allow us to formulate a sort of *separation principle* (Theorem 4.6) in the domain of adaptive regulation. The other solution, the *contraction–exploration principle*, is based on studying systems consisting of compartments of which the dynamics is either contracting or exploring. Contracting compartments reflect relaxation of the system's variables to invariant sets, whereas exploring compartments correspond to a searching mechanism of an adaptive system looking for the most optimal conditions of operation. We will study how these two compartments must interact so that weakly attracting sets emerge at the optimal conditions of the system's operation (Theorem 4.7, or the non-uniform small-gain theorem). We will analyze the asymptotic properties of such systems (Corollary 4.2) and provide simple criteria connecting parameters of the input–output and input–state mappings of the interacting contracting-exploring compartments with the estimates of attractor basins (Corollary 4.3). In addition to providing general principles for functional and structural synthesis of adaptive systems, these results lead to a generalized non-uniform small-gain theorem for cascades of integrally input-to-state stable systems with bidirectional connections.

4.1 Operator description of dynamical systems

Similarly to Zadeh and Desoer (1991) and Pupkov *et al.* (1976), let us consider a physical object $\mathcal{O}$. Let us suppose that the following list of assumptions holds:

(1) observation of variables, parameters, and generally, processes of $\mathcal{O}$ does not change the object;
(2) processes of $\mathcal{O}$ are described by real functions of real variables defined on a non-empty interval $\mathcal{T} \subset \mathbb{R}$;
(3) external stimulation $\mathbf{u}$, the influence of the environment $\mathbf{e}$, internal processes $\mathbf{x}$ in the object, and measurements $\mathbf{y}$ of these processes are also real functions of real variables defined on $\mathcal{T}$.

The function $\mathbf{u} : \mathcal{T} \to \mathbb{R}^m$ is referred to as a *control input*, the function $\mathbf{e} : \mathcal{T} \to \mathbb{R}^s$ models the influence of the environment (*perturbations*), the function

$\mathbf{x} : \mathcal{T} \to \mathbb{R}^n$ is the *state* of $\mathcal{O}$, and $\mathbf{y} : \mathcal{T} \to \mathbb{R}^h$ are the observables or outputs of $\mathcal{O}$.

The sets of admissible functions $\mathbf{u}(t)$, $\mathbf{e}(t)$, $\mathbf{x}(t)$ and $\mathbf{y}(t)$, $t \in \mathcal{T}$ are defined by the symbols $\mathcal{U}$, $\mathcal{E}$, $\mathcal{X}$, and $\mathcal{Y}$, respectively.

The intervals $\mathcal{T}$, on which these functions are defined, are referred to as *intervals of existence of* the object. The maximal interval of existence $\mathcal{T}^*(\mathcal{O}, \mathbf{u}, \mathbf{e})$ for some $\mathbf{u}$ and $\mathbf{e}$ will be referred to as the *existence time* of $\mathcal{O}$. It is clear that the existence time of $\mathcal{O}$ depends on $\mathbf{u} \in \mathcal{U}$ and $\mathbf{e} \in \mathcal{E}$. If such dependence is clear from the context, then for the sake of notational compactness we will replace $\mathcal{T}^*(\mathcal{O}, \mathbf{u}, \mathbf{e})$ with $\mathcal{T}^*$.

Example 4.1.1 Solutions of the following equation present an example of objects for which the existence time may be finite, depending on the inputs:

$$\dot{x} = x^2 + u, \; x(t_0) = 0.$$

If $u(t)$ = constant, $u \leq 0$ for all $t \geq t_0$, then the solution to this initial-value problem is defined for all $t \geq t_0$, and hence the existence interval is $[t_0, \infty) \subset \mathbb{R}$. If, however, $u > 0$, then $x(t)$ reaches infinity in finite time T^*. Hence the existence interval in this case is $\mathcal{T}^* = [t_0, T^*)$.

In mathematical systems theory a "system" is understood as a binary *relation* $\mathcal{S}$, defined on $\{\mathcal{U}\times\mathcal{E}\}\times\{\mathcal{X}\times\mathcal{Y}\}$. Let $\mathcal{P} \subseteq \{\mathcal{U}\times\mathcal{E}\}\times\{\mathcal{X}\times\mathcal{Y}\}$ be the graph of this relation. Then a system is defined by a set of pairs $(\{\mathbf{u}, \mathbf{e}\}, \{\mathbf{x}, \mathbf{y}\}) \in \{\mathcal{U} \times \mathcal{E}\} \times \{\mathcal{X} \times \mathcal{Y}\}$. It is allowed that $\mathbf{x}(t)$ and $\mathbf{y}(t)$ need not be defined for some $\mathbf{u} \in \mathcal{U}$ and $\mathbf{e} \in \mathcal{E}$. This general definition, however, does not account for causal relations between inputs and outputs.

In *dynamical systems* causal relations are important. In many cases we are interested in predicting the values of $\mathbf{x}(t)$ and $\mathbf{y}(t)$ in response to $\mathbf{u}(t)$ and $\mathbf{e}(t)$. Thus, in order to introduce such causal relations between elements defining the system (i.e. $\mathbf{u}(t)$, $\mathbf{e}(t)$, $\mathbf{x}(t)$, and $\mathbf{y}(t)$), we need to extend our former definition of a system. A possible way to do so is provided in the definition below.

Definition 4.1.1 Let $\mathcal{U}$, $\mathcal{E}$, $\mathcal{X}$, and $\mathcal{Y}$ be given. A system $\mathcal{S}$, defined on $\mathcal{U}$, $\mathcal{E}$, $\mathcal{X}$, and $\mathcal{Y}$ is a six-tuple $\langle\mathcal{U}, \mathcal{E}, \mathcal{X}, \mathcal{Y}, \mathcal{S}_\mathcal{T}, \mathcal{H}_\mathcal{T}\rangle$, where

$$\mathcal{S}_\mathcal{T} : \; \{\mathbf{u}(t), \mathbf{e}(t)\} \subseteq \mathcal{U} \times \mathcal{E} \; \mapsto \; \{\mathbf{x}(t)\} \subseteq \mathcal{X}, \tag{4.1}$$

$$\mathcal{H}_\mathcal{T} : \; \{\mathbf{u}(t), \mathbf{e}(t), \mathbf{x}(t)\} \subseteq \mathcal{U} \times \mathcal{E} \times \mathcal{X} \; \mapsto \; \{\mathbf{y}(t)\} \subseteq \mathcal{Y}, \tag{4.2}$$

$$\mathcal{T} = [t_0, T] \subseteq \mathcal{T}^*.$$

Mapping $\mathcal{S}_\mathcal{T}$ in (4.1) determines the input-to-state properties of the system, and the mapping $\mathcal{H}_\mathcal{T}$ in (4.2) defines the input/state-to-output relation.

In order to be able to compare state $\mathbf{x}(t)$ and output $\mathbf{y}(t)$ of the system for different $\mathbf{u}(t)$ and $\mathbf{e}(t)$, a metric in the sets of $\mathcal{U}$, $\mathcal{E}$,[1] $\mathcal{X}$, and $\mathcal{Y}$ is needed. Moreover, in many problems the sets of functions $\mathcal{U}$ and $\mathcal{X}$ inherit the structure of *linear space*. In other words, in these sets there is an operation "+" with respect to which they form an Abelian group, and there is an operation "·" of multiplication on a scalar that is distributive w.r.t. the operation "+". Thus in what follows we will suppose that $\mathcal{U}$, $\mathcal{E}$ $\mathcal{X}$, and $\mathcal{Y}$ are normed linear spaces of functions defined over $\mathcal{T}$. In particular we will restrict our attention to $L^n_{\mathrm{p}}[t_0, T]$-spaces, $p \in \mathbb{R}_{\geq 1}$, $\mathbb{R}_{\geq 1} = \{x \in \mathbb{R} | x \geq 1\} \cup \infty$, $n \in \mathbb{N}$. This choice is motivated by the fact that the relevant features of $\mathbf{u}(t)$, $\mathbf{x}(t)$, and $\mathbf{y}(t)$, e.g. energy, power, and maximal amplitude, are the values of the corresponding $\|\cdot\|_{\mathrm{p},[t_0,T]}$-norms. Moreover, some standard performance criteria are naturally defined as $L^n_1[t_0, T]$-, $L^n_2[t_0, T]$-, and $L^n_\infty[t_0, T]$-norms.

The set $\mathcal{E}$ is chosen here to be a *linear normed* space $\mathcal{L}_e$. This is because the set $\mathcal{E}$ will accommodate not only the functions of time but also unknown parameters of $\mathcal{O}$ and initial conditions. An example of $\mathcal{L}_e$ is a direct sum $\mathcal{L}_e = \mathbb{R}^d \oplus L_{\mathrm{p}}[t_0, T]$, where $\|\cdot\|_{\mathcal{L}_e}$ in $\mathcal{L}_e$ is induced by the Euclidean norm in $\mathbb{R}^d$ and $L_{\mathrm{p}}[t_0, T]$, respectively:

$$\forall\, \mathbf{z} \in \mathcal{L}_e,\ \mathbf{z} = \boldsymbol{\xi} \oplus \boldsymbol{\nu},\ \boldsymbol{\xi} \in \mathbb{R}^d,\ \boldsymbol{\nu} \in L_{\mathrm{p}}[t_0, T] \Rightarrow \|\mathbf{z}\|_{\mathcal{L}_e} = \|\boldsymbol{\xi}\| + \|\boldsymbol{\nu}(t)\|_{\mathrm{p},[t_0,T]}.$$

From now on we will consider systems $\mathcal{S}$ of which the input–output properties are defined in the following sense.

Definition 4.1.2 Consider a system $\langle \mathcal{U}, \mathcal{E}, \mathcal{X}, \mathcal{Y}, \mathcal{S}_\mathcal{T}, \mathcal{H}_\mathcal{T} \rangle$, where

$$\begin{aligned}
&\mathcal{X} \subseteq \mathcal{L}_x[t_0, T] \subseteq L^n_q[t_0, T],\\
&\mathcal{U} \subseteq \mathcal{L}_u[t_0, T] \subseteq L^m_{\mathrm{p}}[t_0, T],\\
&\mathcal{Y} \subseteq \mathcal{L}_y[t_0, T] \subseteq L^h_k[t_0, T],\ q,\ p,\ k \in \mathbb{R}_{\geq 1} \cup \{\infty\}.
\end{aligned}$$

The system admits an input–state operator $\mathcal{S}_\mathcal{T}(\mathbf{u}, \mathbf{e})$,

$$\mathcal{S}_\mathcal{T}(\mathbf{u}, \mathbf{e}) :\ \mathcal{L}_u[t_0, T] \times \mathcal{E} \mapsto \mathcal{L}_x[t_0, T], \tag{4.3}$$

and an input–output operator $\mathcal{H}_\mathcal{T}(\mathbf{u}, \mathbf{e})$,

$$\mathcal{H}_\mathcal{T}(\mathbf{u}, \mathbf{e}) :\ \mathcal{L}_u[t_0, T] \times \mathcal{E} \mapsto \mathcal{L}_y[t_0, T], \tag{4.4}$$

on $\mathcal{T} = [t_0, T]$ if and only if

$$\mathbf{u}(t) \in \mathcal{L}_u[t_0, T] \Rightarrow \mathbf{x}(t) \in \mathcal{L}_x[t_0, T],\ \forall\, \mathbf{e} \in \mathcal{E};$$

$$\mathbf{u}(t) \in \mathcal{L}_u[t_0, T] \Rightarrow \mathbf{y}(t) \in \mathcal{L}_y[t_0, T],\ \forall\, \mathbf{e} \in \mathcal{E}.$$

[1] We will see below, however, that introducing a metric for $\mathcal{E}$ is not actually necessary for all results of this chapter.

If mappings $\mathcal{S}_{\mathcal{T}}(\mathbf{u},\mathbf{e})$ and $\mathcal{H}_{\mathcal{T}}(\mathbf{u},\mathbf{e})$ were known and available then we could immediately proceed to the analysis of these objects. The problem is that in the context of adaptation these mappings are rarely known precisely. On the other hand, having some information about input–state and input–output properties of the system is necessary for the goals of analysis. This information should allow comparison of the values of $\mathbf{x}(t)$ for different functions $\mathbf{u}(t)$. Yet, it should be of a rather general type that could easily be provided at the modeling stage. Possible candidates for such additional information about the system are the *input–state* and *input–output margins*.

Definition 4.1.3 Consider the system (4.1) and (4.2) defined on $\mathcal{T}=[t_0,T]$. We say that the system admits input–state and input–output margins $\gamma_{S,\mathcal{L}_x}$ and $\gamma_{H,\mathcal{L}_y}$ if and only if

(1) the system (4.1) and (4.2) admits an input–state operator $\mathcal{S}_{\mathcal{T}}(\mathbf{u},\mathbf{e}):\mathcal{L}_u[t_0,T]\times\mathcal{E}\mapsto\mathcal{L}_x[t_0,T]$ and there exists a function $\gamma_{S,\mathcal{L}_x}:\mathcal{L}_e\times\mathbb{R}_{\geq 0}\times\mathbb{R}\to\mathbb{R}_{\geq 0}$, such that

$$\|\mathbf{x}(t)\|_{\mathcal{L}_x,[t_0,T]}\leq\gamma_{S,\mathcal{L}_x}(\mathbf{e},\|\mathbf{u}(t)\|_{\mathcal{L}_u,[t_0,T]},T),\tag{4.5}$$

where $\gamma_{S,\mathcal{L}_x}(\mathbf{e},\|\mathbf{u}(t)\|_{\mathcal{L}_u,[t_0,T]},T)$ is non-decreasing with respect to the term $\|\mathbf{u}(t)\|_{\mathcal{L}_u,[t_0,T]}$ and is locally bounded;

(2) the system (4.1) and (4.2) admits an input–output operator $\mathcal{H}_{\mathcal{T}}(\mathbf{u},\mathbf{e}):\mathcal{L}_u[t_0,T]\times\mathcal{E}\mapsto\mathcal{L}_y[t_0,T]$ and there is a function $\gamma_{H,\mathcal{L}_y}:\mathcal{L}_e\times\mathbb{R}_{\geq 0}\times\mathbb{R}\to\mathbb{R}_{\geq 0}$, such that

$$\|\mathbf{y}(t)\|_{\mathcal{L}_y,[t_0,T]}\leq\gamma_{H,\mathcal{L}_y}(\mathbf{e},\|\mathbf{u}(t)\|_{\mathcal{L}_u,[t_0,T]},T),\tag{4.6}$$

where the function $\gamma_{H,\mathcal{L}_y}(\mathbf{e},\|\mathbf{u}(t)\|_{\mathcal{L}_u,[t_0,T]},T)$ is non-decreasing with respect to $\|\mathbf{u}(t)\|_{\mathcal{L}_u,[t_0,T]}$ and is locally bounded.

The functions $\gamma_{S,\mathcal{L}_x}$ and $\gamma_{H,\mathcal{L}_y}$ provide us with a reasonably rough estimation of how the system's behavior might change with $\mathbf{u}(t)$ and $\mathbf{e}$. These characterizations allow a substantially high degree of uncertainty about the system. Notice, however, that there may be many such margins defined for one system and in this sense these functions do not define the system uniquely. Moreover, these functions may be defined for different norms. Let us illustrate this with an example.

Example 4.1.2 Consider the following ordinary differential equation of first order:

$$\begin{aligned}\dot{x}&=-x+u,\ x(t_0)\in\mathcal{E}\subset\mathbb{R},\\ y&=x,\ \mathcal{T}=[t_0,T],\end{aligned}\tag{4.7}$$

where $u\in C^0[t_0,T]$ and $u\in L^1_\infty[t_0,T]\cap L^1_2,[t_0,T]$. Clearly (4.7) admits an input–state operator $\mathcal{S}_{\mathcal{T}}:L^1_\infty[t_0,T]\cap L^1_2,[t_0,T]\times\mathcal{E}\mapsto L^1_\infty[t_0,T]\cap L^1_2,[t_0,T]$. It is also easy to see that this system admits at least four different input–state margins, $L_2^1 \mapsto L_\infty^1$, $L_2^1 \mapsto L_2^1$, $L_\infty^1 \mapsto L_2^1$, and $L_\infty^1 \mapsto L_\infty^1$.

Amongst all the possible types of input–state and input–output margins the existence of the margins for $\mathcal{L}_x \subseteq L_\infty^n[t_0, T]$ and $\mathcal{L}_y \subseteq L_\infty^h[t_0, T]$ and correspondingly for the norms $\|\cdot\|_{\mathcal{L}_x} = \|\cdot\|_{\infty,[t_0,T]}$ and $\|\cdot\|_{\mathcal{L}_y} = \|\cdot\|_{\infty,[t_0,T]}$ is of special importance. These margins will be denoted by $\gamma_{S,\infty}$ and $\gamma_{H,\infty}$, respectively. The existence of $\gamma_{S,\infty}$ and $\gamma_{H,\infty}$ implies that there exists a non-empty interval $\mathcal{T}$ on which the state of the system (4.1) and (4.2) is a bounded function. The latter is a manifestation of the physical *realizability* of the system. This motivates the introduction of the following notion.

Definition 4.1.4 System $\mathcal{S}$ is called

(1) realizable if for every $\mathbf{u} \in \mathcal{L}_u$ and $\mathbf{e} \in \mathcal{E}$ there exists a number $T(\mathbf{u}, \mathbf{e}) > t_0$ such that

$$\|\mathbf{x}(t)\|_{\infty,[t_0,T]} < \infty, \tag{4.8}$$

$$\|\mathbf{y}(t)\|_{\infty,[t_0,T]} < \infty; \tag{4.9}$$

(2) realizable with a margin with respect to the norm $\|\cdot\|_{\mathcal{L}_u}$ if for every $\mathbf{e} \in \mathcal{E}$ there exists a number $T(\mathbf{e}) > t_0$ such that the following margins $\gamma_{S,\infty}$ and $\gamma_{H,\infty}$ are defined for $\mathcal{S}$ on $\mathcal{T} = [t_0, T]$:

$$\|\mathbf{x}(t)\|_{\infty,[t_0,T]} \le \gamma_{S,\infty}(\mathbf{e}, \|\mathbf{u}(t)\|_{\mathcal{L}_u,[t_0,T]}, T), \tag{4.10}$$

$$\|\mathbf{y}(t)\|_{\infty,[t_0,T]} \le \gamma_{H,\infty}(\mathbf{e}, \|\mathbf{u}(t)\|_{\mathcal{L}_u,[t_0,T]}, T). \tag{4.11}$$

The interval $\mathcal{T} = [t_0, T]$ is referred to as the realizability interval of the system for the given $\mathbf{u} \in \mathcal{L}_u$ and $\mathbf{e} \in \mathcal{E}$.

It is clear that every system defined by a system of ordinary differential equations with continuous right-hand sides is realizable if $\mathbf{u}(t) \in C^0$. This follows immediately from the Peano existence theorem. Notice that realizability implicitly suggests that the values of $\|\mathbf{u}(t)\|_{\mathcal{L}_u,[t_0,T]}$ are defined for those inputs for which realizability is claimed. Using the notion of realizability of a system on $\mathcal{T}$, let us proceed with defining another important characterization of a dynamical system with inputs, namely *completeness*.

Definition 4.1.5 A system $\mathcal{S}$ is called

(1) complete, if it is realizable and its realizability interval $\mathcal{T} = [t_0, \infty)$;
(2) complete with a margin with respect to the norm $\|\cdot\|_{\mathcal{L}_u}$, if it is complete and there exist $\gamma_{S,\infty}$ and $\gamma_{H,\infty}$ such that inequalities (4.10) and (4.11) hold for all $T \ge t_0$.

Completeness of a system implies that the interval on which the system is defined is $[t_0, \infty)$. For systems modeled by ordinary differential equations completeness implies that solutions are defined for all t, $\mathbf{e} \in \mathcal{E}$ and $\mathbf{u}(t) \in \mathcal{L}_u$.

Often completeness of an elementary system describing the original object in the sense of Definition 4.1.5 does not need a proof. In fact, in many cases we start the analysis assuming that the system is complete (excluding exotic cases such as in Example 4.1.1). Yet, when analyzing more complex objects such as various interconnected compartments comprising an adaptive system, establishing completeness becomes an important step in the analysis. Indeed, in order to tell how the state of a system behaves at $t \to \infty$, one must be sure that it is at least defined for all t. We will discuss this in the next sections of this chapter.

Now we will move on to the input–output analysis of systems. We start by establishing basic input–output and input–state properties of stable systems and then proceed with developing analysis methods for the unstable ones.

4.2 Input–output and input–state characterizations of stable systems

Traditional stability notions invoke the notion of the *phase flow*, or simply flow (LaSalle and Lefschetz 1961). The flow is defined as a mapping

$$\mathbf{x}(t, \mathbf{x}_0, t_0), \ \mathbf{x} : \mathbb{R} \times \mathbb{R}^n \times \mathbb{R} \to \mathbb{R}^n, \tag{4.12}$$

satisfying the following condition:

$$\mathbf{x}(t_0, \mathbf{x}_0, t_0) = \mathbf{x}_0.$$

The arguments of $\mathbf{x}(t, \mathbf{x}_0, t_0)$ are the variable $t \in \mathbb{R}$ representing a time instant, the vector $\mathbf{x}_0 \in \mathbb{R}^n$ defining initial conditions, and $t_0 \in \mathbb{R}$ corresponding to a time instant at which the value of $\mathbf{x}$ is equal to $\mathbf{x}_0$. Notice that, since the values of $\mathbf{x}_0$ and t_0 are fixed, we can consider the pair $(\mathbf{x}_0, t_0)$ as an element of a set $\mathcal{E}_0 \subseteq \mathbb{R}^n \oplus \mathbb{R}$ that is a subset of the space $\mathcal{E}$ modeling environmental (external) factors. Thus the phase flow (4.12) can be thought of as a mapping that relates pairs $(\mathbf{x}_0, t_0) \in \mathcal{E}_0 \subset \mathcal{E}$ to elements from $L^n_\infty[t_0, T]$.

Since we are interested in studying systems, not the flows, let us associate the following system with flow (4.12):

$$\begin{aligned} \mathcal{S}_T &: \mathcal{E} \mapsto \mathcal{L}_x \subseteq L^n_\infty[t_0, T], \\ \mathcal{H}_T &: \mathcal{E} \mapsto \mathcal{L}_y \subseteq L^n_\infty[t_0, T]. \end{aligned} \tag{4.13}$$

The mappings are defined on $\mathcal{T} = [t_0, T]$. For the moment we will consider $\mathcal{E}$ as the space of inputs. Let $\Omega^* \subset \mathbb{R}^n$ be an invariant set with respect to $\mathbf{x}(t, \mathbf{x}_0, t_0)$. In Chapter 2 we provided a number of stability definitions relating to the asymptotic behavior of a flow with respect to an invariant set. Despite some differences in many of these stability definitions, they share important general properties. These properties are continuity of the mapping $\mathcal{S}_\mathcal{T}$ and the existence of the corresponding continuous input–state margins $\gamma_{S,\mathcal{L}_x}(\mathbf{e}, T)$:

$$\gamma_{S,\mathcal{L}_x}(\mathbf{e}, T) = \gamma_{S,\mathcal{L}_x}(\mathbf{x}_0 \oplus t_0, T) = \gamma_{S,\mathcal{L}_\infty}(\|\mathbf{x}_0\|_{\Omega^*}, T). \tag{4.14}$$

While continuity of the function $\gamma_{S,\mathcal{L}_\infty}(\cdot,\cdot)$ does not require any additional comments, we would like to clarify our current use of the term "continuity" with regard to the mapping $\mathcal{S}_\mathcal{T}$. Let us define a closeness measure of an element $\mathbf{e} = \mathbf{x}_0 \oplus t_0$ of $\mathcal{E}$ to Ω^* as

$$\rho_e(\mathbf{e}, \Omega) = \|\mathbf{x}_0\|_{\Omega^*},$$

and a closeness measure of an element $\mathbf{x}$ of $\mathcal{L}_x$ to $\mathcal{S}(\Omega^*) = \{\mathbf{z} \in \mathcal{L}_x | \mathbf{z}(t) = \mathbf{x}(t, \mathbf{p}, t_0),\ \mathbf{p} \in \Omega^*\}$ as

$$\begin{aligned} &\rho_x(\mathbf{x}(t), \mathcal{S}(\Omega^*)) = \|\mathbf{x}(t)\|_{\Omega_s,[t_0,\infty]}, \\ &\Omega_s = \{\zeta_t \in \mathbb{R}^n | \zeta_t = \mathbf{p}(t),\ \mathbf{p}(t) \in \mathcal{S}(\Omega^*),\ t \geq t_0\}. \end{aligned} \tag{4.15}$$

We will call the mapping $\mathcal{S}_\mathcal{T}$ continuous with respect to $\mathbf{x}_0$ at Ω^* iff

$$\forall\, \varepsilon > 0\, \exists\, \delta > 0 : \rho_e(\mathbf{e}, \Omega) < \delta \Rightarrow \rho_x(\mathbf{x}(t), \mathcal{S}(\Omega^*)) < \varepsilon. \tag{4.16}$$

Similarly, it will be called locally bounded with respect to $\mathbf{x}_0$ in a neighborhood of Ω^* iff

$$\rho_e(\mathbf{e}, \Omega) < \delta \Rightarrow \exists\, \varepsilon : \ \rho_x(\mathbf{x}(t), \mathcal{S}(\Omega^*)) < \varepsilon. \tag{4.17}$$

Having these notations in mind, we can now formulate the following statement:

Theorem 4.1 *Consider a complete system of which the input–output properties are defined by (4.13).*

(1) Let Ω^* be an invariant set. Then it is stable in the sense of Lyapunov iff the following alternatives hold:

(1.1) the mapping $\mathcal{S}^*_\mathcal{T}(\mathbf{x}_0, t_0) = \mathcal{S}_\mathcal{T}(\mathbf{x}_0 \oplus t_0)$, $\mathcal{T} = [t_0, \infty]$ from $\mathbb{R}^n \times \mathbb{R}$ to $\mathcal{L}_x \subseteq L^n_\infty[t_0,\infty]$ is continuous with respect to $\mathbf{x}_0$ at Ω^* for all $t_0 \in \mathbb{R}$;

(1.2) the input–state margin $\gamma_{S,\mathcal{L}_x}(\mathbf{e}, T)$ of $\mathcal{S}_\mathcal{T}$ can be represented as in (4.14), where $\gamma_{S,L_\infty}(0, t_0) = 0$, $\forall\, t_0 \in \mathbb{R}$ is a locally bounded with respect to t_0, continuous, and non-decreasing function of $\|\mathbf{x}_0\|_{\Omega^*}$ in a vicinity of zero.

(2) The set Ω^* is (locally) stable in the sense of Lagrange if the following alternatives hold:

(2.1) the mapping $\mathcal{S}^*_\mathcal{T}(\mathbf{x}_0, t_0) = \mathcal{S}_\mathcal{T}(\mathbf{x}_0 \oplus t_0)$, $\mathcal{T} = [t_0, \infty]$ from $\mathbb{R}^n \times \mathbb{R}$ to $\mathcal{L}_x \subseteq L^n_\infty[t_0,\infty]$ is locally bounded with respect to $\mathbf{x}_0$ in a neighborhood of Ω^* for all $t_0 \in \mathbb{R}$;

(2.2) the input–state margin $\gamma_{S,\mathcal{L}_x}(\mathbf{e},T)$ of $\mathcal{S}_T$ can be represented as in (4.14), in which the function $\gamma_{S,L_\infty}(0,t_0)=0$, $\forall\, t_0 \in \mathbb{R}$ is locally bounded with respect to t_0 and $\|\mathbf{x}_0\|_{\Omega^*}$ (in a neighborhood of zero).

(3) A solution is stable in the sense of Lyapunov if $\mathcal{S}_{\mathcal{T}}^*(\mathbf{x}_0,t_0)=\mathcal{S}_{\mathcal{T}}(\mathbf{x}_0\oplus t_0)$ with $\mathcal{T}=[t_0,\infty]$ is continuous with respect to $\mathbf{x}_0$ ($\mathcal{L}_x \subseteq L_\infty^n[t_0,\infty]$ is a space with standard $\|\cdot\|_{\infty,[t_0,\infty]}$ metrics).

As follows from Theorem 4.1, stability of sets and solutions in the sense of Lyapunov are equivalent to continuity of some mappings in the corresponding spaces. In each case, however, the closeness measures in $\mathcal{L}_x$ with respect to which the continuity is defined differ. Stability in the sense of Lagrange is equivalent to local boundedness of $\mathcal{S}_{\mathcal{T}}^*(\mathbf{x}_0,t_0)$. It is also worth mentioning that in the case of stable sets and solutions there exist input–state margins $\gamma_{S,\mathcal{L}_x}(\mathbf{e},T)$ that do not depend on T.

Standard stability notions in the domain of dynamical systems characterize the asymptotic behavior of states without paying much attention to inputs. The latter, however, are important components of an abstract system. An extension of the standard stability notions characterizing the actual input–output and input–state behavior of systems was introduced in a series of works by Zames (1966), Sontag and Wang (1996), Angeli *et al.* (2004) and Krichman *et al.* (2001). These are the notions of input-to-output stability and input-to-state stability.

Consider a system, and let $\mathbf{x}(t)\in L_\infty^n[t_0,T]$, $\mathbf{u}(t)\in\mathcal{L}_u[t_0,T]$, and $\mathcal{E}=\mathbb{R}^n\oplus\mathbb{R}$. Then, according to Definition 4.1.2, we have that the system's input–state and input–output relations are defined as

$$\begin{aligned}\mathcal{S}_{\mathcal{T}}:\ \mathcal{L}_u\times\mathcal{E}\mapsto\mathcal{L}_x\subseteq L_\infty^n[t_0,T],\\ \mathcal{H}_{\mathcal{T}}:\ \mathcal{L}_u\times\mathcal{E}\mapsto\mathcal{L}_y\subseteq L_\infty^n[t_0,T].\end{aligned}\tag{4.18}$$

Suppose that system (4.18) is complete. Now, following Zames (1966) and Khalil (2002), we can introduce the notion of input-to-output stability as follows.

Definition 4.2.1 A system (of which the input–output relations are defined by (4.18)) is called input-to-output stable if there exist $\alpha\in\mathcal{K}$ and $\beta\in\mathbb{R}_{>0}$ such that

$$\|\mathbf{y}(t)\|_{\mathcal{L}_y,[t_0,T]}\leq\alpha(\|\mathbf{u}(t)\|_{\mathcal{L}_y,[t_0,T]})+\beta\tag{4.19}$$

holds for all $\mathbf{u}(t)\in\mathcal{L}_u[t_0,\infty]\cap\mathcal{L}_y[t_0,T]$, $\mathbf{e}\in\mathcal{E}$.

A system is input-to-output stable with a finite input-to-output gain if there exists $\gamma>0$ such that

$$\|\mathbf{y}(t)\|_{\mathcal{L}_y,[t_0,T]}\leq\gamma\|\mathbf{u}(t)\|_{\mathcal{L}_y,[t_0,T]}+\beta.\tag{4.20}$$

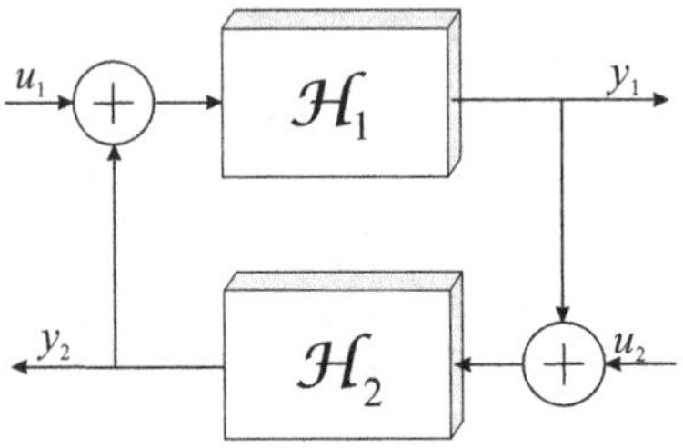

Figure 4.1 A diagram showing feedback interconnection of two systems.

An important property of input-to-output systems is that this characteristic not only allows one to characterize the input–output properties of a single system but also provides a useful tool for the analysis of the existence and stability of interconnections of uncertain systems. Consider, for example, a feedback interconnection of two systems $\mathcal{S}_1$ and $\mathcal{S}_2$ of which the input–state relations are defined by $\mathcal{H}_{1,T}$ and $\mathcal{H}_{2,T}$:

$$\mathcal{H}_{1,T} : \mathbf{y}_1(t) = \mathcal{H}_{1,T}(\mathbf{u}_1(t)),$$
$$\mathcal{H}_{2,T} : \mathbf{y}_2(t) = \mathcal{H}_{2,T}(\mathbf{u}_2(t)).$$

The diagram of this interconnection is shown in Figure 4.1. It is shown in Zames (1966) that if the product $\gamma_1\gamma_2$ (or the composition $\alpha_1 \circ \alpha_2$) is strictly less than 1 then the feedback interconnection is also input-to-output stable. A simple statement of this result, also known as the small-gain theorem, is presented below (Khalil 2002; Zames 1966).

Theorem 4.2 *Let $\mathcal{S}_1$ and $\mathcal{S}_2$ be two input-to-output stable systems:*

$$\|\mathbf{y}_1(t)\|_{\mathcal{L}_y,[t_0,T]} \le \gamma_1 \|\mathbf{u}_1(t)\|_{\mathcal{L}_y,[t_0,T]} + \beta_1,$$

$$\|\mathbf{y}_2(t)\|_{\mathcal{L}_y,[t_0,T]} \le \gamma_2 \|\mathbf{u}_2(t)\|_{\mathcal{L}_y,[t_0,T]} + \beta_2.$$

Then the feedback interconnection of $\mathcal{S}_1$ and $\mathcal{S}_2$ is input-to-output stable (with $(\mathbf{u}_1, \mathbf{u}_2)$ as input and $\mathbf{y}_1$ (or $\mathbf{y}_2$) as output), provided that

$$\gamma_1\gamma_2 < 1. \tag{4.21}$$

The main idea of the proof can be sketched as follows. Given that both systems are input-to-output stable, and that in the feedback interconnection $\mathbf{u}_1$ is replaced with $\mathbf{y}_2 + \mathbf{u}_1$ ($\mathbf{u}_2$ is replaced with $\mathbf{y}_1 + \mathbf{u}_2$), we can write

$$\begin{aligned}\|\mathbf{y}_1(t)\|_{\mathcal{L}_y,[t_0,T]} &\le \gamma_1(\|\mathbf{y}_2(t)\|_{\mathcal{L}_y,[t_0,T]} + \|\mathbf{u}_1(t)\|_{\mathcal{L}_y,[t_0,T]}) + \beta_1 \\ &\le \gamma_1\gamma_2\|\mathbf{y}_1(t)\|_{\mathcal{L}_y,[t_0,T]} + \gamma_1\|\mathbf{u}_1(t)\|_{\mathcal{L}_y,[t_0,T]} \\ &\quad + \gamma_1\gamma_2\|\mathbf{u}_2(t)\|_{\mathcal{L}_y,[t_0,T]} + \beta_1 + \gamma_1\beta_2.\end{aligned}$$

Thus the condition $\gamma_1\gamma_2 < 1$ allows us to express $\|\mathbf{y}_1(t)\|_{\mathcal{L}_y,[t_0,T]}$ as

$$\|\mathbf{y}_1(t)\|_{\mathcal{L}_y,[t_0,T]} \leq \frac{\gamma_1\|\mathbf{u}_1(t)\|_{\mathcal{L}_y,[t_0,T]}}{1-\gamma_1\gamma_2} + \frac{\gamma_1\gamma_2\|\mathbf{u}_2(t)\|_{\mathcal{L}_y,[t_0,T]}}{1-\gamma_1\gamma_2} + \beta_3,$$

where $\beta_3 = (\beta_1 + \gamma_1\beta_2)/(1-\gamma_1\gamma_2)$.

There are generalizations of this result that replace input-to-output stability in the statement of Theorem 4.2 with a less restrictive notion of *practical input-to-output stability* (Jiang *et al.* 1994).

Definition 4.2.2 System (4.18) is called practically input-to-output stable if there exist functions $\alpha \in \mathcal{K}$, $\beta \in \mathcal{KL}$, and $D \in \mathbb{R}_{\geq 0}$ such that

$$\|\mathbf{y}(t)\| \leq \beta(\|\mathbf{x}_0\|, t) + \alpha(\|\mathbf{u}(t)\|_{\mathcal{L}_y,[t_0,T]}) + D, \tag{4.22}$$

for all $\mathbf{u}(t) \in \mathcal{L}_u[t_0,T] \cap \mathcal{L}_y[t_0,T]$, $\mathbf{e} \in \mathcal{E}$.

Input–output stability and small-gain theorems are instrumental for establishing input–output characterizations of larger systems composed of input–output-stable components. They do not, however, immediately allow one to characterize the asymptotic properties of the system's state. A powerful extension of the notion of input–output stability that allows one to characterize both the input–output and the input–state asymptotic behavior of systems is developed in Sontag and Wang (1996), Krichman *et al.* (2001), and Angeli *et al.* (2004). A key element of this extension is constituted by the notions of input-to-state, output-to-state, and input-to-output-state stability.

Definition 4.2.3 Consider a complete system (4.1) and (4.2), and let Ω^* be an invariant set of the system for $\mathbf{u}(t) = 0$. The system is called

(1) globally input-to-state stable w.r.t. $\Omega^* \subset \mathbb{R}^n$ if there are functions $\gamma \in \mathcal{K}$ and $\beta \in \mathcal{KL}$ such that

$$\|\mathbf{x}(t,\mathbf{x}_0,t_0)\|_{\Omega^*} \leq \beta(\|\mathbf{x}_0\|_{\Omega^*}, t) + \gamma(\|\mathbf{u}(\tau)\|_{\infty,[t_0,t]}) \tag{4.23}$$

holds for all $\mathbf{x}_0 \in \mathbb{R}^n$, $t \geq t_0$;

(2) globally integrally input-to-state stable w.r.t. $\Omega^* \subset \mathbb{R}^n$ if there are $\gamma \in \mathcal{K}$ and $\beta \in \mathcal{KL}$ such that for all $\mathbf{x}_0 \in \mathbb{R}^n$ and $t \geq t_0$ the following estimate holds:

$$\|\mathbf{x}(t,\mathbf{x}_0,t_0)\|_{\Omega^*} \leq \beta(\|\mathbf{x}_0\|_{\Omega^*}, t) + \int_{t_0}^{t} \gamma(\|\mathbf{u}(\tau)\|)d\tau; \tag{4.24}$$

(3) globally input–output–state stable w.r.t. $\Omega^* \subset \mathbb{R}^n$ if there are $\gamma_u \in \mathcal{K}$, $\gamma_y \in \mathcal{K}$, and $\beta \in \mathcal{KL}$ such that for all $\mathbf{x}_0 \in \mathbb{R}^n$, $t \geq t_0$ the following holds:

$$\|\mathbf{x}(t, \mathbf{x}_0, t_0)\|_{\Omega^*} \leq \beta(\|\mathbf{x}_0\|_{\Omega^*}, t) + \gamma_u(\|\mathbf{u}(\tau)\|_{\infty,[t_0,t]}) + \gamma_y(\|\mathbf{y}(\tau)\|_{\infty,[t_0,t]}); \tag{4.25}$$

(4) globally output–state stable if there are $\gamma_y \in \mathcal{K}$, $\beta \in \mathcal{KL}$ such that

$$\|\mathbf{x}(t, \mathbf{x}_0, t_0)\|_{\Omega^*} \leq \beta(\|\mathbf{x}_0\|_{\Omega^*}, t) + \gamma_y(\|\mathbf{y}(\tau)\|_{\infty,[t_0,t]}) \tag{4.26}$$

for all $\mathbf{x}_0 \in \mathbb{R}^n$, $t \geq t_0$.

The notion of input-to-state stability is closely related to that of Lyapunov. In particular, in Sontag and Wang (1996) the input-to-state stability is shown to be equivalent to the existence of a Lyapunov function for the original system without inputs. In addition the following obvious properties hold.

Theorem 4.3 *Let (4.18) be a complete system. Then*

(1) if the system is input-to-output stable then there exists a locally bounded w.r.t. $\mathbf{e}$*, continuous in* $\|\mathbf{u}\|_{\mathcal{L}_y,[t_0,T]}$ *and bounded in* T *input–output margin* $\gamma_{H,\mathcal{L}_y}(\mathbf{e}, \|\mathbf{u}(t)\|_{\mathcal{L}_y,[t_0,T]}, T)$*;*

(2) if the system is input–state stable then there exists an input–state margin $\gamma_{S,\mathcal{L}_x}(\mathbf{e}, \|\mathbf{u}(t)\|_{\mathcal{L}_u,[t_0,T]}, T)$*:*

$$\gamma_{S,\mathcal{L}_x}(\mathbf{e}, \|\mathbf{u}(t)\|_{\mathcal{L}_u,[t_0,T]}, T) = \gamma_{S,L_\infty}(\|\mathbf{x}_0\|_{\Omega^*}, t_0, \|\mathbf{u}(t)\|_{\mathcal{L}_u,[t_0,T]}, T),$$

where $\gamma_{S,L_\infty}(\cdot)$ *is a function that is continuous w.r.t.* $\|\mathbf{x}_0\|_{\Omega^*}$ *and* $\|\mathbf{u}(t)\|_{\mathcal{L}_u,[t_0,T]}$*, which is locally bounded in* t_0 *and bounded in* T*. Moreover,* $\gamma_{S,L_\infty}(0, t_0, 0, T) = 0$ *for all* t_0*,* $T \geq t_0$*.*

The theorem follows immediately from Definition 4.2.3.

Similar properties can be established for integral variants of input-to-state stability. The general properties of stable systems formulated in Theorems 4.1 and 4.3 allow us to establish specific properties of the corresponding input–output and input–state relations in the definitions of these systems. These results will be used in the next section to determine what the general invariants of unstable systems are, and what kind of analysis tools may be made available for studying them.

4.3 Input–output and input–state analysis of uncertain unstable systems

The analysis in the previous section showed that popular and widely used notions of stability are very closely related to continuity of the corresponding input–output and input–state characterizations of the system. As has been mentioned earlier, stability in the sense of Lyapunov is often a criterion of the fitness and robustness of a system. Thus establishing continuity of $\mathcal{S}_{\mathcal{T}}^{*}(\mathbf{x}_0, t_0)$ w.r.t. $\mathbf{x}_0$ or continuity of γ_{S,L_∞} would automatically ensure the robustness and fitness of the system.

Continuity and the availability of certain input–output or input–state characterizations of a system allow one to employ a wealth of mathematical resources when studying the asymptotic properties of such systems and their interconnections. In particular, knowledge of a continuous input–output margin is necessary for conventional small-gain-based analysis.

The main issue, however, is that tight, albeit continuous, input–output and input–state characterizations of a system are not always available. A more realistic requirement is that these characterizations are modeled by some locally bounded rather than continuous functions. Thus an extension of the theory to this class of systems is needed.

In order to be able to proceed with a formal statement of the problem, let us first define three basic types of interconnections: serial, parallel, and feedback. For this purpose consider two systems $\mathcal{S}_1$ and $\mathcal{S}_2$:

$$\begin{aligned} \mathcal{S}_{1,\mathcal{T}} &: \mathcal{L}_{u_1} \times \mathcal{E}_1 \mapsto \mathcal{L}_{x_1} \subseteq L_\infty^{n_1}[t_0, T], \\ \mathcal{H}_{1,\mathcal{T}} &: \mathcal{L}_{u_1} \times \mathcal{E}_1 \mapsto \mathcal{L}_{y_1} \subseteq L_\infty^{m_1}[t_0, T], \end{aligned} \tag{4.27}$$

$$\begin{aligned} \mathcal{S}_{2,\mathcal{T}} &: \mathcal{L}_{u_2} \times \mathcal{E}_2 \mapsto \mathcal{L}_{x_2} \subseteq L_\infty^{n_2}[t_0, T], \\ \mathcal{H}_{2,\mathcal{T}} &: \mathcal{L}_{u_2} \times \mathcal{E}_2 \mapsto \mathcal{L}_{y_2} \subseteq L_\infty^{m_2}[t_0, T]. \end{aligned} \tag{4.28}$$

Definition 4.3.1 Let systems $\mathcal{S}_1$ and $\mathcal{S}_2$ be given, and $\mathcal{L}_{y_1} \subseteq \mathcal{L}_{u_2}$. Serial interconnection of $\mathcal{S}_1$ and $\mathcal{S}_2$ is the system $\mathcal{S}$:

$$\begin{aligned} \mathcal{S}_{\mathcal{T}} &: \mathcal{L}_u \times \mathcal{E} \mapsto \mathcal{L}_x \subseteq L_\infty^{n_1+n_2}[t_0, T], \\ \mathcal{H}_{\mathcal{T}} &: \mathcal{L}_u \times \mathcal{E} \mapsto \mathcal{L}_y \subseteq L_\infty^{m_2}[t_0, T], \end{aligned} \tag{4.29}$$

where

$$\begin{aligned} &\mathcal{E} = \mathcal{E}_1 \oplus \mathcal{E}_2, \quad \mathcal{L}_x = \mathcal{L}_{x_1} \oplus \mathcal{L}_{x_2}, \quad \mathcal{L}_y = \mathcal{L}_{y_2}, \quad \mathcal{L}_u = \mathcal{L}_{u_1}, \\ &\mathcal{S}_{\mathcal{T}}(\mathbf{u}, \mathbf{e}_1 \oplus \mathbf{e}_2) = \mathcal{S}_{1,\mathcal{T}}(\mathbf{u}, \mathbf{e}_1) \oplus \mathcal{S}_{2,\mathcal{T}}(\mathcal{H}_{1,\mathcal{T}}(\mathbf{u}, \mathbf{e}_1), \mathbf{e}_2), \\ &\mathcal{H}_{\mathcal{T}}(\mathbf{u}, \mathbf{e}_1 \oplus \mathbf{e}_2) = \mathcal{H}_{2,\mathcal{T}}(\mathcal{H}_{1,\mathcal{T}}(\mathbf{u}, \mathbf{e}_1), \mathbf{e}_2). \end{aligned} \tag{4.30}$$

Definition 4.3.2 Let systems $\mathcal{S}_1$ and $\mathcal{S}_2$ be given, and $\mathcal{L}_u = \mathcal{L}_{u_1} \cap \mathcal{L}_{u_2} \neq 0$. Parallel interconnection of $\mathcal{S}_1$ and $\mathcal{S}_2$ is the following system $\mathcal{S}$:

$$\begin{aligned} \mathcal{S}_{\mathcal{T}} &: \mathcal{L}_u \times \mathcal{E} \mapsto \mathcal{L}_x \subseteq L_\infty^{n_1+n_2}[t_0, T], \\ \mathcal{H}_{\mathcal{T}} &: \mathcal{L}_u \times \mathcal{E} \mapsto \mathcal{L}_y \subseteq L_\infty^{m_1+m_2}[t_0, T], \end{aligned} \tag{4.31}$$

where

$$\begin{aligned}
&\mathcal{E} = \mathcal{E}_1 \oplus \mathcal{E}_2, \ \ \mathcal{L}_x = \mathcal{L}_{x_1} \oplus \mathcal{L}_{x_2}, \ \ \mathcal{L}_y = \mathcal{L}_{y_1} \oplus \mathcal{L}_{y_2}, \\
&\mathcal{S}_\mathcal{T}(\mathbf{u}, \mathbf{e}_1 \oplus \mathbf{e}_2) = \mathcal{S}_{1,\mathcal{T}}(\mathbf{u}, \mathbf{e}_1) \oplus \mathcal{S}_{2,\mathcal{T}}(\mathbf{u}, \mathbf{e}_2), \\
&\mathcal{H}_\mathcal{T}(\mathbf{u}, \mathbf{e}_1 \oplus \mathbf{e}_2) = \mathcal{H}_{1,\mathcal{T}}(\mathbf{u}, \mathbf{e}_1) \oplus \mathcal{H}_{2,\mathcal{T}}(\mathbf{u}, \mathbf{e}_2).
\end{aligned} \tag{4.32}$$

Definition 4.3.3 Let $\mathcal{S}_1$ and $\mathcal{S}_2$ be given, and $\mathcal{L}_{y_1} \subseteq \mathcal{L}_{u_2}$ and $\mathcal{L}_{y_2} \subseteq \mathcal{L}_{u_1}$. Feedback interconnection of $\mathcal{S}_1$ and $\mathcal{S}_2$ is the system $\mathcal{S}$:

$$\begin{aligned}
\mathcal{S}_\mathcal{T} &: \ \mathcal{L}_u \times \mathcal{E} \mapsto \mathcal{L}_x \subseteq L_\infty^{n_1+n_2}[t_0, T], \\
\mathcal{H}_\mathcal{T} &: \ \mathcal{L}_u \times \mathcal{E} \mapsto \mathcal{L}_y \subseteq L_\infty^{m_1+m_2}[t_0, T],
\end{aligned} \tag{4.33}$$

where

$$\mathcal{E} = \mathcal{E}_1 \oplus \mathcal{E}_2, \ \ \mathcal{L}_x = \mathcal{L}_{x_1} \oplus \mathcal{L}_{x_2}, \ \ \mathcal{L}_y = \mathcal{L}_{y_1} \oplus \mathcal{L}_{y_2}, \ \ \mathcal{L}_u = \mathcal{L}_{u_1} \oplus \mathcal{L}_{u_2},$$

$$\begin{aligned}
\mathbf{y}_1 &= \mathcal{H}_{1,\mathcal{T}}(\mathbf{u}_1 + \mathbf{y}_2, \mathbf{e}_1), \\
\mathbf{y}_2 &= \mathcal{H}_{2,\mathcal{T}}(\mathbf{u}_2 + \mathbf{y}_1, \mathbf{e}_2), \\
\mathcal{S}_\mathcal{T}(\mathbf{u}_1 \oplus \mathbf{u}_2, \mathbf{e}_1 \oplus \mathbf{e}_2) &= \mathcal{S}_{1,\mathcal{T}}(\mathbf{u}_1 + \mathbf{y}_2, \mathbf{e}_1) \oplus \mathcal{S}_{2,\mathcal{T}}(\mathbf{u}_2 + \mathbf{y}_1, \mathbf{e}_2), \\
\mathcal{H}_\mathcal{T}(\mathbf{u}_1 \oplus \mathbf{u}_2, \mathbf{e}_1 \oplus \mathbf{e}_2) &= \mathcal{H}_{1,\mathcal{T}}(\mathbf{u}_1 + \mathbf{y}_2, \mathbf{e}_1) \oplus \mathcal{H}_{2,\mathcal{T}}(\mathbf{u}_2 + \mathbf{y}_1, \mathbf{e}_2).
\end{aligned} \tag{4.34}$$

One of the basic questions about interconnections is whether they are realizable and complete. While establishing the realizability and completeness of serial and parallel interconnections of two or more systems is a rather straightforward exercise, finding a satisfactory answer for the feedback interconnections is non-trivial. This is because the definition of feedback interconnections, (4.34), is implicit. Thus $\mathcal{S}_\mathcal{T}$ and $\mathcal{H}_\mathcal{T}$ are not explicitly defined. Moreover, we wish to allow that the input–output and input–state characterization of each system involved be known up to a locally bounded function. Thus the questions of realizability and completeness become even more complicated since the functions $\mathcal{H}_{i,\mathcal{T}}$ and $\mathcal{S}_{i,\mathcal{T}}$ in (4.34) are not known precisely.

Realizability and completeness are not the only properties one needs to know when dealing with interconnections of uncertain systems. In particular, having estimates of the norms of the system's state and output (in $\mathcal{L}_x$ and $\mathcal{L}_y$) for the given input environmental factors from $\mathcal{E}$ is often desirable. In addition, estimating the limit sets (ω-limit sets) of the interconnection is sometimes required.

Thus the following problems are relevant.

Problem 4.1 *Realizability and completeness analysis*. Let $\mathcal{S}$, defined by

$$\begin{aligned}
\mathcal{S}_\mathcal{T} &: \ \mathcal{L}_u \times \mathcal{E} \mapsto \mathcal{L}_x \subseteq L_\infty^{n}[t_0, T], \\
\mathcal{H}_\mathcal{T} &: \ \mathcal{L}_u \times \mathcal{E} \mapsto \mathcal{L}_y \subseteq L_\infty^{n}[t_0, T],
\end{aligned} \tag{4.35}$$

be an interconnection (serial, parallel, feedback) of two systems $\mathcal{S}_1$ and $\mathcal{S}_2$.

Determine

(1) conditions[2] when the system $\mathcal{S}$ is realizable;
(2) conditions when the system $\mathcal{S}$ is complete.

Problem 4.2 *Local boundedness of input–state/input–output characterizations of interconnections.* Let $\mathcal{S}$ be a complete system defined by (4.35).

(1) Determine conditions ensuring that input–state and input–output mappings of $\mathcal{S}$ are locally bounded for all $\mathbf{u} \in \mathcal{L}_u$.
(2) Estimate the input–state, $\gamma_{S,\mathcal{L}_x}(\cdot,\cdot,\cdot)$, and input–output, $\gamma_{H,\mathcal{L}_x}(\cdot,\cdot,\cdot)$, margins of the corresponding input–state and input–output mappings:

$$\|\mathbf{x}(t)\|_{\mathcal{L}_x,[t_0,T]} \leq \gamma_{S,\mathcal{L}_x}(\mathbf{e}, \|\mathbf{u}(t)\|_{\mathcal{L}_u,[t_0,T]}, T),$$
$$\|\mathbf{y}(t)\|_{\mathcal{L}_y,[t_0,T]} \leq \gamma_{H,\mathcal{L}_y}(\mathbf{e}, \|\mathbf{u}(t)\|_{\mathcal{L}_u,[t_0,T]}, T).$$

Problem 4.3 *Asymptotic analysis of interconnections of systems.* Let $\mathcal{S}$ be a complete system defined by (4.35), and let its input–state and input–output mappings be locally bounded.

Determine estimates of the ω-limit sets of state $\mathbf{x}(t)$ and output $\mathbf{y}(t)$ of the system. In addition, derive the bounds for

$$\lim_{T\to\infty} \|\mathbf{x}(t)\|_{\mathcal{L}_x,[T,\infty]}, \quad \lim_{T\to\infty} \|\mathbf{y}(t)\|_{\mathcal{L}_y,[T,\infty]} \tag{4.36}$$

as functions $\mathbf{e} \in \mathcal{E}$ and $\mathbf{u} \in \mathcal{L}_u$.

Solutions to these three problems and other related questions are provided in the next sections.

4.3.1 Realizability of interconnections of systems with locally bounded operators

Let us consider the realizability and completeness problem, and let us start with the simplest cases of serial and parallel interconnections. Completeness and realizability conditions of the interconnections in these cases are provided in the theorem below.

Theorem 4.4 *Consider (4.27) and (4.28), and let (4.31) and (4.32) be realizable (complete).*

[2] Here the term conditions refers to specific properties of $\mathcal{S}_1$ and $\mathcal{S}_2$.

Then

(1) parallel interconnection of (4.31) and (4.32) is realizable (complete).
In addition, if systems $\mathcal{S}_1$ and $\mathcal{S}_2$ admit input–state and input–output margins w.r.t. the norms $\|\cdot\|_{\mathcal{L}_{u_1}}$ and $\|\cdot\|_{\mathcal{L}_{u_2}}$ on $[t_0, T_1]$, $[t_0, T_2]$, and $\gamma_{H,\mathcal{L}_{y_1}}$,

$$\|\mathbf{y}_1(t)\|_{\mathcal{L}_{u_2},[t_0,T_1]} \leq \gamma_{H,\mathcal{L}_{y_1}\subseteq\mathcal{L}_{u_2}}(\mathbf{e}_1, \|\mathbf{u}_1(t)\|_{\mathcal{L}_{u_1}}, T_1),$$

then

(2) serial interconnection (4.29) and (4.30) of $\mathcal{S}_1$ and $\mathcal{S}_2$ is also realizable (complete), and its input–state and input–output margins can be defined as

$$\begin{aligned}\|\mathbf{x}(t)\|_{\infty,[t_0,T]} &\leq \gamma_{S_1,\infty}(\mathbf{e}_1, \|\mathbf{u}_1(t)\|_{\mathcal{L}_{u_1},[t_0,T]}, T)\\ &\quad + \gamma_{S_2,\infty}(\mathbf{e}_2, \gamma_{H,\mathcal{L}_{y_1}}(\mathbf{e}_1, \|\mathbf{u}_1(t)\|_{\mathcal{L}_{u_1}}, T), T), \qquad (4.37)\\ \|\mathbf{y}(t)\|_{\infty,[t_0,T]} &\leq \gamma_{H_2,\infty}(\mathbf{e}_2, \gamma_{H,\mathcal{L}_{y_1}}(\mathbf{e}_1, \|\mathbf{u}_1(t)\|_{\mathcal{L}_{u_1}}, T), T).\end{aligned}$$

In the case in which the systems are not complete, the existence time $T = T^*(\mathcal{S})$ can be estimated as

$$T^*(\mathcal{S}) \leq \min\{T^*(\mathcal{S}_1), T^*(\mathcal{S}_2)\}. \qquad (4.38)$$

Realizability and completeness of serial and parallel interconnections together with estimates (4.37) will be used in the analysis of feedback interconnections of systems. Before we proceed to these results, let us first introduce an additional useful notion.

Definition 4.3.4 Consider a system $\mathcal{S}$ given by (4.1) and (4.2) for which the input–state and input–output mappings are $\mathcal{S}_\mathcal{T}$ and $\mathcal{H}_\mathcal{T}$, (4.3) and (4.4), and $\mathcal{T} = [t_0, T]$. Let, in addition, the following mapping be defined for $\mathcal{S}$:

$$\boldsymbol{\psi} : \mathcal{L}_x[t_0, T] \times \mathcal{E}_\psi \mapsto \mathcal{L}_\psi[t_0, T], \qquad (4.39)$$

where $\mathcal{L}_\psi[t_0, T]$ is a linear normed space with the norm $\|\cdot\|_{\mathcal{L}_\psi,[t_0,T]}$, and $\mathcal{E}_\psi$ is a linear space.

We will say that the mapping $\boldsymbol{\psi}(\mathbf{x}(t), \mathbf{e}_\psi) \in \mathcal{L}_\psi[t_0, T]$ majorizes state $\mathbf{x}(t)$ and output $\mathbf{y}(t)$ with respect to $\|\cdot\|_{\mathcal{L}_\psi}$, iff there exist functions $\mu_{S,\infty} : \mathcal{E} \times \mathbb{R}_{\geq 0} \to \mathbb{R}_{\geq 0}$ and $\mu_{H,\infty} : \mathcal{E} \times \mathbb{R}_{\geq 0} \to \mathbb{R}_{\geq 0}$ such that

$$\|\mathbf{x}(t)\|_{\infty,[t_0,T]} \leq \mu_{S,\infty}(\mathbf{e}, \|\boldsymbol{\psi}(\mathbf{x}(t), \mathbf{e}_\psi)\|_{\mathcal{L}_\psi,[t_0,T]}), \qquad (4.40)$$

$$\|\mathbf{y}(t)\|_{\infty,[t_0,T]} \leq \mu_{H,\infty}(\mathbf{e}, \|\boldsymbol{\psi}(\mathbf{x}(t), \mathbf{e}_\psi)\|_{\mathcal{L}_\psi,[t_0,T]}). \qquad (4.41)$$

Without loss of generality we will suppose that $\mu_{S,\infty}(\cdot, p)$ and $\mu_{H,\infty}(\cdot, p)$ are non-decreasing functions of a real variable p.

Majorization in general and majorizing functions in particular have been used successfully in various areas of adaptive regulation and control.[3] The utility of using majorizing functions is that they allow us to tell how a number of (potentially unaccessible) variables of a system vary by observing just a few available quantities (majorizing functions).

Let $\mathcal{X} \subseteq \mathbb{R}^n$, then an example of a majorizing function state is any estimate of the following type:

$$\|\mathbf{x}\| \leq \mu(\|\boldsymbol{\psi}(\mathbf{x})\|). \tag{4.42}$$

Notice that standard conditions (2.23) specifying the choice of a Lyapunov function in the analysis of stability $V(\mathbf{x}, t)$,

$$\alpha_1(\|\mathbf{x}\|) \leq V(\mathbf{x}, t) \leq \alpha_2(\|\mathbf{x}\|),\ \alpha_1, \alpha_2 \in \mathcal{K}_\infty,$$

can be viewed as a majorization.

Despite the fact that (2.23) and (4.40)–(4.42) may indeed look very similar, they are different in that the latter trio relates norms of functions in the corresponding spaces while the former expression compares the values of functions at a point. This difference is further illustrated in the example below.

Example 4.3.1 Consider the following system of equations describing the motion of a point mass attached to a spring:

$$\begin{aligned} \dot{x}_1 &= x_2, \\ \dot{x}_2 &= k_0 x_1 + f(x_2, t) + u(t),\ k_0 < 0. \end{aligned} \tag{4.43}$$

The function $f : \mathbb{R} \times \mathbb{R}_{\geq 0} \to \mathbb{R}$, $f \in C^1$ models the influence of nonlinear damping. Let $\psi(\mathbf{x}(t), \lambda) = \lambda x_1(t) + x_2(t)$, $\lambda \in \mathbb{R}_{>0}$, and consider

$$\dot{x}_1 = -\lambda x_1 + \psi(t).$$

The function $\psi(\mathbf{x}(t), \lambda) \in L^1_\infty[t_0, T]$ majorizes the state of (4.43) w.r.t. the uniform norm $\|\cdot\|_{\infty,[t_0,T]}$. We notice that $x_1(t)$ satisfies the following equation:

$$x_1(t) = e^{-\lambda(t-t_0)} x_1(t_0) + e^{-\lambda t} \int_{t_0}^{t} e^{\lambda \tau} \psi(\tau) d\tau \Rightarrow$$

$$|x_1(t)| \leq |x_1(t_0)| + \frac{1}{\lambda} \|\psi(t)\|_{\infty,[t_0,t]}.$$

Then, using

$$|x_2(t)| \leq |\psi(t)| + \lambda |x_1(t)|,$$

[3] See, for example, Putov (1993), where the concept is used for solving problems of robust adaptive control. In Lin and Qian (2002a,b) majorizing functions are employed to ensure adaptive compensation of nonlinearly parametrized perturbations.

we can obtain that

$$\|\mathbf{x}(t)\|_{\infty,[t_0,t]} \leq \left(1+\frac{1}{\lambda}\right)\|\psi(t)\|_{\infty,[t_0,t]} + (1+\lambda)|x_1(t_0)|. \tag{4.44}$$

Moreover, in addition to proving that (4.44) holds, it is possible to show that the function $\psi(\mathbf{x}(t),\lambda) \in L^1_\infty[t_0,T]$ majorizes $x_1(t)$ in (4.43) w.r.t. the norms $L^1_p[t_0,T]$, $p \geq 1$. Indeed, using Hölder's inequality and the triangle inequality (Kolmogorov and Fomin 1976) and also that

$$e^{-\lambda t}\left(\int_{t_0}^{t} |e^{\lambda\tau}|^q \, d\tau\right)^{\frac{1}{q}} \leq \left(\frac{1}{\lambda q}\right)^{\frac{1}{q}},$$

we can obtain

$$\|x_1(t)\|_{\infty,[t_0,t]} \leq |x_1(t_0)| + \left(\frac{1}{\lambda q}\right)^{\frac{1}{q}} \|\psi(t)\|_{\mathrm{p},[t_0,t]}, \quad \frac{1}{p}+\frac{1}{q}=1.$$

Hence variable $x_1(t)$ of (4.43) is majorized by $\psi(t)$ w.r.t. the norms $\|\cdot\|_{\mathrm{p},[t_0,t]}$ too. These examples imply that, to keep the values of $x_1(t)$ and $x_2(t)$ within some given bounds, it is sufficient to ensure that the value of $\|\psi(t)\|_{\infty,[t_0,t]}$ is bounded from above. If boundedness of $x_1(t)$ is needed, then ensuring that $\|\psi(t)\|_{p,[t_0,t]}$, $p \geq 1$ is bounded suffices. Moreover, precise knowledge of the nonlinear damping $f(x_2,t)$ is not required in order to draw such a conclusion.

Let $\mathcal{S}$ be a system and $\boldsymbol{\psi}(\mathbf{x},\mathbf{e}_\psi)$ be a majorizing mapping for this system. Given that the state of the system can be expressed as $\mathbf{x} = \mathcal{S}_T(\mathbf{u},\mathbf{e})$, the following mapping can now be defined:

$$\begin{aligned} &\mathcal{P}_T(\mathbf{u},\mathbf{e},\mathbf{e}_\psi) : \mathcal{L}_u[t_0,T] \times \mathcal{E} \times \mathcal{E}_\psi \mapsto \mathcal{L}_\psi[t_0,T], \\ &\mathcal{P}_T(\mathbf{u},\mathbf{e},\mathbf{e}_\psi) = \boldsymbol{\psi}(\mathcal{S}_T(\mathbf{u},\mathbf{e}),\mathbf{e}_\psi). \end{aligned} \tag{4.45}$$

Similarly to input–state and input–output margins, we can introduce a margin function $\gamma_{P,\mathcal{L}_\psi}$ for $\mathcal{P}_T(\mathbf{u},\mathbf{e},\mathbf{e}_\psi)$:

$$\|\psi(t)\|_{\mathcal{L}_\psi,[t_0,T]} \leq \gamma_{P,\mathcal{L}_\psi}(\mathbf{e},\mathbf{e}_\psi,\|\mathbf{u}(t)\|_{\mathcal{L}_u,[t_0,T]},T),$$

where $\gamma_{P,\mathcal{L}_\psi}$ is a function that is locally bounded and non-decreasing in $\|\mathbf{u}(t)\|_{\mathcal{L}_u,[t_0,T]}$.

General input–output, input–state, and majorizing characterizations of systems define the behavior of systems for all inputs $\mathbf{u}$ and $\mathbf{e}$. In addition to this general description, it is also useful to know the system's behavior relative to a certain class of inputs. For example, it may happen that the system's inputs can be decomposed as $\mathbf{u}(t) = \mathbf{u}^*(t) + \delta(t)$, where $\mathbf{u}^*$ is an element of $\mathcal{L}_u$ and δ is an element of a subset

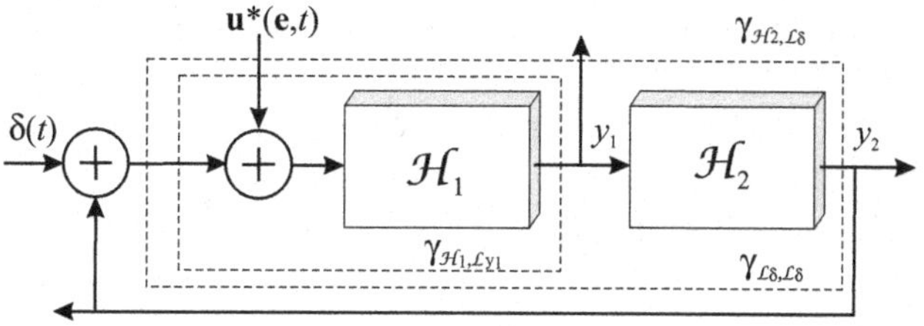

Figure 4.2 Feedback interconnection of systems $\mathcal{S}_1$ and $\mathcal{S}_2$.

of $\mathcal{L}_u$ with some special properties. In order to be able to capture these possibilities, let us summarize this as an assumption.

Assumption 4.1 *Consider a system $\mathcal{S}$ and suppose that for every $\mathbf{e} \in \mathcal{E}$ there is a function $\mathbf{u}^*(\mathbf{e}, t)$ and a subspace $\mathcal{L}_\delta[t_0, T] \subseteq \mathcal{L}_u[t_0, T]$ such that the following relations hold:*

$$\|\psi(t)\|_{\mathcal{L}_\psi,[t_0,T]} \le \gamma_{P,\mathcal{L}_\psi}(\mathbf{e}, \mathbf{e}_\psi, \|\delta(t)\|_{\mathcal{L}_\delta,[t_0,T]}, T), \tag{4.46}$$

$$\|\mathbf{y}(t)\|_{\mathcal{L}_y,[t_0,T]} \le \gamma_{H^*,\mathcal{L}_y}(\mathbf{e}, \|\delta(t)\|_{\mathcal{L}_\delta,[t_0,T]}, T). \tag{4.47}$$

The functions $\gamma_{P,\mathcal{L}_\psi}$ and $\gamma_{H^*,\mathcal{L}_y}$ are the margins for the corresponding majorizing and input–state mappings $\mathcal{L}_\delta[t_0, T] \mapsto \mathcal{L}_\psi[t_0, T]$ and $\mathcal{L}_\delta[t_0, T] \mapsto \mathcal{L}_y[t_0, T]$:

$$\mathcal{P}_\mathcal{T}(\mathbf{u}^*(\mathbf{e}, t) + \delta(t), \mathbf{e}, \mathbf{e}_\psi),$$
$$\mathcal{S}_\mathcal{T}(\mathbf{u}^*(\mathbf{e}, t) + \delta(t)).$$

Let us now consider a feedback interconnection of a system $\mathcal{S}_1$ satisfying Assumption 4.1 and a system $\mathcal{S}_2$. The diagram of this interconnection is shown in Figure 4.2. As has been mentioned earlier, feedback interconnections (4.34) are not explicitly defined in terms of the input–output mappings of $\mathcal{S}_1$ and $\mathcal{S}_2$, hence determining the realizability and completeness of the interconnection is generally an issue. The problem becomes even more challenging if merely input–state and input–output margins are known. Yet, in a number of cases, the realizability and completeness of feedback interconnections can be established relatively easily. These cases are specified in the theorem below.

Theorem 4.5 Existence of small gain, or the bottle-neck theorem. *Let $\mathcal{S}_1$ be a realizable (complete) system of which the input–output and input–state mappings $\mathcal{S}_{1,\mathcal{T}}$ and $\mathcal{H}_{1,\mathcal{T}}$ are defined as (4.27), and let $\mathcal{T} = [t_0, T]$ be the realizability interval of $\mathcal{S}_1$. Let us suppose that*

(1) for the system $\mathcal{S}_1$ there exists a majorizing mapping $\boldsymbol{\psi}(\mathbf{x}_1(t), \mathbf{e}_\psi)$ and a function $\mathbf{u}^*(\mathbf{e}_1, t) \in \mathcal{L}_u[t_0, T]$ satisfying Assumption 4.1.

Consider $\mathcal{S}_2$, (4.28), of which the input–state and input–output mappings are

$$\mathcal{S}_{2,T}: \mathcal{L}_{y_1}[t_0,T] \times \mathcal{E}_2 \mapsto \mathcal{L}_{x_2}[t_0,T],$$

$$\mathcal{H}_{2,T}: \mathcal{L}_{y_1}[t_0,T] \times \mathcal{E}_2 \mapsto \mathcal{L}_{\delta}[t_0,T],$$

and let system $\mathcal{S}_2$

(2) be realizable (complete) w.r.t. the norm $\|\cdot\|_{\mathcal{L}_{y_1},[t_0,T]}$

(3) and have an input–output margin $\gamma_{H_2,\mathcal{L}_\delta}$ that is globally bounded w.r.t. $\|\cdot\|_{\mathcal{L}_{y_1},[t_0,T]}$; i.e. there exists a continuous function $\gamma^*_{H_2,\mathcal{L}_\delta}(\mathbf{e}_2,T) : \mathcal{E}_2 \times \mathbb{R}_{\geq 0} \to \mathbb{R}_{\geq 0}$ such that the inequality

$$\gamma_{H_2,\mathcal{L}_\delta}(\mathbf{e}_2, \|\mathbf{y}_1(t)\|_{\mathcal{L}_{y_1},[t_0,T]}, T) \leq \gamma^*_{H_2,\mathcal{L}_\delta}(\mathbf{e}_2,T) \tag{4.48}$$

holds for all $\mathbf{y}_1(t) \in \mathcal{L}_{y_1}[t_0,T]$.

Then the feedback interconnection (4.34) of systems $\mathcal{S}_1$ and $\mathcal{S}_2$ is realizable (complete) for all $\mathbf{u}(t)$:

$$\mathbf{u}(t) = \mathbf{u}^*(\mathbf{e}_1,t) + \delta(t), \ \delta(t) \in \mathcal{L}_\delta[t_0,T]. \tag{4.49}$$

Moreover, if the functions $\gamma_{P,\mathcal{L}_\psi}(\cdot)$ and $\gamma^*_{H_2,\mathcal{L}_\delta}(\cdot)$ are bounded in T,

$$\sup_{T\geq t_0} \gamma_{P,\mathcal{L}_\psi}(\mathbf{e}_1,\mathbf{e}_\psi,d,T) = \Delta_P(\mathbf{e}_1,\mathbf{e}_\psi,d),$$

$$\sup_{T\geq t_0} \gamma^*_{H_2,\mathcal{L}_\delta}(\mathbf{e}_2,T) = \Delta_C(\mathbf{e}_2),$$

then $\mathbf{x}_1(t)$ and $\mathbf{y}_1(t)$ are bounded, and hence the following estimates hold:

$$\begin{aligned} \|\mathbf{x}_1(t)\|_{\infty,[t_0,T]} &\leq \mu_{S_1,\infty}(\mathbf{e}_1, \Delta_P(\mathbf{e}_1,\mathbf{e}_\psi,\Delta_C(\mathbf{e}_2) + \|\delta(t)\|_{\mathcal{L}_\delta,[t_0,T]})),\\ \|\mathbf{y}_1(t)\|_{\infty,[t_0,T]} &\leq \mu_{H_1,\infty}(\mathbf{e}_1, \Delta_P(\mathbf{e}_1,\mathbf{e}_\psi,\Delta_C(\mathbf{e}_2) + \|\delta(t)\|_{\mathcal{L}_\delta,[t_0,T]})). \end{aligned} \tag{4.50}$$

According to Theorem 4.5, which we will refer to as the *existence of small gain theorem* or the *bottle-neck theorem*, realizability (completeness) of the feedback interconnections of systems $\mathcal{S}_1$ and $\mathcal{S}_2$ with locally bounded input–state and input–output mappings follows automatically if an input–output margin $\gamma_{H_2,\mathcal{L}_\delta}$ corresponding to the input–output mapping $\mathcal{L}_{y_1}[t_0,T]$ in $\mathcal{L}_\delta[t_0,T]$ is a bounded function of $\|\mathbf{y}_1(t)\|_{\mathcal{L}_{y_1},[t_0,T]}$ for all $\mathbf{y}_1(t) \in \mathcal{L}_{y_1}[t_0,T]$. In contrast to the standard small-gain results (see e.g. Theorem 4.2), Theorem 4.5 does not require the existence and availability of input–output gains γ_1 and γ_2 for $\mathcal{H}_1$ and $\mathcal{H}_2$, and does not require that the product $\gamma_1\gamma_2 < 1$. Instead, the mere *existence* of an input–output margin of mapping (4.48) that is bounded w.r.t. $\|\mathbf{y}_1(t)\|_{\mathcal{L}_{y_1},[t_0,T]}$ is required. The

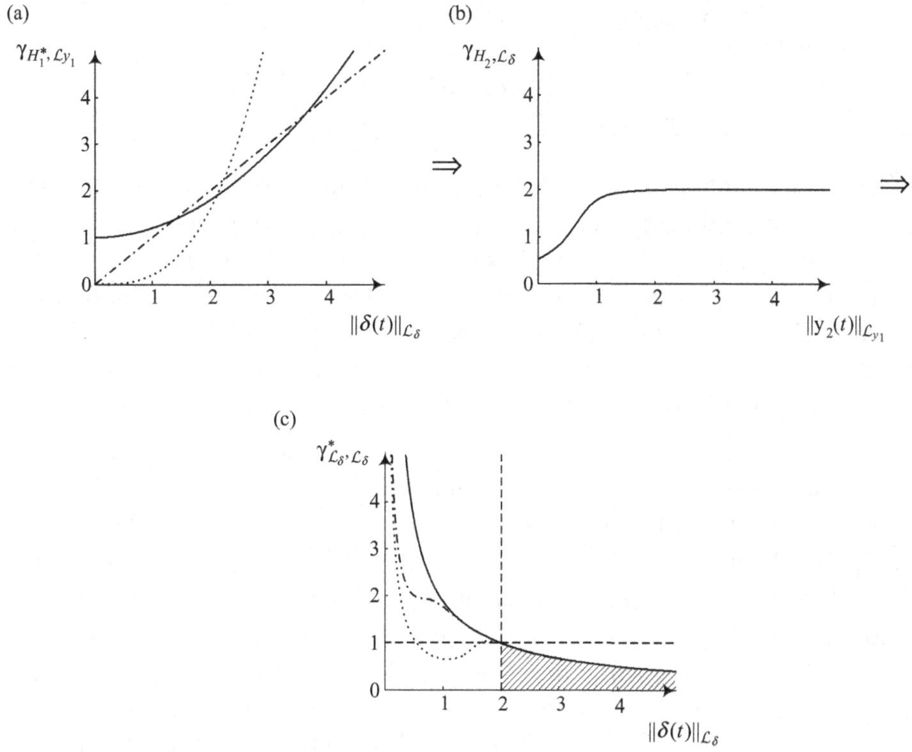

Figure 4.3 Illustration of the conditions of Theorem 4.5.

input–output gains for both systems,

$$\gamma_{\mathcal{L}_{y_1},\mathcal{L}_\delta} = \sup_{\delta(t)\in\mathcal{L}_\delta[t_0,T]} \frac{\|\mathcal{H}_{1,T}(\mathbf{u}^*(\mathbf{e},t)+\delta(t),\mathbf{e}_1)\|_{\mathcal{L}_{y_1},[t_0,T]}}{\|\delta(t)\|_{\mathcal{L}_\delta[t_0,T]}},$$

$$\gamma_{\mathcal{L}_\delta,\mathcal{L}_{y_1}} = \sup_{\mathbf{y}_1(t)\in\mathcal{L}_{y_1}[t_0,T]} \frac{\|\mathcal{H}_{2,T}(\mathbf{y}_1(t),\mathbf{e}_2)\|_{\mathcal{L}_\delta,[t_0,T]}}{\|\mathbf{y}_1(t)\|_{\mathcal{L}_{y_1}[t_0,T]}}, \tag{4.51}$$

$$\gamma_{\mathcal{L}_\delta,\mathcal{L}_\delta} = \sup_{\delta(t)\in\mathcal{L}_\delta[t_0,T]} \frac{\|\mathcal{H}_{2,T}(\mathcal{H}_{1,T}(\mathbf{u}^*(\mathbf{e}_1,t)+\delta(t),\mathbf{e}_1),\mathbf{e}_2)\|_{\mathcal{L}_\delta,[t_0,T]}}{\|\delta(t)\|_{\mathcal{L}_\delta,[t_0,T]}},$$

are allowed to be undefined for certain inputs (see Figure 4.3(c)), grow unboundedly (Figure 4.3(a)), and do not necessarily satisfy the smallness condition $\gamma_{\mathcal{L}_\delta,\mathcal{L}_\delta} = \gamma_{\mathcal{L}_{y_1},\mathcal{L}_\delta} \cdot \gamma_{\mathcal{L}_\delta,\mathcal{L}_{y_1}} < 1$ for all $\mathbf{y}_1(t) \in \mathcal{L}_{y_1}[t_0,T]$ and $\delta(t) \in \mathcal{L}_\delta[t_0,T]$ (Figure 4.3(b)). Figure 4.3(a) depicts graphs of $\gamma_{H_1^*,\mathcal{L}_{y_1}}(\mathbf{e}_1, \|\delta(t)\|_{\mathcal{L}_\delta,[t_0,T]}, T)$ for three different values of $\mathbf{e}_1$. Figure 4.3(b) shows $\gamma_{H_2,\mathcal{L}_\delta}(\mathbf{e}_2, \|\mathbf{y}_1(t)\|_{\mathcal{L}_{y_1},[t_0,T]}, T)$, and Figure 4.3(c) depicts the corresponding input–output gains $\gamma^*_{\mathcal{L}_\delta,\mathcal{L}_\delta}$ at $\delta(t) \in \mathcal{L}_\delta[t_0, T]$:

$$\gamma^*_{\mathcal{L}_\delta,\mathcal{L}_\delta}(\|\delta(t)\|_{\mathcal{L}_\delta,[t_0,T]}, T) = \frac{\gamma_{H_2,\mathcal{L}_\delta}(\mathbf{e}_2, \gamma_{H_1^*,\mathcal{L}_{y_1}}(\mathbf{e}_1, \|\delta(t)\|_{\mathcal{L}_\delta,[t_0,T]}, T), T)}{\|\delta(t)\|_{\mathcal{L}_\delta,[t_0,T]}}.$$

Given that condition (4.48) holds, there exists a non-increasing function the graph of which is above the graph of $\gamma^*_{\mathcal{L}_\delta,\mathcal{L}_\delta}(\|\delta(t)\|_{\mathcal{L}_\delta,[t_0,T]}, T)$ for every $\mathbf{e}_1 \in \mathcal{E}_1$ and T fixed (see Figure 4.3(c), in which plots of $\gamma^*_{\mathcal{L}_\delta,\mathcal{L}_\delta}(\|\delta(t)\|_{\mathcal{L}_\delta,[t_0,T]}, T)$ corresponding to various values of $\mathbf{e}_1$ are shown in different styles). It is also clear that for every such $\mathbf{e}_1 \in \mathcal{E}_1$ there is $r(\mathbf{e}_1) \in \mathbb{R}_{\geq 0}$ such that

$$\gamma_{\mathcal{L}_\delta,\mathcal{L}_\delta}(r(\mathbf{e}_1)) = \sup_{\|\delta(t)\|_{\mathcal{L}_\delta,[t_0,T]} > r(\mathbf{e}_1)} \gamma^*_{\mathcal{L}_\delta,\mathcal{L}_\delta}(\|\delta(t)\|_{\mathcal{L}_\delta,[t_0,T]}, T) < 1 \tag{4.52}$$

(the leftmost boundary of the shaded domain in Figure 4.3(c)). Hence, for all $\mathbf{e}_1 \in \mathcal{E}_1$ there would exist $r(\mathbf{e}_1) \geq 0$, such that the loop gain $\gamma_{\mathcal{L}_\delta,\mathcal{L}_\delta}$ in (4.51) is strictly smaller than 1 for $\delta(t) \in \mathcal{L}_\delta[t_0, T]$ of which the norm $\|\delta(t)\|_{\mathcal{L}_\delta,[t_0,T]}$ is sufficiently large: $\|\delta(t)\|_{\mathcal{L}_\delta,[t_0,T]} > r(\mathbf{e}_1)$. If, in addition, the function $\gamma^*_{H_2,\mathcal{L}_\delta}(\mathbf{e}_2, T)$ in (4.48) is globally bounded in T then the loop gain in the feedback interconnection of $\mathcal{S}_1$ and $\mathcal{S}_2$ will be strictly smaller than 1 for all $T > t_0$ and $\|\delta(t)\|_{\mathcal{L}_\delta,[t_0,T]} > r(\mathbf{e}_1)$. This property is very similar to the standard small-gain condition. The difference is that in our case this condition is satisfied for signals $\delta(t)$ of which the norm is sufficiently large.

Notice also that, on denoting $\delta_e(t) = \delta(t) + \mathbf{y}_2(t)$ and using (4.52), one can derive the following estimate:

$$\begin{aligned}\|\mathbf{y}_2(t)\|_{\mathcal{L}_\delta,[t_0,T]} &\leq \gamma_{\mathcal{L}_\delta,\mathcal{L}_\delta}(r(\mathbf{e}_1))\|\delta_e(t)\|_{\mathcal{L}_\delta,[t_0,T]} \\ &\leq \gamma_{\mathcal{L}_\delta,\mathcal{L}_\delta}(r(\mathbf{e}_1))\|\mathbf{y}_2(t)\|_{\mathcal{L}_\delta,[t_0,T]} + \gamma_{\mathcal{L}_\delta,\mathcal{L}_\delta}(r(\mathbf{e}_1))\|\delta(t)\|_{\mathcal{L}_\delta,[t_0,T]} \Rightarrow \\ \|\mathbf{y}_2(t)\|_{\mathcal{L}_\delta,[t_0,T]} &\leq \frac{\gamma_{\mathcal{L}_\delta,\mathcal{L}_\delta}(r(\mathbf{e}_1))}{1 - \gamma_{\mathcal{L}_\delta,\mathcal{L}_\delta}(r(\mathbf{e}_1))}\|\delta(t)\|_{\mathcal{L}_\delta,[t_0,T]}. \end{aligned} \tag{4.53}$$

Inequality (4.53), in turn, can be referred to as *input–output stability "for large inputs,"* where the term "large inputs" refers to $\delta : \|\delta(t)\|_{\mathcal{L}_\delta,[t_0,T]} > r(\mathbf{e}_1)$.[4]

The term input–output stability for large inputs introduced above complements the notion of *input–output stability in small* (Khalil (2002), Definition 5.2, p. 201):

$$\|\mathbf{y}_2(t)\|_{\mathcal{L}_\delta,[t_0,T]} \leq \alpha_r(\|\delta(t)\|_{\mathcal{L}_\delta,[t_0,T]}) + \beta,$$

where $\alpha_r \in \mathcal{K}$ is defined for $\|\delta(t)\|_{\mathcal{L}_\delta,[t_0,T]} \leq r$. Indeed, input–output stability in small assumes that all singularities and resonances occur outside a bounded domain

[4] See also Definition 4.2.1, in which the notion of input–output stability is specified.

$\mathcal{D}_0(r) \subseteq \mathcal{L}_\delta[t_0, T] : \|\delta(t)\|_{\mathcal{L}_\delta,[t_0,T]} \leq r$ in $\mathcal{L}_\delta[t_0, T]$. Input–output stability for large inputs, (4.53), however, implies that singularities and resonances can occur only for those inputs δ of which the amplitude is sufficiently small. For every $\mathbf{e}_1$ the domain within which these singularities and resonances occur is bounded by the sphere $\mathcal{D}_0(r(\mathbf{e}_1))$.

In the next section we will use the results of the analysis of realizability and completeness of serial, parallel, and feedback interconnections summarized in Theorems 4.4 and 4.5 for stating and solving the problem of input–output (functional) adaptive regulation.

4.3.2 Functional synthesis of adaptive systems: the separation principle

In order to be able to make a step forward from the analysis of interconnections of systems of which the models bear a degree of uncertainty to developing basic principles of organization of systems of which the goals include adaptive regulation we need to specify the notions of a controlled or regulated process (object) and of a controlling or regulating process (controller), and the goals and objectives of regulation/control, including a performance measure. We will suppose that an object to be controlled can be defined as the following system $\mathcal{S}_\mathrm{p}$:

$$\begin{aligned} \mathcal{S}_{\mathrm{p},T} &: \mathcal{L}_u[t_0, T] \times \mathcal{E}_\mathrm{p} \mapsto \mathcal{L}_{x_\mathrm{p}}[t_0, T], \\ \mathcal{H}_{\mathrm{p},T} &: \mathcal{L}_u[t_0, T] \times \mathcal{E}_\mathrm{p} \mapsto \mathcal{L}_{y_\mathrm{p}}[t_0, T], \\ \mathcal{P}_{\mathrm{p},T} &: \mathcal{L}_{x_\mathrm{p}}[t_0, T] \times \mathcal{E}_\mathrm{p} \times \mathcal{E}_\psi \mapsto \mathcal{L}_{\psi_\mathrm{p}}[t_0, T]. \end{aligned} \tag{4.54}$$

Let the space of inputs be denoted as $\mathcal{L}_u[t_0, T]$, and let $\mathcal{U}^* \subseteq \mathcal{L}_u[t_0, T]$ be the set of *admissible* inputs. We allow that the inputs $\mathbf{u}^*$ can be functions of $\mathbf{x}_\mathrm{p}(t)$ and $\mathbf{e}_\mathrm{p}$ as long as these functions are defined. Keeping this dependence in mind, we write:

$$\mathbf{u}^*(t) = \mathbf{u}^*(\mathbf{e}_\mathrm{p}, \mathbf{x}_\mathrm{p}(t), t).$$

Notice that $\mathcal{L}_u[t_0, T]$ is a linear space. Hence, an element $\mathbf{u}(t) \in \mathcal{L}_u[t_0, T]$ can always be represented as the following sum:

$$\mathbf{u}(t) = \mathbf{u}^*(\mathbf{e}_\mathrm{p}, \mathbf{x}_\mathrm{p}(t), t) + \delta(t), \quad \delta(t) \in \mathcal{L}_\delta[t_0, T] \subseteq \mathcal{L}_u[t_0, T], \tag{4.55}$$

where $\mathbf{u}(t) \in \mathcal{U}^*$ is given and $\mathcal{L}_\delta[t_0, T]$ is a subspace of $\mathcal{L}_u[t_0, T]$.[5] The decomposition (4.55) allows us to view the original system (4.54) as a new system $\mathcal{S}_\mathrm{p}^*$ of which the inputs are from $\mathcal{L}_\delta[t_0, T] \subseteq \mathcal{L}_u[t_0, T]$. "Modified" input–output and

[5] A trivial example of $\mathcal{L}_\delta[t_0, T]$ is $\mathcal{L}_u[t_0, T]$.

input–state mappings of (4.54) can be obtained by replacing $\mathbf{u}(t)$ in (4.54) with $\mathbf{u}^*(\mathbf{e}_\mathrm{p}, \mathbf{x}_\mathrm{p}(t), t) + \delta(t)$, $\delta(t) \in \mathcal{L}_\delta[t_0, T] \subseteq \mathcal{L}_u[t_0, T]$:

$$\begin{aligned} \mathcal{S}^*_{\mathrm{p},\mathcal{T}}(\delta, \mathbf{e}) &= \mathcal{S}_{\mathrm{p},\mathcal{T}}(\mathbf{u}^* + \delta, \mathbf{e}_\mathrm{p}), \\ \mathcal{H}^*_{\mathrm{p},\mathcal{T}}(\delta, \mathbf{e}) &= \mathcal{H}_{\mathrm{p},\mathcal{T}}(\mathbf{u}^* + \delta, \mathbf{e}_\mathrm{p}), \\ \mathcal{P}^*_{\mathrm{p},\mathcal{T}}(\delta, \mathbf{e}, \mathbf{e}_\psi) &= \mathcal{P}_{\mathrm{p},\mathcal{T}}(\mathbf{u}^* + \delta, \mathbf{e}_\mathrm{p}, \mathbf{e}_\psi). \end{aligned} \tag{4.56}$$

A minimal control objective at this stage is to ensure that the state $\mathbf{x}_\mathrm{p}(t)$ and output $\mathbf{y}_\mathrm{p}(t)$ of $\mathcal{S}_\mathrm{p}$ are defined for all $t \in [t_0, T]$. If $\mathbf{u}^*$, $\mathbf{x}_\mathrm{p}$ and $\mathbf{e}_\mathrm{p}$ were known then this goal could easily be attained by choosing the input appropriately. The problem, however, is that this is not always the case, and hence a procedure for feeding the system $\mathcal{S}_\mathrm{p}$ with admissible inputs is needed.

A controller is a system $\mathcal{S}_\mathrm{c}$,

$$\begin{aligned} \mathcal{S}_{\mathrm{c},\mathcal{T}} &: \mathcal{L}_{y_\mathrm{p}}[t_0, T] \times \mathcal{E}_\mathrm{c} \mapsto \mathcal{L}_{x_\mathrm{c}}[t_0, T], \\ \mathcal{H}_{\mathrm{c},\mathcal{T}} &: \mathcal{L}_{y_\mathrm{p}}[t_0, T] \times \mathcal{E}_\mathrm{c} \mapsto \mathcal{L}_{y_\mathrm{c}}[t_0, T] \subseteq \mathcal{L}_u[t_0, T], \end{aligned} \tag{4.57}$$

producing the estimates of $\mathbf{y}_\mathrm{c}(t) \in \mathcal{L}_{y_\mathrm{c}}[t_0, T] \subseteq \mathcal{L}_u[t_0, T]$, $\mathbf{u}^*(\mathbf{e}_\mathrm{p}, \mathbf{x}_\mathrm{p}(t), t) \in \mathcal{U}^*$ from $\mathbf{y}_\mathrm{p}(t) \in \mathcal{L}_{y_\mathrm{p}}[t_0, T]$.

Given that $\mathbf{u}^*(\mathbf{e}_\mathrm{p}, \mathbf{x}(t), t) \in \mathcal{U}^*$ depends on $\mathbf{x}_\mathrm{p}(t)$, $\mathbf{e}_\mathrm{p} \in \mathcal{E}_\mathrm{p}$, and that these functions might not be available, reconstruction of relevant information about the values of $\mathbf{x}_\mathrm{p}(t)$ and $\mathbf{e}_\mathrm{p}$ is necessary. For this purpose we introduce an additional observer system $\mathcal{S}_\mathrm{o}$,

$$\begin{aligned} \mathcal{S}_{\mathrm{o},\mathcal{T}} &: \mathcal{L}_{y_\mathrm{p}}[t_0, T] \times \mathcal{E}_\mathrm{o} \mapsto \mathcal{L}_{x_\mathrm{o}}[t_0, T], \\ \mathcal{H}_{\mathrm{o},\mathcal{T}} &: \mathcal{L}_{y_\mathrm{p}}[t_0, T] \times \mathcal{E}_\mathrm{o} \mapsto \mathcal{L}_{y_\mathrm{o}}[t_0, T], \end{aligned} \tag{4.58}$$

an *adaptation* system $\mathcal{S}_\mathrm{a}$,

$$\begin{aligned} \mathcal{S}_{\mathrm{a},\mathcal{T}} &: \mathcal{L}_{y_\mathrm{p}}[t_0, T] \oplus \mathcal{L}_{y_\mathrm{o}}[t_0, T] \times \mathcal{E}_\mathrm{a} \mapsto \mathcal{L}_{x_\mathrm{a}}[t_0, T], \\ \mathcal{H}_{\mathrm{a},\mathcal{T}} &: \mathcal{L}_{y_\mathrm{p}}[t_0, T] \oplus \mathcal{L}_{y_\mathrm{o}}[t_0, T] \times \mathcal{E}_\mathrm{a} \mapsto \mathcal{L}_{y_\mathrm{a}}[t_0, T], \end{aligned} \tag{4.59}$$

and the functions

$$\begin{aligned} \mathbf{u}_\mathrm{o} &: \mathcal{E}_\mathrm{p} \times \mathcal{L}_{y_\mathrm{o}}[t_0, T] \times \mathbb{R}_{\geq 0} \to \mathcal{L}_u[t_0, T], \\ \mathbf{u}_\mathrm{a} &: \mathcal{L}_{y_\mathrm{a}}[t_0, T] \times \mathcal{L}_{y_\mathrm{o}}[t_0, T] \times \mathbb{R}_{\geq 0} \to \mathcal{L}_u[t_0, T], \end{aligned}$$

where $\mathbf{y}_\mathrm{o}(t)$ is the estimate of $\mathbf{x}_\mathrm{p}(t)$ at $\mathbf{e}_\mathrm{p}$ as a function(al) of $\mathbf{y}_\mathrm{p}$, and $\mathbf{y}_\mathrm{a}(t)$ is the estimate of $\mathbf{e}_\mathrm{p}$ as a function(al) of $\mathbf{y}_\mathrm{p}(t)$ and $\mathbf{y}_\mathrm{o}(t)$. Notice that the function $\mathbf{u}_\mathrm{o}(\mathbf{e}_\mathrm{p}, \mathbf{y}_\mathrm{o}(t), t)$ can be thought of as an estimate of $\mathbf{u}^*(\mathbf{e}_\mathrm{p}, \mathbf{x}_\mathrm{p}(t), t)$ from the values of $\mathbf{y}_\mathrm{p}(t)$; knowledge of the values of $\mathbf{e}_\mathrm{p}$ is assumed. The function $\mathbf{u}_\mathrm{a}(\mathbf{y}_\mathrm{a}(t), \mathbf{y}_\mathrm{o}(t), t)$, in turn, is an estimate of $\mathbf{u}_o(\mathbf{e}_p, \mathbf{y}_o(t), t)$ from just $\mathbf{y}_p(t)$ and $\mathbf{y}_o(t)$. Thus the function $\mathbf{u}_a(\mathbf{y}_a(t), \mathbf{y}_o(t), t)$ is an estimate of $\mathbf{u}^*(\mathbf{e}_p, \mathbf{x}_p(t), t)$ from $\mathbf{y}_p(t)$. We will therefore assign that $\mathbf{u}_a(\mathbf{y}_a(t), \mathbf{y}_o(t), t)$ is the output $\mathbf{y}_c(t)$ of the controller $\mathcal{S}_c$.

In general, the performance of a system is measured in terms of how far the system's behavior is from the desired one. Since the desired behavior in our case is mere boundedness of state and outputs, accessing the performance of the system in terms of the closeness of the actual inputs to $\mathcal{S}_p$ to the desired ones is a plausible option. In particular, the distances separating $\mathbf{u}_o(\mathbf{e}_p, \mathbf{y}_o(t), t)$, $\mathbf{u}_a(\mathbf{y}_a(t), \mathbf{y}_o(t), t)$, and $\mathbf{u}^*(\mathbf{e}_p, \mathbf{x}_p(t), t)$ in $\mathcal{L}_\delta[t_0, T]$ can be chosen as indicators of the overall performance of an adaptive system:

$$\mathcal{J}_x[t_0, T] = \|\mathbf{u}^*(\mathbf{e}_p, \mathbf{x}_p(t), t) - \mathbf{u}_o(\mathbf{e}_p, \mathbf{y}_o(t), t)\|_{\mathcal{L}_\delta, [t_0, T]}, \tag{4.60}$$

$$\mathcal{J}_e[t_0, T] = \|\mathbf{u}_o(\mathbf{e}_p, \mathbf{y}_o(t), t) - \mathbf{u}_a(\mathbf{y}_a(t), \mathbf{y}_o(t), t)\|_{\mathcal{L}_\delta, [t_0, T]}. \tag{4.61}$$

Taking these notations into account, the problem of functional synthesis of an adaptive system (or adaptive controller) can be stated as follows.

Problem 4.4 *Functional synthesis of adaptive controllers.* Let system $\mathcal{S}_p$ be realizable (complete) with input–output and input–state mappings defined as (4.54). Determine the functions $\mathbf{u}^*(\mathbf{e}_p, \mathbf{x}_p(t), t)$ and systems $\mathcal{S}_o$ and $\mathcal{S}_a$ in (4.58) and (4.59) such that for all $\mathbf{e}_p \in \mathcal{E}_p$

(1) the feedback interconnection of system $\mathcal{S}_p^*$ (4.56) and $\mathcal{S}_o$ and $\mathcal{S}_a$ is realizable (complete);
(2) the states and outputs of $\mathcal{S}_p$ and the controller are bounded.
In addition, determine conditions when
(3) $\mathcal{J}_x$ and $\mathcal{J}_e$, (4.60) and (4.61), and $\|\mathbf{x}_p(t)\|_{\mathcal{L}_{xp}, [t_0, T]}$ and $\|\mathbf{y}_p(t)\|_{\mathcal{L}_{yp}, [t_0, T]}$ can be estimated from the above as functions of $\mathbf{e}_p$, $\mathbf{e}_o$, and $\mathbf{e}_a$.[6]

Sufficient conditions specifying classes of systems $\mathcal{S}_o$ and $\mathcal{S}_a$ for which this problem is solvable follow from the next theorem.

Theorem 4.6 *Consider system (4.54), of which the input–output and input–state properties are defined by mappings $\mathcal{S}_{p,\mathcal{T}}$, $\mathcal{H}_{p,\mathcal{T}}$, and $\mathcal{P}_{p,\mathcal{T}}$. Suppose that*

(1) there is an input (4.55) such that the system (4.56) satisfies Assumption 4.1;
(2) there are realizable (complete) systems $\mathcal{S}_o$ (4.58) and $\mathcal{S}_a$ (4.59) with input–state and input–output mappings $\mathcal{S}_{o,\mathcal{T}}$, $\mathcal{H}_{o,\mathcal{T}}$ and $\mathcal{S}_{a,\mathcal{T}}$, $\mathcal{H}_{a,\mathcal{T}}$, respectively;
(3) the values of $\mathcal{J}_x[t_0, T]$ and $\mathcal{J}_e[t_0, T]$ defined by (4.60) and (4.61) are bounded,

$$\mathcal{J}_x \le \Delta_{J_x}, \qquad \mathcal{J}_e \le \Delta_{J_e}. \tag{4.62}$$

[6] This requirement will be needed for comparing behaviors of the system in various environments.

Then the feedback interconnection of system $\mathcal{S}_{\mathrm{p}}$ with $\mathcal{S}_{\mathrm{c}}$,

$$\begin{aligned}\mathcal{S}_{\mathrm{c},\mathcal{T}}(\mathbf{y}_{\mathrm{p}},\mathbf{e}_{\mathrm{o}}\oplus\mathbf{e}_{\mathrm{a}}) &= \mathcal{S}_{\mathrm{a},\mathcal{T}}(\mathbf{y}_{\mathrm{p}},\mathcal{H}_{\mathrm{o},\mathcal{T}}(\mathbf{y}_{\mathrm{p}},\mathbf{e}_{\mathrm{o}}),\mathbf{e}_{\mathrm{a}})\oplus\mathcal{S}_{\mathrm{o}}(\mathbf{y}_{\mathrm{p}},\mathbf{e}_{\mathrm{o}}),\\ \mathcal{H}_{\mathrm{c},\mathcal{T}}(\mathbf{y}_{\mathrm{p}},\mathbf{e}_{\mathrm{o}}0\oplus\mathbf{e}_{\mathrm{a}}) &= \mathbf{u}_{\mathrm{a}}(\mathcal{H}_{\mathrm{a}}(\mathbf{y}_{\mathrm{p}},\mathcal{H}_{\mathrm{o}}(\mathbf{y}_{\mathrm{p}},\mathbf{e}_{\mathrm{o}}),\mathbf{e}_{\mathrm{a}}),\mathcal{H}_{\mathrm{o}}(\mathbf{y}_{\mathrm{p}},\mathbf{e}_{\mathrm{o}}),t),\end{aligned}\tag{4.63}$$

is realizable (complete) for all $\delta(t)\in\mathcal{L}_{\delta}[t_0,T]$. In addition, if the function $\gamma_{P,\mathcal{L}_{\psi}}(\cdot)$ is bounded w.r.t. T,

$$\sup_{T\geq t_0}\gamma_{P,\mathcal{L}_{\psi}}(\mathbf{e}_{\mathrm{p}},\mathbf{e}_{\psi},d,T)=\Delta_P(\mathbf{e}_{\mathrm{p}},\mathbf{e}_{\psi},d),\tag{4.64}$$

then $\mathbf{x}_{\mathrm{p}}(t)$ is bounded:

$$\|\mathbf{x}_{\mathrm{p}}(t)\|_{\infty,[t_0,T]}\leq\mu_{S_{\mathrm{p}},\infty}(\mathbf{e}_{\mathrm{p}},\Delta_P(\mathbf{e}_{\mathrm{p}},\mathbf{e}_{\psi},\Delta_{J_x}+\Delta_{J_e}+\|\delta(t)\|_{\mathcal{L}_{\delta},[t_0,T]})).\tag{4.65}$$

Theorem 4.6 specifies conditions ensuring the solvability of task (1) of Problem 4.4. The solution to task (2) follows automatically if systems $\mathcal{S}_{\mathrm{a}}$ and $\mathcal{S}_{\mathrm{o}}$ share the bounded-input–bounded-output property. Indeed, according to Theorem 4.6, condition (4.64) implies that $\mathbf{x}_{\mathrm{p}}(t)$ is bounded. Obtaining solutions to task (3) clearly requires the availability of estimates of the input–output mappings of $\mathcal{S}_{\mathrm{o}}$ and $\mathcal{S}_{\mathrm{a}}$; this information, however, is often available for specific systems.

According to Theorem 4.6, the synthesis problem can be split into the following sequence of elementary tasks:

(1) determine the class of functions $\mathbf{u}^*(\mathbf{e}_{\mathrm{p}},\mathbf{x}_{\mathrm{p}}(t),t)$ (for $\mathbf{x}_{\mathrm{p}}$ and $\mathbf{e}_{\mathrm{p}}$ known) transforming $\mathcal{S}_{\mathrm{p}}$ into system (4.56) satisfying Assumption 4.1;
(2) derive system $\mathcal{S}_{\mathrm{o}}$ (observer) ensuring that $\mathcal{J}_x[t_0,T]$ is bounded for all $\mathbf{u}^*\in\mathcal{U}^*$ (the values of $\mathbf{e}_{\mathrm{p}}$ are assumed to be known);
(3) derive system $\mathcal{S}_{\mathrm{a}}$ (adaptation to $\mathbf{e}_{\mathrm{p}}$) ensuring that $\mathcal{J}_e[t_0,T]$ is bounded for all $\mathbf{y}_{\mathrm{o}}\in\mathcal{L}_{y_{\mathrm{o}}}$.

Notice that the conditions of realizability and completeness of the overall system are not required to depend on any knowledge of how solutions to the particular problems (1)–(3) above are obtained. For example, solution of (1) does not depend on the outcomes of tasks (2) and (3). One needs just to find $\mathbf{u}^*\in\mathcal{U}$ satisfying Assumption 4.1. On the other hand, solving tasks (2) and (3) is equivalent to ensuring that (4.62) holds. If these tasks are solved for classes of $\mathbf{u}^*\in\mathcal{U}$ and $\mathbf{y}_{\mathrm{o}}\in\mathcal{L}_{y_{\mathrm{o}}}$ then the overall solution does not depend on the particular choice of functions $\mathbf{u}^*$ and $\mathbf{y}_{\mathrm{o}}$. Thus the theorem can be viewed as a *separation principle* too.

4.4 Asymptotic properties of systems with locally bounded input–output and input–state mappings

In the previous sections of this chapter we considered analysis and, partially, synthesis problems of systems and their interconnections. Input–state and input–output mappings of these systems were assumed to be locally bounded. Theorems 4.4, 4.5, and 4.6 establish conditions ensuring that serial, parallel, and feedback interconnections of systems with locally bounded mappings are also systems of which the input–state and input–output mappings are defined; these mappings are also locally bounded. These results provide us with solutions to Problems 4.1 and 4.2.

Finding a solution to Problem 4.3 not only involves showing that the system's state is bounded but also requires specification of domains to which the state belongs (e.g. (4.36)). Rough estimates of these domains can be derived immediately from (4.65) or (4.50), provided that the upper bounds for $\|\delta(t)\|_{\mathcal{L}_\delta,[t_0,\infty]}$ are known and that the functions $\mu_{S,\infty}(\cdot)$, $\mu_{H,\infty}(\cdot)$, and $\mu_{S_{\mathrm{p}},\infty}(\cdot)$ are available together with the bounds for $\mathbf{e}_{\mathrm{p}}$ and $\mathcal{J}_x[t_0,\infty]$ and $\mathcal{J}_e[t_0,\infty]$. More accurate estimates can be obtained by using the notions of invariant (see Definition 2.1.1) and limit sets (Birkhoff 1927; Guckenheimer and Holmes 2002).

Definition 4.4.1 Consider a flow $\mathbf{x}(t,\mathbf{x}_0,t_0)$, $\mathbf{x}_0 \in \mathbb{R}^n$, $t \in \mathbb{R}$. A point $p \in \mathbb{R}^n$ is called an ω-limit point of $\mathbf{x}_0$ w.r.t. the flow $\mathbf{x}(t,\mathbf{x}_0,t_0)$, iff there exists an infinite sequence of $t_1 < t_2 < \cdots < t_i < \ldots$, $\lim_{i\to\infty} t_i = \infty$ such that

$$\lim_{i\to\infty} \mathbf{x}(t_i,\mathbf{x}_0,t_0) = p.$$

The set of all ω-limit points p of $\mathbf{x}_0$ is called the ω-limit set of $\mathbf{x}_0$.

The literature on the analysis of limit sets of (dynamical) systems is huge. We do not wish to provide a review of these results here, but rather concentrate on the simple case of an autonomous system

$$\dot{\mathbf{x}} = \mathbf{f}(\mathbf{x}), \tag{4.66}$$

in which the function $\mathbf{f}(\cdot)$ is locally Lipschitz. Fundamental properties of limit sets of these systems are formulated in the lemma below (see Birkhoff (1927), p. 198, or Khalil (2002), p. 127, Lemma 4.1).

Lemma 4.1 *Suppose that solutions of (4.66) are bounded for all* $\mathbf{x}_0 \in \mathcal{D}$ *in the interval* $[0,\infty)$. *Then the set* $\Omega(\mathcal{D}) = \bigcup_{\mathbf{x}_0\in\mathcal{D}} \omega(\mathbf{x}_0)$ *is closed and invariant, and*

$$\lim_{t\to\infty} \|\mathbf{x}(t,\mathbf{x}_0,t_0)\|_{\Omega(\mathcal{D})} = 0.$$

The following corollary follows immediately from Lemma 4.1 and Theorem 4.6.

Corollary 4.1 *Let $\mathcal{S}_\mathrm{p}$ and $\mathcal{S}_\mathrm{c}$ be systems satisfying the conditions of Theorem 4.6. In addition, suppose that the input–state and input–output mappings of systems $\mathcal{S}_\mathrm{a}$ and $\mathcal{S}_\mathrm{o}$ are locally bounded and that $\delta(t) \equiv 0$. Moreover, let the feedback interconnection be described by (4.66) in the domain specified by (4.65).*[7] *Then the state of the interconnection converges asymptotically to the maximal invariant set contained in*

$$\begin{aligned} \|\mathbf{x}_\mathrm{p}\| &\leq \max_{\mathbf{e}_\mathrm{p}\in\mathcal{E}_\mathrm{p}, d\leq M} \mu_{\mathcal{S}_\mathrm{p},\infty}(\mathbf{e}_\mathrm{p}, d), \\ M &= \max_{\mathbf{e}_\mathrm{p}\in\mathcal{E}_\mathrm{p}, \mathbf{e}_\psi\in\mathcal{E}_\psi, d\leq\Delta_{J_x}+\Delta_{J_e}} \Delta_P(\mathbf{e}_\mathrm{p}, \mathbf{e}_\psi, d). \end{aligned} \tag{4.67}$$

In order to apply the corollary, one needs to make sure that the description of the overall system (including the process to be controlled or regulated, the controllers, and the environment) as a system of ordinary differential equations is plausible. If this is the case, then the next key assumption is knowledge of the system's invariant sets. In some cases the allocation and topology of these sets can be easily estimated (e.g. equilibria and periodic orbits of systems in the plane). In general, however, finding the invariant sets of a dynamical system is a rather difficult and technically involved task. In addition to the genuine theoretical difficulty of this task, in the context of adaptive control and regulation it gains additional complications. The first is the presence of uncertainties in any system in which adaptation is needed. The second complication is that the dimension of the overall system (and hence the difficulty of finding all of the invariant sets) increases substantially due to the need to incorporate the state vectors of the controllers into the extended state space of the system.

A possible remedy that would facilitate the finding of a solution to this problem is to derive mechanisms and schemes of adaptation that do not induce new undesired invariant sets in the extended state space. The ultimate goal, of course, is that the invariant sets of the extended system correspond to the system's target motions.

In conclusion, we would like to emphasize the differences between the estimates (4.67) which follow from the input–state and input–output analysis of systems and the standard estimates of asymptotic behavior adopted in the theory of adaptive control. These differences are illustrated with Figure 4.4. Our conditions for realizability, completeness, and state boundedness involved majorizing mappings $\boldsymbol{\psi}(\mathbf{x}, \mathbf{e}_\psi)$. For simplicity, consider the case in which $\boldsymbol{\psi} : \mathbb{R}^n \to \mathbb{R}$ is a scalar and

[7] In this case $\mathbf{x}$ is the generalized state vector combining the states of the controlled process, controller, and variables of the environment.

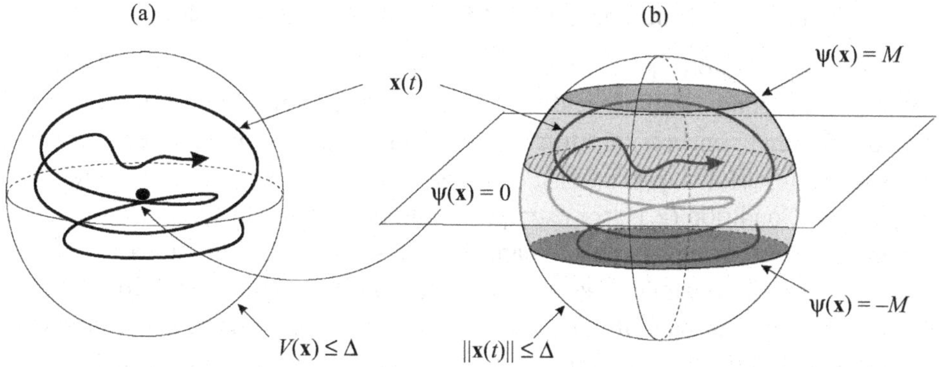

Figure 4.4 (a) Conventional Lyapunov-function-based estimates. (b) Estimates following from (4.67).

differentiable function that does not depend on $\mathbf{e}_\psi$. The set $\Omega_\psi = \{\mathbf{x} \in \mathbb{R}^n | \boldsymbol{\psi}(\mathbf{x}) = 0\}$ defines a target set in $\mathbb{R}^n$. In standard formulations of adaptive regulation and control problems the target sets are defined as sets on which a known positive definite and radially unbounded function $V : \mathbb{R}^n \to \mathbb{R}_{\geq 0}$ is zero (Figure 4.4(a)). This condition implies that these target sets need to be known a priori. If these sets are to be modified then the function V may need to be altered too, and likewise the adaptation mechanisms. In the case of target-set assignment based on the input–state and input–output analysis these restrictions may be relaxed. Indeed, the boundedness of the system's state follows from Theorems 4.5 and 4.6 (the sphere $\Omega_\mathbf{x} = \{\mathbf{x} \in \mathbb{R}^n | \mathbf{x} : \|\mathbf{x}(t)\| \leq \Delta\}$ in Figure 4.4(b)). On the other hand, $\mathbf{x}(t) \in \Omega_\psi^M = \{\mathbf{x} \in \mathbb{R}^n | \mathbf{x} : |\psi(\mathbf{x})| \leq M\}$. Thus, the trajectory $\mathbf{x}(t, \mathbf{x}_0, t_0, \boldsymbol{\theta}, \mathbf{u}(t))$ will necessarily be in $\Omega_\mathbf{x} \cap \Omega_\psi^M$ for all $t \geq t_0$ (the shaded domain in Figure 4.4(b)). If, in addition, the extended system can be described by (4.66), then its solutions will have to converge to the maximal invariant set in $\Omega_\mathbf{x} \cap \Omega_\psi^M$. Knowledge of this set is not required a priori. If, however, $\psi(\mathbf{x}) \in C^1$ and it is known that $\psi \to 0$ as $t \to \infty$, then one can easily see that the solutions will converge to the maximal invariant set in Ω_ψ.

Figure 4.4 also illustrates the differences between the possible target-set assignments in the realm of input–state and input–output approaches and those for approaches based on geometric representations (Astolfi and Ortega 2003). Results based on coordinate transformation around the target manifold (5.4) are applicable only in a subset of $\mathbb{R}^n$ where $\psi(\mathbf{x}, t)$ does not depend explicitly on t and the rank of $\psi(\mathbf{x}, t)$ is constant. In this respect these results are local. Theorems 4.5 and 4.6 do not require constant-rank conditions and allow time-varying $\psi(\mathbf{x}, t)$. They may therefore replace the conventional ones for systems with non-stationary target dynamics or ones that are far away from the target manifolds.

4.5 Asymptotic properties of a class of unstable systems

Analysis of asymptotic properties of dynamical systems is often based on the notions of Lyapunov stability and attracting sets (see Definition 2.1.3). Although the conventional concepts of attracting sets and Lyapunov stability are a powerful tandem in various applications, some problems cannot be solved within this framework. Condition (2.1), for example, could be violated in systems with intermittent, itinerant, or meta-stable dynamics. In general the condition does not hold when the system's dynamics, loosely speaking, is exploring rather than contracting. Such systems appear naturally in the context of global optimization. For instance, in Shang and Wah (1996) finding the global minimum of a differentiable cost function $Q : \mathbb{R}^n \rightarrow \mathbb{R}_{\geq 0}$ in a bounded subset $\Omega_x \subset \mathbb{R}^n$ is achieved by splitting the search procedure into a locally attracting gradient $\mathcal{S}_{\mathrm{a}}$, and a wandering part $\mathcal{S}_{\mathrm{w}}$:

$$\begin{aligned} \mathcal{S}_{\mathrm{a}} &: \dot{\mathbf{x}} = -\mu_x \frac{\partial Q(\mathbf{x})}{\partial \mathbf{x}} + \mu_t T(t), \ \mu_x, \mu_t \in \mathbb{R}_{\geq 0}, \\ \mathcal{S}_{\mathrm{w}} &: T(t) = h\{t, \mathbf{x}(t)\}, \ h : \mathbb{R}_{\geq 0} \times L^n_\infty[t_0, t] \rightarrow L^n_\infty[t_0, t]. \end{aligned} \tag{4.68}$$

The trace function, $T(t)$, in (4.68) is supposed to cover (i.e. be dense in) the whole searching domain Ω_x. Even though the results in Shang and Wah (1996) are purely simulation studies, they illustrate the superior performance of algorithms (4.68) compared with standard local minimizers and classical methods of global optimization in a variety of benchmark problems. Abandoning Lyapunov stability is likewise advantageous in problems of identification and adaptation in the presence of general nonlinear parametrization (Tyukin and van Leeuwen 2005), in maneuvering and path searching (Suemitsu and Nara 2004), and in decision making in intelligent systems (van Leeuwen and Raffone 2001; van Leeuwen *et al.* 2000). Systems with attracting, yet unstable, invariant sets are relevant for modeling complex behavior in biological and physical systems (Ashwin and Timme 2005). Last but not least, Lyapunov-unstable attracting sets are relevant in problems of synchronization (Bischi *et al.* 1998; Ott and Sommerer 1994; Timme *et al.* 2002).[8]

Even when it is appropriate to consider a system to be stable, we may be limited in our success in meeting the requirement to identify a proper Lyapunov function. This is the case, for instance, when the system's dynamics is only partially known. Trading stability requirements for the sake of convergence might be a possible

[8] See also Pogromsky *et al.* (2003), where the striking difference between stable and "almost stable" synchronization in terms of the coupling strengths for a pair of Lorentz oscillators is demonstrated analytically.

remedy. Known results in this direction can be found in Ilchman (1997) and Pomet (1992).[9]

In all the cases which are problematic under condition (2.1) of Definition 2.1.3, condition (2.2), namely the convergence of $\mathbf{x}(t, \mathbf{x}_0)$ to an invariant set $\mathcal{A}$, is still a requirement that has to be met. In order to treat these cases analytically we shall, first of all, move from the standard concept of attracting sets in Definition 2.1.3 to one that does not assume that the basin of attraction is necessarily a neighborhood of the invariant set $\mathcal{A}$. In other words, we shall allow convergence that is not uniform in initial conditions. This requirement is captured by the concept of weak, or Milnor, attraction (Milnor 1985) (Definition 2.1.4).

Conventional methods such as La Salle's invariance principle (La Salle 1976) or center-manifold theory (Carr 1981) can, in principle, address the issue of convergence to weak equilibria. They do so, however, at the expense of requiring detailed knowledge of the vector fields of the ordinary differential equations of the model. When such information is not available, the system can be thought of as a mere interconnection of input–output maps. Small-gain theorems (Zames 1966; Jiang *et al.* 1994) are usually efficient in this case. These results, however, apply only under the assumption of stability of each component in the interconnection. If stability is not explicitly required, as is the case for Theorem 4.5, global boundedness of input–output mappings is needed.

In this section we will provide a result that would allow us to analyze interconnections of systems that are neither necessarily stable nor required to have globally bounded input–output mappings. The systems we consider here, however, are of a rather special class. The object of our study is a class of systems that can be decomposed into an attracting, or stable, component $\mathcal{S}_{\mathrm{a}}$ and an exploratory, generally unstable, part $\mathcal{S}_{\mathrm{w}}$. Typical systems of this class are nonlinear systems in cascaded form,

$$\begin{aligned} \mathcal{S}_{\mathrm{a}} &: \ \dot{\mathbf{x}} = \mathbf{f}(\mathbf{x}, \mathbf{z}), \\ \mathcal{S}_{\mathrm{w}} &: \ \dot{\mathbf{z}} = \mathbf{q}(\mathbf{z}, \mathbf{x}), \end{aligned} \tag{4.69}$$

where the zero solution of the $\mathbf{x}$-subsystem is asymptotically stable in the absence of input $\mathbf{z}$, and the state of the $\mathbf{z}$-subsystem is a function of the integral $\int_{t_0}^{t} \|\mathbf{x}(\tau)\| d\tau$. Even when both subsystems in (4.69) are stable and the $\mathbf{x}$-subsystem does not depend on state $\mathbf{z}$, the cascade can still be unstable (Arcak *et al.* 2002). We show, however, that for unstable interconnections (4.69), under certain conditions that involve only input-to-state properties of $\mathcal{S}_{\mathrm{a}}$ and $\mathcal{S}_{\mathrm{w}}$, there is a set $\mathcal{V}$ in the system's

[9] In Chapter 8 we demonstrate how explorative dynamics can solve the problem of simultaneous state and parameter observation for a system that cannot be transformed into a canonical adaptive-observer form (Bastin and Gevers 1988).

state space such that trajectories starting in $\mathcal{V}$ remain bounded. The result is formally stated in Theorem 4.7. When an additional measure of invariance is defined for $\mathcal{S}_a$ (in our case a steady-state characteristic), a weak, Milnor attracting set emerges. Its location is completely determined by the zeros of the steady-state response of system $\mathcal{S}_a$.

Consider a system that can be decomposed into two interconnected subsystems, $\mathcal{S}_a$ and $\mathcal{S}_w$:

$$\begin{aligned} \mathcal{S}_a &: (u_a, \mathbf{x}_0) \mapsto \mathbf{x}(t), \\ \mathcal{S}_w &: (u_w, \mathbf{z}_0) \mapsto \mathbf{z}(t), \end{aligned} \tag{4.70}$$

where $u_a \in \mathcal{U}_a \subseteq L_\infty[t_0, \infty]$ and $u_w \in \mathcal{U}_w \subseteq L_\infty[t_0, \infty]$ are the spaces of inputs to $\mathcal{S}_a$ and $\mathcal{S}_w$, respectively, $\mathbf{x}_0 \in \mathbb{R}^n$ and $\mathbf{z}_0 \in \mathbb{R}^m$ represent the initial conditions, and $\mathbf{x}(t) \in \mathcal{X} \subseteq L^n_\infty[t_0, \infty]$ and $\mathbf{z}(t) \in \mathcal{Z} \subseteq L^m_\infty[t_0, \infty]$ are the system states.

System $\mathcal{S}_a$ represents the contracting dynamics. More precisely, we require that $\mathcal{S}_a$ is input-to-state stable[10] (Sontag 1990) with respect to a compact set $\mathcal{A}$.

Assumption 4.2

$$\begin{aligned} \mathcal{S}_a : \quad & \|\mathbf{x}(t)\|_{\mathcal{A}} \le \beta(\|\mathbf{x}(t_0)\|_{\mathcal{A}}, t - t_0) + c\|u_a(t)\|_{\infty,[t_0,t]}, \\ & \forall\, t_0 \in \mathbb{R}_{\ge 0},\ t \ge t_0, \end{aligned} \tag{4.71}$$

where the function $\beta(\cdot,\cdot) \in \mathcal{KL}$ and $c > 0$ is some positive constant.

The function $\beta(\cdot,\cdot)$ in (4.71) specifies the contraction property of the unperturbed dynamics of $\mathcal{S}_a$. In other words, it models the rate at which the system forgets its initial conditions $\mathbf{x}_0$, if left unperturbed. Propagation of the input to output is estimated in terms of a continuous mapping, $c\|u_a(t)\|_{\infty,[t_0,t]}$, which, in our case, is chosen for simplicity to be linear. Notice that this mapping should not necessarily be contracting. In what follows we will assume that the function $\beta(\cdot,\cdot)$ and constant c are known or can be estimated a priori.

For systems $\mathcal{S}_a$ of which a model is given by a system of ordinary differential equations

$$\dot{\mathbf{x}} = \mathbf{f}_x(\mathbf{x}, u_a),\ \mathbf{f}_x(\cdot,\cdot) \in \mathcal{C}^1, \tag{4.72}$$

Assumption 4.2 is equivalent, for instance, to the combination of the following properties:[11]

[10] In general, as will be demonstrated with examples, our analysis can be carried out for (integral) input-to-output/state stable systems as well.

[11] For a comprehensive characterization of the input-to-state stability and detailed mathematical arguments we refer the reader to the paper by E. D. Sontag and Y. Wang (Sontag and Wang 1996).

(1) let $u_a(t) \equiv 0$ for all t, then set $\mathcal{A}$ is Lyapunov stable and globally attracting for (4.72);
(2) for all $u_a \in \mathcal{U}_a$ and $\mathbf{x}_0 \in \mathbb{R}^n$ there exists a non-decreasing function $\kappa : \mathbb{R}_{\geq 0} \rightarrow \mathbb{R}_{\geq 0} : \kappa(0) = 0$ such that

$$\inf_{t \in [0,\infty)} \|\mathbf{x}(t)\|_{\mathcal{A}} \leq \kappa\left(\|u_a(t)\|_{\infty,[t_0,\infty]}\right).$$

The system $\mathcal{S}_w$ stands for the searching or wandering dynamics. We will consider $\mathcal{S}_w$ subject to the following conditions.

Assumption 4.3 *The system $\mathcal{S}_w$ is forward-complete:*

$$u_w(t) \in \mathcal{U}_w \Rightarrow \mathbf{z}(t) \in \mathcal{Z}, \ \forall\, t \geq t_0, \ t_0 \in \mathbb{R}_{\geq 0},$$

and there exists an "output" function $h : \mathbb{R}^m \rightarrow \mathbb{R}$ and two "bounding" functions, $\gamma_0 \in \mathcal{K}_{\infty,e}$ and $\gamma_1 \in \mathcal{K}_{\infty,e}$, such that the following integral inequality holds:

$$\mathcal{S}_w : \int_{t_0}^{t} \gamma_1(u_w(\tau))d\tau \leq h(\mathbf{z}(t_0)) - h(\mathbf{z}(t)) \leq \int_{t_0}^{t} \gamma_0(u_w(\tau))d\tau,$$
$$\forall\, t \geq t_0, \ t_0 \in \mathbb{R}_{\geq 0}. \tag{4.73}$$

When system $\mathcal{S}_w$ is specified in terms of vector fields

$$\dot{\mathbf{z}} = \mathbf{f}_z(\mathbf{z}, u_w), \ \mathbf{f}_z(\cdot,\cdot) \in \mathcal{C}^1. \tag{4.74}$$

Assumption 4.3 can be viewed, for example, as postulating the existence of a function $h : \mathbb{R}^m \rightarrow \mathbb{R}_{\geq 0}$ of which the evolution in time is a mere integration of the input $u_w(t)$. In general, for $u_w : u_w(t) \geq 0 \ \forall\, t \in \mathbb{R}_{\geq 0}$, inequality (4.73) implies *monotonicity* of function $h(\mathbf{z}(t))$ in t. Regarding the function $\gamma_0(\cdot)$ in (4.73), we assume that for any $M \in \mathbb{R}_{\geq 0}$ there exists a function $\gamma_{0,1} : \mathbb{R}_{\geq 0} \rightarrow \mathbb{R}_{\geq 0}$ and a *non-decreasing* function $\gamma_{0,2} : \mathbb{R}_{\geq 0} \rightarrow \mathbb{R}_{\geq 0}$ such that

$$\gamma_0(a \cdot b) \leq \gamma_{0,1}(a) \cdot \gamma_{0,2}(b), \quad \forall\, a, b \in [0, M]. \tag{4.75}$$

Requirement (4.75) is a technical assumption that will be used in the formulation and proof of the main results which follow. Yet, it is not too restrictive; it holds, for instance, for a wide class of locally Lipschitz functions $\gamma_0(\cdot) : \gamma_0(a \cdot b) \leq L_0(M) \cdot (a \cdot b)$, $L_0(M) \in \mathbb{R}_{\geq 0}$. Another example for which the assumption holds is the class of polynomial functions $\gamma_0(\cdot) : \gamma_0(a \cdot b) = (a \cdot b)^p = a^p \cdot b^p$, $p > 0$. No further restrictions will be imposed a priori on $\mathcal{S}_a$, $\mathcal{S}_w$.

Now consider the interconnection of (4.71) and (4.73) with coupling $u_a(t) = h(\mathbf{z}(t))$, and $u_s(t) = \|\mathbf{x}(t)\|_{\mathcal{A}}$. The equations for the combined system can be

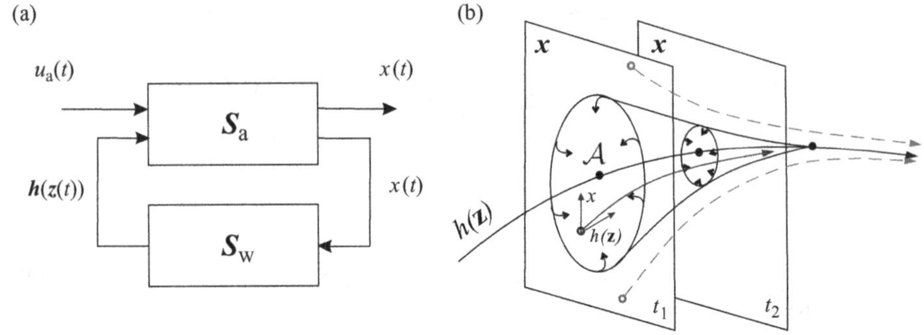

Figure 4.5 (a) The class of interconnected systems S_a and S_w. System $\mathcal{S}_a$, the "contracting system," has an attracting invariant set $\mathcal{A}$ in its state space, whereas system $\mathcal{S}_w$ does not necessarily have an attracting set. It represents the "wandering" dynamics. A typical example of such behavior is the dynamics of the flow in a neighborhood of a saddle point in three-dimensional space (b).

written as:

$$\begin{aligned} \|\mathbf{x}(t)\|_{\mathcal{A}} &\le \beta(\|\mathbf{x}(t_0)\|_{\mathcal{A}}, t - t_0) + c\|h(\mathbf{z}(t))\|_{\infty,[t_0,t]}, \\ \int_{t_0}^{t} \gamma_1(\|\mathbf{x}(\tau)\|_{\mathcal{A}})d\tau &\le h(\mathbf{z}(t_0)) - h(\mathbf{z}(t)) \le \int_{t_0}^{t} \gamma_0(\|\mathbf{x}(\tau)\|_{\mathcal{A}})d\tau. \end{aligned} \tag{4.76}$$

A diagram illustrating the general structure of the entire system (4.76) is given in Figure 4.5.

Equations (4.76) capture the relevant interplay between contracting, $\mathcal{S}_a$, and wandering, $\mathcal{S}_w$, dynamics inherent to a variety of searching strategies in the realm of optimization, (4.68), and interconnections (4.69) in general systems theory. In addition, this kind of interconnection describes the behavior of systems that undergo transcritical or saddle-node bifurcations.

Example 4.5.1 Consider for instance the following system:

$$\begin{aligned} \dot{x}_1 &= -x_1 + x_2, \\ \dot{x}_2 &= \varepsilon + \gamma x_1^2, \quad \gamma > 0, \end{aligned} \tag{4.77}$$

where the parameter ε varies from negative to positive values. At $\varepsilon = 0$ stable and unstable equilibria collide, leading to the cascade satisfying equations (4.76). An alternative bifurcation scenario could be represented by the system

$$\begin{aligned} \dot{x}_1 &= -x_1 + x_2, \\ \dot{x}_2 &= \varepsilon + \gamma x_2^2, \quad \gamma > 0. \end{aligned} \tag{4.78}$$

In this case, however, the dynamics of the variable x_2 is *independent* of x_1, and the analysis of the asymptotic behavior of (4.78) reduces to the analysis of each equation separately. Thus systems like (4.78) are easier to deal with than (4.77). This constitutes an additional motivation for the present approach.

When analyzing the asymptotic behavior of interconnection (4.76) we will address the following set of questions. Is there a set (a weak trapping set in the system's state space) such that the trajectories which start in this set are bounded? It is natural to expect that the existence of such a set depends on the specific functions $\gamma_0(\cdot)$ and $\gamma_1(\cdot)$ in (4.76), on the properties of $\beta(\cdot,\cdot)$, and on the values of c. If such a set exists and could be defined, the next questions are, therefore, where will the trajectories converge and how can these domains be characterized?

We formulate conditions ensuring that there exists a point $\mathbf{x}_0 \oplus \mathbf{z}_0$ such that the *ω-limit set* of $\mathbf{x}_0 \oplus \mathbf{z}_0$[12] is bounded in the following sense:

$$\|\omega_{\mathbf{x}}(\mathbf{x}_0 \oplus \mathbf{z}_0)\|_{\mathcal{A}} < \infty, \quad |h(\omega_{\mathbf{z}}(\mathbf{x}_0 \oplus \mathbf{z}_0))| < \infty. \tag{4.79}$$

These conditions and also a specification of the set Ω_γ of points $\mathbf{x}' \oplus \mathbf{z}'$ for which the ω-limit set satisfies property (4.79) are provided in Theorem 4.7.

In order to verify whether an attracting set exists in $\omega(\Omega_\gamma)$ that is a subset of Ω_γ, we use an additional characterization of the contracting system $\mathcal{S}_a$. In particular, we introduce the intuitively clear notion of the input-to-state *steady-state characteristics*[13] of a system. It is possible to show that, when system $\mathcal{S}_a$ has a steady-state characteristic, there exists an attracting set in $\omega(\Omega_\gamma)$ and this set is uniquely defined by the zeros of the steady-state characteristics of $\mathcal{S}_a$. A diagram illustrating the steps of our analysis is provided in Figure 4.6, together with the sequence of conditions leading to the emergence of the attracting set in (4.76).

4.5.1 Small-gain theorems for the analysis of non-uniform convergence

Before we formulate the main results of this subsection, let us first comment briefly on the machinery of our analysis. First of all, we introduce three sequences

$$\mathcal{S} = \{\sigma_i\}_{i=0}^{\infty}, \ \sigma_i \in \mathbb{R}_{\geq 0},$$
$$\Xi = \{\xi_i\}_{i=0}^{\infty}, \ \xi_i \in \mathbb{R}_{\geq 0},$$
$$\mathcal{T} = \{\tau_i\}_{i=0}^{\infty}, \ \tau_i \in \mathbb{R}_{\geq 0}.$$

12 Recall that in our current notation a point $\mathbf{p} \in \mathbb{R}^{m+n}$ is an ω-limit point of $\mathbf{x}' \oplus \mathbf{z}'$ if there exists a sequence $\{t_i\}$, $i = 1, 2, \ldots$ such that $\lim_{i\to\infty} t_i = \infty$ and $\lim_{t_i\to\infty} \mathbf{x}(t_i, \mathbf{x}' \oplus \mathbf{z}') \oplus \mathbf{z}(t_i, \mathbf{x}' \oplus \mathbf{z}') = \mathbf{p}$, where $\mathbf{x}(t, \mathbf{x}' \oplus \mathbf{z}') \oplus \mathbf{z}(t, \mathbf{x}' \oplus \mathbf{z}')$ denotes the flow of interconnection (4.76). A set of all ω-limit points of $\mathbf{x}' \oplus \mathbf{z}'$ is an ω-limit set of $\mathbf{x}' \oplus \mathbf{z}'$.

13 A more precise definition of the steady-state characteristics is given in Section 4.5.2.

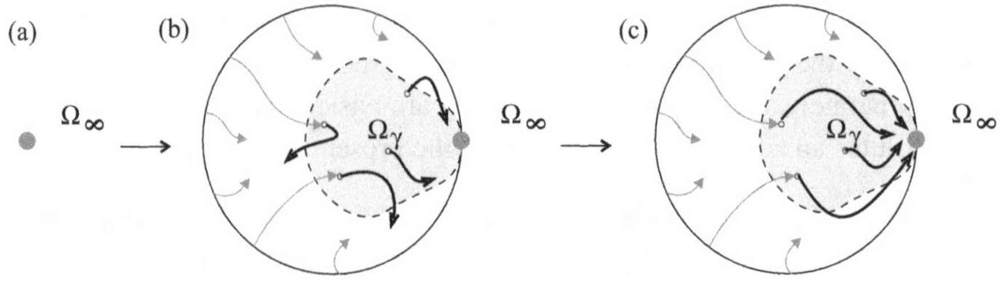

Figure 4.6 The emergence of a weak (Milnor) attracting set Ω_∞. In (a) the target invariant set Ω_∞ is depicted as a filled circle. First (Theorem 4.7), we investigate whether a domain $\Omega_\gamma \subset \mathbb{R}^n \times \mathbb{R}^m$ exists such that $\|\mathbf{x}(t)\|_{\mathcal{A}}$ and $h(\mathbf{z}(t))$ are bounded for all $\mathbf{x}_0 \oplus \mathbf{z}_0 \in \Omega_\gamma$. In the text we refer to this set as a weak trapping region or simply a trapping region. The trapping region is shown as a gray domain in (b) In principle, the system's states can eventually leave the domain Ω_γ. They must, however, satisfy (4.79), ensuring boundedness of $\|\mathbf{x}(t)\|_{\mathcal{A}}$ and $h(\mathbf{z}(t))$. As a result they will dwell within the region shown as a circle in (b). Notice that neither this domain nor the previous one need necessarily be neighborhoods of Ω_∞. Second (Lemmas 4.2 and 4.3, Corollary 4.2), we show in (c) the conditions which lead to the emergence of a weak attracting set in the trapping region Ω_γ.

The first sequence, $\mathcal{S}$, partitions the interval $[0, h(\mathbf{z}_0)]$, $h(\mathbf{z}_0) > 0$ into the union of shrinking subintervals H_i:

$$[0, h(\mathbf{z}_0)] = \cup_{i=0}^{\infty} H_i, \;\; H_i = [\sigma_{i+1} h(\mathbf{z}_0), \sigma_i h(\mathbf{z}_0)]. \tag{4.80}$$

For the sake of transparency, let us define this property formally in the form of Property 4.1.

Property 4.1 *The sequence $\mathcal{S}$ is strictly monotone and converging,*

$$\{\sigma_n\}_{n=0}^{\infty} : \lim_{n\to\infty} \sigma_n = 0, \; \sigma_0 = 1. \tag{4.81}$$

Sequences Ξ and $\mathcal{T}$ will specify the desired rates $\xi_i \in \Xi$ of the contracting dynamics (4.71) in terms of the function $\beta(\cdot,\cdot)$ and $\tau_i \in \mathcal{T}$. Let us, therefore, impose the following constraint on the choice of Ξ and $\mathcal{T}$.

Property 4.2 *Sequences Ξ and $\mathcal{T}$ are such that for the given function $\beta(\cdot,\cdot) \in \mathcal{KL}$ in (4.71) the following inequality holds:*

$$\beta(\cdot, T) \le \xi_i \beta(\cdot, 0), \; \forall\, T \ge \tau_i. \tag{4.82}$$

Property 4.2 states that, for the given, yet arbitrary, factor ξ_i and time instant t_0, the amount of time τ_i is sufficient for the state $\mathbf{x}$ to reach the domain:

$$\|\mathbf{x}\|_{\mathcal{A}} \le \xi_i \beta(\|\mathbf{x}(t_0)\|_{\mathcal{A}}, 0).$$

in the absence of u_a.

In order to specify the desired convergence rates ξ_i, it will be necessary to define another measure in addition to (4.82). This is a measure of the propagation of initial conditions $\mathbf{x}_0$ and input $h(\mathbf{z}_0)$ to the state $\mathbf{x}(t)$ of the contracting dynamics (4.71) when the system travels in $h(\mathbf{z}(t)) \in [0, h(\mathbf{z}_0)]$. For this reason we introduce two systems of functions, Φ and Υ:

$$\Phi: \begin{array}{l} \phi_j(s) = \phi_{j-1} \circ \rho_{\phi,j}(\xi_{i-j} \cdot \beta(s,0)), \; j = 1, \dots, i, \\ \phi_0(s) = \beta(s,0), \end{array} \tag{4.83}$$

$$\Upsilon: \begin{array}{l} \upsilon_j(s) = \phi_{j-1} \circ \rho_{\upsilon,j}(s), \; j = 1, \dots, i, \\ \upsilon_0(s) = \beta(s,0), \end{array} \tag{4.84}$$

where the functions $\rho_{\phi,j}$, $\rho_{\upsilon,j} \in \mathcal{K}$ satisfy the following inequality:

$$\phi_{j-1}(a+b) \leq \phi_{j-1} \circ \rho_{\phi,j}(a) + \phi_{j-1} \circ \rho_{\upsilon,j}(b). \tag{4.85}$$

Notice that in the case $\beta(\cdot,0) \in \mathcal{K}_\infty$ the functions $\rho_{\phi,j}(\cdot)$ and $\rho_{\upsilon,j}(\cdot)$ will always exist (Jiang *et al.* 1994). The properties of the sequence Ξ which ensure the desired propagation rate of the influence of the initial condition $\mathbf{x}_0$ and input $h(\mathbf{z}_0)$ to the state $\mathbf{x}(t)$ are specified in Property 4.3.

Property 4.3 *The sequences*

$$\sigma_n^{-1} \cdot \phi_n(\|\mathbf{x}_0\|_\mathcal{A}), \;\; \sigma_n^{-1} \cdot \left(\sum_{i=0}^{n} \upsilon_i(c|h(\mathbf{z}_0)|\sigma_{n-i}) \right), \; n = 0, \dots, \infty$$

are bounded from above, i.e. there exist functions $B_1(\|\mathbf{x}_0\|)$ and $B_2(|h(\mathbf{z}_0)|, c)$ such that

$$\sigma_n^{-1} \cdot \phi_n(\|\mathbf{x}_0\|_\mathcal{A}) \leq B_1(\|\mathbf{x}_0\|_\mathcal{A}), \tag{4.86}$$

$$\sigma_n^{-1} \cdot \left(\sum_{i=0}^{n} \upsilon_i(c|h(\mathbf{z}_0)|\sigma_{n-i}) \right) \leq B_2(|h(\mathbf{z}_0)|, c) \tag{4.87}$$

for all $n = 0, 1, \dots, \infty$.

For a large class of functions $\beta(s,0)$, for instance those that are Lipschitz in s, these conditions are reduced to more transparent ones that can always be satisfied by an appropriate choice of sequences Ξ and $\mathcal{S}$. This case is considered in detail as a corollary of Theorem 4.7 in Section 4.5.3.

In order to prove the emergence of the trapping region, we consider the following collection of volumes induced by the sequence $\mathcal{S}_i$ and the corresponding partition (4.80) of the interval $[0, h(\mathbf{z}_0)]$:

$$\Omega_i = \{\mathbf{x} \in \mathcal{X}, \mathbf{z} \in \mathcal{Z} | h(\mathbf{z}(t)) \in H_i\}. \tag{4.88}$$

For the given initial conditions $\mathbf{x}_0 \in \mathcal{X}$ and $\mathbf{z}_0 \in \mathcal{Z}$ two alternative possibilities exist. First, there exists an i such that the trajectory $\mathbf{x}(t, \mathbf{x}_0) \oplus \mathbf{z}(t, \mathbf{z}_0)$ enters Ω_i and stays there forever. Hence, for $t \to \infty$ the state will converge into

$$\Omega_{\mathrm{a}} = \{\mathbf{x} \in \mathcal{X},\ \mathbf{z} \in \mathcal{Z} |\ \|\mathbf{x}\|_{\mathcal{A}} \leq c \cdot h(\mathbf{z}_0),\ \mathbf{z} :\ h(\mathbf{z}) \in [0, h(\mathbf{z}_0)]\}. \tag{4.89}$$

The second alternative is that for each $i = 0, 1, \dots$ the trajectory $\mathbf{x}(t, \mathbf{x}_0) \oplus \mathbf{z}(t, \mathbf{z}_0)$ enters Ω_i and leaves some time later. Let t_i be the time instants when it hits the hypersurfaces $h(\mathbf{z}(t)) = h(\mathbf{z}_0)\sigma_i$. Then the state of the coupled system stays in $\cup_{i=0}^{\infty} \Omega_i$ only if the sequence $\{t_i\}_{i=0}^{\infty}$ diverges. Theorem 4.7 provides sufficient conditions specifying the latter case in terms of the properties of sequences $\mathcal{S}$, Ξ, and $\mathcal{T}$ and the function $\gamma_0(\cdot)$ in (4.76). For a large class of interconnections (4.76) it is possible to formulate these conditions in terms of the input–output properties of systems $\mathcal{S}_{\mathrm{a}}$ and $\mathcal{S}_{\mathrm{w}}$ explicitly, i.e. in terms of functions $\beta(\cdot, \cdot)$ and $\gamma_0(\cdot)$ and the values of c. The results are presented as immediate corollaries of Theorem 4.7.

A diagram illustrating the main ideas of the proof is provided in Figure 4.7.

Theorem 4.7 *Let systems $\mathcal{S}_{\mathrm{a}}$ and $\mathcal{S}_{\mathrm{w}}$ be given and let them satisfy Assumptions 4.2 and 4.3. Consider interconnection (4.76) and suppose that there exist sequences $\mathcal{S}$, Ξ, and $\mathcal{T}$ satisfying Properties 4.1–4.3. In addition, suppose that the following conditions hold:*

(1) There exists a positive number $\Delta_0 > 0$ such that

$$\frac{1}{\tau_i} \frac{\sigma_i - \sigma_{i+1}}{\gamma_{0,1}(\sigma_i)} \geq \Delta_0 \quad \forall\, i = 0, 1, \dots, \infty. \tag{4.90}$$

(2) The set Ω_γ of all points $\mathbf{x}_0$ and $\mathbf{z}_0$ satisfying the inequality

$$\gamma_{0,2}(B_1(\|\mathbf{x}_0\|_{\mathcal{A}}) + B_2(|h(\mathbf{z}_0)|, c) + c|h(\mathbf{z}_0)|) \leq h(\mathbf{z}_0)\Delta_0 \tag{4.91}$$

is not empty.

(3) Partial sums of elements from $\mathcal{T}$ diverge:

$$\sum_{i=0}^{\infty} \tau_i = \infty. \tag{4.92}$$

Then for all $\mathbf{x}_0$, $\mathbf{z}_0 \in \Omega_\gamma$ the state $\mathbf{x}(t, \mathbf{z}_0) \oplus \mathbf{z}(t, \mathbf{z}_0)$ of system (4.76) converges into the set specified by (4.89),

$$\Omega_{\mathrm{a}} = \{\mathbf{x} \in \mathcal{X},\ \mathbf{z} \in \mathcal{Z} |\ \|\mathbf{x}\|_{\mathcal{A}} \leq c \cdot h(\mathbf{z}_0),\ \mathbf{z} :\ h(\mathbf{z}) \in [0, h(\mathbf{z}_0)]\}.$$

The major difference between the conditions of Theorem 4.7 and those of conventional small-gain theorems (Zames 1966; Jiang *et al.* 1994) is that the latter involve

<table>
<tr><th>Standard</th><th>Proposed</th></tr>
<tr><td>1) Domain of attraction is a neighborhood

2) Implies Lyapunov stability</td><td>1) Domain of attraction is a set of positive measure (not necessarily a neighborhood)

2) Allows one to analyze convergence in Lyapunov-unstable systems</td></tr>
<tr><td>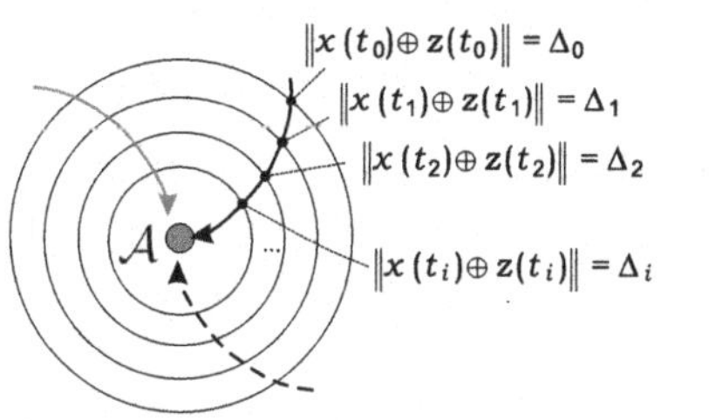
</td><td>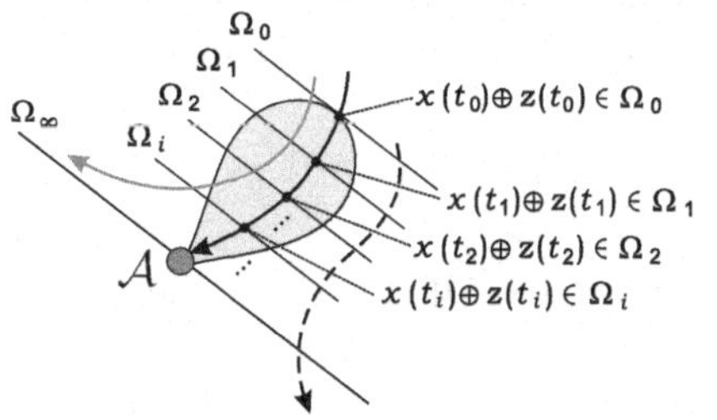
</td></tr>
<tr><td>Given: a sequence of diverging time instants t_i</td><td>Given: a sequence of sets Ω_i whose distance Δ_i to $\mathcal{A}$ is converging to zero</td></tr>
<tr><td>Prove: convergence of norms $\|\mathbf{x}(t_i) \oplus \mathbf{z}(t_i)\| = \Delta_i$ to zero</td><td>Prove: divergence of $\{t_i\}$, where t_i : $\mathbf{x}(t_i) \oplus \mathbf{z}(t_i) \in \Omega_i$</td></tr>
</table>

Figure 4.7 Key differences between the conventional concept of convergence (left panel) and the concept of weak, non-uniform, convergence (right panel). In the uniform case, trajectories that start in a neighborhood of $\mathcal{A}$ remain in a neighborhood of $\mathcal{A}$ (solid and dashed lines). In the non-uniform case, only a fraction of the initial conditions in a neighborhood of $\mathcal{A}$ will produce trajectories that remain in a neighborhood of $\mathcal{A}$ (solid bold line). In the most general case a necessary condition for this to happen is that the sequence $\{t_i\}$ diverges. In our current problem statement divergence of $\{t_i\}$ implies boundedness of $\|\mathbf{x}(t)\|_{\mathcal{A}}$. To show state boundedness and convergence of $\mathbf{x}(t)$ to $\mathcal{A}$, additional information on the system dynamics will be required.

only input–output or input–state mappings. Formulating conditions for state boundedness of the interconnection in terms of input–output or input–state mappings is possible in the traditional case because the interconnected systems are assumed to be input-to-state stable. Hence their internal dynamics can be neglected. In our case, however, the dynamics of $\mathcal{S}_w$ is generally unstable in the Lyapunov sense. Hence, in order to ensure boundedness of $\mathbf{x}(t, \mathbf{x}_0)$ and $h(\mathbf{z}(t, \mathbf{z}_0))$, the rate/degree of stability of $\mathcal{S}_a$ should be taken into account. Roughly speaking, system $\mathcal{S}_a$ should ensure a sufficiently high degree of contraction in $\mathbf{x}_0$ while the input–output response of $\mathcal{S}_w$ should be sufficiently small. The rate of contraction in $\mathbf{x}_0$ of $\mathcal{S}_a$, according to (4.71), is specified in terms of the function $\beta(\cdot,\cdot)$. Properties of this function that are relevant for convergence are explicitly accounted for in Property 4.3 and (4.92). The domain of admissible initial conditions and actually the small-gain condition (input–state–output properties of $\mathcal{S}_w$ and $\mathcal{S}_a$) are defined by (4.90) and (4.91), respectively. Notice also that Ω_γ is not necessarily a neighborhood of Ω_a, thus the convergence ensured by Theorem 4.7 is allowed to be non-uniform in $\mathbf{x}_0$ and $\mathbf{z}_0$.

In addition, notice that the theorem remains valid if instead of interconnection (4.76) the following is considered:

$$
\begin{aligned}
&\|\mathbf{x}(t)\|_{\mathcal{A}} \leq \beta(\|\mathbf{x}(t_0)\|_{\mathcal{A}}, t - t_0) + c\|h(\mathbf{z}(t))\|_{\infty,[t_0,t]} + \|\varepsilon(t)\|_{\infty,[t_0,t]},\\
&\int_{t_0}^{t} \gamma_1(\|\mathbf{x}(\tau)\|_{\mathcal{A}})d\tau \leq h(\mathbf{z}(t_0)) - h(\mathbf{z}(t)) \leq \int_{t_0}^{t} \gamma_0(\|\mathbf{x}(\tau)\|_{\mathcal{A}})d\tau,
\end{aligned}
\tag{4.93}
$$

where $\varepsilon(t)$ is an asymptotically decaying perturbation that satisfies

$$
|\varepsilon(t)| \leq M \cdot h(\mathbf{z}_0) \cdot \sigma_i, \; t \geq \sum_{j=0}^{i} \tau_i - \tau_0.
$$

In this case condition (4.91) transforms into

$$
\gamma_{0,2}(B_1(\|\mathbf{x}_0\|_{\mathcal{A}}) + B_2(|h(\mathbf{z}_0)|, c + M) + (c + M)|h(\mathbf{z}_0)|) \leq h(\mathbf{z}_0)\Delta_0. \tag{4.94}
$$

This enables one to apply Theorem 4.7 for systems of which the state is known up to an asymptotically decaying error.

4.5.2 Estimates of Milnor attracting sets in the system's state space

Even for interconnections of Lyapunov-stable systems, small-gain conditions usually are effective merely for establishing boundedness of states or outputs. Yet, even in the setting of Theorem 4.7, it is still possible to derive estimates (such as, for instance, (4.89)) of the domains to which the state will converge. These estimates, however, are often too conservative. If a more precise characterization of these domains is required, additional information on the dynamics of systems $\mathcal{S}_a$ and $\mathcal{S}_w$ will be needed. The question, therefore, is how detailed should this information be? It appears that some additional knowledge of the steady-state characteristics of system $\mathcal{S}_a$ is sufficient to improve the estimates (4.89) substantially.

Let us formally introduce the notion of a steady-state characteristic as follows.

Definition 4.5.1 We say that system (4.71) has steady-state characteristic $\chi : \mathbb{R} \to \mathcal{S}\{\mathbb{R}_{\geq 0}\}$ with respect to the norm $\|\mathbf{x}\|_{\mathcal{A}}$ if and only if for each constant $\bar{u}_a$ the following holds:

$$\forall\, u_a(t) \in \mathcal{U}_a : \lim_{t\to\infty} u_a(t) = \bar{u}_a \Rightarrow \lim_{t\to\infty} \|\mathbf{x}(t)\|_{\mathcal{A}} \in \chi(\bar{u}_a). \tag{4.95}$$

The key property captured by Definition 4.5.1, is that there exists a limit of $\|\mathbf{x}(t)\|_{\mathcal{A}}$ as $t \to \infty$, provided that the limit for $u_a(t)$, $t \to \infty$ is defined and constant. Notice that the mapping χ is set-valued. This means that for each $\bar{u}_a$ there is a set $\chi(\bar{u}_a) \subset \mathbb{R}_{\geq 0}$ such that $\|\mathbf{x}(t)\|_{\mathcal{A}}$ converges to an element of $\chi(\bar{u}_a)$ as $t \to \infty$. Therefore, our definition allows a fairly large amount of uncertainty for $\mathcal{S}_a$. It will be of essential importance, however, that such a characterization exists for the system $\mathcal{S}_a$.

Clearly, not every system obeys a steady-state characteristic $\chi(\cdot)$ of Definition 4.5.1. There are relatively simple systems for which the state does not converge even in the "norm" sense for constant converging inputs (condition (4.95)). In mechanics, physics, and biology such systems encompass the large class of nonlinear oscillators that can be excited by constant inputs. In order to take such systems into consideration, we introduce a weaker notion, that of a steady-state characteristic *on average*.

Definition 4.5.2 We say that system (4.71) has a steady-state characteristic on average $\chi_T : \mathbb{R} \to \mathcal{S}\{\mathbb{R}_{\geq 0}\}$ with respect to the norm $\|\mathbf{x}\|_{\mathcal{A}}$ if and only if for each constant $\bar{u}_a$ and some $T > 0$ the following holds:

$$\forall\, u_a(t) \in \mathcal{U}_a : \lim_{t\to\infty} u_a(t) = \bar{u}_a \Rightarrow \lim_{t\to\infty} \int_t^{t+T} \|\mathbf{x}(\tau)\|_{\mathcal{A}}\, d\tau \in \chi_T(\bar{u}_a). \tag{4.96}$$

Steady-state characterization of system $\mathcal{S}_a$ allows further specification of the asymptotic behavior of interconnection (4.76). These results are summarized in Lemmas 4.2 and 4.3 below.

Lemma 4.2 *Let system (4.76) be given and let $h(\mathbf{z}(t, \mathbf{z}_0))$ be bounded for some $\mathbf{x}_0$ and $\mathbf{z}_0$. Let, furthermore, system (4.71) have steady-state characteristic $\chi(\cdot) : \mathbb{R} \to \mathcal{S}\{\mathbb{R}_{\geq 0}\}$. Then the following limiting relations hold:*[14]

$$\lim_{t\to\infty} \|\mathbf{x}(t, \mathbf{x}_0)\|_{\mathcal{A}} = 0, \qquad \lim_{t\to\infty} h(\mathbf{z}(t, \mathbf{z}_0)) \in \chi^{-1}(0). \tag{4.97}$$

As follows from Lemma 4.2, in a case in which the steady-state characteristic of $\mathcal{S}_a$ is defined, the asymptotic behavior of interconnection (4.76) is characterized by

[14] The symbol $\chi^{-1}(0)$ in (4.97) denotes the set $\chi^{-1}(0) = \bigcup_{\bar{u}_a \in \mathbb{R}_{\geq 0}} \bar{u}_a : \chi(\bar{u}_a) \ni 0$.

the zeros of the steady-state mapping $\chi(\cdot)$. For the steady-state characteristics on average a slightly modified conclusion can be derived.

Lemma 4.3 *Let system (4.76) be given, let $h(\mathbf{z}(t,\mathbf{z}_0))$ be bounded for some $\mathbf{x}_0, \mathbf{z}_0$, $h(\mathbf{z}(t,\mathbf{z}_0)) \in [0, h(\mathbf{z}_0)]$, and let system (4.71) have steady-state characteristic $\chi_T(\cdot) : \mathbb{R} \to \mathcal{S}\{\mathbb{R}_{\geq 0}\}$ on average. Furthermore, let there exist a positive constant $\bar{\gamma}$ such that the function $\gamma_1(\cdot)$ in (4.73) satisfies the following constraint:*

$$\gamma_1(s) \geq \bar{\gamma} \cdot s, \quad \forall s \in [0, \bar{s}], \bar{s} \in \mathbb{R}_{\geq 0} : \ \bar{s} > c \cdot h(\mathbf{z}_0). \tag{4.98}$$

In addition, suppose that $\chi_T(\cdot)$ has no zeros in the positive domain, i.e. $0 \notin \chi_T(\bar{u}_{\mathrm{a}})$ for all $\bar{u}_{\mathrm{a}} > 0$. Then

$$\lim_{t\to\infty} \|\mathbf{x}(t,\mathbf{x}_0)\|_{\mathcal{A}} = 0, \qquad \lim_{t\to\infty} h(\mathbf{z}(t,\mathbf{z}_0)) = 0. \tag{4.99}$$

An immediate outcome of Lemmas 4.2 and 4.3 is that when the conditions of Theorem 4.7 are satisfied and system (4.71) has steady-state characteristic $\chi(\cdot)$ or $\chi_T(\cdot)$ the domain of convergence Ω_{a} becomes

$$\Omega_{\mathrm{a}} = \{\mathbf{x} \in \mathcal{X},\ \mathbf{z} \in \mathcal{Z} |\ \|\mathbf{x}\|_{\mathcal{A}} = 0,\ \mathbf{z} :\ h(\mathbf{z}) \in [0, h(\mathbf{z}_0)]\}. \tag{4.100}$$

It is possible, however, to improve estimate (4.100) further under additional hypotheses on the dynamics of systems $\mathcal{S}_{\mathrm{a}}$ and $\mathcal{S}_{\mathrm{w}}$. This result is formulated in the corollary below.

Corollary 4.2 *Let system (4.76) be given and let it satisfy the assumptions of Theorem 4.7. In addition,*

(C1) let the flow $\mathbf{x}(t,\mathbf{x}_0) \oplus \mathbf{z}(t,\mathbf{z}_0)$ be generated by a system of autonomous differential equations with locally Lipschitz right-hand side;
(C2) let subsystem $\mathcal{S}_{\mathrm{w}}$ be practically integral-input-to-state stable:

$$\|\mathbf{z}(\tau)\|_{\infty,[t_0,t]} \leq C_z + \int_0^t \gamma_1(u_{\mathrm{w}}(\tau))d\tau, \tag{4.101}$$

and let the function $h(\cdot) \in C^0$ in (4.73);
(C3) let system $\mathcal{S}_{\mathrm{a}}$ have steady-state characteristic $\chi(\cdot)$.

Then for all $\mathbf{x}_0, \mathbf{z}_0 \in \Omega_\gamma$ the state of the interconnection converges to the set

$$\Omega_{\mathrm{a}} = \{\mathbf{x} \in \mathcal{X}, \mathbf{z} \in \mathcal{Z} |\ \|\mathbf{x}\|_{\mathcal{A}} = 0,\ h(\mathbf{z}) \in \chi^{-1}(0)\}. \tag{4.102}$$

As follows from Corollary 4.2, the zeros of the steady-state characteristic of system $\mathcal{S}_{\mathrm{a}}$ actually "control" the domains to which the state of interconnection (4.76) might potentially converge. This is illustrated in Figure 4.8. Notice also that in this case condition (C3) in Corollary 4.2 is replaced with the following alternative:

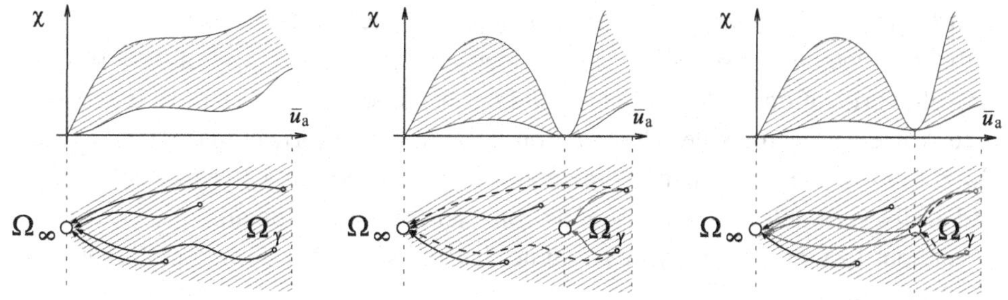

Figure 4.8 Control of the attracting set by means of the system's steady-state characteristics

(C3′) *if system $\mathcal{S}_a$ has a steady-state characteristic on average $\chi_T(\cdot)$, condition (4.98) holds, and $\chi_T(\cdot)$ has no zeros in the positive domain, then it is possible to show that the state converges to*

$$\Omega_a = \{\mathbf{x} \in \mathcal{X}, \mathbf{z} \in \mathcal{Z} | \ \|\mathbf{x}\|_{\mathcal{A}} = 0, \ h(\mathbf{z}) = 0\}. \tag{4.103}$$

The proof follows straightforwardly from the proof of Corollary 4.2 and is therefore omitted.

4.5.3 Systems with separable dynamics in space-time

So far we have presented convergence tests and estimates of the trapping region, and also characterized the attracting sets of interconnection (4.76) under the assumptions of uniform asymptotic stability of $\mathcal{S}_a$ and input–output properties (4.73) and (4.101) of system $\mathcal{S}_w$. The conditions are given for rather general functions $\beta(\cdot,\cdot) \in \mathcal{KL}$ in (4.71) and $\gamma_0(\cdot)$ and $\gamma_1(\cdot)$ in (4.73). It appears, however, that these conditions can be substantially simplified if additional properties of $\beta(\cdot,\cdot)$ and $\gamma_0(\cdot)$ are available. This information is, in particular, the separability of the function $\beta(\cdot,\cdot)$ or, equivalently, the possibility of factorization:

$$\beta(\|\mathbf{x}\|_{\mathcal{A}}, t) \leq \beta_x(\|\mathbf{x}\|_{\mathcal{A}}) \cdot \beta_t(t), \tag{4.104}$$

where $\beta_x(\cdot) \in \mathcal{K}$ and $\beta_t(\cdot) \in C^0$ is strictly decreasing[15] with

$$\lim_{t \to \infty} \beta_t(t) = 0. \tag{4.105}$$

In principle, as shown in Grune *et al.* (1999), the factorization (4.104) is achievable for a large class of uniformly asymptotically stable systems under an appropriate

[15] If $\beta_t(\cdot)$ is not strictly monotone, it can always be majorized by a strictly decreasing function.

coordinate transformation. An immediate consequence of the factorization (4.104) is that the elements of sequence Ξ in Property 4.2 are independent of $\|\mathbf{x}(t_i)\|_{\mathcal{A}}$. As a result, verification of Properties 4.2 and 4.3 becomes easier. The most interesting case, however, occurs when the function $\beta_x(\cdot)$ in the factorization (4.104) is Lipschitz. For this class of functions the conditions of Theorem 4.7 reduce to a single and easily verifiable inequality. Let us consider this case in detail.

Without loss of generality, we assume that the state $\mathbf{x}(t)$ of system $\mathcal{S}_a$ satisfies the following equation:

$$\|\mathbf{x}(t)\|_{\mathcal{A}} \leq \|\mathbf{x}(t_0)\|_{\mathcal{A}} \cdot \beta_t(t - t_0) + c \cdot \|h(\mathbf{z}(\tau, \mathbf{z}_0))\|_{\infty,[t_0,t]}, \tag{4.106}$$

where $\beta_t(0)$ is greater than or equal to unity. Given that $\beta_t(t)$ is strictly decreasing, the mapping $\beta_t : [0, \infty] \mapsto [0, \beta_t(0)]$ is injective. Moreover, since $\beta_t(t)$ is continuous, it is surjective and, therefore, bijective. In other words, there is a (continuous) mapping $\beta_t^{-1} : [0, \beta_t(0)] \mapsto \mathbb{R}_{\geq 0}$:

$$\beta_t^{-1} \circ \beta_t(t) = t, \ \forall\, t > 0. \tag{4.107}$$

Conditions for the emergence of the trapping region for interconnection (4.76) with the dynamics of system $\mathcal{S}_a$ governed by (4.106) are summarized below.

Corollary 4.3 *Let the interconnection (4.76) be given, let system $\mathcal{S}_a$ satisfy (4.106), and let the function $\gamma_0(\cdot)$ in (4.73) be Lipschitz:*

$$|\gamma_0(s)| \leq D_{\gamma,0} \cdot |s|. \tag{4.108}$$

Furthermore, the domain

$$\Omega_\gamma : \ D_{\gamma,0} \leq \left(\beta_t^{-1}\left(\frac{d}{\kappa}\right)\right)^{-1} \frac{\kappa - 1}{\kappa} \times \frac{h(\mathbf{z}_0)}{\beta_t(0)\|\mathbf{x}_0\|_{\mathcal{A}} + \beta_t(0) \cdot c \cdot |h(\mathbf{z}_0)|(1 + \kappa/(1 - d)) + c|h(\mathbf{z}_0)|} \tag{4.109}$$

is not empty for some $d < 1$, $\kappa > 1$. Then for all initial conditions $\mathbf{x}_0$, $\mathbf{z}_0 \in \Omega_\gamma$ the state $\mathbf{x}(t, \mathbf{x}_0) \oplus \mathbf{z}(t, \mathbf{z}_0)$ of interconnection (4.76) converges into the set Ω_a specified by (4.89). If, in addition, conditions (C1)–(C3) of Corollary 4.2 hold, then the domain of convergence is given by (4.100).

A practically important consequence of this corollary concerns systems $\mathcal{S}_a$ that are exponentially stable:

$$\|\mathbf{x}(t)\|_{\mathcal{A}} \leq \|\mathbf{x}(t_0)\|_{\mathcal{A}} D_\beta \exp(-\lambda(t - t_0)) + c \cdot \|h(\mathbf{z}(t, \mathbf{z}_0))\|_{\infty,[t_0,t]},$$
$$\lambda > 0, \ D_\beta \geq 1. \tag{4.110}$$

In this case the domain (4.109) of initial conditions ensuring convergence into Ω_{a} is defined as

$$D_{\gamma,0} \leq \max_{\kappa>1,\, d\in(0,1)} -\lambda \left(\ln\left(\frac{d}{\kappa}\right)\right)^{-1} \frac{\kappa-1}{\kappa} \times \frac{h(\mathbf{z}_0)}{D_\beta \|\mathbf{x}_0\|_{\mathcal{A}} + D_\beta \cdot c \cdot |h(\mathbf{z}_0)|(1+\kappa/(1-d)) + c|h(\mathbf{z}_0)|}.$$

We would also like to mention that statements of Theorem 4.7 and Corollaries 4.2–4.3 constitute additional theoretical tools for the analysis of the asymptotic behavior of systems in cascaded form. In particular, they are complementary to the results of Arcak *et al.* (2002), where *asymptotic stability* of systems of the type

$$\begin{aligned}
\dot{\mathbf{x}} &= \mathbf{f}(\mathbf{x}),\\
\dot{\mathbf{z}} &= \mathbf{q}(\mathbf{x},\mathbf{z}),\ \mathbf{f}:\mathbb{R}^n \to \mathbb{R}^n,\ \mathbf{q}:\mathbb{R}^n \times \mathbb{R}^m \to \mathbb{R}^m
\end{aligned}$$

was considered under the assumption that the $\mathbf{x}$-subsystem is globally asymptotically stable and the $\mathbf{z}$-subsystem is integral-input-to-state stable. In contrast to this, our results apply to establishing *asymptotic convergence* for systems with the structure

$$\begin{aligned}
\dot{\mathbf{x}} &= \mathbf{f}(\mathbf{x},\mathbf{z}),\\
\dot{\mathbf{z}} &= \mathbf{q}(\mathbf{x},\mathbf{z}),\ \mathbf{f}:\mathbb{R}^n \times \mathbb{R}^m \to \mathbb{R}^n,\ \mathbf{q}:\mathbb{R}^n \times \mathbb{R}^m \to \mathbb{R}^m,
\end{aligned}$$

where the $\mathbf{x}$-subsystem is input-to-state stable, and the $\mathbf{z}$-subsystem could be practically integral-input-to-state stable (see Corollary 4.2), although in general no stability assumptions are imposed on it.

Corollary 4.3 can be easily modified to tackle interconnections in which a stable subsystem, $\mathcal{S}_{\mathrm{a}}$, is perturbed by a bounded additive disturbance. In particular, the following result holds.

Corollary 4.4 *Consider an interconnection of systems that is governed by the following set of equations:*

$$\begin{aligned}
&\|\mathbf{x}(t)\| \leq \beta_t(t-t_0)(\|\mathbf{x}(t_0)\| + \|\mathbf{y}(t_0)\|) + c\|h(\mathbf{z}(t))\|_{\infty,[t_0,t]} + \Delta,\ \Delta \in \mathbb{R}_{\geq 0},\\
&-\int_{t_0}^{t} \gamma_0(\|\mathbf{x}(\tau)\|_\varepsilon)d\tau \leq h(\mathbf{z}(t)) - h(\mathbf{z}(t_0)) \leq h(\mathbf{z}(t_0)),\ \gamma_0 \in \mathcal{K}, \quad (4.111)
\end{aligned}$$

where $\beta_t : \mathbb{R} \to \mathbb{R}$ is a strictly monotone function asymptotically decreasing to zero, and γ_0 satisfies (4.108). Then trajectories $\mathbf{x}(t)$ and $h(\mathbf{z}(t))$ passing through $\mathbf{x}(t_0) = \mathbf{x}_0$ and $h(\mathbf{z}(t_0)) = h_0$ at $t = t_0$ are bounded in forward time, provided that

$$D_{\gamma,0} \leq \frac{\kappa-1}{\kappa}\left[\beta_t^{-1}\left(\frac{d}{\kappa}\right)\right]^{-1} \frac{h_0}{\beta_t(0)\|\mathbf{x}(t_0)\| + ch_0(1+\kappa\beta_t(0)/(1-d))}$$

and

$$\varepsilon \geq \left(\beta_t(0)\left(1-\frac{d}{\kappa}\right)^{-1}+1\right)\Delta \tag{4.112}$$

for some of $d \in (0,1)$, $\kappa \in (1,\infty)$.

The conditions specifying state boundedness formulated in Theorem 4.7 and Corollaries 4.2 and 4.3 depend explicitly on the initial conditions $\mathbf{x}(t_0)$ and $\mathbf{z}(t_0)$. Such a dependence is inevitable when the convergence is allowed to be non-uniform. But if the mere existence of a trapping region is asked for, the dependence on initial conditions may be removed from the statements of the results. The next corollary presents such modified conditions.

Corollary 4.5 *Consider interconnection (4.76) where the system* $\mathcal{S}_a$ *satisfies inequality (4.106) and the function* $\gamma_0(\cdot)$ *obeys (4.108). Then there exists a set* Ω_γ *of initial conditions corresponding to the trajectories converging to* Ω_a *if the following condition is satisfied:*

$$D_{\gamma,0}\cdot c\cdot \mathcal{G} < 1, \tag{4.113}$$

where

$$\mathcal{G} = \beta_t^{-1}\left(\frac{d}{\kappa}\right)\frac{k}{k-1}\left(\beta_t(0)\left(1+\frac{\kappa}{1-d}\right)+1\right)$$

for some $d \in (0,1)$, $\kappa \in (1,\infty)$. *In particular,* Ω_γ *contains the following domain:*

$$\|\mathbf{x}(t_0)\|_{\mathcal{A}} \leq \frac{h(\mathbf{z}(t_0))}{\beta_t(0)}\left[\frac{1}{D_{\gamma,0}}\left(\beta_t^{-1}\left(\frac{d}{\kappa}\right)\right)^{-1}\frac{k-1}{k} - c\left(\beta_t(0)\left(1+\frac{\kappa}{1-d}\right)+1\right)\right].$$

If the function $h(\mathbf{z})$ *in (4.76) is continuous, the volume of the set* Ω_γ *is non-zero in* $\mathbb{R}^n \oplus \mathbb{R}^m$.

Notice that when the dynamics of the contracting subsystem $\mathcal{S}_a$ is exponentially stable, i.e. it satisfies inequality (4.110), the term $\mathcal{G}$ in condition (4.113) reduces to

$$\mathcal{G} = \frac{1}{\lambda}\cdot \ln\left(\frac{\kappa}{d}\right)\frac{k}{k-1}\left(D_\beta\left(1+\frac{\kappa}{1-d}\right)+1\right). \tag{4.114}$$

For $D_\beta = 1$ the minimal value of $\mathcal{G}$ in (4.114) can be estimated as

$$\mathcal{G}^* = \frac{1}{\lambda}\min_{d\in(0,1),\ \kappa\in(1,\infty)} \ln\left(\frac{\kappa}{d}\right)\frac{k}{k-1}\left(2+\frac{\kappa}{1-d}\right) \approx \frac{15.6886}{\lambda} < \frac{16}{\lambda}, \tag{4.115}$$

which leads to an even simpler formulation of (4.114):

$$D_{\gamma,0} \cdot \frac{c}{\lambda} \leq \frac{1}{16}.$$

Corollary 4.5 provides an explicit and easy-to-check condition for the existence of a trapping region in the state space of a class of Lyapunov-unstable systems. In addition, it allows one to specify explicitly the points $\mathbf{x}(t_0)$ and $\mathbf{z}(t_0)$ which belong to the emergent trapping region. Notice also that the existence condition, inequality (4.113), has the flavor of conventional small-gain constraints. Yet, it is substantially different from these classical results. This is because the input–output gain for the wandering subsystem, $\mathcal{S}_{\mathrm{w}}$, need not be finite or need not even be defined.

To elucidate these differences as well as the similarities between conditions of conventional small-gain theorems and those formulated in Corollary 4.5 we provide an example.

Example 4.5.2 Consider the following systems:

$$\begin{cases} \dot{x}_1 = -\lambda_1 x_1 + c_1 x_2, \\ \dot{x}_2 = -\lambda_2 x_2 - c_2 |x_1|, \end{cases} \tag{4.116}$$

$$\begin{cases} \dot{x}_1 = -\lambda_1 x_1 + c_1 x_2, \\ \dot{x}_2 = -c_2 |x_1|. \end{cases} \tag{4.117}$$

System (4.116) can be viewed as an interconnection of two input-to-state stable systems, x_1 and x_2, with input–output L_∞-gains c_1/λ_1 and c_2/λ_2, respectively. Therefore, in order to prove state boundedness of (4.116) we can, in principle, invoke the conventional small-gain theorem. The small-gain condition in this case is as follows:

$$\frac{c_1}{\lambda_1} \cdot \frac{c_2}{\lambda_2} < 1.$$

The theorem, however, does not apply to system (4.117) because the input–output gain of its second subsystem, x_2, is infinite. Yet, by invoking Corollary 4.5, it is still possible to show the existence of a weak attracting set in the state space of system (4.117) and specify its basin of attraction. As follows from Corollary 4.5, the condition

$$\frac{c_1}{\lambda_1} \cdot \frac{c_2}{\lambda_1} < \frac{1}{16}$$

ensures the existence of the trapping region, and the trapping region itself is given by

$$|x_1(t_0)| \leq \left[\frac{1}{c_2} \lambda_1 \left(\ln \left(\frac{\kappa}{d} \right) \right)^{-1} \frac{k-1}{k} - \frac{c_1}{\lambda_1} \left(2 + \frac{\kappa}{1-d} \right) \right] x_2(t_0).$$

Results provided in this section play an important role in the development of various adaptation laws and procedures which we present in the next chapters. These include adaptive observers and identification algorithms for systems with nonlinear parametrization and models of which the equations are not in adaptive-observer canonical forms (Bastin and Gevers 1988). The latter algorithms and procedures will form a basis for explaining adaptation mechanisms in fixed-weight recurrent neural networks and adaptation algorithms for solving an invariant template-matching problem.

Appendix to Chapter 4

A4.1 Proof of Theorem 4.1

Let us first prove statement (3) of the theorem. Consider $\mathcal{S}_{\mathcal{T}}^*(\mathbf{x}_0, t_0) : \mathbb{R}^n \times \mathbb{R} \mapsto \mathcal{L}_x$, where $\mathcal{L}_x$ is a linear normed space with a norm $\|\cdot\|_{\mathcal{L}_x}$, and let $\|\cdot\|_{\mathbb{R}^n}$ stand for the Euclidean norm. According to the standard $\varepsilon - \delta$ definition of continuity of functions in metric spaces, continuity of $\mathcal{S}_{\mathcal{T}}^*(\mathbf{x}_0, t_0)$ with respect to $\mathbf{x}_0$ is equivalent to the following:

$$\begin{aligned}\forall\, \varepsilon > 0\, \exists\, \delta(\varepsilon, t_0) > 0:\ & \|\mathbf{x}_0' - \mathbf{x}_0''\|_{\mathbb{R}^n} < \delta(\varepsilon, t_0) \Rightarrow \\ & \|\mathcal{S}_{\mathcal{T}}^*(\mathbf{x}_0', t_0) - \mathcal{S}_{\mathcal{T}}^*(\mathbf{x}_0'', t_0)\|_{\mathcal{L}_x} < \varepsilon.\end{aligned} \tag{A4.1}$$

Taking into account that $\mathcal{S}_{\mathcal{T}}^*(\mathbf{x}_0, t_0) = \mathcal{S}_{\mathcal{T}}(\mathbf{x}_0 \oplus t_0)$ we can conclude that (A4.1) is equivalent to

$$\begin{aligned}\forall\, \varepsilon > 0\, \exists\, \delta(\varepsilon, t_0) > 0:\ & \|\mathbf{x}_0' - \mathbf{x}_0''\|_{\mathbb{R}^n} < \delta(\varepsilon, t_0) \Rightarrow \\ & \|\mathcal{S}_{\mathcal{T}}(\mathbf{x}_0' \oplus t_0) - \mathcal{S}_{\mathcal{T}}(\mathbf{x}_0'' \oplus t_0)\|_{\mathcal{L}_x} < \varepsilon.\end{aligned} \tag{A4.2}$$

On replacing $\mathcal{S}_{\mathcal{T}}(\mathbf{x}_0' \oplus t_0)$ and $\mathcal{S}_{\mathcal{T}}(\mathbf{x}_0'' \oplus t_0)$ in (A4.2) with $\mathbf{x}(t, \mathbf{x}_0', t_0)$ and $\mathbf{x}(t, \mathbf{x}_0'', t_0)$ we obtain that

$$\begin{aligned}\forall\, \varepsilon > 0\, \exists\, \delta(\varepsilon, t_0) > 0:\ & \|\mathbf{x}_0' - \mathbf{x}_0''\|_{\mathbb{R}^n} < \delta(\varepsilon, t_0) \Rightarrow \\ & \|\mathbf{x}(t, \mathbf{x}_0', t_0) - \mathbf{x}(t, \mathbf{x}_0'', t_0)\|_{\mathcal{L}_x} < \varepsilon.\end{aligned} \tag{A4.3}$$

Clearly (A4.3) is equivalent to the definition of stability of a solution $\mathbf{x}(t, \mathbf{x}_0', t_0)$ in the sense of Lyapunov. Thus statement (3) is obvious.

Let us now prove statement (1). The set Ω^* is forward-invariant, and hence one can easily see that $\Omega_s \subseteq \Omega^*$ (Ω_s is defined in (4.15)). On the other hand, since $\mathbf{x}_0(t_0, \mathbf{x}_0, t_0) = \mathbf{x}_0$, $\Omega^* \subseteq \Omega_s$, and hence $\Omega^* = \Omega_s$. Thus, according to (4.16), continuity of $\mathcal{S}_{\mathcal{T}}^*(\mathbf{x}_0, t_0)$ implies that

$$\begin{aligned}\forall\, \varepsilon > 0\, \exists\, \delta(\varepsilon, t_0) > 0:\ & \|\mathbf{x}_0\|_{\Omega^*} < \delta(\varepsilon, t_0) \Rightarrow \\ & \|\mathcal{S}_{\mathcal{T}}^*(\mathbf{x}_0, t_0)\|_{\Omega^*, [t_0, \infty]} < \varepsilon.\end{aligned} \tag{A4.4}$$

Given that $\mathbf{x}(t, \mathbf{x}_0, t_0) = \mathcal{S}_{\mathcal{T}}^*(\mathbf{x}_0, t_0) = \mathcal{S}_{\mathcal{T}}(\mathbf{x}_0 \oplus t_0)$, we can conclude that inequality (A4.4) is identical to that used in the definition of stability of a set in the sense of Lyapunov.

Let us prove property (1.2). We start by showing sufficiency of existence of a function $\gamma_{S,L_\infty}(\cdot)$ defined as in (4.14). According to (4.10) and (4.14) the following holds:

$$\begin{aligned}\|\mathbf{x}(t)\|_{\infty,[t_0,\infty]} &\leq \gamma_{S,\mathcal{L}_x}(\mathbf{e}, T) \\ &= \gamma_{S,\mathcal{L}_x}(\mathbf{x}_0 \oplus t_0, T) = \gamma_{S,L_\infty}(\|\mathbf{x}_0\|_{\Omega^*}, t_0),\end{aligned} \tag{A4.5}$$

where $\gamma_{S,L_\infty}(\|\mathbf{x}_0\|_{\Omega^*}, t_0)$ is non-decreasing w.r.t. $\|\mathbf{x}_0\|_{\Omega^*}$, and $\gamma_{S,L_\infty}(0, t_0) = 0$. It is clear that for every t_0 the function $\gamma_{S,L_\infty}(\|\mathbf{x}_0\|_{\Omega^*}, t_0)$ can be majorized by a continuous and strictly monotone function $\gamma_s : \mathbb{R}_{\geq 0} \times \mathbb{R} \to \mathbb{R}_{\geq 0}$ such that

$$\lim_{r\to\infty} \gamma_s(r, t_0) = \infty.$$

Taking (A4.5) into account, we can conclude that

$$\|\mathbf{x}(t)\|_{\infty,[t_0,\infty]} \leq \gamma_s(\|\mathbf{x}_0\|_{\Omega^*}, t_0). \tag{A4.6}$$

The function $\gamma_s(r, t_0)$ has an inverse $\gamma_s^{-1}(r, t_0)$:

$$\gamma_s(\gamma_s^{-1}(r, t_0), t_0) = r,$$

and the function $\gamma_s^{-1}(r, t_0)$ is a strictly monotone and non-decreasing function w.r.t. r. Thus the following inequalities hold:

$$\begin{aligned}\gamma_s(a, t_0) &\leq \gamma_s(b, t_0)\ \forall\, a \leq b,\ a, b \in \mathbb{R}_{\geq 0}; \\ \gamma_s^{-1}(a, t_0) &\leq \gamma_s^{-1}(b, t_0)\ \forall\, a \leq b,\ a, b \in \mathbb{R}_{\geq 0}.\end{aligned}$$

Hence for all $\mathbf{x}_0 \in \mathcal{V}(\Omega^*, \delta)$, $\mathcal{V}(\Omega^*, \delta) = \{\mathbf{x} \in \mathbb{R}^n |\ \|\mathbf{x}\|_{\Omega^*} \leq \delta\}$, where $\delta = \gamma_s^{-1}(\varepsilon, t_0)$, $\varepsilon \in \mathbb{R}_{\geq 0}$ the following holds:

$$\|\mathbf{x}(t)\|_{\infty,[t_0,\infty]} \leq \gamma_s(\|\mathbf{x}_0\|_{\Omega^*}, t_0) \leq \gamma_s(\gamma_s^{-1}(\varepsilon, t_0), t_0) = \varepsilon. \tag{A4.7}$$

Notice that, since the value of ε was arbitrary, inequality (A4.7) is true for all $\varepsilon > 0$, provided that $\mathbf{x}_0$ is in a δ-neighborhood of Ω^*, $\delta = \gamma_s^{-1}(\varepsilon, t_0)$. Thus

$$\forall\, \varepsilon > 0\ \exists\, \delta(\varepsilon, t_0) = \gamma_s^{-1}(\varepsilon, t_0) :\ \|\mathbf{x}_0\|_{\Omega^*} < \delta(\varepsilon, t_0) \Rightarrow \|\mathbf{x}(t)\|_{\Omega^*} < \varepsilon,$$

which proves sufficiency.

Let's prove the necessity part of property (1.2). Consider ε_1 as a function of δ and t_0:

$$\varepsilon_1(\delta, t_0) = \sup_{\mathbf{x}_0 \in \mathcal{V}(\Omega^*, \delta)} \|\mathbf{x}(t, \mathbf{x}_0, t_0)\|_{\Omega^*,[t_0,\infty]}.$$

It is clear that $\varepsilon_1(\delta, t_0) \geq \delta$. Moreover, $\varepsilon_1(0, t_0) = 0$ because the set Ω^* is invariant. In addition

$$\varepsilon_1(\delta_1, t_0) \geq \varepsilon_1(\delta_2, t_0), \ \forall\, \delta_1, \delta_2 \in \mathbb{R}_{\geq 0}, \ \delta_1 \geq \delta_2,$$

because $\mathcal{V}(\Omega^*, \delta_1) \supseteq \mathcal{V}(\Omega^*, \delta_2)$. Thus the function $\varepsilon_1(\delta, t_0)$ is non-decreasing w.r.t. δ. Finally, observe that the function $\varepsilon_1(\delta, t_0)$ is continuous w.r.t. δ at $\delta = 0$. Setting

$$\gamma_{S,L_\infty}(\|\mathbf{x}_0\|_{\Omega^*}, t_0) = \varepsilon_1(\|\mathbf{x}_0\|_{\Omega^*}, t_0)$$

completes the proof of necessity.

Statement (2.1) of the theorem can be proven along the same lines as in the proof of (1.1) and (3). According to the definition (of local boundedness, (4.17)) we have that

$$\forall\, \varepsilon > 0\, \exists\, \delta(\varepsilon, \mathbf{x}_0, t_0) : \ \|\mathbf{x}_0\|_{\Omega^*} < \varepsilon \Rightarrow$$
$$\|\mathcal{S}_{\mathcal{T}}^*(\mathbf{x}_0, t_0)\|_{\Omega^*,[t_0,\infty]} = \|\mathbf{x}(t, \mathbf{x}_0, t_0)\|_{\Omega^*,[t_0,\infty]} < \delta(\varepsilon, \mathbf{x}_0, t_0).$$

Let us prove (2.2). Sufficiency follows immediately from the definition of the input–state margin,

$$\|\mathbf{x}(t)\|_{\Omega^*,[t_0,\infty]} \leq \gamma_{S,\mathcal{L}_x}(\mathbf{e}, \infty) = \gamma_{S,L_\infty}(\|\mathbf{x}_0\|_{\Omega^*}, t_0),$$

and the local boundedness of $\gamma_{S,L_\infty}(\|\mathbf{x}_0\|_{\Omega^*}, t_0)$ w.r.t. $\|\mathbf{x}_0\|_{\Omega^*}$:

$$\forall\, \varepsilon > 0\, \exists\, \delta(\varepsilon, t_0) : \ \|\mathbf{x}_0\|_{\Omega^*} < \varepsilon \Rightarrow$$
$$\|\mathbf{x}(t)\|_{\Omega^*,[t_0,\infty]} \leq \gamma_{S,L_\infty}(\|\mathbf{x}_0\|_{\Omega^*}, t_0) < \delta(\varepsilon, t_0).$$

Let us show the necessity part. Stability in the sense of Lagrange implies that $\|\mathbf{x}(t)\|_{\Omega^*,[t_0,\infty]}$ is defined. Consider

$$\delta_1(\varepsilon, t_0) = \sup_{\mathbf{x}_0 \in \mathcal{V}(\Omega^*, \varepsilon)} \|\mathbf{x}(t, \mathbf{x}_0, t_0)\|_{\Omega^*,[t_0,\infty]}.$$

The function $\delta_1(\varepsilon, t_0)$ is non-negative and locally bounded w.r.t. t_0 and ε. Moreover, $\delta_1(0, t_0) = 0$ because the set Ω^* is invariant. Hence choosing

$$\gamma_{S,L_\infty}(\|\mathbf{x}_0\|_{\Omega^*}, t_0) = \delta_1(\|\mathbf{x}_0\|_{\Omega^*}, t_0)$$

completes the proof of (2.2) and the theorem. □

A4.2 Proof of Theorem 4.4

Consider parallel interconnection of $\mathcal{S}_1$ and $\mathcal{S}_2$ specified by (4.31) and (4.32). Systems $\mathcal{S}_1$ and $\mathcal{S}_2$ are realizable. Thus for every pair $(\mathbf{u}, \mathbf{e}_1) \in \mathcal{L}_u \times \mathcal{E}_1$ and $(\mathbf{u}, \mathbf{e}_2) \in \mathcal{L}_u \times \mathcal{E}_2$ there exist positive numbers $T_1(\mathbf{u}, \mathbf{e}_1)$ and $T_2(\mathbf{u}, \mathbf{e}_2)$ such that

$$\|\mathcal{S}_1(\mathbf{u}, \mathbf{e}_1)\|_{\infty,[t_0,T_1]} < \infty, \qquad \|\mathcal{H}_1(\mathbf{u}, \mathbf{e}_1)\|_{\infty,[t_0,T_1]} < \infty;$$

$$\|\mathcal{S}_2(\mathbf{u},\mathbf{e}_2)\|_{\infty,[t_0,T_2]} < \infty, \qquad \|\mathcal{H}_2(\mathbf{u},\mathbf{e}_2)\|_{\infty,[t_0,T_2]} < \infty.$$

Let

$$T = \min\{T_1, T_2\}. \tag{A4.8}$$

Then, taking (4.32) into account we can produce the following estimate:

$$\begin{aligned}
\|\mathcal{S}(\mathbf{u},\mathbf{e}_1 \oplus \mathbf{e}_2)\|_{\infty,[t_0,T]} &= \|\mathcal{S}_1(\mathbf{u},\mathbf{e}_1)\|_{\infty,[t_0,T]} + \|\mathcal{S}_2(\mathbf{u},\mathbf{e}_2)\|_{\infty,[t_0,T]} < \infty,\\
\|\mathcal{H}(\mathbf{u},\mathbf{e}_1 \oplus \mathbf{e}_2)\|_{\infty,[t_0,T]} &= \|\mathcal{H}_1(\mathbf{u},\mathbf{e}_1)\|_{\infty,[t_0,T]} + \|\mathcal{H}_2(\mathbf{u},\mathbf{e}_2)\|_{\infty,[t_0,T]} < \infty.
\end{aligned} \tag{A4.9}$$

This shows the realizability of the parallel interconnection. Given that the values of T_1 and T_2 can be chosen arbitrarily large for complete systems $\mathcal{S}_1$ and $\mathcal{S}_2$, completeness of parallel interconnections of complete systems now follows immediately from (A4.9) and (A4.8).

Let us consider serial interconnection, (4.29) and (4.30), of $\mathcal{S}_1$ and $\mathcal{S}_2$. Systems $\mathcal{S}_1$ and $\mathcal{S}_2$ are realizable. Thus for every $\mathbf{e}_1 \in \mathcal{E}_1$ and $\mathbf{e}_2 \in \mathcal{E}_2$ there exist positive numbers $T_1(\mathbf{e}_1) > t_0$ and $T_2(\mathbf{e}_2) > t_0$ such that

$$\begin{aligned}
\|\mathcal{S}_1(\mathbf{u}_1,\mathbf{e}_1)\|_{\infty,[t_0,T_1]} &\leq \gamma_{S_1,\infty}(\mathbf{e}_1, \|\mathbf{u}_1(t)\|_{\mathcal{L}_{u_1},[t_0,T_1]}, T_1),\\
\|\mathcal{H}_1(\mathbf{u}_1,\mathbf{e}_1)\|_{\infty,[t_0,T_1]} &\leq \gamma_{H_1,\infty}(\mathbf{e}_1, \|\mathbf{u}_1(t)\|_{\mathcal{L}_{u_1},[t_0,T_1]}, T_1),\\
\|\mathcal{S}_2(\mathbf{u}_2,\mathbf{e}_2)\|_{\infty,[t_0,T_2]} &\leq \gamma_{S_2,\infty}(\mathbf{e}_2, \|\mathbf{u}_2(t)\|_{\mathcal{L}_{u_2},[t_0,T_2]}, T_2),\\
\|\mathcal{H}_2(\mathbf{u}_2,\mathbf{e}_2)\|_{\infty,[t_0,T_2]} &\leq \gamma_{H_2,\infty}(\mathbf{e}_2, \|\mathbf{u}_2(t)\|_{\mathcal{L}_{u_2},[t_0,T_2]}, T_2).
\end{aligned} \tag{A4.10}$$

According to the assumptions of the theorem the following inequality holds:

$$\|\mathcal{H}_1(\mathbf{u}_1,\mathbf{e}_1)\|_{\mathcal{L}_{u_2}} \leq \gamma_{H,\mathcal{L}_{y_1}}(\mathbf{e}_1, \|\mathbf{u}_1(t)\|_{\mathcal{L}_{u_1}}, T_1).$$

Thus, on choosing the value of T according to (A4.8) and taking into account that the functions $\gamma_{S_2,\infty}$ and $\gamma_{H_2,\infty}$ are monotone w.r.t. $\|\mathbf{u}_2(t)\|_{\mathcal{L}_{u_2}}$, we can derive that

$$\begin{aligned}
\|\mathcal{S}_2(\mathcal{H}_1(\mathbf{u}_1,\mathbf{e}_1),\mathbf{e}_2)\|_{\infty,[t_0,T]} &\leq \gamma_{S_2,\infty}(\mathbf{e}_2, \gamma_{H,\mathcal{L}_{y_1}}(\mathbf{e}_1, \|\mathbf{u}_1(t)\|_{\mathcal{L}_{u_1}}, T), T),\\
\|\mathcal{H}_2(\mathcal{H}_1(\mathbf{u}_1,\mathbf{e}_1),\mathbf{e}_2)\|_{\infty,[t_0,T]} &\leq \gamma_{H_2,\infty}(\mathbf{e}_2, \gamma_{H,\mathcal{L}_{y_1}}(\mathbf{e}_1, \|\mathbf{u}_1(t)\|_{\mathcal{L}_{u_1}}, T), T).
\end{aligned} \tag{A4.11}$$

Finally, we notice that (4.29), (4.30), and (A4.10) imply

$$\begin{aligned}
\|\mathbf{x}(t)\|_{\infty,[t_0,T]} &= \|\mathbf{x}_1(t) \oplus \mathbf{x}_2(t)\|_{\infty,[t_0,T]}\\
&= \|\mathcal{S}_1(\mathbf{u}_1,\mathbf{e}_1) \oplus \mathcal{S}_2(\mathcal{H}_1(\mathbf{u}_1,\mathbf{e}_1),\mathbf{e}_2)\|_{\infty,[t_0,T]}
\end{aligned}$$

$$
\begin{aligned}
&\leq \|\mathcal{S}_1(\mathbf{u}_1, \mathbf{e}_1)\|_{\infty,[t_0,T]} + \|\mathcal{S}_2(\mathcal{H}_1(\mathbf{u}_1, \mathbf{e}_1), \mathbf{e}_2)\|_{\infty,[t_0,T]} \\
&\leq \gamma_{S_1,\infty}(\mathbf{e}_1, \|\mathbf{u}_1(t)\|_{\mathcal{L}_{u_1},[t_0,T]}, T) \\
&\quad + \gamma_{S_2,\infty}(\mathbf{e}_2, \gamma_{H,\mathcal{L}_{y_1}}(\mathbf{e}_1, \|\mathbf{u}_1(t)\|_{\mathcal{L}_{u_1}}, T), T),
\end{aligned}
$$

$$
\begin{aligned}
\|\mathbf{y}(t)\|_{\infty,[t_0,T]} &= \|\mathbf{y}_2(t)\|_{\infty,[t_0,T]} = \|\mathcal{H}_2(\mathcal{H}_1(\mathbf{u}_1, \mathbf{e}_1), \mathbf{e}_2)\|_{\infty,[t_0,T]} \\
&\leq \gamma_{H_2,\infty}(\mathbf{e}_2, \gamma_{H,\mathcal{L}_{y_1}}(\mathbf{e}_1, \|\mathbf{u}_1(t)\|_{\mathcal{L}_{u_1}}, T), T). \qquad \text{(A4.12)}
\end{aligned}
$$

Inequality (A4.12), in turn, assures that serial interconnection (4.29) and (4.30) is realizable (complete). Moreover, (A4.8) guarantees that (4.38) holds. □

A4.3 Proof of Theorem 4.5

Consider serial interconnection, (4.30), of systems $\mathcal{S}_1$ and $\mathcal{S}_2$, provided that the class of inputs $\mathcal{U}_1$ for system $\mathcal{S}_1$ is restricted to (4.49). System $\mathcal{S}_1$ in this case can be redefined as a system $\tilde{\mathcal{S}}_1$ mapping signals from $\mathcal{L}_\delta[t_0, T_1] \times \mathcal{E}_1$ (where $\mathcal{L}_\delta[t_0, T_1] \subseteq \mathcal{L}_u[t_0, T])$ into $\mathcal{L}_{x_1}[t_0, T_1]$ and $\mathcal{L}_{y_1}[t_0, T_1]$:

$$
\begin{aligned}
\tilde{\mathcal{S}}_{1,\mathcal{T}} &= \mathcal{S}_{1,\mathcal{T}}(\mathbf{u}^*(\mathbf{e}_1, t) + \delta(t), \mathbf{e}_1), \\
\tilde{\mathcal{H}}_{1,\mathcal{T}} &= \mathcal{H}_{1,\mathcal{T}}(\mathbf{u}^*(\mathbf{e}_1, t) + \delta(t), \mathbf{e}_1).
\end{aligned} \qquad \text{(A4.13)}
$$

According to Theorem 4.4, serial interconnection of two systems $\tilde{\mathcal{S}}_1$ and $\mathcal{S}_2$ is realizable (complete) if (1) they are realizable (complete) and admit input–state and input–output mappings w.r.t. the norms $\|\cdot\|_{\mathcal{L}_\delta,[t_0,T_1]}$ and $\|\cdot(t)\|_{\mathcal{L}_{y_1},[t_0,T_2]}$ respectively, and (2) system $\tilde{\mathcal{S}}_1$ admits the following input–output margin:

$$
\|\mathbf{y}_1(t)\|_{\mathcal{L}_{y_1},[t_0,T_1]} \leq \gamma_{\tilde{H}_1,\mathcal{L}_\delta}(\mathbf{e}_1, \|\delta(t)\|_{\mathcal{L}_\delta,[t_0,T_1]}, T_1). \qquad \text{(A4.14)}
$$

According to assumption (2) of Theorem 4.5, system $\mathcal{S}_2$ is realizable (complete) and admits input–state and input–output mappings w.r.t. $\|\cdot\|_{\mathcal{L}_{y_1},[t_0,T_2]}$. Let us show that $\tilde{\mathcal{S}}_1$ is realizable (complete) too. Condition (1) of the theorem implies the existence of a majorizing mapping $\boldsymbol{\psi}(\mathbf{x}(t), \mathbf{e}_\psi)$ for $\mathcal{S}_1$. In addition, the following holds for all $\mathbf{u}(t)$ in (4.49) (see (4.46), Assumption 4.1):

$$
\|\psi(t)\|_{\mathcal{L}_\psi,[t_0,T_1]} \leq \gamma_{P,\mathcal{L}_\psi}(\mathbf{e}, \mathbf{e}_\psi, \|\delta(t)\|_{\mathcal{L}_\delta,[t_0,T_1]}, T_1).
$$

In accordance with (4.40), (4.41) (Definition 4.3.4), and (4.46), we obtain that

$$
\|\mathbf{x}_1(t)\|_{\infty,[t_0,T_1]} \leq \mu_{S_1,\infty}(\mathbf{e}_1, \gamma_{P,\mathcal{L}_\psi}(\mathbf{e}_1, \mathbf{e}_\psi, \|\delta(t)\|_{\mathcal{L}_\delta,[t_0,T_1]}, T_1)), \qquad \text{(A4.15)}
$$

$$
\|\mathbf{y}_1(t)\|_{\infty,[t_0,T_1]} \leq \mu_{H_1,\infty}(\mathbf{e}_1, \gamma_{P,\mathcal{L}_\psi}(\mathbf{e}_1, \mathbf{e}_\psi, \|\delta(t)\|_{\mathcal{L}_\delta,[t_0,T_1]}, T_1)). \qquad \text{(A4.16)}
$$

Inequalities (A4.15) and (A4.16) imply that system $\tilde{\mathcal{S}}_1$, defined by (A4.13), is complete. Notice that (4.47) in Assumption 4.1 implies that (A4.14) holds. Hence serial interconnection of $\tilde{\mathcal{S}}_1$ and $\mathcal{S}_2$ is realizable (complete).

Let us show that feedback interconnection of these systems is also realizable (complete). In order to do so, pick $T = \min\{T_1, T_2\}$.[16] Notice that the conditions of the theorem (inequality (4.48)) ensure that the following estimate holds:

$$\|\mathbf{y}_2(t)\|_{\mathcal{L}_\delta,[t_0,T]} \leq \gamma^*_{H_2,\mathcal{L}_\delta}(\mathbf{e}_2, T).$$

Hence, on taking (A4.15) and (A4.16) into account and denoting $\upsilon(t) = \mathbf{y}_2(t) + \delta(t)$, we obtain

$$\begin{aligned}\|\mathbf{x}_1(t)\|_{\infty,[t_0,T]} &\leq \mu_{S_1,\infty}(\mathbf{e}_1, \gamma_{P,\mathcal{L}_\psi}(\mathbf{e}_1, \mathbf{e}_\psi, \|\upsilon(t)\|_{\mathcal{L}_\delta,[t_0,T]}, T)) \\ &\leq \mu_{S_1,\infty}(\mathbf{e}_1, \gamma_{P,\mathcal{L}_\psi}(\mathbf{e}_1, \mathbf{e}_\psi, \|\delta(t)\|_{\mathcal{L}_\delta,[t_0,T]} + \gamma^*_{H_2,\mathcal{L}_\delta}(\mathbf{e}_2, T), T)),\end{aligned} \tag{A4.17}$$

$$\begin{aligned}\|\mathbf{y}_1(t)\|_{\infty,[t_0,T]} &\leq \mu_{H_1,\infty}(\mathbf{e}_1, \gamma_{P,\mathcal{L}_\psi}(\mathbf{e}_1, \mathbf{e}_\psi, \|\upsilon(t)\|_{\mathcal{L}_\delta,[t_0,T]}, T)) \\ &\leq \mu_{H_1,\infty}(\mathbf{e}_1, \gamma_{P,\mathcal{L}_\psi}(\mathbf{e}_1, \mathbf{e}_\psi, \|\delta(t)\|_{\mathcal{L}_\delta,[t_0,T]} + \gamma^*_{H_2,\mathcal{L}_\delta}(\mathbf{e}_2, T), T)).\end{aligned} \tag{A4.18}$$

Using (A4.14) and invoking similar arguments, the norm $\|\mathbf{y}_1(t)\|_{\mathcal{L}_{y_1},[t_0,T]}$ can be estimated as follows:

$$\begin{aligned}\|\mathbf{y}_1(t)\|_{\mathcal{L}_{y_1},[t_0,T]} &\leq \gamma_{\tilde{H}_1,\infty}(\mathbf{e}_1, \|\upsilon(t)\|_{\mathcal{L}_\delta,[t_0,T]}, T) \\ &\leq \gamma_{\tilde{H}_1,\infty}(\mathbf{e}_1, \|\delta(t)\|_{\mathcal{L}_\delta,[t_0,T]} + \gamma^*_{H_2,\mathcal{L}_\delta}(\mathbf{e}_2, T), T).\end{aligned}$$

Therefore, realizability (completeness) of $\mathcal{S}_2$ implies that

$$\begin{aligned}\|\mathbf{x}_2(t)\|_{\infty,[t_0,T]} &\leq \gamma_{S_2,\infty}(\mathbf{e}_2, \|\mathbf{y}_1(t)\|_{\mathcal{L}_{y_1},[t_0,T]}, T) \\ &\leq \gamma_{S_2,\infty}(\mathbf{e}_2, \gamma_{\tilde{H}_1,\infty}(\mathbf{e}_1, \|\delta(t)\|_{\mathcal{L}_\delta,[t_0,T]} + \gamma^*_{H_2,\mathcal{L}_\delta}(\mathbf{e}_2, T), T), T),\end{aligned} \tag{A4.19}$$

$$\begin{aligned}\|\mathbf{y}_2(t)\|_{\infty,[t_0,T]} &\leq \gamma_{H_2,\infty}(\mathbf{e}_2, \|\mathbf{y}_1(t)\|_{\mathcal{L}_{y_1},[t_0,T]}, T) \\ &\leq \gamma_{H_2,\infty}(\mathbf{e}_2, \gamma_{\tilde{H}_1,\infty}(\mathbf{e}_1, \|\delta(t)\|_{\mathcal{L}_\delta,[t_0,T]} + \gamma^*_{H_2,\mathcal{L}_\delta}(\mathbf{e}_2, T), T), T).\end{aligned} \tag{A4.20}$$

[16] $T = T_2$ since system $\tilde{\mathcal{S}}_1$ is complete.

Inequalities (A4.17)–(A4.20), in turn, guarantee that

$$\|\mathbf{x}(t)\|_{\infty,[t_0,T]} \leq \mu_{S_1,\infty}(\mathbf{e}_1, \gamma_{P,\mathcal{L}_\psi}(\mathbf{e}_1, \mathbf{e}_\psi, \|\delta(t)\|_{\mathcal{L}_\delta,[t_0,T]} + \gamma^*_{H_2,\mathcal{L}_\delta}(\mathbf{e}_2, T), T)) + \gamma_{S_2,\infty}(\mathbf{e}_2, \gamma_{\tilde{H}_1,\infty}(\mathbf{e}_1, \|\delta(t)\|_{\mathcal{L}_\delta,[t_0,T]} + \gamma^*_{H_2,\mathcal{L}_\delta}(\mathbf{e}_2, T), T), T), \quad \text{(A4.21)}$$

$$\|\mathbf{y}(t)\|_{\infty,[t_0,T]} \leq \mu_{H_1,\infty}(\mathbf{e}_1, \gamma_{P,\mathcal{L}_\psi}(\mathbf{e}_1, \mathbf{e}_\psi, \|\delta(t)\|_{\mathcal{L}_\delta,[t_0,T]} + \gamma^*_{H_2,\mathcal{L}_\delta}(\mathbf{e}_2, T), T)) + \gamma_{H_2,\infty}(\mathbf{e}_2, \gamma_{\tilde{H}_1,\infty}(\mathbf{e}_1, \|\delta(t)\|_{\mathcal{L}_\delta,[t_0,T]} + \gamma^*_{H_2,\mathcal{L}_\delta}(\mathbf{e}_2, T), T), T). \quad \text{(A4.22)}$$

Equations (A4.21) and (A4.22), clearly, assure the realizability of the interconnection. Furthermore, if system $\mathcal{S}_2$ is complete then the interconnection $\mathcal{S}$ is complete as well. Inequalities (A4.17) and (A4.18) automatically imply that (4.50) holds. □

A4.4 Proof of Theorem 4.6

Notice that system $\mathcal{S}_{\text{c}}$ is serial interconnection of $\mathcal{S}_{\text{o}}$ and $\mathcal{S}_{\text{a}}$ (Definition 4.3.1). According to Theorem 4.4, realizability (completeness) of $\mathcal{S}_{\text{a}}$ and $\mathcal{S}_{\text{o}}$ (condition (2)) automatically implies that $\mathcal{S}_{\text{c}}$ is realizable (complete). Thus realizability (completeness) of the overall system is guaranteed if feedback interconnection of $\mathcal{S}_{\text{c}}$ and $\mathcal{S}_{\text{p}}$ is realizable (complete).

According to (4.63) and condition (1) of the theorem, feedback interconnection of $\mathcal{S}_{\text{p}}$ and $\mathcal{S}_{\text{c}}$ can be represented as feedback interconnection of systems $\mathcal{S}^*_{\text{p}}$ and $\tilde{\mathcal{S}}_{\text{c}}$, where

$$\begin{aligned} \tilde{\mathcal{S}}_{\text{c},\mathcal{T}}(\mathbf{y}_{\text{p}} \oplus \mathbf{u}^*, \mathbf{e}_{\text{o}} \oplus \mathbf{e}_{\text{a}}) &= \mathcal{S}_{a,\mathcal{T}}(\mathbf{y}_{\text{p}}, \mathcal{H}_{o,\mathcal{T}}(\mathbf{y}_{\text{p}}, \mathbf{e}_{\text{o}}), \mathbf{e}_{\text{a}}) \oplus \mathcal{S}_{\text{o}}(\mathbf{y}_{\text{p}}, \mathbf{e}_{\text{o}}), \\ \tilde{\mathcal{H}}_{\text{c},\mathcal{T}}(\mathbf{y}_{\text{p}} \oplus \mathbf{u}^*, \mathbf{e}_{\text{o}} \oplus \mathbf{e}_{\text{a}}) &= \mathbf{u}_{\text{a}}(\mathcal{H}_{\text{a}}(\mathbf{y}_{\text{p}}, \mathcal{H}_{\text{o}}(\mathbf{y}_{\text{p}}, \mathbf{e}_{\text{o}}), \mathbf{e}_{\text{a}}), \mathcal{H}_{\text{o}}(\mathbf{y}_{\text{p}}, \mathbf{e}_{\text{o}}), t) \\ &\quad - \mathbf{u}^*(t). \end{aligned} \quad \text{(A4.23)}$$

Condition (3) of the theorem implies that the mapping $\tilde{\mathcal{H}}_{\text{c},\mathcal{T}}$ in (A4.23) is bounded:

$$\|\tilde{\mathcal{H}}_{\text{c},\mathcal{T}}(\mathbf{y}_{\text{p}} \oplus \mathbf{u}^*, \mathbf{e}_{\text{o}} \oplus \mathbf{e}_{\text{a}})\|_{\mathcal{L}_\delta,[t_0,T]} \leq \Delta_{J_x} + \Delta_{J_e}.$$

Therefore in accordance with Theorem 4.5 feedback interconnection of systems $\mathcal{S}^*_{\text{p}}$ and $\tilde{\mathcal{S}}_{\text{c}}$ is realizable (complete). Hence interconnection of $\mathcal{S}_{\text{p}}$ and $\mathcal{S}_{\text{c}}$ is also realizable (complete). Estimates (4.65) now follow from (4.50). □

A4.5 Proof of Theorem 4.7

Let the conditions of the theorem be satisfied for given $t_0 \in \mathbb{R}_{\geq 0} : \mathbf{x}(t_0) = \mathbf{x}_0$, $\mathbf{z}(t_0) = \mathbf{z}_0$. Notice that in this case $h(\mathbf{z}_0) \geq 0$, otherwise requirement (4.91) will be violated. Consider the sequence (4.88) of volumes Ω_i induced by $\mathcal{S}$:

$$\Omega_i = \{\mathbf{x} \in \mathcal{X},\ \mathbf{z} \in \mathcal{Z} | h(\mathbf{z}(t)) \in H_i\}.$$

To prove the theorem we show that $0 \leq h(\mathbf{z}(t)) \leq h(\mathbf{z}_0)$ for all $t \geq t_0$. For the given partition (4.88) we consider two alternatives.

First, in the degenerative case, the state $\mathbf{x}(t) \oplus \mathbf{z}(t)$ enters some Ω_j, $j \geq 0$ and stays there afterward, which automatically guarantees that $0 \leq |h(\mathbf{z})| \leq h(\mathbf{z}_0)$. Then, according to (4.71), the trajectory $\mathbf{x}(t)$ satisfies the following inequality:

$$\begin{aligned} \|\mathbf{x}(t)\|_{\mathcal{A}} &\leq \beta(\|\mathbf{x}_0\|_{\mathcal{A}}, t - t_0) + c\|h(\mathbf{z}(t))\|_{\infty,[t_0,t]} \\ &\leq \beta(\|\mathbf{x}_0\|_{\mathcal{A}}, t - t_0) + c|h(\mathbf{z}_0)|. \end{aligned} \tag{A4.24}$$

Taking into account that $\beta(\cdot,\cdot) \in \mathcal{KL}$ we can conclude that (A4.24) implies that

$$\lim\sup_{t\to\infty} \|\mathbf{x}(t)\|_{\mathcal{A}} \leq c|h(\mathbf{z}_0)|. \tag{A4.25}$$

Therefore the statements of the theorem hold.

Let us consider the second alternative, where the state $\mathbf{x}(t) \oplus \mathbf{z}(t)$ enters each Ω_j and leaves later. Given that $h(\mathbf{z}(t))$ is monotone and non-increasing in t, this implies that there exists an ordered sequence of time instants t_j:

$$t_0 < t_1 < t_2 \cdots t_j < t_{j+1} \cdots \tag{A4.26}$$

such that

$$h(\mathbf{z}(t_i)) = \sigma_i h(\mathbf{z}_0). \tag{A4.27}$$

Hence, in order to prove the theorem we must show that the sequence $\{t_i\}_{i=0}^{\infty}$ does not converge. In other words, the boundary $\sigma_\infty h(\mathbf{z}_0) = 0$ will not be reached in finite time.

In order to do this, let us estimate the upper bounds for the following differences:

$$T_i = t_{i+1} - t_i.$$

Taking into account inequality (4.73) and the fact that $\gamma_0(\cdot) \in \mathcal{K}_e$, we can derive that

$$\begin{aligned} h(\mathbf{z}(t_i)) - h(\mathbf{z}(t_{i+1})) &\leq T_i \max_{\tau\in[t_i,t_{i+1}]} \gamma_0(\|\mathbf{x}(\tau)\|_{\mathcal{A}}) \\ &\leq T_i \gamma_0(\|\mathbf{x}(\tau)\|_{\mathcal{A}\infty,[t_i,t_{i+1}]}). \end{aligned} \tag{A4.28}$$

According to the definition of t_i in (A4.27) and noticing that the sequence $\mathcal{S}$ is strictly decreasing, we have

$$h(\mathbf{z}(t_i)) - h(\mathbf{z}(t_{i+1})) = (\sigma_i - \sigma_{i+1})h(\mathbf{z}_0) > 0.$$

Hence $h(\mathbf{z}_0) > 0$ implies that $\gamma_0(\|\mathbf{x}(\tau)\|_{\mathcal{A}_\infty,[t_i,t_{i+1}]}) > 0$ and, therefore, (A4.28) results in the following estimate of T_i:

$$T_i \geq \frac{h(\mathbf{z}(t_i)) - h(\mathbf{z}(t_{i+1}))}{\gamma_0(\|\mathbf{x}(\tau)\|_{\mathcal{A}_\infty,[t_i,t_{i+1}]})} = \frac{h(\mathbf{z}_0)(\sigma_i - \sigma_{i+1})}{\gamma_0(\|\mathbf{x}(\tau)\|_{\mathcal{A}_\infty,[t_i,t_{i+1}]})}. \tag{A4.29}$$

Taking into account that $h(\mathbf{z}(t))$ is non-increasing over $[t_i, t_{i+1}]$ and using (4.71), we can bound the norm $\|\mathbf{x}(\tau)\|_{\mathcal{A}_\infty,[t_i,t_{i+1}]}$ as follows:

$$\begin{aligned}\|\mathbf{x}(\tau)\|_{\mathcal{A}_\infty,[t_i,t_{i+1}]} &\leq \beta(\|\mathbf{x}(t_i)\|_{\mathcal{A}}, 0) + c\|h(\mathbf{z}(\tau))\|_{\infty,[t_i,t_{i+1}]} \\ &\leq \beta(\|\mathbf{x}(t_i)\|_{\mathcal{A}}, 0) + c \cdot \sigma_i h(\mathbf{z}_0).\end{aligned} \tag{A4.30}$$

Hence, on combining (A4.29) and (A4.30) we obtain that

$$T_i \geq \frac{h(\mathbf{z}_0)(\sigma_i - \sigma_{i+1})}{\gamma_0(\sigma_i(\sigma_i^{-1}\beta(\|\mathbf{x}(t_i)\|_{\mathcal{A}}, 0) + c \cdot h(\mathbf{z}_0)))}.$$

Then, using property (4.75) of function γ_0, we can derive that

$$T_i \geq \frac{h(\mathbf{z}_0)(\sigma_i - \sigma_{i+1})}{\gamma_{0,1}(\sigma_i)} \frac{1}{\gamma_{0,2}(\sigma_i^{-1}\beta(\|\mathbf{x}(t_i)\|_{\mathcal{A}}, 0) + c \cdot h(\mathbf{z}_0))}. \tag{A4.31}$$

Taking into account condition (4.92) of the theorem, the theorem will be proven if we assure that

$$T_i \geq \tau_i \tag{A4.32}$$

for all $i = 0, 1, 2, \ldots, \infty$. We prove this claim by induction with respect to the index $i = 0, 1, \ldots, \infty$. We start with $i = 0$, and then show that for all $i > 0$ the implication

$$T_i \geq \tau_i \Rightarrow T_{i+1} \geq \tau_{i+1} \tag{A4.33}$$

holds. Let us prove that (A4.32) holds for $i = 0$. To this end, consider the term

$$(\sigma_i - \sigma_{i+1})/\gamma_{0,1}(\sigma_i).$$

As follows immediately from the conditions of the theorem, (4.90), we have that

$$\frac{\sigma_i - \sigma_{i+1}}{\gamma_{0,1}(\sigma_i)} \geq \tau_i \Delta_0 \ \forall\, i \geq 0. \tag{A4.34}$$

In particular

$$\frac{\sigma_0 - \sigma_1}{\gamma_{0,1}(\sigma_0)} \geq \tau_0 \Delta_0.$$

Therefore, inequality (A4.31) reduces to

$$T_0 \geq \tau_0 \Delta_0 \frac{h(\mathbf{z}_0)}{\gamma_{0,2}(\sigma_0^{-1}\beta(\|\mathbf{x}(t_0)\|_{\mathcal{A}}, 0) + c \cdot h(\mathbf{z}_0))}. \tag{A4.35}$$

Moreover, taking into account Property 4.3 and (4.83) and (4.84), we can derive the following estimate:

$$\begin{aligned}\sigma_0^{-1}\beta(\|\mathbf{x}(t_0)\|_{\mathcal{A}},0) &\le \sigma_0^{-1}\phi_0(\|\mathbf{x}(t_0)\|_{\mathcal{A}}) + \sigma_0^{-1}\upsilon_0(c\cdot|h(\mathbf{z}_0)|\sigma_0)\\ &\le B_1(\|\mathbf{x}_0\|_{\mathcal{A}}) + B_2(|h(\mathbf{z}_0)|,c).\end{aligned}$$

According to the theorem conditions $\mathbf{x}_0$ and $\mathbf{z}_0$ satisfy inequality (4.91). This in turn implies that

$$\begin{aligned}&\gamma_{0,2}(\sigma_0^{-1}\beta(\|\mathbf{x}(t_0)\|_{\mathcal{A}},0) + c\cdot h(\mathbf{z}_0))\\ &\quad\le \gamma_{0,2}(B_1(\|\mathbf{x}_0\|_{\mathcal{A}}) + B_2(|h(\mathbf{z}_0)|,c) + c\cdot h(\mathbf{z}_0)) \le \Delta_0\cdot h(\mathbf{z}_0).\end{aligned} \tag{A4.36}$$

On combining (A4.35) and (A4.36) we obtain the desired inequality

$$T_0 \ge \tau_0\Delta_0\frac{h(\mathbf{z}_0)}{\gamma_{0,2}(\sigma_0^{-1}\beta(\|\mathbf{x}(t_0)\|_{\mathcal{A}},0) + c\cdot h(\mathbf{z}_0))} \ge \tau_0\frac{\Delta_0 h(\mathbf{z}_0)}{\Delta_0 h(\mathbf{z}_0)} = \tau_0.$$

Thus the basis of induction is proven.

Let us assume that (A4.32) holds for all $i = 0,\ldots,n$, $n \ge 0$. We shall prove now that implication (A4.33) holds for $i = n+1$. Consider the term $\beta(\|\mathbf{x}(t_{n+1})\|_{\mathcal{A}},0)$:

$$\begin{aligned}\beta(\|\mathbf{x}(t_{n+1})\|_{\mathcal{A}},0) &\le \beta(\beta(\|\mathbf{x}(t_n)\|_{\mathcal{A}},T_n) + c\|h(\mathbf{z}(\tau))\|_{\infty,[t_n,t_{n+1}]},0)\\ &\le \beta(\beta(\|\mathbf{x}(t_n)\|_{\mathcal{A}},T_n) + c\cdot\sigma_n\cdot h(\mathbf{z}_0),0).\end{aligned}$$

Taking into account Property 4.2 (specifically, inequality (4.82)) and (4.83)–(4.85), we can derive that

$$\begin{aligned}\beta(\|\mathbf{x}(t_{n+1})\|_{\mathcal{A}},0) &\le \beta(\xi_n\cdot\beta(\|\mathbf{x}(t_n)\|_{\mathcal{A}},0) + c\cdot\sigma_n\cdot h(\mathbf{z}_0),0)\\ &\le \phi_1(\|\mathbf{x}(t_n)\|_{\mathcal{A}}) + \upsilon_1(c\cdot|h(\mathbf{z}_0)|\cdot\sigma_n).\end{aligned} \tag{A4.37}$$

Notice that, according to the inductive hypothesis ($T_i \ge \tau_i$), the following holds:

$$\begin{aligned}\|\mathbf{x}(t_{i+1})\|_{\mathcal{A}} &\le \beta(\|\mathbf{x}(t_i)\|_{\mathcal{A}},T_i) + c\cdot\sigma_i\cdot h(\mathbf{z}_0)\\ &\le \xi_i\beta(\|\mathbf{x}(t_i)\|_{\mathcal{A}},0) + c\cdot\sigma_i\cdot h(\mathbf{z}_0)\end{aligned} \tag{A4.38}$$

for all $i = 0, \ldots, n$. Then (A4.37), (A4.38), and (4.83)–(4.85) imply that

$$\begin{aligned}\beta(\|\mathbf{x}(t_{n+1})\|_{\mathcal{A}}, 0) &\leq \phi_1(\xi_i \beta(\|\mathbf{x}(t_{n-1})\|_{\mathcal{A}}, 0) + c \cdot \sigma_{n-1} \cdot h(\mathbf{z}_0)) \\ &\quad + \upsilon_1(c \cdot |h(\mathbf{z})_0| \cdot \sigma_n) \\ &\leq \phi_2(\|\mathbf{x}(t_{n-1})\|_{\mathcal{A}}) + \upsilon_2(c \cdot |h(\mathbf{z}_0)| \cdot \sigma_{n-1}) \\ &\quad + \upsilon_1(c \cdot |h(\mathbf{z}_0)| \cdot \sigma_n) \\ &\leq \phi_{n+1}(\|\mathbf{x}_0\|_{\mathcal{A}}) + \sum_{i=1}^{n+1} \upsilon_i(c \cdot |h(\mathbf{z}_0)| \sigma_{n+1-i}) \\ &\leq \phi_{n+1}(\|\mathbf{x}_0\|_{\mathcal{A}}) + \sum_{i=0}^{n+1} \upsilon_i(c \cdot |h(\mathbf{z}_0)| \sigma_{n+1-i}). \end{aligned} \tag{A4.39}$$

According to Property 4.3, the term

$$\sigma_{n+1}^{-1} \left(\phi_{n+1}(\|\mathbf{x}_0\|_{\mathcal{A}}) + \sum_{i=0}^{n+1} \upsilon_i(c \cdot |h(\mathbf{z}_0)| \sigma_{n+1-i}) \right)$$

is bounded from above by the sum

$$B_1(\|\mathbf{x}_0\|_{\mathcal{A}}) + B_2(|h(\mathbf{z}_0)|, c).$$

Therefore, the monotonicity of $\gamma_{0,2}$, estimate (A4.39), and inequality (4.91) lead to the following inequality:

$$\begin{aligned}\gamma_{0,2}(\sigma_{n+1}^{-1}\beta(\|\mathbf{x}(t_{n+1})\|_{\mathcal{A}}, 0) + c \cdot h(\mathbf{z}_0)) &\leq \gamma_{0,2}(B_1(\|\mathbf{x}_0\|_{\mathcal{A}}) + B_2(|h(\mathbf{z}_0)|, c) \\ &\quad + c \cdot h(\mathbf{z}_0)) \\ &\leq h(\mathbf{z}_0)\Delta_0. \end{aligned}$$

Hence, according to (A4.31) and (A4.34) we have

$$\begin{aligned} T_{n+1} &\geq \frac{\sigma_{n+1} - \sigma_{n+2}}{\gamma_{0,1}(\sigma_{n+1})} \frac{h(\mathbf{z}_0)}{\gamma_{0,2}(\sigma_{n+1}^{-1}\beta(\|\mathbf{x}(t_{n+1})\|_{\mathcal{A}}, 0) + c \cdot h(\mathbf{z}_0))} \\ &\geq \tau_{n+1} \frac{\Delta_0 h(\mathbf{z}_0)}{\Delta_0 h(\mathbf{z}_0)} = \tau_{n+1}. \end{aligned}$$

Thus implication (A4.33) is proven. This implies that $h(\mathbf{z}(t)) \in [0, h(\mathbf{z}_0)]$ for all $t \geq t_0$ and, consequently, that (A4.25) holds. □

A4.6 Proof of Lemma 4.2

As follows from the assumptions, $h(\mathbf{z}(t, \mathbf{z}_0))$ is bounded. Assume it belongs to the interval $[a, h(\mathbf{z}_0)]$, $a \leq h(\mathbf{z}_0)$. Taking into account that $h(\mathbf{z}(t, \mathbf{z}_0))$ is bounded and monotone in t (every subsequence of it is again monotone) and applying the Bolzano–Weierstrass theorem, we can conclude that $h(\mathbf{z}(t, \mathbf{z}_0))$ converges in $[a, h(\mathbf{z}_0)]$. In particular, there exists $\bar{h} \in [a, h(\mathbf{z}_0)]$ such that

$$\lim_{t\to\infty} h(\mathbf{z}(t, \mathbf{z}_0)) = \bar{h}. \tag{A4.40}$$

Therefore, as follows from (4.73), we can conclude that

$$\begin{aligned} 0 &\leq \lim_{t\to\infty} \int_{t_0}^{t} \gamma_1(\|\mathbf{x}(\tau, \mathbf{x}_0)\|_{\mathcal{A}})d\tau \leq \lim_{t\to\infty} (h(\mathbf{z}_0) - h(\mathbf{z}(t, \mathbf{z}_0))) \\ &= h(\mathbf{z}_0) - \bar{h} \leq h(\mathbf{z}_0) - a < \infty. \end{aligned} \tag{A4.41}$$

According to the lemma assumptions, system $\mathcal{S}_{\mathrm{a}}$ has steady-state characteristics. This means that there exists a constant $\bar{x} \in \mathbb{R}_{\geq 0}$ such that

$$\lim_{t\to\infty} \|\mathbf{x}(t, \mathbf{x}_0)\|_{\mathcal{A}} = \bar{x}. \tag{A4.42}$$

Suppose that $\bar{x} > 0$. Then it follows from (A4.42) that there exists a time instant t_1, $t_0 \leq t_1 < \infty$, and some constant $0 < \delta < \bar{x}$ such that

$$\|\mathbf{x}(t)\|_{\mathcal{A}} \geq \delta \; \forall\, t \geq t_1.$$

Hence, using (A4.41) and noticing that $\gamma_1 \in \mathcal{K}_e$, we obtain

$$\infty > h(\mathbf{z}_0) - \bar{h} \geq \lim_{t\to\infty} \int_{t_0}^{t} \gamma_1(\|\mathbf{x}(\tau, \mathbf{x}_0)\|_{\mathcal{A}})d\tau \geq \lim_{t\to\infty} \int_{t_1}^{t} \gamma_1(\delta)d\tau = \infty.$$

Thus we have obtained a contradiction. Hence, $\bar{x} = 0$ and, consequently,

$$\lim_{t\to\infty} \|\mathbf{x}(t)\|_{\mathcal{A}} = 0.$$

Then, according to the notion of a steady-state characteristic in Definition 4.5.1, this is possible only if $\bar{h} \in \chi^{-1}(0)$. □

A4.7 Proof of Lemma 4.3

Analogously to the proof of Lemma 4.2, we notice that (A4.41) holds. This, however, implies that for any constant and positive T the limit

$$\lim_{t\to\infty} \int_{t}^{t+T} \gamma_1(\|\mathbf{x}(\tau)\|_{\mathcal{A}})d\tau$$

exists and equals zero. Furthermore, $h(\mathbf{z}(t, \mathbf{z}_0)) \in [0, h(\mathbf{z}_0)]$ for all $t \geq t_0$. Hence, there exists a time instant t' such that

$$\|\mathbf{x}(t)\|_{\mathcal{A}} \leq c \cdot h(\mathbf{z}_0) + \varepsilon, \; \forall \, t \geq t',$$

where $\varepsilon > 0$ is arbitrarily small. Then, taking into account (4.98) we can conclude that

$$\lim_{t \to \infty} \int_t^{t+T} \gamma_1(\|\mathbf{x}(\tau)\|_{\mathcal{A}}) d\tau \geq \bar{\gamma} \int_t^{t+T} \|\mathbf{x}(\tau)\|_{\mathcal{A}} \, d\tau = 0. \tag{A4.43}$$

Given that (A4.40) holds, that system (4.71) has the steady-state characteristic on average and that $\chi_T(\cdot)$ has no zeros in the positive domain, the limiting relation (A4.43) is possible only if $\bar{h} = 0$. Then, according to (4.71), $\lim_{t \to \infty} \|\mathbf{x}(t)\|_{\mathcal{A}} = 0$. □

A4.8 Proof of Corollary 4.2

As follows from Theorem 4.7, the state $\mathbf{x}(t, \mathbf{x}_0) \oplus \mathbf{z}(t, \mathbf{z}_0)$ converges to the set Ω_{a} specified by (4.89). Hence $h(\mathbf{z}(t, \mathbf{z}_0))$ is bounded. Then, according to (4.73), estimate (A4.41) holds. This, in combination with condition (4.101), implies that $\mathbf{z}(t, \mathbf{z}_0)$ is bounded. In other words

$$\mathbf{x}(t, \mathbf{x}_0) \oplus \mathbf{z}(t, \mathbf{z}_0) \in \Omega' \; \forall \, t \geq t_0,$$

where Ω' is a bounded subset in $\mathbb{R}^n \times \mathbb{R}^m$. By applying the Bolzano–Weierstrass theorem we can conclude that for every point $\mathbf{x}_0 \oplus \mathbf{z}_0 \in \Omega_\gamma$ there is an ω-limit set $\omega(\mathbf{x}_0 \oplus \mathbf{z}_0) \subseteq \Omega'$ (non-empty).

As follows from (C3) and Lemma 4.2, the following holds:

$$\lim_{t \to \infty} h(\mathbf{z}(t, \mathbf{z}_0)) \in \chi^{-1}(0).$$

Therefore, given that $h(\cdot) \in \mathcal{C}^0$, we can obtain that

$$\lim_{t_i \to \infty} h(\mathbf{z}(t_i, \mathbf{z}_0)) = h\left(\lim_{t_i \to \infty} \mathbf{z}(t_i, \mathbf{z}_0) \right) = h(\omega_z(\mathbf{x}_0 \oplus \mathbf{z}_0)) \in \chi^{-1}(0).$$

In other words,

$$\omega_z(\mathbf{x}_0 \oplus \mathbf{z}_0) \subseteq \Omega_h = \{\mathbf{x} \in \mathbb{R}^n, \; \mathbf{z} \in \mathbb{R}^m | h(\mathbf{z}) \in \chi^{-1}(0)\}.$$

Moreover,

$$\omega_x(\mathbf{x}_0 \oplus \mathbf{z}_0) \subseteq \Omega_{\mathrm{a}} = \{\mathbf{x} \in \mathbb{R}^n, \; \mathbf{z} \in \mathbb{R}^m | \; \|\mathbf{x}\|_{\mathcal{A}} = 0\}.$$

According to assumption (C1), the flow $\mathbf{x}(t, \mathbf{x}_0) \oplus \mathbf{z}(t, \mathbf{z}_0)$ is generated by a system of autonomous differential equations with locally Lipschitz right-hand sides. Then, as follows from Khalil (2002) (Lemma 4.1, page 127),

$$\lim_{t \to \infty} \text{dist}(\mathbf{x}(t, \mathbf{x}_0) \oplus \mathbf{z}(t, \mathbf{z}_0), \omega(\mathbf{x}_0 \oplus \mathbf{z}_0)) = 0.$$

Noticing that

$$\text{dist}(\mathbf{x}(t, \mathbf{x}_0) \oplus \mathbf{z}(t, \mathbf{z}_0), \omega(\mathbf{x}_0 \oplus \mathbf{z}_0)) \geq \text{dist}(\mathbf{x}(t, \mathbf{x}_0), \Omega_{\text{a}}) + \text{dist}(\mathbf{z}(t, \mathbf{z}_0), \Omega_h)$$

we can finally obtain that

$$\lim_{t \to \infty} \text{dist}(\mathbf{x}(t, \mathbf{x}_0), \Omega_{\text{a}}) = 0, \qquad \lim_{t \to \infty} \text{dist}(\mathbf{z}(t, \mathbf{z}_0), \Omega_h) = 0.$$

□

A4.9 Proof of Corollary 4.3

As follows from Theorem 4.7, the corollary will be proven if Properties 4.1–4.3 are satisfied and also (4.90)–(4.92) hold. In order to satisfy Property 4.1, we select the following sequence $\mathcal{S}$:

$$\mathcal{S} = \{\sigma_i\}_{i=0}^{\infty}, \; \sigma_i = \frac{1}{\kappa^i}, \; \kappa \in \mathbb{R}_{\geq 0}, \; \kappa > 1. \tag{A4.44}$$

Let us choose sequences $\mathcal{T}$ and Ξ as follows:

$$\mathcal{T} = \{\tau_i\}_{i=0}^{\infty}, \; \tau_i = \tau^*, \tag{A4.45}$$

$$\Xi = \{\xi_i\}_{i=0}^{\infty}, \; \xi_i = \xi^*, \tag{A4.46}$$

where τ^* and ξ^* are positive constants that have yet to be defined. Notice that choosing $\mathcal{T}$ as in (A4.45) automatically fulfills condition (4.92) of Theorem 4.7. On the other hand, taking into account (4.82) and (4.106) and that $\beta_t(t)$ is monotonically decreasing in t, this choice defines a constant ξ^* as follows:

$$\beta_t(\tau^*) \leq \xi^* \beta_t(0) < \beta_t(0), \; 0 \leq \xi^* < 1. \tag{A4.47}$$

Given that the inverse β_t^{-1} exists, (4.107), this choice is always possible. In particular, (A4.47) will be satisfied for the following values of τ^*:

$$\tau^* \geq \beta_t^{-1}(\xi^* \beta_t(0)). \tag{A4.48}$$

Let us now find the values for τ^* and ξ^* such that Property 4.3 is also satisfied. To this end, consider systems of functions Φ and Υ specified by (4.83) and (4.84). Notice that the function $\beta(s, 0)$ in (4.83) and (4.84) is linear for system (4.106),

$$\beta(s, 0) = s \cdot \beta_t(0),$$

and therefore the functions $\rho_{\phi,j}(\cdot)$ and $\rho_{\upsilon,j}(\cdot)$ are identity maps. Hence, Φ and Υ reduce to the following:

$$\Phi: \begin{array}{l} \phi_j(s) = \phi_{j-1} \cdot \xi^* \cdot \beta(s,0) = \xi^* \cdot \beta_t(0) \cdot \phi_{j-1}(s), \\ \phi_0(s) = \beta_t(0) \cdot s, \ j = 1, \ldots, i \end{array} \tag{A4.49}$$

$$\Upsilon: \begin{array}{l} \upsilon_j(s) = \phi_{j-1}(s), \\ \upsilon_0(s) = \beta_t(0) \cdot s, \ j = 1, \ldots, i. \end{array} \tag{A4.50}$$

Taking into account (A4.44), (A4.49), and (A4.50), let us explicitly formulate requirements (4.86) and (4.87) in Property 4.3. These conditions are equivalent to the boundedness of the following functions:

$$\|\mathbf{x}(t_0)\|_{\mathcal{A}} \cdot \beta_t(0) \cdot \kappa^n (\xi^* \cdot \beta_t(0))^n, \tag{A4.51}$$

$$\begin{aligned} &\kappa^n \left(\beta_t(0) \frac{c|h(\mathbf{z}_0)|}{\kappa^n} + \frac{\beta_t(0) c |h(\mathbf{z}_0)|}{\kappa^{n-1}} + \beta_t(0) \sum_{i=2}^{n} c|h(\mathbf{z}_0)| \frac{1}{k^{n-i}} (\xi^* \cdot \beta_t(0))^{i-1} \right) \\ &= \beta_t(0) c |h(\mathbf{z}_0)| + \beta_t(0) c |h(\mathbf{z}_0)| \kappa \left(1 + \sum_{i=2}^{n} \kappa^{i-1} (\xi^* \cdot \beta_t(0))^{i-1} \right). \end{aligned} \tag{A4.52}$$

Boundedness of the functions $B_1(\|\mathbf{x}_0\|_{\mathcal{A}})$ and $B_2(|h(\mathbf{z}_0)|, c)$ is ensured if ξ^* satisfies the inequality

$$\xi^* \leq \frac{d}{\kappa \cdot \beta_t(0)} \tag{A4.53}$$

for some $0 \leq d < 1$. Notice that $\kappa > 1$ and $\beta_t(0) \geq 1$ imply that $\xi^* \leq 1$ and therefore constant τ^* satisfying (A4.48) will always be defined. Hence, according to (A4.51) and (A4.52), the functions $B_1(\|\mathbf{x}_0\|_{\mathcal{A}})$ and $B_2(|h(\mathbf{z}_0)|, c)$ satisfying Property 4.3 can be chosen as

$$\begin{aligned} B_1(\|\mathbf{x}_0\|_{\mathcal{A}}) &= \beta_t(0) \, \|\mathbf{x}_0\|_{\mathcal{A}}, \\ B_2(|h(\mathbf{z}_0)|, c) &= \beta_t(0) \cdot c \cdot |h(\mathbf{z}_0)| \left(1 + \frac{\kappa}{1-d} \right). \end{aligned} \tag{A4.54}$$

In order to apply Theorem 4.7 we have to check the remaining conditions (4.90) and (4.91). This requires the possibility of factorization (4.75) for the function $\gamma_0(\cdot)$. According to assumption (4.108) of the corollary the function $\gamma_0(\cdot)$ is Lipschitz:

$$|\gamma_0(s)| \leq D_{\gamma,0} \cdot |s|.$$

This allows us to choose the functions $\gamma_{0,1}(\cdot)$ and $\gamma_{0,2}(\cdot)$ as follows:

$$\gamma_{0,1}(s) = s, \qquad \gamma_{0,2}(s) = D_{\gamma,0} \cdot s. \tag{A4.55}$$

Condition (4.90), therefore, is equivalent to solvability of the following inequality:

$$\left(\frac{1}{\kappa^i} - \frac{1}{\kappa^{i+1}}\right)\frac{\kappa^i}{\tau^*} \geq \Delta_0. \qquad \text{(A4.56)}$$

Taking into account inequalities (A4.48) and (A4.53), we can derive that solvability of

$$\Delta_0 = \left(\beta_t^{-1}\left(\frac{d}{\kappa}\right)\right)^{-1}\frac{\kappa - 1}{\kappa} \qquad \text{(A4.57)}$$

implies the existence of $\Delta_0 > 0$ satisfying (A4.56) and, consequently, condition (4.90) of Theorem 4.7. Given that $d < 1$, $\kappa > 1$, and $\beta_t(0) \geq 1$, a positive solution to (A4.57) is always defined. Hence, the proof will be complete and the claim is non-vacuous if the domain

$$D_{\gamma,0} \leq \left(\beta_t^{-1}\left(\frac{d}{\kappa}\right)\right)^{-1}\frac{\kappa - 1}{\kappa} \times \frac{h(\mathbf{z}_0)}{\beta_t(0)\|\mathbf{x}_0\|_{\mathcal{A}} + \beta_t(0)\cdot c\cdot |h(\mathbf{z}_0)|(1 + \kappa/(1-d)) + c|h(\mathbf{z}_0)|} \qquad \text{(A4.58)}$$

is not empty. □

A4.10 Proof of Corollary 4.4

Similarly to the proof of Theorem 4.7, we introduce a strictly decreasing sequence

$$\{\sigma_i\},\ i = 0, 1, \ldots,$$

such that $\sigma_0 = 1$, and σ_i asymptotically converge to zero. Let

$$\{t_i\},\ i = 1, \ldots$$

be an ordered sequence of time instants such that[17]

$$h(t_i) = \sigma_i h(t_0).$$

We wish to show that the amount of time needed to reach the set specified by $\|\mathbf{x}(t)\| = 0$ from the given initial condition is infinite.

[17] When this equality does not hold, nothing remains to be proven; for $\|h(t)\|$ will always be separated away from zero for all $t \geq t_0$.

Consider time differences $T_i = t_i - t_{i-1}$. It is clear that

$$T_i \|\mathbf{x}(\tau)\|_{\varepsilon\infty,[t_{i-1},t_i]} \geq \frac{h_0(\sigma_{i-1}-\sigma_i)}{D_{\gamma,0}},$$

$$T_i \geq \begin{cases} \dfrac{h_0(\sigma_{i-1}-\sigma_i)}{D_{\gamma,0}} \dfrac{1}{\|\mathbf{x}(\tau)\|_{\infty,[t_{i-1},t_i]} - \varepsilon}, \\ \|\mathbf{x}(\tau)\|_{\infty,[t_{i-1},t_i]} - \varepsilon > 0, \\ \infty, \quad \|\mathbf{x}(\tau)\|_{\infty,[t_{i-1},t_i]} - \varepsilon \leq 0. \end{cases} \tag{A4.59}$$

Consider the case when $\|\mathbf{x}(\tau)\|_{\infty,[t_{i-1},t_i]} - \varepsilon > 0$ for all i, and introduce the sequence $\{\tau_i\}$, $\tau_i = \tau^*$, $\tau^* \in \mathbb{R}_{>0}$, $i = 1, \ldots$ The sequence $\{\tau_i = \tau^*\}$ gives rise to the series with divergent partial sums $\sum_i \tau_i = \sum_i \tau^*$. Hence proving that

$$T_i \geq \tau^* \Rightarrow T_{i+1} \geq \tau^* \; \forall\, i$$

will constitute the proof that $\mathbf{x}(t)$ and $h(t)$ are bounded for all $t \geq t_0$. Let $T_j \geq \tau^*$ for all $1 \leq j \leq i-1$ and consider

$$\begin{aligned} \|\mathbf{x}(\tau)\|_{\infty,[t_{i-1},t_i]} &\leq \beta_t(0)\, \|\mathbf{x}(t_{i-1})\| + ch_0\sigma_{i-1} + \Delta \\ &\leq \beta_t(0)[\beta_t(T_{i-1})\, \|\mathbf{x}(t_{i-2})\| + ch_0\sigma_{i-2}] + ch_0\sigma_{i-1} + \beta_t(0)\Delta + \Delta \\ &\leq \beta_t(0)\beta_t^2(\tau^*)\, \|\mathbf{x}(t_{i-3})\| + P_2, \end{aligned}$$

where

$$\begin{aligned} P_2 = \beta_t(0)\big[\beta_t(\tau^*)c\sigma_{i-3} + c\sigma_{i-2}\big]h_0 + c\sigma_{i-1}h_0 \\ + \beta_t(0)\left[\beta_t(\tau^*)\Delta + \Delta\right] + \Delta. \end{aligned}$$

Repeating this iteration with respect to i leads to

$$\begin{aligned} &\|\mathbf{x}(\tau)\|_{\infty,[t_{i-1},t_i]} \leq \beta_t(0)\beta_t^3(\tau^*)\|\mathbf{x}(t_{i-4})\| + P_3, \\ &P_3 = ch_0\beta_t(0)\left[\beta_t^2(\tau^*)\sigma_{i-4} + \beta_t(\tau^*)\sigma_{i-3} + \sigma_{i-2}\right] + \sigma_{i-1}ch_0 \\ &\quad + \Delta\beta_t(0)\left[\beta_t^2(\tau^*) + \beta_t(\tau^*) + 1\right] + \Delta \\ &\quad = ch_0\beta_t(0)\left[\sum_{j=0}^{2} \beta_t^j(\tau^*)\sigma_{i-j-2}\right] + ch_0\sigma_{i-1} + \Delta\beta_t(0)\left[\sum_{j=0}^{2}\beta_t^j(\tau^*)\right] + \Delta, \end{aligned}$$

and after $i-1$ steps we obtain

$$\|\mathbf{x}(\tau)\|_{\infty,[t_{i-1},t_i]} \leq \beta_t(0)\beta_t(\tau^*)^{i-1}\|\mathbf{x}(t_0)\| + P_{i-1},$$

$$P_{i-1} = ch_0\beta_t(0)\left[\sum_{j=0}^{i-2}\beta_t^j(\tau^*)\sigma_{i-j-2}\right] + ch_0\sigma_{i-1} + \Delta\beta_t(0)\left[\sum_{j=0}^{i-2}\beta_t^j(\tau^*)\right] + \Delta. \tag{A4.60}$$

From (A4.59) it follows that

$$T_i \geq \frac{\sigma_{i-1}-\sigma_i}{\sigma_{i-1}}\frac{h_0}{D_{\gamma,0}}\frac{1}{\sigma_{i-1}^{-1}\left(\|\mathbf{x}(\tau)\|_{\infty,[t_{i-1},t_i]} - \varepsilon\right)}.$$

Hence, if we can show that there exist $\mathbf{x}_0$ such that for some $\Delta_0 \in \mathbb{R}_{\geq 0}$ and ε,

$$\frac{\sigma_{i-1}-\sigma_i}{\sigma_{i-1}}\frac{h_0}{\tau^*} \geq \Delta_0, \tag{A4.61}$$

we have

$$D_{\gamma,0}\sigma_{i-1}^{-1}\left(\|\mathbf{x}(\tau)\|_{\infty,[t_{i-1},t_i]} - \varepsilon\right) \leq D_{\gamma,0}B(\mathbf{x}_0) \leq \Delta_0 \ \forall\, i, \tag{A4.62}$$

where $B(\cdot)$ is a function of $\mathbf{x}_0$, then boundedness of trajectories will follow. Consider the term $\sigma_{i-1}^{-1}\|\mathbf{x}(\tau)\|_{\infty,[t_{i-1},t_i]}$, and let

$$\sigma_i = \frac{1}{\kappa^i},\ \kappa > 1.$$

According to (A4.60) we have

$$\begin{aligned}
\sigma_{i-1}^{-1}\left(\|\mathbf{x}(\tau)\|_{\infty,[t_{i-1},t_i]} - \varepsilon\right) &\leq \beta_t(0)\left[\sigma_{i-1}^{-1}\beta_t(\tau^*)^{i-1}\right]\|\mathbf{x}(t_0)\| \\
&\quad + \sigma_{i-1}^{-1}P_{i-1} - \sigma_{i-1}^{-1}\varepsilon \\
&= \beta_t(0)(\kappa\beta_t(\tau^*))^{i-1}\|\mathbf{x}(t_0)\| + \kappa^{i-1}P_{i-1} - \kappa^{i-1}\varepsilon \\
&= \beta_t(0)(\kappa\beta_t(\tau^*))^{i-1}\|\mathbf{x}(t_0)\| \\
&\quad + ch_0\beta_t(0)\kappa\left[\sum_{j=0}^{i-2}\beta_t^j(\tau^*)\kappa^j\right] \\
&\quad + ch_0 + \kappa^{i-1}\left(\beta_t(0)\Delta\sum_{j=0}^{i-2}\beta_t^j(\tau^*) + \Delta - \varepsilon\right).
\end{aligned}$$

Hence choosing the value of τ^* as

$$\kappa\beta_t(\tau^*) \leq d,\ d \in (0,1) \tag{A4.63}$$

results in the following estimate:

$$\sigma_{i-1}^{-1}\left(\|\mathbf{x}(\tau)\|_{\infty,[t_{i-1},t_i]}-\varepsilon\right)\leq B(\mathbf{x}_0)$$
$$=\beta_t(0)\|\mathbf{x}(t_0)\|+ch_0\left(1+\frac{\beta_t(0)\kappa}{1-d}\right)+\kappa^{i-1}\left(\Delta\left[\frac{\beta_t(0)}{1-d/k}+1\right]-\varepsilon\right).$$

Condition (4.112) implies that

$$\kappa^{i-1}\left(\Delta\left[\frac{\beta_t(0)}{1-d/k}+1\right]-\varepsilon\right)\leq 0.$$

Hence

$$\sigma_{i-1}^{-1}\left(\|\mathbf{x}(\tau)\|_{\infty,[t_{i-1},t_i]}-\varepsilon\right)\leq B(\mathbf{x}_0)$$
$$=\beta_t(0)\|\mathbf{x}(t_0)\|+ch_0\left(1+\frac{\beta_t(0)\kappa}{1-d}\right).$$

Solving (A4.63) and (A4.61) with respect to Δ_0 results in

$$\Delta_0=\frac{\kappa-1}{\kappa}\left[\beta_t^{-1}\left(\frac{d}{\kappa}\right)\right]^{-1}h_0.$$

This in turn implies that for all $\mathbf{x}_0$ and h_0 such that

$$D_{\gamma,0}\leq\frac{\kappa-1}{\kappa}\left[\beta_t^{-1}\left(\frac{d}{\kappa}\right)\right]^{-1}\frac{h_0}{\beta_t(0)\|\mathbf{x}(t_0)\|+ch_0(1+\kappa\beta_t(0)/(1-d))}$$

the following implication must hold: $T_i\geq\tau^*\Rightarrow T_{i+1}\geq\tau^*$. Therefore, trajectories $\mathbf{x}(t)$ and $h(t)$ passing through $\mathbf{x}(t_0)=\mathbf{x}_0$ and $h(t_0)=h_0$ at $t=t_0$ are bounded in forward time. □

A4.11 Proof of Corollary 4.5

It follows from Corollary 4.3 that the state of the interconnection converges into Ω_a for all initial conditions $\mathbf{x}_0$ and $\mathbf{z}_0$ satisfying (A4.58). In other words the following inequality should hold:

$$D_{\gamma,0}\left(\beta_t(0)\|\mathbf{x}_0\|_{\mathcal{A}}+\beta_t(0)\cdot c\cdot|h(\mathbf{z}_0)|\left(1+\frac{\kappa}{1-d}\right)+c|h(\mathbf{z}_0)|\right)$$
$$\leq\left(\beta_t^{-1}\left(\frac{d}{\kappa}\right)\right)^{-1}\frac{\kappa-1}{\kappa}\cdot h(\mathbf{z}_0). \tag{A4.64}$$

Hence, assuming that $h(\mathbf{z}_0) > 0$, we can rewrite (A4.64) in the following way:

$$D_{\gamma,0}\cdot\beta_t(0)\|\mathbf{x}_0\|_{\mathcal{A}} \leq \left(\left(\beta_t^{-1}\left(\frac{d}{\kappa}\right)\right)^{-1}\frac{\kappa-1}{\kappa} - D_{\gamma,0}\cdot c\left(\beta_t(0)\cdot\left(1+\frac{\kappa}{1-d}\right)+1\right)\right)h(\mathbf{z}_0). \tag{A4.65}$$

Solutions to (A4.65) exist, however, if the inequality

$$\left(\beta_t^{-1}\left(\frac{d}{\kappa}\right)\right)^{-1}\frac{\kappa-1}{\kappa} \geq D_{\gamma,0}\cdot c\left(\beta_t(0)\cdot\left(1+\frac{\kappa}{1-d}\right)+1\right)$$

or, equivalently,

$$D_{\gamma,0}\cdot c\cdot\left(\beta_t(0)\cdot\left(1+\frac{\kappa}{1-d}\right)+1\right)\cdot\beta_t^{-1}\left(\frac{d}{\kappa}\right)\frac{\kappa}{\kappa-1} < 1 \tag{A4.66}$$

is satisfied. The estimate of the trapping region follows from (A4.65).

Let us finally show that continuity of $h(\mathbf{z})$ implies that the volume of Ω_γ is nonzero in $\mathbb{R}^n \oplus \mathbb{R}^m$. For the sake of compactness we rewrite inequality (A4.65) in the following form:

$$\|\mathbf{x}_0\|_{\mathcal{A}} \leq C_\gamma h(\mathbf{z}_0), \tag{A4.67}$$

where C_γ is a constant depending on $d, \kappa, \beta_t(0)$, and $D_{\gamma,0}$. Given that (A4.66) holds, we can conclude that $C_\gamma > 0$. According to (A4.67), the domain Ω_γ contains the following set:

$$\{\mathbf{x}_0 \in \mathbb{R}^n,\ \mathbf{z}_0 \in \mathbb{R}^m | h(\mathbf{z}_0) > D_z \in \mathbb{R}_{\geq 0},\ \|\mathbf{x}_0\|_{\mathcal{A}} \leq C_\gamma D_z\}.$$

Consider the following domain: $\Omega_{\mathbf{x},\gamma} = \{\mathbf{x}_0 \in \mathbb{R}^n |\ \|\mathbf{x}_0\|_{\mathcal{A}} \leq C_\gamma D_z\}$. Clearly, it contains a point $\mathbf{x}_{0,1} \in \mathbb{R}^n :\ \left\|\mathbf{x}_{0,1}\right\|_{\mathcal{A}} = C_\gamma D_z/2$. For the point $\mathbf{x}_{0,1}$ and for all $\boldsymbol{\varepsilon}_1 \in \mathbb{R}^n :\ \|\boldsymbol{\varepsilon}_1\| \leq C_\gamma D_z/4$ we have that $\left\|\mathbf{x}_{0,1}+\boldsymbol{\varepsilon}_1\right\|_{\mathcal{A}} = \inf_{\mathbf{q}\in\mathcal{A}}\|\mathbf{x}_{0,1}+\boldsymbol{\varepsilon}_1-\mathbf{q}\| \leq \inf_{\mathbf{q}\in\mathcal{A}}\{\|\mathbf{x}_{0,1}-\mathbf{q}\| + \|\boldsymbol{\varepsilon}_1\|\} \leq 3C_\gamma D_z/4$. On the other hand, $\left\|\mathbf{x}_{0,1}+\boldsymbol{\varepsilon}_1\right\|_{\mathcal{A}} = \inf_{\mathbf{q}\in\mathcal{A}}\|\mathbf{x}_{0,1}+\boldsymbol{\varepsilon}_1-\mathbf{q}\| \geq \inf_{\mathbf{q}\in\mathcal{A}}\{\|\mathbf{x}_{0,1}-\mathbf{q}\| - \|\boldsymbol{\varepsilon}_1\|\} \geq C_\gamma D_z/4$. This implies that there exists a set of points $\mathbf{x}_{0,2} = \mathbf{x}_{0,1}+\boldsymbol{\varepsilon}_1 \in \mathbb{R}^n$: $\|\mathbf{x}_{0,1}-\mathbf{x}_{0,2}\| \leq C_\gamma D_z/4$, $\mathbf{x}_{0,2} \notin \mathcal{A}$, $\left\|\mathbf{x}_{0,2}\right\|_{\mathcal{A}} \leq C_\gamma D_z$.

Consider now the following domain: $\Omega_{\mathbf{z},\gamma} = \{\mathbf{z}_0 \in \mathbb{R}^m | h(\mathbf{z}_0) > D_z\}$. Let us pick $\mathbf{z}_{0,1} \in \Omega_{\mathbf{z},\gamma}$: $h(\mathbf{z}_{0,1}) = 2D_z$. Because $h(\cdot)$ is continuous we have that

$$\forall\,\varepsilon > 0,\ \exists\,\delta > 0:\ \|\mathbf{z}_{0,1}-\mathbf{z}_{0,2}\| < \delta \Rightarrow |h(\mathbf{z}_{0,1})-h(\mathbf{z}_{0,2})| < \varepsilon.$$

Let $\varepsilon = D_z$, then $-D_z < h(\mathbf{z}_{0,1}) - h(\mathbf{z}_{0,2}) < D_z$ and therefore $h(\mathbf{z}_{0,2}) > D_z$. Hence there exists a set of points $\mathbf{z}_{0,2} \in \mathbb{R}^m$: $\|\mathbf{z}_{0,1}-\mathbf{z}_{0,2}\| < \delta$, $\mathbf{z}_{0,2} \in \Omega_{\mathbf{z},\gamma}$.

Consider the following set:

$$\Omega_{\mathbf{xz},\gamma} = \Big\{\mathbf{x}' \in \mathbb{R}^n,\ \mathbf{z}' \in \mathbb{R}^m \,|\ \|\mathbf{x}_{0,1} - \mathbf{x}'\|^2 + \|\mathbf{z}_{0,1} - \mathbf{z}'\|^2 \leq r^2,$$

$$r = \min\left\{\delta, \frac{C_\gamma D_z}{4}\right\}\Big\}.$$

For all $\mathbf{x}_0, \mathbf{z}_0 \in \Omega_{\mathbf{xz},\gamma}$ we have that $\mathbf{x}_0 \in \Omega_{\mathbf{x},\gamma}$ and $\mathbf{z}_0 \in \Omega_{\mathbf{z},\gamma}$. Hence, inequality (A4.67) holds, and $\mathbf{x}_0 \oplus \mathbf{z}_0 \in \Omega_\gamma$. The volume of the set $\Omega_{\mathbf{xz},\gamma}$ is defined by the volume of the interior of a sphere in $\mathbb{R}^{n+m}$ with non-zero radius. Thus the volume of $\Omega_\gamma \supset \Omega_{\mathbf{xz},\gamma}$ is also non-zero. □

5

Algorithms of adaptive regulation and adaptation in dynamical systems in the presence of nonlinear parametrization and/or possibly unstable target dynamics

In this chapter we discuss a range of synthesis problems in the domain of adaptive control and regulation for dynamical systems with nonlinear parametrization and, possibly, unstable target dynamics. Results presented in the previous chapters, such as e.g. the bottle-neck principle and (non-uniform) small-gain theorems from Chapter 4, will play important roles in the development of suitable formal statements of these problems. In particular, when specifying the target dynamics of an adapting system, we will exploit input–output characterizations such as input–state and input–output margins, and majorizing of mappings and functions. No stability requirements will be imposed on the target motions in the adapting system a priori. This will offer us greater flexibility and thus will create opportunities to overcome certain limitations of standard approaches (see Chapter 3) with regard to the target dynamics and nonlinear parametrization.

We begin by stating the general problem of adaptation and adaptive regulation and providing a set of solutions to this problem. Having developed these solutions, we will proceed by considering several specific problems of adaptation. These problems are

(1) adaptive regulation to invariant sets;
(2) adaptive control of interconnected nonlinear systems;
(3) parametric identification of systems of ordinary differential equations with monotone nonlinear parametrization;
(4) non-dominating adaptive control and identification for systems with general nonlinear parametrization of uncertainties.

In order to provide particular solutions to the general problem of adaptation and also to the specific problems (1)–(4) we introduce a synthesis method – the method of the *virtual adaptation algorithm*. The method is explained in detail in Section 5.2.1. The non-uniform small-gain theorem will be applied in problem (4), and the bottle-neck principle will be used in finding solutions to problems (1)–(3).

5.1 Problems of adaptive control of nonlinear systems in the presence of nonlinear parametrization

Consider models the dynamics of which are governed by the following system of ordinary differential equations:

$$\begin{aligned}\dot{\mathbf{x}}_1 &= \mathbf{f}_1(\mathbf{x}, t) + \mathbf{g}_1(\mathbf{x}, t)u,\\ \dot{\mathbf{x}}_2 &= \mathbf{f}_2(\mathbf{x}, \boldsymbol{\theta}, t) + \mathbf{g}_2(\mathbf{x}, t)u,\end{aligned} \tag{5.1}$$

where

$$\begin{aligned}\mathbf{x}_1 &= (x_{11}, \ldots, x_{1q})^{\mathrm{T}} \in \mathbb{R}^q,\\ \mathbf{x}_2 &= (x_{21}, \ldots, x_{2p})^{\mathrm{T}} \in \mathbb{R}^p,\\ \mathbf{x} &= (x_{11}, \ldots, x_{1q}, x_{21}, \ldots, x_{2p})^{\mathrm{T}} \in \mathbb{R}^n.\end{aligned}$$

The symbol $\boldsymbol{\theta} \in \Omega_\theta \subset \mathbb{R}^d$ in (5.1) stands for a vector of unknown parameters; for simplicity we will suppose that Ω_θ is a closed bounded subset of $\mathbb{R}^d$; $u \in \mathbb{R}$ is the control input; knowledge of the boundaries of Ω_θ is not required, unless specified otherwise or they are clear from the context. The vectors $\mathbf{x}_{1,0}$ and $\mathbf{x}_{2,0}$ are the initial conditions. The functions

$$\begin{aligned}&\mathbf{f}_1 : \mathbb{R}^n \times \mathbb{R} \to \mathbb{R}^q,\ \mathbf{f}_2 : \mathbb{R}^n \times \mathbb{R}^d \times \mathbb{R} \to \mathbb{R}^p,\\ &\mathbf{g}_1 : \mathbb{R}^n \times \mathbb{R} \to \mathbb{R}^q,\ \mathbf{g}_2 : \mathbb{R}^n \times \mathbb{R} \to \mathbb{R}^p\end{aligned}$$

are locally bounded and continuous in $\mathbf{x}$ (if not stated otherwise) and globally bounded and continuous in t.[1] The vector $\mathbf{x} \in \mathbb{R}^n$ is a state vector.

Unless stated otherwise, we will suppose that the values of functions $\mathbf{f}_1(\mathbf{x}, t)$ and $\mathbf{g}_1(\mathbf{x}, t)$ are known for every $\mathbf{x}$ and t, whereas the values of function $\mathbf{f}_2(\mathbf{x}, \boldsymbol{\theta}, t)$ may be unknown because e.g. of the dependence of $\mathbf{f}_2$ on the unknown $\boldsymbol{\theta}$. Thus, vectors $\mathbf{x}_1$ and $\mathbf{x}_2$ are referred to as *uncertainty-independent* and *uncertainty-dependent* partitions of $\mathbf{x}$, respectively. For the sake of compactness we will also use the following description of (5.1):

$$\dot{\mathbf{x}} = \mathbf{f}(\mathbf{x}, \boldsymbol{\theta}, t) + \mathbf{g}(\mathbf{x}, t)u, \tag{5.2}$$

where

$$\begin{aligned}\mathbf{g}(\mathbf{x}, t) &= (g_{11}(\mathbf{x}, t), \ldots, g_{1q}(\mathbf{x}, t), g_{21}(\mathbf{x}, t), \ldots, g_{2p}(\mathbf{x}, t))^{\mathrm{T}},\\ \mathbf{f}(\mathbf{x}, \boldsymbol{\theta}, t) &= (f_{11}(\mathbf{x}, t), \ldots, f_{1q}(\mathbf{x}, t), f_{21}(\mathbf{x}, \boldsymbol{\theta}, t), \ldots, f_{2p}(\mathbf{x}, \boldsymbol{\theta}, t))^{\mathrm{T}},\end{aligned}$$

[1] In principle the continuity requirement can be lifted, provided that solutions of the system remain defined, at least locally in t.

so all of the assumptions and constraints we imposed on the right-hand side of (5.1) apply automatically to the functions $\mathbf{f}(\mathbf{x}, \boldsymbol{\theta}, t)$ as well.

Our general objective here is to derive the control input as a function of the state variable $\mathbf{x}(t)$, controller parameters $\hat{\boldsymbol{\theta}}(t) : \mathbb{R}_{\geq 0} \rightarrow \mathbb{R}^d$, and time t, e.g. $u = u(\mathbf{x}(t), \hat{\boldsymbol{\theta}}(t), t)$ such that, for a class of nonlinearly parametrized $\mathbf{f}(\mathbf{x}, \boldsymbol{\theta}, t)$, all $\boldsymbol{\theta} \in \Omega_\theta$ and $\mathbf{x}_{1,0} \oplus \mathbf{x}_{2,0} \in \mathbb{R}^n$, (1) all trajectories of the system with $u = u(\mathbf{x}(t), \hat{\boldsymbol{\theta}}(t), t)$ are bounded and (2) state $\mathbf{x}(t)$ converges to a given bounded target set, (3) ensuring, if possible, that the estimate $\hat{\boldsymbol{\theta}}(t)$ converges to unknown $\boldsymbol{\theta} \in \Omega_\theta$ asymptotically.

Let us now proceed by providing a formal statement of the general synthesis problem of adaptive regulation. In the previous chapters we have seen the benefits of describing the behavior of complex and uncertain dynamical systems in terms of the majorizing mappings and corresponding input–output margins (Definition 4.3.4).[2] In particular, this description allows us to reduce the amount of prior knowledge about the system, which is especially useful when mathematical models of the system are uncertain. Therefore, bringing in elements of such description into the statement of the *synthesis* problem of adaptive regulation and control is a desirable option.

To this end, we introduce and consider the function $\psi : \mathbb{R}^n \times \mathbb{R}_{\geq 0} \rightarrow \mathbb{R}$, $\psi \in \mathcal{C}^1$ satisfying the following assumption.

Assumption 5.1 *For a given function $\psi(\mathbf{x}, t) \in \mathcal{C}^1$ and all $T \geq t_0$ the following property holds:*[3]

$$\|\mathbf{x}(t)\|_{\infty,[t_0,T]} \leq \tilde{\gamma}\left(\mathbf{x}_0, \boldsymbol{\theta}, \|\psi(\mathbf{x}(t), t)\|_{\infty,[t_0,T]}\right), \tag{5.3}$$

where $\tilde{\gamma}\left(\mathbf{x}_0, \boldsymbol{\theta}, \|\psi(\mathbf{x}(t), t)\|_{\infty,[t_0,T]}\right)$ is a non-negative, locally bounded function that is non-decreasing in $\|\psi(\mathbf{x}(t), t)\|_{\infty,[t_0,T]}$.

According to Definition 4.3.4, the function $\psi(\mathbf{x}, t)$ majorizes state $\mathbf{x}(t)$ of system (5.1), hence the distance from $\mathbf{x}(t)$ to

$$\Omega_0 = \{\mathbf{x}(t) \in \mathbb{R}^n | \psi(\mathbf{x}(t), t) = 0\} \tag{5.4}$$

is one of the natural performance criteria in this case. The fact that $\mathbf{x}(t)$ stays within an ε-neighborhood of Ω_0 implies that $\mathbf{x}(t)$ is bounded, and that the upper bound for $\|\mathbf{x}(t)\|$ is a function of the initial conditions $\mathbf{x}_0$, the unknown parameter vector $\boldsymbol{\theta}$, and ε. In what follows the set Ω_0 will be termed the *target set*.

In a number of standard statements of the problem of adaptive regulation and control the function $\psi(\mathbf{x}, t)$ should additionally satisfy some (*algebraic*) metric

[2] These are also known as macro-variables in the terminology of Kolesnikov (1994).
[3] Without substantial loss of generality we will suppose that $t_0 \geq 0$.

restrictions. In particular, it is required that $\psi(\mathbf{x}, t)$ be (positive) definite with respect to the target set Ω_0. For example, in problems involving regulating the state $\mathbf{x}(t)$ to $\boldsymbol{\xi} \in \mathbb{R}^n$, these restrictions lead to the following constraints:

$$\nu_1(\|\mathbf{x} - \boldsymbol{\xi}\|) \leq \psi(\mathbf{x}, t) \leq \nu_2(\|\mathbf{x} - \boldsymbol{\xi}\|), \quad \nu_1, \nu_2 \in \mathcal{K}_\infty. \tag{5.5}$$

In addition, the function $\psi(\mathbf{x}, t)$ is required to be a Lyapunov candidate for the closed-loop system at $\hat{\boldsymbol{\theta}} = \boldsymbol{\theta}$. Knowledge of $\psi(\mathbf{x}, t)$ is explicitly used in standard certainty-equivalence approaches (Fomin *et al.* 1981; Fradkov 1990; Fradkov *et al.* 1999). Finding such a Lyapunov candidate is not a trivial task. For systems with globally Lyapunov-unstable target dynamics no appropriate goal function $\psi(\mathbf{x}, t)$ could be given. That is why, instead of confining ourselves to the realm of stable systems and motions in the sense of Lyapunov, we opted to specify conservative target sets as in Assumption 5.1.

Specification of target sets as in Assumption 5.1 applies to a wider class of systems. Indeed, no stability constraints are imposed a priori, and there is no need to demand that the function $\psi(\mathbf{x}, t)$ in the definition of the target set (5.4) should additionally satisfy condition (5.5). The standard *algebraic* restriction (5.5) is replaced in Assumption 5.1 with *operator* relations. Standard norms $\|\cdot\|$ in $\mathbb{R}^n$ in (5.5) are replaced with *functional norms* $\|\mathbf{x}(t)\|_{p,[t_0,T]}$, $T \geq t_0$ in the *functional spaces* $L_p[t_0, T]$, $T \geq t_0$, $p \in \mathbb{R}_{\geq 1} \cup \infty$. This allows us to keep the function $\psi(\mathbf{x}, t)$ as a measure of the closeness of trajectories $\mathbf{x}(t)$ to the desired set $\Omega_{0,t}$ without imposing state-metric or definiteness restrictions (5.5) on $\psi(\mathbf{x}, t)$.

Assumption 5.1 can be interpreted as the *unbounded observability* (Jiang *et al.* 1994) of system (5.1) with respect to the "output" $\psi(\mathbf{x}, t)$. Clearly, it includes conventional definiteness requirements on $\psi(\mathbf{x}, t)$ as a special case. A large class of systems obeys Assumption 5.1 without requiring the function $\psi(\mathbf{x}, t)$ to be definite (see e.g. Example 4.3.1 in Chapter 4.). For a more general class of systems

$$\begin{aligned} \dot{x}_i &= f_i(x_1, \ldots, x_i) + x_{i+1}, \; i = \{1, \ldots, n-1\}, \\ \dot{x}_n &= f_n(x_1, \ldots, x_n) + \nu(\mathbf{x}, \boldsymbol{\theta}) + u \end{aligned} \tag{5.6}$$

and a class of non-definite functions $\psi(\mathbf{x}, t) = x_n - p(x_1, \ldots, x_{n-1})$, $p(\cdot) \in \mathcal{C}^1$, checking Assumption 5.1 amounts to establishing a *bounded-input–bounded-state* property for the following cascade:

$$\begin{aligned} \dot{x}_i &= f_i(x_1, \ldots, x_i) + x_{i+1}, \; i = \{1, \ldots, n-2\}, \\ \dot{x}_{n-1} &= p(x_1, \ldots, x_{n-1}) + f_{n-1}(x_1, \ldots, x_{n-1}) + \upsilon. \end{aligned}$$

In some systems the state $\mathbf{x}(t)$ may be bounded for all $\boldsymbol{\theta}$. Clearly, in these cases the choice of the function $\psi(\mathbf{x}, t)$ for determining the conservative target set (5.4) should not be restricted by the need to satisfy Assumption 5.1. These, however, are rather exceptional situations, and in what follows we will presume that there exists a set of initial conditions and values of $\boldsymbol{\theta}$ such that the corresponding solutions of the original system grow unboundedly with t. Thus, it is necessary to find a proper control input u.

Let us specify a class of control inputs u that can ensure boundedness of solutions $\mathbf{x}(t, \mathbf{x}_0, t_0, \boldsymbol{\theta}, u)$ for every $\boldsymbol{\theta} \in \Omega_\theta$ and $\mathbf{x}_0 \in \mathbb{R}^n$. According to (5.3), boundedness of $\mathbf{x}(t, \mathbf{x}_0, t_0, \boldsymbol{\theta}, u)$ is ensured if we find a control input u such that $\psi(\mathbf{x}(t), t) \in L^1_\infty[t_0, \infty]$. To this end, consider the dynamics of system (5.2) with respect to $\psi(\mathbf{x}, t)$:

$$\dot{\psi} = L_{\mathbf{f}(\mathbf{x},\boldsymbol{\theta},t)}\psi(\mathbf{x}, t) + L_{\mathbf{g}(\mathbf{x},t)}\psi(\mathbf{x}, t)u + \frac{\partial\psi(\mathbf{x}, t)}{\partial t}. \tag{5.7}$$

In what follows we will consider control inputs u from the class of functions $\mathcal{U}^*$ such that

$$\begin{aligned} u \in \mathcal{U}^* \Rightarrow L_{\mathbf{f}(\mathbf{x},\boldsymbol{\theta},t)}\psi(\mathbf{x}, t) &+ L_{\mathbf{g}(\mathbf{x},t)}\psi(\mathbf{x}, t)u + \frac{\partial\psi(\mathbf{x}, t)}{\partial t} \\ &= f(\mathbf{x}, \boldsymbol{\theta}, t) - f(\mathbf{x}, \hat{\boldsymbol{\theta}}, t) - \varphi(\psi, \boldsymbol{\omega}, t) + \varepsilon(t), \end{aligned} \tag{5.8}$$

provided that solutions of the closed-loop system are defined locally. According to (5.7) the dynamics of $\psi(\mathbf{x}, t)$ is affected by unknown $\boldsymbol{\theta}$ through the following term:

$$L_{\mathbf{f}(\mathbf{x},\boldsymbol{\theta},t)}\psi(\mathbf{x}, t) = f(\mathbf{x}, \boldsymbol{\theta}, t). \tag{5.9}$$

We require a feedback $u(\mathbf{x}, \hat{\boldsymbol{\theta}}, t)$ that is capable of annihilating the influence of uncertainty $f(\mathbf{x}, \boldsymbol{\theta}, t)$ on the dynamics of $\psi(\mathbf{x}, t)$ and, in addition, ensures boundedness of $\psi(\mathbf{x}, t)$. Assuming, for example, that the inverse $\left(L_{\mathbf{g}(\mathbf{x},t)}\psi(\mathbf{x}, t)\right)^{-1}$ exists everywhere[4] and using the notation (5.9), we define the control input u as follows:

$$\begin{aligned} u(\mathbf{x}, \hat{\boldsymbol{\theta}}, \boldsymbol{\omega}, t) &= (L_{\mathbf{g}(\mathbf{x},t)}\psi(\mathbf{x}, t))^{-1}\left[- f(\mathbf{x}, \hat{\boldsymbol{\theta}}, t) - \varphi(\psi, \boldsymbol{\omega}, t) - \frac{\partial\psi(\mathbf{x}, t)}{\partial t} + \varepsilon(t)\right], \\ \varphi &: \mathbb{R} \times \mathbb{R}^w \times \mathbb{R}_{\geq 0} \to \mathbb{R}, \end{aligned} \tag{5.10}$$

where $\boldsymbol{\omega} \in \Omega_\omega \subset \mathbb{R}^w$ is a vector of *known* parameters of the function $\varphi(\psi, \boldsymbol{\omega}, t)$, and the function $\varepsilon : \mathbb{R}_{\geq 0} \to \mathbb{R}$ stands for disturbances due to measurement

[4] This assumption limits our choice of functions $\psi(\mathbf{x}, t)$ to ones that satisfy the following constraint: $\operatorname{sign}(\sum_{i=1}^n g_i(\mathbf{x}, t)\partial\psi(\mathbf{x}, t)/\partial x_i) = \text{constant}$. Even though invertibility of $L_{\mathbf{g}(\mathbf{x},t)}\psi(\mathbf{x}, t)$ is not at all necessary for our approach and the specific choice of feedback $u(\mathbf{x}, \boldsymbol{\theta}, t)$ is not the central topic of our present contribution, we sacrifice generality for the sake of constructive design.

noise, *unmodeled dynamics* etc. Unless stated otherwise, we assume that $\varepsilon \in L_2^1[t_0, \infty] \cap \mathcal{C}^0$.

Feedback (5.10) renders (5.7) into the *error-model* form[5] (Narendra and Annaswamy 1989):

$$\dot{\psi} = f(\mathbf{x}, \boldsymbol{\theta}, t) - f(\mathbf{x}, \hat{\boldsymbol{\theta}}, t) - \varphi(\psi, \boldsymbol{\omega}, t) + \varepsilon(t), \tag{5.11}$$

or, when $\varepsilon(t) \equiv 0$,

$$\dot{\psi} = f(\mathbf{x}, \boldsymbol{\theta}, t) - f(\mathbf{x}, \hat{\boldsymbol{\theta}}, t) - \varphi(\psi, \boldsymbol{\omega}, t). \tag{5.12}$$

The vector $\hat{\boldsymbol{\theta}}$ in (5.10) and (5.12) can be interpreted as an estimate of the unknown parameter vector $\boldsymbol{\theta}$.

Let us now specify the desired properties of the function $\varphi(\psi, \boldsymbol{\omega}, t)$ in (5.10)–(5.12). Often (see e.g. (Sastry and Bodson 1989; Narendra and Annaswamy 1989; Krstić *et al.* 1995; Fradkov 1990), and Fomin *et al.* (1981)) the desired, or target, dynamics of the controlled system at $\hat{\boldsymbol{\theta}} = \boldsymbol{\theta}$ is expected to be stable in the sense of Lyapunov. An example of such a standard requirement in our setting is that $\psi(\mathbf{x}, t) = 0$ is a globally asymptotically stable equilibrium of (5.12) at $\boldsymbol{\theta} \equiv \hat{\boldsymbol{\theta}}$. This requirement, however, might not always be natural. On the other hand, a wide class of processes satisfies a passivity property: an input signal of which the energy is finite produces finite (bounded) state deviations. Thus, instead of global Lyapunov stability of system (5.11) for $\boldsymbol{\theta} \equiv \hat{\boldsymbol{\theta}}$, we propose that finite energy of the signal $f(\mathbf{x}(t), \boldsymbol{\theta}, t) - f(\mathbf{x}(t), \hat{\boldsymbol{\theta}}(t), t)$, defined for example by its $L_2^1[t_0, \infty]$-norm with respect to the variable t, results in boundedness of $\psi(\mathbf{x}(t), t)$ and hence, by virtue of Assumption 5.1, of the state $\mathbf{x}(t)$. Formally this requirement is introduced in Assumption 5.2.

Assumption 5.2 *Consider the following system:*

$$\dot{\psi} = -\varphi(\psi, \boldsymbol{\omega}, t) + \zeta(t), \ \psi_0 \in \mathbb{R}, \tag{5.13}$$

where $\zeta : \mathbb{R}_{\geq 0} \to \mathbb{R}$ *and* $\varphi(\psi, \boldsymbol{\omega}, t)$ *is from (5.11). For every* $\boldsymbol{\omega} \in \Omega_\omega$ *system (5.13) has* $L_2^1[t_0, \infty] \mapsto L_\infty^1[t_0, \infty]$ *margin with respect to input* $\zeta(t)$*. In other words, there exists a locally bounded function* $\gamma_{\infty,2} : \mathbb{R} \times \mathbb{R}^\omega \times \mathbb{R}_{\geq 0} \to \mathbb{R}_{\geq 0}$ *such that*

$$\|\psi(t)\|_{\infty,[t_0,T]} \leq \gamma_{\infty,2}(\psi_0, \boldsymbol{\omega}, \|\zeta(t)\|_{2,[t_0,T]}), \quad \forall\, \zeta(t) \in L_2^1[t_0, T] \tag{5.14}$$

and the function $\gamma_{\infty,2}(\psi_0, \boldsymbol{\omega}, \|\zeta(t)\|_{2,[t_0,T]})$ *is non-decreasing in* $\|\zeta(t)\|_{2,[t_0,T]}$.

[5] Error models (5.12) have been shown to be convenient when solving adaptive control, regulation, and identification problems for systems with nonlinear parametrization (Loh *et al.* 1999; Tyukin *et al.* 2003a; Cao *et al.* 2003).

Assumption 5.2 does not require global *asymptotic stability* of the origin of the unperturbed system (5.13), i.e. for $\zeta(t) = 0$. System (5.13) is allowed to have Lyapunov-unstable equilibria, multiple attractors, or no equilibria at all. When the target dynamics is stable, a benefit of Assumption 5.2 is that there is no need to know the *particular Lyapunov function* of the unperturbed system.

In addition to the problem of proper selection of target sets and target dynamics there is another important issue we wish to mention here. This issue is the possibility of achieving control objectives by using control signals that are small in some meaningful sense. In the framework of adaptive control it is not very clear how to characterize the smallness of control signals in the most natural fashion. This is because classes of admissible feedbacks are often presumed to be given, and the problem is restricted to finding a suitable adaptation algorithm for tuning the parameters of these feedbacks (Fomin *et al.* 1981; Fradkov 1990; Narendra and Annaswamy 1989; Krstić *et al.* 1995). Here we do not wish to dispute this strategy. Yet, we would still like to introduce a reasonable characterization of control smallness in the context of adaption. This characterization is the property of *non-domination*, of which the formal definition is provided below.

Definition 5.1.1 The adaptation law $\hat{\boldsymbol{\theta}}(\mathbf{x}, t) : \mathbb{R}^n \times \mathbb{R}_{\geq 0} \rightarrow \mathbb{R}^d$ is called non-dominating of class $\mathcal{C}_\varphi$, iff the control objective is reached for all $\mathbf{x}_0 \in \mathbb{R}^n$, $\boldsymbol{\theta} \in \Omega_\theta$ and $\varphi \in \mathcal{C}_\varphi$.

Let an adaptation law be non-dominating, for example, in the class of functions $\mathcal{C}_\varphi = \{\varphi : \mathbb{R} \rightarrow \mathbb{R} | |\varphi(\cdot)| \in \mathcal{K},\ \varphi(\sigma)\sigma \leq 0,\ \sigma \in \mathbb{R}\}$. Then, according to Definition 5.1.1, the goal of the regulation should be achievable in this system for all $\varphi(\psi) = k\psi$, $k \in \mathbb{R}_{>0}$, irrespective of how small the value of k is.

So far we have discussed the main constraints on the classes of target sets, target dynamics, and control functions. Let us now shift our attention to describing the class of functions $f(\mathbf{x}, \boldsymbol{\theta}, t)$ in (5.11). It is obvious that the larger the class of admissible functions $f(\mathbf{x}, \boldsymbol{\theta}, t)$, the wider the spectrum of potential applications. We will allow two classes of nonlinearities $f(\mathbf{x}, \boldsymbol{\theta}, t)$: one is the class of locally Lipschitz functions with respect to $\boldsymbol{\theta}$ (which is termed here *general*), and the other is somewhat more restricted (termed here *monotone*).

With regard to the second class of functions, instead of considering general nonlinear parametrization of $f(\mathbf{x}, \boldsymbol{\theta}, t)$ we find that a broad range of physical, mechanical, and biological phenomena can be described by a specific class of nonlinear functions, in particular

$$f(\mathbf{x}, \boldsymbol{\theta}, t) = \lambda(\mathbf{x}, t) f_{\mathrm{m}}(\mathbf{x}, \boldsymbol{\phi}(\mathbf{x}, t)^{\mathrm{T}} \boldsymbol{\theta}, t), \tag{5.15}$$

where $\lambda : \mathbb{R}^n \times \mathbb{R}_{\geq 0} \rightarrow \mathbb{R}$, $\boldsymbol{\phi} : \mathbb{R}^n \times \mathbb{R}_{\geq 0} \rightarrow \mathbb{R}^d$, and $f_{\mathrm{m}} : \mathbb{R}^n \times \mathbb{R} \times \mathbb{R}_{\geq 0} \rightarrow \mathbb{R}$ are continuous functions and $f_{\mathrm{m}}(\mathbf{x}, \boldsymbol{\phi}(\mathbf{x}, t)^{\mathrm{T}} \boldsymbol{\theta}, t)$ is monotone in $\boldsymbol{\phi}(\mathbf{x}, t)^{\mathrm{T}} \boldsymbol{\theta}$. Functions

Table 5.1. *Examples of nonlinearities satisfying Assumption 5.3. The parameter* Δ_θ *is a positive constant*

Physical meaning	Mathematical model of uncertainty $f(\mathbf{x},\boldsymbol{\theta},t)$	Domain of physical relevance	$\boldsymbol{\alpha}(\mathbf{x},t)^{\mathrm{T}}$
Stiction forces	$f:\ \theta_0 e^{-x_2^2\theta_1}$ $= e^{-x_2^2\theta_1+\ln(\theta_0)}$ $\mathbf{x}=(x_1,x_2)$	$\Delta_\theta>\theta_0,\theta_1>0$	$(-x_2^2,1)$
Tyre–road friction	$f:\ F_n\mathrm{sign}(x_1-rx_2)\dfrac{\theta G\frac{\sigma_0}{L}\frac{x_3}{1-x_3}}{\theta G+\frac{\sigma_0}{L}\frac{x_3}{1-x_3}}$ $G=(\mu_C+(\mu_S-\mu_C)$ $\times e^{-\frac{\lvert rx_2x_3\rvert}{\lvert 1-x_3\rvert v_S}})$ $\mathbf{x}=(x_1,x_2,x_3),\ \mu_S>\mu_C$ $F_n,L,r,\sigma_0,\mu_S,\mu_C>0$ – parameters	$\Delta_\theta>\theta>0,$ $x_1,x_2\geq 0,$ $x_3\in(0,1)$	$x_3/(1-x_3)$
Force supported by hydraulic emulsion in suspension dampers	$f:$ $\dfrac{K_o(\theta+1)A_p(x_1-x_2)}{L(\theta+K_oP_o/x_3^2)}$ $\mathbf{x}=(x_1,x_2,x_3)$ $K_o,A_p,P_o,L>0$ – parameters	$\Delta_\theta>\theta>0$ $x_3>0$	$A_p(x_1-x_2)$
Nonlinearities in Monod's growth model of microorganisms	$f:\ \dfrac{x_1x_2}{\theta_0+\theta_1x_1},\ \dfrac{x_1x_2}{\theta_0+\theta_1x_2}$ $\mathbf{x}=(x_1,x_2)$	$\Delta_\theta>\theta_0,\theta_1>0$ $x_1,x_2>0$	$x_1x_2(1,x_1)$ $x_1x_2(1,x_2)$
Blur distortion model in problems of processing visual information	$f:$ $\int_{\Omega_x} e^{-\lVert\mathbf{x}-\boldsymbol{\xi}\rVert^2/\theta}s(\boldsymbol{\xi})d\boldsymbol{\xi},$ $\Omega_x\subset\mathbb{R}^2$ $\mathbf{x}=(x_1,x_2),$ $s:\Omega_x\to\mathbb{R}_{\geq 0}$ – image	$\Delta_\theta>\theta>0,$ $s(\mathbf{x})$ is bounded and non-negative for all $\mathbf{x}\in\Omega_x$	1

(5.15) naturally extend from linear to nonlinear parametrizations, covering a wide range of practically relevant models (Armstrong-Helouvry 1993; Canudas de Wit and Tsiotras 1999; Kitching *et al.* 2000; Boskovic 1995), as illustrated in Table 5.1.

These observations motivated us to consider functions $f(\mathbf{x},\boldsymbol{\theta},t)$ satisfying the following assumption.

Assumption 5.3 *For the given function $f(\mathbf{x},\boldsymbol{\theta},t)$ in (5.11) there exists a function $\boldsymbol{\alpha}(\mathbf{x},t) : \mathbb{R}^n \times \mathbb{R}_{\geq 0} \to \mathbb{R}^d$, $\boldsymbol{\alpha}(\mathbf{x},t) \in \mathcal{C}^1$ and positive constant $D > 0$ such that*

$$(f(\mathbf{x},\hat{\boldsymbol{\theta}},t) - f(\mathbf{x},\boldsymbol{\theta},t))(\boldsymbol{\alpha}(\mathbf{x},t)^{\mathrm{T}}(\hat{\boldsymbol{\theta}} - \boldsymbol{\theta})) \geq 0, \tag{5.16}$$

$$|f(\mathbf{x},\hat{\boldsymbol{\theta}},t) - f(\mathbf{x},\boldsymbol{\theta},t)| \leq D|\boldsymbol{\alpha}(\mathbf{x},t)^{\mathrm{T}}(\hat{\boldsymbol{\theta}} - \boldsymbol{\theta})|. \tag{5.17}$$

Inequality (5.16) in Assumption 5.3 holds, for instance, for all functions (5.15). In this case the function $\boldsymbol{\alpha}(\mathbf{x},t)$ can be given as follows: $\boldsymbol{\alpha}(\mathbf{x},t) = (-1)^p \lambda(\mathbf{x},t)\boldsymbol{\phi}(\mathbf{x},t)$, where $p = 0$ if $f_{\mathrm{m}}(\mathbf{x},\boldsymbol{\phi}(\mathbf{x},t)^{\mathrm{T}}\boldsymbol{\theta},t)$ is non-decreasing in $\boldsymbol{\phi}(\mathbf{x},t)^{\mathrm{T}}\boldsymbol{\theta}$ and $p = 1$ if $f_{\mathrm{m}}(\mathbf{x},\boldsymbol{\phi}(\mathbf{x},t)^{\mathrm{T}}\boldsymbol{\theta},t)$ is non-increasing in $\boldsymbol{\phi}(\mathbf{x},t)^{\mathrm{T}}\boldsymbol{\theta}$.

Inequality (5.17) is satisfied if $f_{\mathrm{m}}(\mathbf{x},\boldsymbol{\phi}(\mathbf{x},t)^{\mathrm{T}}\boldsymbol{\theta},t)$ does not grow faster than a linear function in the variable $\boldsymbol{\phi}(\mathbf{x},t)^{\mathrm{T}}\boldsymbol{\theta}$ for every $\mathbf{x} \in \mathbb{R}^n$. This requirement holds, for example, for functions $f_{\mathrm{m}}(\mathbf{x},\boldsymbol{\phi}(\mathbf{x},t)^{\mathrm{T}}\boldsymbol{\theta},t)$ that are globally Lipschitz in $\boldsymbol{\phi}(\mathbf{x},t)^{\mathrm{T}}\boldsymbol{\theta}$:

$$|f_{\mathrm{m}}(\mathbf{x},\boldsymbol{\phi}(\mathbf{x},t)^{\mathrm{T}}\boldsymbol{\theta},t) - f_{\mathrm{m}}(\mathbf{x},\boldsymbol{\phi}(\mathbf{x},t)^{\mathrm{T}}\boldsymbol{\theta}',t)| \leq D_\theta(\mathbf{x},t)|\boldsymbol{\phi}(\mathbf{x},t)^{\mathrm{T}}(\boldsymbol{\theta} - \boldsymbol{\theta}')|.$$

In this case, inequalities (5.16) and (5.17) hold for the functions $f(\mathbf{x},\boldsymbol{\theta},t)$ defined by (5.15) with $\boldsymbol{\alpha}(\mathbf{x},t) = (-1)^p D_\theta(\mathbf{x},t)\lambda(\mathbf{x},t)\boldsymbol{\phi}(\mathbf{x},t)$.

Figure 5.1 illustrates possible choices of the function $\boldsymbol{\alpha}(\mathbf{x},t)$. Examples of the relevant parametrizations in Table 5.1 also satisfy Assumption 5.3. Their corresponding functions $\boldsymbol{\alpha}(\mathbf{x},t)$ are listed in the right-hand column.

In general, when $f(\mathbf{x},\boldsymbol{\theta},t) \in \mathcal{C}^1$, validation of Assumption 5.3 amounts, according to Hadamard's lemma, to finding $\boldsymbol{\alpha}(\mathbf{x},t)$ and D such that the following holds for all $\mathbf{x},\boldsymbol{\theta}$, and $\hat{\boldsymbol{\theta}}$:

$$0 \leq \boldsymbol{\alpha}(\mathbf{x},t)\int_0^1 \frac{\partial f(\mathbf{x},\mathbf{s}(\boldsymbol{\theta},\hat{\boldsymbol{\theta}},\sigma),t)}{\partial \mathbf{s}(\boldsymbol{\theta},\hat{\boldsymbol{\theta}},\sigma)}d\sigma \leq D\boldsymbol{\alpha}(\mathbf{x},t)\boldsymbol{\alpha}(\mathbf{x},t)^{\mathrm{T}},$$
$$\mathbf{s}(\boldsymbol{\theta},\hat{\boldsymbol{\theta}},\sigma) = \sigma\boldsymbol{\theta} + (1-\sigma)\hat{\boldsymbol{\theta}}. \tag{5.18}$$

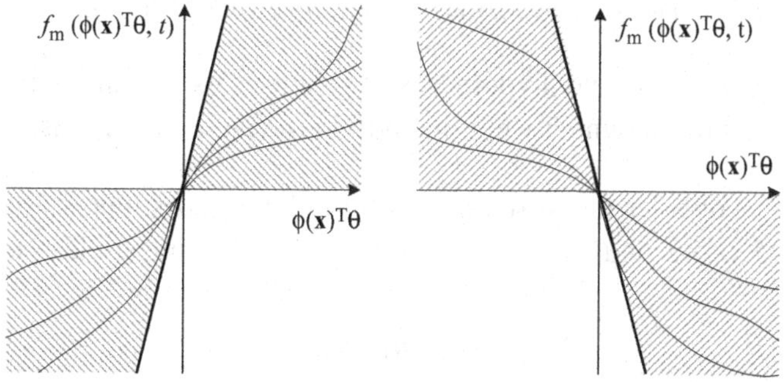

Figure 5.1 An illustration of the conditions of Assumption 5.3 for functions $f(\mathbf{x},\boldsymbol{\theta},t)$ that belong to the class (5.15). Thick lines represent the function $D\boldsymbol{\phi}(\mathbf{x},t)^{\mathrm{T}}\boldsymbol{\theta}$, $D = \max_{\mathbf{x},t}|D_\theta(\mathbf{x},t)\lambda(\mathbf{x},t)|$ in each block, respectively.

Assumption 5.3 bounds the growth rate of the difference $|f(\mathbf{x},\boldsymbol{\theta},t) - f(\mathbf{x},\hat{\boldsymbol{\theta}},t)|$ by the functional $D|\boldsymbol{\alpha}(\mathbf{x},t)^{\mathrm{T}}(\hat{\boldsymbol{\theta}}-\boldsymbol{\theta})|$. This will help us to find a parameter-estimation algorithm such that the estimates converge to $\boldsymbol{\theta}$ sufficiently fast for the solutions of (5.1) and (5.11) to remain bounded with non-dominating feedback (5.10). On the other hand, parametric error $\hat{\boldsymbol{\theta}} - \boldsymbol{\theta}$ can be inferred from the changes in the variable $\psi(\mathbf{x},t)$, according to (5.11), only by means of the difference $f(\mathbf{x},\boldsymbol{\theta},t) - f(\mathbf{x},\hat{\boldsymbol{\theta}},t)$. Therefore, as long as convergence of the estimates $\hat{\boldsymbol{\theta}}$ to $\boldsymbol{\theta}$ is expected, it is useful to have the estimate of $|f(\mathbf{x},\boldsymbol{\theta},t) - f(\mathbf{x},\hat{\boldsymbol{\theta}},t)|$ from below, as specified in Assumption 5.4.

Assumption 5.4 *For a given function $f(\mathbf{x},\boldsymbol{\theta},t)$ in (5.11) and a function $\boldsymbol{\alpha}(\mathbf{x},t)$ satisfying Assumption 5.3, there exists a positive constant $D_1 > 0$ such that*

$$|f(\mathbf{x},\hat{\boldsymbol{\theta}},t) - f(\mathbf{x},\boldsymbol{\theta},t)| \geq D_1|\boldsymbol{\alpha}(\mathbf{x},t)^{\mathrm{T}}(\hat{\boldsymbol{\theta}} - \boldsymbol{\theta})|. \tag{5.19}$$

In problems of parameter estimation, the effectiveness of the algorithms often depends on how "good" the nonlinearity $f(\mathbf{x},\boldsymbol{\theta},t)$ is, and how predictable locally the system's behavior is. As measures of goodness and predictability, usually smoothness and boundedness are considered. Likewise, in our study, we distinguish several specific properties of the functions $f(\mathbf{x},\boldsymbol{\theta},t)$ and $\varphi(\psi,\boldsymbol{\omega},t)$:

H 1. *The function $f(\mathbf{x},\boldsymbol{\theta},t)$ is locally bounded with respect to $\mathbf{x}$ and $\boldsymbol{\theta}$ uniformly in t.*

H 2. *The function $f(\mathbf{x},\boldsymbol{\theta},t) \in C^1$, and $\partial f(\mathbf{x},\boldsymbol{\theta},t)/\partial t$ is locally bounded with respect to $\mathbf{x}$ and $\boldsymbol{\theta}$ uniformly in t.*

H 3. *Let $U_x \subset \mathbb{R}^n$ and $U_\theta \subset \mathbb{R}^d$ be bounded. There exists constant $D_{U_x,U_\theta} > 0$ such that for every $\mathbf{x} \in U_x$ and $\boldsymbol{\theta}, \hat{\boldsymbol{\theta}} \in U_\theta$ Assumption 5.4 is satisfied with $D_1 = D_{U_x,U_\theta}$.*

H 4. *The function $\varphi(\psi,\boldsymbol{\omega},t)$ is locally bounded in ψ and $\boldsymbol{\omega}$ uniformly in t.*

Assumptions 5.1–5.4 and hypotheses 1–4 allow us to state several specific problems of adaptation, which we will study in the next paragraphs.

Problem 5.1 Consider the system (5.1) and (5.8), and suppose that it satisfies Assumptions 5.1–5.3 and, possibly, Assumption 5.4. Find a class of adaptation algorithms

$$\hat{\boldsymbol{\theta}} = \hat{\boldsymbol{\theta}}(\mathbf{x},t) \tag{5.20}$$

ensuring that

(1) the system is forward-complete;

(2) the signal $f(\mathbf{x}(t), \boldsymbol{\theta}, t) - f(\mathbf{x}(t), \hat{\boldsymbol{\theta}}(t), t)$ has bounded $L_2[t_0, \infty]$-norm,

$$f(\mathbf{x}(t), \boldsymbol{\theta}, t) - f(\mathbf{x}(t), \hat{\boldsymbol{\theta}}(t), t) \in L_2^1[t_0, \infty],$$
$$\|f(\mathbf{x}(t), \boldsymbol{\theta}, t) - f(\mathbf{x}(t), \hat{\boldsymbol{\theta}}(t), t)\|_{2,[t_0,\infty]} < \infty, \tag{5.21}$$

for all $\mathbf{x}_0 \in \mathbb{R}^n$ and $\boldsymbol{\theta} \in \Omega_\theta$;

(3) the influence of uncertainty on the target dynamics vanishes asymptotically as $t \to \infty$:

$$\lim_{t\to\infty} f(\mathbf{x}(t), \boldsymbol{\theta}, t) - f(\mathbf{x}(t), \hat{\boldsymbol{\theta}}(t), t) = 0. \tag{5.22}$$

Finding an adaptation algorithm that is a solution to Problem 5.1 allows one to ensure that solutions of the combined system exist and are bounded on any bounded interval (property (5.21) and see Assumptions 5.1 and 5.2). The limiting relation (5.22) ensures that the influence of uncertainties on the target dynamics is compensated for asymptotically; this is in full agreement with what is expected from an adaptive system (Polderman and Pait 2003).

Developing adaptation laws with more delicate and "certain" asymptotic behavior is a natural step forward after solutions to Problem 5.1 have been found. Suppose that $\boldsymbol{\theta}$ is a perturbation and the system has an invariant set at $\boldsymbol{\theta} = 0$. Then an interesting question is that of whether we can steer the system's state to this set when the value of $\boldsymbol{\theta} \neq 0$ and is uncertain. This constitutes Problem 5.2.

Problem 5.2 Consider system (5.1), and let Ω_0 be its invariant set at $\boldsymbol{\theta} = 0, u = 0$. Find the functions

$$u = u(\mathbf{x}, \hat{\boldsymbol{\theta}}, t), \ \hat{\boldsymbol{\theta}} = \hat{\boldsymbol{\theta}}(\mathbf{x}, t), \tag{5.23}$$

such that for all $\boldsymbol{\theta} \in \Omega_\theta$ trajectories $\mathbf{x}(t)$ of the combined system (5.1) and (5.23) converge to Ω_0 asymptotically.

Equations (5.1) in Problems 5.1 and 5.2 describe a single system with uncertainties. Yet, in a range of situations it is worthwhile to decompose the system into an interconnection of smaller ones. Each smaller subsystem will have its own inputs, outputs, and states, and a state of one subsystem may not be generally accessible from another. The question is how can adaptation and adaptive regulation be realized in this case? Thus we converge to Problem 5.3.

Problem 5.3 Consider an interconnection of (not necessarily identical) systems S_1 and S_2. Suppose that the dynamics of each system is governed by (5.1), (5.8), and (5.20), and that these control laws are solutions of Problem 5.1 for each of the subsystems.

(1) Determine conditions ensuring that
  (1.1) the interconnection is complete;

(1.2) the state of the combined system is bounded for all $\boldsymbol{\theta}_1 \in \Omega_{\theta_1}$ and $\boldsymbol{\theta}_2 \in \Omega_{\theta_2}$; $\boldsymbol{\theta}_1$ and $\boldsymbol{\theta}_2$ are unknown parameters of systems $\mathcal{S}_1$ and $\mathcal{S}_2$ respectively; $\boldsymbol{\theta} = \boldsymbol{\theta}_1 \oplus \boldsymbol{\theta}_2$;

(1.3) the limiting relation (5.22) holds for each subsystem.

Consider interconnection of systems $\mathcal{S}_1$ and $\mathcal{D}_2$, where $\mathcal{D}_2$ is complete.

(2) Determine conditions ensuring that
   (2.1) the interconnection is complete;
   (2.2) the state of the interconnection is bounded for all $\boldsymbol{\theta} \in \Omega_\theta$.

(3) Find solutions of (1) and (2) for interconnections of an arbitrary (finite) number of systems.

Notice that the decentralized adaptive control problem is a special case of Problem 5.3.

Problems 5.1–5.3 focused on the issues of control and regulation of uncertain dynamical systems, albeit in the framework of adaptation. Let us now introduce two more problems in which estimation of unknown parameters of dynamical systems is necessary.

Problem 5.4 Consider (5.1) and (5.8), and suppose that Assumptions 5.1–5.4 hold. Find an adaptation algorithm (5.20) such that

(1) the combined system is complete;

(2) the property

$$\lim_{t\to\infty} \hat{\boldsymbol{\theta}}(t) = \boldsymbol{\theta} \tag{5.24}$$

holds for all $\mathbf{x}_0 \in \mathbb{R}^n$ and $\boldsymbol{\theta} \in \Omega_\theta$, preferably providing the rates of convergence of $\hat{\boldsymbol{\theta}}$ to $\boldsymbol{\theta}$.

Unlike other problems considered so far in this chapter, Problem 5.4 does not require that solutions of the combined system remain bounded or converge to some specified set. Since the function u is fixed, we can interpret this problem as a kind of adaptive bifurcation control should the dynamics of the combined system change dramatically at $\boldsymbol{\theta} = \hat{\boldsymbol{\theta}}$. The fact that nonlinear parametrization is allowed here makes this formulation different from other known statements (see e.g. Moreau *et al.* (2003) and Moreau and Sontag (2003)).

Notice that, despite the fact that the models in Problem 5.4 are allowed to be nonlinearly parametrized, they are restricted by the monotonicity constraints formulated in Assumptions 5.3 and 5.4. The main reason for introducing this constraint is that in Problem 5.4 we aim at finding algorithms by solving the estimation problem globally. That is, we require that (5.24) holds for all $\mathbf{x}_0 \in \mathbb{R}^n$ and $\boldsymbol{\theta} \in \Omega_\theta$. In the case of a more general class of parametrizations it may be impossible to establish global results. Yet, one can still pose a question about finding parameter-estimation algorithms for systems in which the uncertainty models are not limited to nonlinearities that are monotone in parameters. For this reason we introduce Problem 5.5.

Problem 5.5 Consider system (5.1). Find a control input u, as a function of $\mathbf{x}$ and t, adaptation algorithm (5.20), and domains Ω_x, $\Omega'_\theta \subset \Omega_\theta$ such that state $\mathbf{x}$ of (5.1) is bounded and (5.24) holds for all $\mathbf{x}_0 \in \Omega_x$ and $\boldsymbol{\theta} \in \Omega'_\theta$.

Solutions of Problems 5.1–5.5 are provided in the following sections.

5.2 Direct adaptive control

In this section we will present constructive solutions to Problem 5.1. We start by introducing a synthesis method, the method of the *virtual adaptation algorithm*. The method is described in Section 5.2.1, and its applicability conditions for system (5.1) are presented in Theorem 5.1. One of the most critical conditions is an integrability constraint. This constraint allows us to specify a class of systems for which the method can be applied directly. In Section 5.2.2 we present a technique enabling us to expand the results to a wider class of systems, which do not satisfy the conditions of Theorem 5.1 explicitly. The technique is based on reducing the number of independent variables in the integrability constraint so that the integrability condition will eventually be satisfied. This is achieved by a purposeful extension of the state space of the original (5.1). In Section 5.2.3 we demonstrate how the results of Sections 5.2.1 and 5.2.2 can be applied to deal with the cascades of (5.1).

5.2.1 Virtual adaptation algorithms. Sufficient conditions of realizability

Standard approaches in parameter-estimation and adaptation problems usually assume feedback and a parameter-adjustment algorithm in the following form:

$$u = u(\mathbf{x}, \hat{\boldsymbol{\theta}}, t), \quad \dot{\hat{\boldsymbol{\theta}}} = \mathcal{A}_{\text{lg}}(\psi, \mathbf{x}, t), \tag{5.25}$$

where $\mathcal{A}_{\text{lg}} : \mathbb{R} \times \mathbb{R}^n \times \mathbb{R}_{\geq 0} \to \mathbb{R}^d$ is a function of the error function $\psi(\mathbf{x}, t)$, state $\mathbf{x}$, and time t. The favorite strategy for finding these is a two-stage design prescription, known as the *certainty-equivalence principle*. First, construct uncertainty-dependent feedback $u(\mathbf{x}, \boldsymbol{\theta}, t)$, $\boldsymbol{\theta} \in \Omega_\theta$ which ensures boundedness of the trajectories $\mathbf{x}(t)$. Second, replace $\boldsymbol{\theta}$ with $\hat{\boldsymbol{\theta}}$ in $u(\mathbf{x}, \boldsymbol{\theta}, t)$ and, given the constraints (e.g. $\boldsymbol{\theta}$ and $\dot{\mathbf{x}}$ cannot be measured explicitly while state $\mathbf{x}$ is available), design a function $\mathcal{A}_{\text{lg}}(\cdot)$ that guarantees (5.21)–(5.24), and/or $\psi(\mathbf{x}(t), t) \to 0$.

With this strategy, the design of the feedback $u(\mathbf{x},\boldsymbol{\theta},t)$ is generally independent[6] of the specific design of the parameter-estimation algorithm $\mathcal{A}_{\mathrm{lg}}(\psi,\mathbf{x},t)$. This allows one to obtain the full benefit of nonlinear control theory in designing the feedback $u(\mathbf{x},\boldsymbol{\theta},t)$. On the other hand, this strategy equally benefits from conventional parameter-estimation and adaptation theories, which provide a list of ready-to-be-implemented algorithms under the *assumption* that the feedback $u(\mathbf{x},\boldsymbol{\theta},t)$ ensures stability of the system.

Ironically, the power of the certainty-equivalence principle – simplicity and independence of the design stages – is also its Achilles' heel. It ignores the possibility of advantageous interactions between control and parameter estimation procedures. There are numerous reports (Stotsky 1993; Ortega *et al.* 2002; Astolfi and Ortega 2003; Tyukin *et al.* 2003a) that an additional interaction term

$$\hat{\boldsymbol{\theta}}_P(\mathbf{x},t) : \mathbb{R}^n \times \mathbb{R}_{\geq 0} \to \mathbb{R}^d \tag{5.26}$$

added to the parameters $\hat{\boldsymbol{\theta}}$ in the function $u(\mathbf{x},\hat{\boldsymbol{\theta}},t)$,

$$u(\mathbf{x},\hat{\boldsymbol{\theta}} + \hat{\boldsymbol{\theta}}_P(\mathbf{x},t),t), \tag{5.27}$$

introduces new properties to the system. Unfortunately, straightforward introduction of this term as a new variable of the design negatively affects its simplicity and hence the so-much-favored independence of the design stages.

An alternative strategy is proposed in Tyukin (2003) and Tyukin *et al.* (2003b). It introduces a simple design paradigm for choosing (5.26) and (5.27), in which the adaptation algorithms in (5.25) are initially allowed to depend on unmeasurable variables $\dot{\psi}$, $\dot{\mathbf{x}}$, and $\boldsymbol{\theta}$:

$$\dot{\hat{\boldsymbol{\theta}}} = \mathcal{A}^*_{\mathrm{lg}}(\psi,\dot{\psi},\mathbf{x},\dot{\mathbf{x}},\boldsymbol{\theta},t). \tag{5.28}$$

For this reason such algorithms are called *virtual algorithms*. If the desired properties (5.21)–(5.24) are ensured with (5.28) then the unrealizable algorithm (5.28) is converted into *integro-differential*, or *finite, form* (Fradkov 1986):

$$\begin{aligned} \hat{\boldsymbol{\theta}} &= \Gamma(\hat{\boldsymbol{\theta}}_P(\mathbf{x},t) + \hat{\boldsymbol{\theta}}_I(t)), \\ \dot{\hat{\boldsymbol{\theta}}}_I &= \mathcal{A}_{\mathrm{lg}}(\psi,\mathbf{x},t), \ \Gamma \in \mathbb{R}^{d\times d}, \ \Gamma > 0. \end{aligned} \tag{5.29}$$

Under the following condition, the finite-form representation (5.29) is equivalent to (5.28):

$$\mathcal{A}^*_{\mathrm{lg}}(\psi,\dot{\psi},\mathbf{x},\dot{\mathbf{x}},\boldsymbol{\theta},t) = \Gamma\left(\frac{\partial \hat{\boldsymbol{\theta}}_P(\mathbf{x},t)}{\partial \mathbf{x}}\dot{\mathbf{x}} + \frac{\partial \hat{\boldsymbol{\theta}}_P(\mathbf{x},t)}{\partial t} + \mathcal{A}_{\mathrm{lg}}(\psi,\mathbf{x},t)\right). \tag{5.30}$$

[6] In particular, it is a standard requirement that the function $u(\mathbf{x},\boldsymbol{\theta},t)$ should guarantee Lyapunov stability of the system for $\hat{\boldsymbol{\theta}} = \boldsymbol{\theta}$, whereas parameter-adjustment algorithms use this property in order to ensure stability of the whole system. No other properties are required from the function $u(\mathbf{x},\boldsymbol{\theta},t)$.

Thus, instead of attempting to solve a synthesis problem subject to a given set of constraints, we propose to start by finding solutions to an unconstrained problem. At a later stage, the initial constraints will be taken into account, and we shall retain only those solutions which satisfy these constraints.

According to this strategy, design of the adaptation algorithms requires, first, finding appropriate virtual algorithms $\mathcal{A}^*_{\mathrm{lg}}(\psi, \dot{\psi}, \mathbf{x}, \dot{\mathbf{x}}, \boldsymbol{\theta}, t)$ and, second, solving (5.30) for $\hat{\boldsymbol{\theta}}_P(\mathbf{x}, t)$ and $\mathcal{A}_{\mathrm{lg}}(\psi, \mathbf{x}, t)$. This approach preserves the convenience of the certainty-equivalence principle, since the feedback $u(\mathbf{x}, \boldsymbol{\theta}, t)$ could, in principle, be built independently of the subsequent parameter adjustment procedure. At the same time, it provides in a systematic way the necessary interaction term $\hat{\boldsymbol{\theta}}_P(\mathbf{x}, t)$, ensuring the required properties (5.21)–(5.24) of the closed-loop system even if the function $f(\mathbf{x}, \boldsymbol{\theta}, t)$ in (5.11) is nonlinear in $\boldsymbol{\theta}$.

Clearly, the choice of equations describing *virtual adaptation algorithms* (5.28) must be adequate to the requirements of Problems 5.1–5.5. In particular, the algorithms must be able to cope with nonlinear parametrization without invoking any unnecessary domination of the unknown nonlinearities (this contrasts sharply with a more conservative design philosophy in Putov (1993), Loh *et al.* (1999), and Lin and Qian (2002a,b)). Second, they have to guarantee that property (5.24) (Problems 5.4 and 5.5) holds. Finally, in order to be able to find a solution to Problem 5.2, it is natural to require that no unwanted invariant sets emerge as a result of adaptation.

Taking these requirements into account, we consider the following class of *virtual adaptation algorithms*:[7]

$$\dot{\hat{\boldsymbol{\theta}}} = \Gamma(\dot{\psi} + \varphi(\psi, \boldsymbol{\omega}, t))\boldsymbol{\alpha}(\mathbf{x}, t) + \mathcal{Q}(\mathbf{x}, \hat{\boldsymbol{\theta}}, t)(\boldsymbol{\theta} - \hat{\boldsymbol{\theta}}), \quad \Gamma \in \mathbb{R}^{d\times d}, \ \Gamma > 0, \tag{5.31}$$

where $\mathcal{Q}(\mathbf{x}, \hat{\boldsymbol{\theta}}, t) : \mathbb{R}^n \times \mathbb{R}^d \times \mathbb{R}_{\geq 0} \to \mathbb{R}^{d\times d}$, $\mathcal{Q}(\cdot) \in \mathcal{C}^0$. Indeed, depending on the choice of $\mathcal{Q}$ and $\boldsymbol{\alpha}$, adaptation can be switched off either along the trajectories corresponding to the target dynamics $\dot{\psi} = -\varphi(\psi, \boldsymbol{\omega}, t)$ or when $\hat{\boldsymbol{\theta}} = \boldsymbol{\theta}$. As a candidate for the finite-form realization (5.29) of algorithms of type (5.31) we select the following set of equations:

$$\begin{aligned} \hat{\boldsymbol{\theta}}(\mathbf{x}, t) &= \Gamma(\hat{\boldsymbol{\theta}}_P(\mathbf{x}, t) + \hat{\boldsymbol{\theta}}_I(t)), \ \Gamma \in \mathbb{R}^{d\times d}, \ \Gamma > 0, \\ \hat{\boldsymbol{\theta}}_P(\mathbf{x}, t) &= \psi(\mathbf{x}, t)\boldsymbol{\alpha}(\mathbf{x}, t) - \Psi(\mathbf{x}, t), \\ \dot{\hat{\boldsymbol{\theta}}}_I &= \varphi(\psi(\mathbf{x}, t), \boldsymbol{\omega}, t)\boldsymbol{\alpha}(\mathbf{x}, t) + \mathcal{R}(\mathbf{x}, \hat{\boldsymbol{\theta}}, u(\mathbf{x}, \hat{\boldsymbol{\theta}}, t), t), \end{aligned} \tag{5.32}$$

where the function $\Psi(\mathbf{x}, t) : \mathbb{R}^n \times \mathbb{R}_{\geq 0} \to \mathbb{R}_d$, $\Psi(\mathbf{x}, t) \in \mathcal{C}^1$ satisfies the following integrability assumption.

[7] This choice is motivated by our previous study of derivative-dependent algorithms for systems with nonlinearly parametrized uncertainties (Prokhorov *et al.* 2002b; Tyukin *et al.* 2003a).

Assumption 5.5 *There exists a function* $\Psi(\mathbf{x},t)$ *such that*

$$\frac{\partial \Psi(\mathbf{x},t)}{\partial \mathbf{x}_2} - \psi(\mathbf{x},t)\frac{\partial \alpha(\mathbf{x},t)}{\partial \mathbf{x}_2} = \mathcal{B}(\mathbf{x},t), \tag{5.33}$$

where $\mathcal{B}(\mathbf{x},t) : \mathbb{R}^n \times \mathbb{R}_{\geq 0} \rightarrow \mathbb{R}^{d\times p}$ *is either zero or, if* $\mathbf{f}_2(\mathbf{x},\boldsymbol{\theta},t)$ *is differentiable in* $\boldsymbol{\theta}$*, satisfies the condition*

$$\mathcal{B}(\mathbf{x},t)\mathcal{F}(\mathbf{x},\boldsymbol{\theta},\boldsymbol{\theta}',t) \leq 0 \ \forall\, \boldsymbol{\theta},\boldsymbol{\theta}' \in \Omega_\theta,\ \mathbf{x} \in \mathbb{R}^n,$$

$$\mathcal{F}(\mathbf{x},\boldsymbol{\theta},\boldsymbol{\theta}',t) = \int_0^1 \frac{\partial \mathbf{f}_2(\mathbf{x},\mathbf{s}(\lambda),t)}{\partial \mathbf{s}} d\lambda, \quad \mathbf{s}(\lambda) = \boldsymbol{\theta}'\lambda + \boldsymbol{\theta}(1-\lambda).$$

The function $\mathcal{R}(\mathbf{x},\hat{\boldsymbol{\theta}},u(\mathbf{x},\hat{\boldsymbol{\theta}},t),t) : \mathbb{R}^n \times \mathbb{R}^d \times \mathbb{R} \times \mathbb{R}_{\geq 0} \rightarrow \mathbb{R}^d$ in (5.32) is given as follows:

$$\begin{aligned}\mathcal{R}(\mathbf{x},\hat{\boldsymbol{\theta}},u(\mathbf{x},\hat{\boldsymbol{\theta}},t),t) = {} & \frac{\partial \Psi(\mathbf{x},t)}{\partial t} - \psi(\mathbf{x},t)\frac{\partial \alpha(\mathbf{x},t)}{\partial t} \\ & - (\psi(\mathbf{x},t)L_{\mathbf{f}_1}\alpha(\mathbf{x},t) - L_{\mathbf{f}_1}\Psi(\mathbf{x},t)) \\ & - (\psi(\mathbf{x},t)L_{\mathbf{g}_1}\alpha(\mathbf{x},t) - L_{\mathbf{g}_1}\Psi(\mathbf{x},t))u(\mathbf{x},\hat{\boldsymbol{\theta}},t) \\ & + \mathcal{B}(\mathbf{x},t)(\mathbf{f}_2(\mathbf{x},\hat{\boldsymbol{\theta}},t) + \mathbf{g}_2(\mathbf{x},t)u(\mathbf{x},\hat{\boldsymbol{\theta}},t)).\end{aligned} \tag{5.34}$$

The functions $\Psi(\mathbf{x},t)$ and $\mathcal{R}(\mathbf{x},\hat{\boldsymbol{\theta}},u(\mathbf{x},\hat{\boldsymbol{\theta}},t),t)$ are introduced into (5.32) in order to shape the derivative $\dot{\hat{\boldsymbol{\theta}}}(\mathbf{x},t)$ to fit (5.31). The role of the function $\Psi(\mathbf{x},t)$ in (5.32) is to compensate for the uncertainty-dependent term $\psi(\mathbf{x},t)L_{\mathbf{f}_2(\mathbf{x},\boldsymbol{\theta},t)}\alpha(\mathbf{x},t)$, and (5.33) is the condition for such a compensation to be possible.[8] With the function $\mathcal{R}(\mathbf{x},\hat{\boldsymbol{\theta}},u(\mathbf{x},\hat{\boldsymbol{\theta}},t),t)$ we eliminate the influence of the uncertainty-independent vector fields $\mathbf{f}_1(\mathbf{x},t)$, $\mathbf{g}_1(\mathbf{x},t)$, and $\mathbf{g}_2(\mathbf{x},t)$ on the desired form of the time-derivative $\dot{\hat{\boldsymbol{\theta}}}(\mathbf{x},t)$. In this sense Assumption 5.5 specifies the condition for solvability of (5.30) for the class of virtual algorithms (5.31).

The properties of the system (5.1) and (5.11) with adaptation algorithm (5.32) and (5.34) are summarized in Theorem 5.1.

Theorem 5.1 *Let the system (5.1), (5.11), (5.32), and (5.34) be given and let Assumptions 5.3–5.5 be satisfied. Then the following properties hold.*

(1) Let, for the given initial conditions $\mathbf{x}(t_0)$, $\hat{\boldsymbol{\theta}}_I(t_0)$, *and parameter vector* $\boldsymbol{\theta}$, *the the interval* $[t_0, T^*]$, $T^* \geq t_0$ *be the (maximal) time-interval of existence of*

[8] Its technical relevance and issues related to validation of Assumption 5.5 are discussed after the formulation of Theorem 5.1.

solutions in the closed-loop system (5.1), (5.11), (5.32), and (5.34). Then

$$f(\mathbf{x}(t),\boldsymbol{\theta},t) - f(\mathbf{x}(t),\hat{\boldsymbol{\theta}}(t),t)) \in L_2^1[t_0,T^*], \tag{5.35}$$

$$\|\boldsymbol{\theta} - \hat{\boldsymbol{\theta}}(t)\|^2_{\Gamma^{-1}} \le \|\hat{\boldsymbol{\theta}}(t_0) - \boldsymbol{\theta}\|^2_{\Gamma^{-1}} + \frac{D}{2D_1^2}\|\varepsilon(t)\|^2_{2,[t_0,T^*]}. \tag{5.36}$$

In particular,

$$\|f(\mathbf{x}(t),\boldsymbol{\theta},t) - f(\mathbf{x}(t),\hat{\boldsymbol{\theta}}(t),t))\|_{2,[t_0,T^*]} \le D_f(\boldsymbol{\theta},t_0,\Gamma,\|\varepsilon(t)\|_{2,[t_0,T^*]}), \tag{5.37}$$

where

$$D_f(\boldsymbol{\theta},t_0,\Gamma,\|\varepsilon(t)\|_{2,[t_0,T^*]}) = \left(\frac{D}{2}\|\boldsymbol{\theta} - \hat{\boldsymbol{\theta}}(t_0)\|^2_{\Gamma^{-1}}\right)^{0.5} + \frac{D}{D_1}\|\varepsilon(t)\|_{2,[t_0,T^*]}. \tag{5.38}$$

In addition, if Assumptions 5.1 and 5.2 are satisfied then

(2) $\psi(\mathbf{x}(t),t) \in L^1_\infty[t_0,\infty]$, $\mathbf{x}(t) \in L^n_\infty[t_0,\infty]$, *and*

$$\|\psi(\mathbf{x}(t),t)\|_{\infty,[t_0,\infty]} \le \gamma_{\infty,2}\left(\psi(\mathbf{x}_0,t_0),\boldsymbol{\omega},D^*\right), \tag{5.39}$$

where $D^* = D_f(\boldsymbol{\theta},t_0,\Gamma,\|\varepsilon(t)\|_{2,[t_0,\infty]}) + \|\varepsilon(t)\|_{2,[t_0,\infty]}$.

(3) if properties H1 and H4 hold, and the system (5.13) admits an $L_2^1[t_0,\infty] \mapsto L_p^1[t_0,\infty]$, $p > 1$ *margin with respect to input* $\zeta(t)$ *and output* ψ, *then*

$$\varepsilon(t) \in L_2^1[t_0,\infty] \cap L^1_\infty[t_0,\infty] \Rightarrow \lim_{t\to\infty} \psi(\mathbf{x}(t),t) = 0. \tag{5.40}$$

If, in addition, property H2 holds, and the functions $\boldsymbol{\alpha}(\mathbf{x},t)$ *and* $\partial\psi(\mathbf{x},t)/\partial t$ *are locally bounded with respect to* $\mathbf{x}$ *uniformly in* t, *then*

(4) the following limiting relation holds:

$$\lim_{t\to\infty}(f(\mathbf{x}(t),\boldsymbol{\theta},t) - f(\mathbf{x}(t),\hat{\boldsymbol{\theta}}(t),t)) = 0. \tag{5.41}$$

Theorem 5.1 provides a set of conditions ensuring that algorithms (5.32) are solutions of Problem 5.1. In addition to assumptions introduced earlier, the theorem requires that an extra assumption holds, namely Assumption 5.5. Therefore, prior to discussing the results of Theorem 5.1, we wish to comment on Assumption 5.5. Because the function $\psi(\mathbf{x},t)$ specifies the desired target set and $\boldsymbol{\alpha}(\mathbf{x},t)$ is determined by $\mathbf{f}_2(\mathbf{x},\boldsymbol{\theta},t)$ in (5.1) and the function $\psi(\mathbf{x},t)$, the uncertainty models $\mathbf{f}_2(\mathbf{x},\boldsymbol{\theta},t)$ and the choice of the goal function $\psi(\mathbf{x},t)$ are interrelated through the conditions for existence of a function $\Psi(\mathbf{x},t)$ satisfying (5.33). When $\mathcal{B}(\mathbf{x},t) \in \mathcal{C}^1$, $\mathcal{B}(\mathbf{x},t) = \mathrm{col}(\mathcal{B}_1(\mathbf{x},t),\ldots,\mathcal{B}_d(\mathbf{x},t))$, and $\boldsymbol{\alpha}(\mathbf{x},t) \in \mathcal{C}^2$, $\boldsymbol{\alpha}(\mathbf{x},t) = \mathrm{col}(\alpha_1(\mathbf{x},t),\ldots,\alpha_d(\mathbf{x},t))$, these conditions follow from

$$\frac{\partial}{\partial \mathbf{x}_2}\left(\psi(\mathbf{x},t)\frac{\partial \alpha_i(\mathbf{x},t)}{\partial \mathbf{x}_2} + \mathcal{B}_i(\mathbf{x},t)\right) = \left(\frac{\partial}{\partial \mathbf{x}_2}\left(\psi(\mathbf{x},t)\frac{\partial \alpha_i(\mathbf{x},t)}{\partial \mathbf{x}_2} + \mathcal{B}_i(\mathbf{x},t)\right)\right)^{\mathrm{T}}. \tag{5.42}$$

As a condition for the existence of $\Psi(\mathbf{x},t)$, this relation takes into account the structural properties of the system (5.1) and (5.11). Indeed, let $\mathcal{B}(\mathbf{x},t) = 0$, and consider partial derivatives $\partial\alpha_i(\mathbf{x},t)/\partial\mathbf{x}_2$ and $\partial\psi(\mathbf{x},t)/\partial\mathbf{x}_2$ with respect to the vector $\mathbf{x}_2 = (x_{21},\ldots,x_{2p})^{\mathrm{T}}$. Let for some $k \in \{1,\ldots,p\}$ the following hold:

$$\begin{aligned}\frac{\partial\psi(\mathbf{x},t)}{\partial\mathbf{x}_2} &= \Big(\underbrace{0\,0\,\cdots\,0}_{k-1}\ \ *\ \ \underbrace{0\,\cdots\,0}_{p-k}\Big),\\ \frac{\partial\alpha_i(\mathbf{x},t)}{\partial\mathbf{x}_2} &= \Big(\underbrace{0\,0\,\cdots\,0}_{k-1}\ \ *\ \ \underbrace{0\,\cdots\,0}_{p-k}\Big),\end{aligned} \tag{5.43}$$

where the symbol $*$ denotes a scalar function of $\mathbf{x}$ and t. Then condition (5.43) guarantees that the equality (5.42) and, consequently, Assumption 5.5, hold. Hence, whether Assumption 5.5 holds depends, roughly speaking, on how the partition $\mathbf{x}_2$ enters the arguments of the functions $\psi(\mathbf{x},t)$ and $\boldsymbol{\alpha}(\mathbf{x},t)$. When $\partial\boldsymbol{\alpha}(\mathbf{x}_1 \oplus \mathbf{x}_2,t)/\partial\mathbf{x}_2 = 0$, Assumption 5.5 holds for arbitrary $\psi(\mathbf{x},t) \in \mathcal{C}^1$. If $\psi(\mathbf{x},t)$ and $\boldsymbol{\alpha}(\mathbf{x},t)$ depend on just a single component of $\mathbf{x}_2$, for instance x_{2k}, conditions (5.43) hold and the function $\Psi(\mathbf{x},t)$ can be derived explicitly by taking the indefinite integral

$$\Psi(\mathbf{x},t) = \int \psi(\mathbf{x},t)\frac{\boldsymbol{\alpha}(\mathbf{x},t)}{\partial x_{2k}}dx_{2k}, \tag{5.44}$$

or, numerically,

$$\Psi(\mathbf{x},t) = \int_{x_{2k}(t_0)}^{x_{2k}(t)} \psi(\mathbf{x},t)\frac{\boldsymbol{\alpha}(\mathbf{x},t)}{\partial x_{2k}}dx_{2k}. \tag{5.45}$$

In all other cases, the existence of the required function $\Psi(\mathbf{x},t)$ follows from (5.42).

The necessity to satisfy Assumption 5.5 may, at first, look like a strong constraint. Yet, we notice that Assumption 5.5 holds in the relevant problem settings for arbitrary $\boldsymbol{\alpha}(\mathbf{x},t), \psi(\mathbf{x},t) \in \mathcal{C}^1$. Consider for instance Cao *et al.* (2003), where the class of systems is restricted to (5.46):

$$\dot{x} = -\varrho(x,u)x + f(\boldsymbol{\theta},u,x),\ x \in \mathbb{R}, \varrho(x,u) > \varrho_{\min} > 0. \tag{5.46}$$

The system state in (5.46) has dimension $\dim\{\mathbf{x}\} = \dim\{\mathbf{x}_2\} = 1$. Hence, according to (5.44), and in the case of functions $\psi(x,t)$, $\boldsymbol{\alpha}(x,t) \in \mathcal{C}^1$, there will always exist a function $\Psi(x,t)$ satisfying equality (5.33) with $\mathcal{B}(x,t) = 0$.

In a large variety of situations, when $\dim\{\mathbf{x}_2\} > 1$, the problem of finding $\Psi(\mathbf{x},t)$ satisfying condition (5.33) can be avoided (or converted into one with an already known solution such as (5.42) or (5.44)) by the *embedding* technique proposed in Tyukin *et al.* (2003b) and Tyukin and Prokhorov (2004). We will describe it in detail in Section 5.2.2.

For the time being, however, let us proceed with a brief discussion of the results stated in Theorem 5.1. The theorem ensures a set of relevant properties for both control (properties 2 and 3) and parameter-estimation problems (properties 1 and 4). These properties, as illustrated with (5.35)–(5.41), provide conditions for boundedness of the solutions $\mathbf{x}(t,\mathbf{x}_0,t_0,\boldsymbol{\theta},u(t))$, zeroing of the goal function $\psi(\mathbf{x},t)$, and exact compensation of the uncertainty term $f(\mathbf{x},\boldsymbol{\theta},t)$ even in the presence of unknown disturbances $\varepsilon(t) \in L_2^1[t_0,\infty] \cap L_\infty^1[t_0,\infty]$. All this follows from the fact that $(f(\mathbf{x}(t),\boldsymbol{\theta},t) - f(\mathbf{x}(t),\hat{\boldsymbol{\theta}}(t),t))) \in L_2^1[t_0,\infty]$ (see also (5.21) in Problem 5.1)), which in turn is guaranteed by properties (5.16), (5.17), and (5.19) of the function $f(\mathbf{x},\boldsymbol{\theta},t)$ in Assumptions 5.3 and 5.4. Estimate (5.19) in Assumption 5.4 is particularly important by virtue of allowing potentially unbounded disturbances from $L_2^1[t_0,\infty]$. When no disturbances are present, it is possible to show that properties 1–4 hold without involving Assumption 5.4.

Corollary 5.1 *Let the system (5.1), (5.11), (5.32), and (5.34) be given, let* $\varepsilon(t) \equiv 0$, *and let Assumptions 5.3 and 5.5 hold. Then*

(5) the norm $\|\boldsymbol{\theta} - \hat{\boldsymbol{\theta}}(t)\|^2_{\Gamma^{-1}}$ *is non-increasing and properties 1–4*[9] *of Theorem 5.1 hold with* $\varepsilon(t) \equiv 0$.

In addition to the fact that $|f(\mathbf{x},\boldsymbol{\theta},t) - f(\mathbf{x},\hat{\boldsymbol{\theta}},t)|$ is not required to be bounded from below as in (5.19), Corollary 5.1 ensures that $\|\boldsymbol{\theta} - \hat{\boldsymbol{\theta}}(t)\|^2_{\Gamma^{-1}}$ is *not growing* with time when $\varepsilon(t) \equiv 0$. Its practical relevance is that it guarantees the desired convergence (5.24) with a much weaker, local, version of Assumption 5.4. It will also help us to establish conditions for exponential stability in the unperturbed system. However, before we start analyzing these new properties of algorithms (5.32), let us first address the question of how to avoid complications arising due to the integrability assumption, Assumption 5.5, for (5.1).

[9] In this case the bound for $\|\psi(\mathbf{x}(t),t)\|_{\infty,[t_0,\infty]}$ will be different from the one given by (5.39) in Theorem 5.1.

5.2.2 Embedding problem

When $\dim\{\mathbf{x}_2\} > 1$, the problem of finding $\Psi(\mathbf{x},t)$ satisfying condition (5.33) can be avoided (or converted into one with an already known solution such as (5.42) and (5.44)) by the *embedding* technique proposed in Tyukin *et al.* (2003b). The main idea of this method is to introduce an auxiliary (forward-complete) system

$$\begin{aligned}\dot{\boldsymbol{\xi}} &= \mathbf{f}_{\xi}(\mathbf{x},\boldsymbol{\xi},t),\ \boldsymbol{\xi} \in \mathbb{R}^z;\\ \mathbf{h}_{\xi} &= \mathbf{h}_{\xi}(\boldsymbol{\xi},t) : \ \mathbb{R}^z \times \mathbb{R}_{\geq 0} \to \mathbb{R}^h,\end{aligned} \tag{5.47}$$

such that for all $\boldsymbol{\theta} \in \mathbb{R}^d$

$$\varepsilon_{\theta}(t) = f(\mathbf{x}(t),\boldsymbol{\theta},t) - f(\mathbf{x}_1(t) \oplus \mathbf{h}_{\xi}(t) \oplus \mathbf{x}_2'(t),\boldsymbol{\theta},t) \in L_2^1[t_0,\infty] \tag{5.48}$$

and $\dim\{\mathbf{h}_{\xi}\} + \dim\{\mathbf{x}_2'\} = p$. Then (5.11) can be rewritten as[10]

$$\dot{\psi} = f(\mathbf{x}_1 \oplus \mathbf{h}_{\xi} \oplus \mathbf{x}_2',\boldsymbol{\theta},t) - f(\mathbf{x}_1 \oplus \mathbf{h}_{\xi} \oplus \mathbf{x}_2',\hat{\boldsymbol{\theta}},t) - \varphi(\psi,\boldsymbol{\omega},t) + \varepsilon_{\xi}(t), \tag{5.49}$$

where $\varepsilon_{\xi}(t) = \varepsilon_{\theta}(t) + \varepsilon(t) \in L_2^1[t_0,\infty]$, and $\dim\{\mathbf{x}_2'\} = p-h < p$. In principle, the dimension of $\mathbf{x}_2'$ could be reduced to 1 or 0. As soon as this is ensured, Assumption 5.5 will be satisfied, and the results of Theorem 5.1 follow.

The desired properties of system (5.47) are summarized in the following assumption.

Assumption 5.6 *System (5.47)*

(1) is forward-complete,

$$\mathbf{x} \in L_{\infty}^n[t_0,T] \Rightarrow \boldsymbol{\xi} \in L_{\infty}^z[t_0,T]; \tag{5.50}$$

(2) there exists a locally bounded function $\Delta_{\xi} : \mathbb{R}^d \times \mathbb{R}^n \to \mathbb{R}_{\geq 0}$ *such that for all* $\boldsymbol{\theta} \in \Omega_{\theta}$ *and* $\mathbf{x}(t_0)$ *the following holds along the solutions of (5.47):*

$$\|f(\mathbf{x},\boldsymbol{\theta},t) - f(\mathbf{x}_1 \oplus \mathbf{x}_2' \oplus \mathbf{h}_{\xi},\boldsymbol{\theta},t)\|_{2,[t_0,T^*]} \leq \Delta_{\xi}(\boldsymbol{\theta},\mathbf{x}_0), \tag{5.51}$$

where T^* *is the maximal interval of existence of the solution* $\mathbf{x}(t,\mathbf{x}_0)$.

Extending the state space of the original system (5.1) by including extra variables of which the dynamics is governed by (5.47) transforms (5.11) into (5.49), where

$$\|\varepsilon_{\xi}(t)\|_{2,[t_0,T]} \leq \Delta_{\xi}(\boldsymbol{\theta},\mathbf{x}_0).$$

[10] In general, the $L_2^1[t_0,\infty]$-norm of $\varepsilon_{\theta}(t)$ depends on $\boldsymbol{\theta}$. Therefore, given that bounds of $\hat{\boldsymbol{\theta}}$ might not be available a priori, it is not always possible to prove that the $L_2^1[t_0,\infty]$-norm of $\varepsilon_{\hat{\theta}}(t)$ is bounded. In this case a modified control (5.10), where the term $f(\mathbf{x},\hat{\boldsymbol{\theta}},t)$ is replaced with $f(\mathbf{x}_1(t) \oplus \mathbf{h}_{\xi}(t) \oplus \mathbf{x}_2'(t),\hat{\boldsymbol{\theta}},t)$, could be used to render (5.11) into (5.49) with $\varepsilon_{\xi}(t) = \varepsilon_{\theta}(t) + \varepsilon(t) \in L_2^1[t_0,\infty]$.

Let us now introduce the following adaptation algorithm for the extended system (5.1) and (5.47):

$$\begin{aligned}\hat{\boldsymbol{\theta}}(\mathbf{x},t) &= \Gamma(\hat{\boldsymbol{\theta}}_P(\mathbf{x},t)+\hat{\boldsymbol{\theta}}_I(t)),\ \Gamma\in\mathbb{R}^{d\times d},\ \Gamma>0,\\ \hat{\boldsymbol{\theta}}_P(\mathbf{x},t) &= \psi(\mathbf{x},t)\boldsymbol{\alpha}(\mathbf{x}_1\oplus\mathbf{x}_2'\oplus\mathbf{h}_\xi,t)-\Psi(\mathbf{x}_1\oplus\mathbf{x}_2'\oplus\mathbf{h}_\xi,t),\\ \dot{\hat{\boldsymbol{\theta}}}_I &= \varphi(\psi(\mathbf{x},t),\boldsymbol{\omega},t)\boldsymbol{\alpha}(\mathbf{x}_1\oplus\mathbf{x}_2'\oplus\mathbf{h}_\xi,t)+\mathcal{R}(\mathbf{x},\hat{\boldsymbol{\theta}},u(\mathbf{x},\hat{\boldsymbol{\theta}},t),t),\end{aligned} \tag{5.52}$$

where the function $\Psi(\mathbf{x}_1\oplus\mathbf{x}_2'\oplus\mathbf{h}_\xi,t)$ satisfies Assumption 5.7.

Assumption 5.7 *There exists a function* $\Psi(\mathbf{x}_1\oplus\mathbf{x}_2'\oplus\mathbf{h}_\xi,t)\in\mathcal{C}^1$, *such that the following holds:*

$$\frac{\partial\Psi(\mathbf{x}_1\oplus\mathbf{x}_2'\oplus\mathbf{h}_\xi,t)}{\partial\mathbf{x}_2'}=\psi(\mathbf{x},t)\frac{\partial\boldsymbol{\alpha}(\mathbf{x}_1\oplus\mathbf{x}_2'\oplus\mathbf{h}_\xi,t)}{\partial\mathbf{x}_2'}. \tag{5.53}$$

The function $\mathcal{R}(\mathbf{x},\hat{\boldsymbol{\theta}},u(\mathbf{x},\hat{\boldsymbol{\theta}},t),t):\mathbb{R}^n\times\mathbb{R}^d\times\mathbb{R}\times\mathbb{R}_{\geq 0}\to\mathbb{R}^d$ in (5.52) is defined as

$$\begin{aligned}\mathcal{R}(\mathbf{x},u(\mathbf{x},\hat{\boldsymbol{\theta}},t),t) &= \frac{\partial\Psi(\mathbf{x}_1\oplus\mathbf{x}_2'\oplus\mathbf{h}_\xi,t)}{\partial t}-\psi(\mathbf{x},t)\frac{\partial\boldsymbol{\alpha}(\mathbf{x}_1\oplus\mathbf{x}_2'\oplus\mathbf{h}_\xi,t)}{\partial t}\\ &\quad+\left(\frac{\partial\Psi(\mathbf{x}_1\oplus\mathbf{x}_2'\oplus\mathbf{h}_\xi,t)}{\partial\mathbf{h}_\xi}-\psi(\mathbf{x},t)\frac{\partial\boldsymbol{\alpha}(\mathbf{x}_1\oplus\mathbf{x}_2'\oplus\mathbf{h}_\xi,t)}{\partial\mathbf{h}_\xi}\right)\\ &\quad\cdot\frac{\partial\mathbf{h}_\xi(\boldsymbol{\xi},t)}{\partial\boldsymbol{\xi}}\mathbf{f}_\xi(\mathbf{x},\boldsymbol{\xi},t)-(\psi(\mathbf{x},t)L_{\mathbf{f}_1}\boldsymbol{\alpha}(\mathbf{x}_1\oplus\mathbf{x}_2'\oplus\mathbf{h}_\xi,t)\\ &\quad-L_{\mathbf{f}_1}\Psi(\mathbf{x}_1\oplus\mathbf{x}_2'\oplus\mathbf{h}_\xi,t))\\ &\quad-(\psi(\mathbf{x},t)L_{\mathbf{g}_1}\boldsymbol{\alpha}(\mathbf{x}_1\oplus\mathbf{x}_2'\oplus\mathbf{h}_\xi,t)\\ &\quad-L_{\mathbf{g}_1}\Psi(\mathbf{x}_1\oplus\mathbf{x}_2'\oplus\mathbf{h}_\xi,t))u(\mathbf{x},\hat{\boldsymbol{\theta}},t).\end{aligned} \tag{5.54}$$

The properties of the extended system (5.1), (5.47), (5.52), and (5.54) are formulated in Theorem 5.2.

Theorem 5.2 *Consider the system (5.1), (5.47), (5.52), and (5.54), and suppose that system (5.47) satisfies Assumption 5.6. Then the statements of Theorem 5.1 hold for trajectories* $\mathbf{x}$ *and* $\hat{\boldsymbol{\theta}}$ *of the extended system, provided that Assumption 5.5 in the formulation of Theorem 5.1 is replaced with Assumption 5.7.*

Moreover, if Assumption 5.2 holds for systems (5.1) and (5.13) then the extended system is forward-complete and its solutions are bounded.

Theorem 5.2 enables us to replace the integrability constraint specified in Assumption 5.5 with a much weaker requirement formulated in Assumption 5.7.

Indeed, the existence of system (5.47) satisfying Assumption 5.6 with $\dim\{\mathbf{h}_\xi\} = p$ implies that Assumption 5.7 holds. Hence we can conclude that Theorem 5.2 reduces Problem 5.1 to that of finding a suitable embedding of the original (5.1) into a higher-dimensional system (5.1) and (5.47), with $\dim\{\mathbf{h}_\xi\} = p$ satisfying Assumption 5.6.

Let us now describe examples of dynamical systems for which it is possible to find such an embedding. We start with (5.1) in which the functions $\mathbf{f}_2(\mathbf{x}, \boldsymbol{\theta}, t)$ satisfy the following assumption.

Assumption 5.8 *There is a function* $\delta_f(\mathbf{x}, t) : \mathbb{R}^n \times \mathbb{R}_{\geq 0} \to \mathbb{R}^p$, $\delta_{f,i}(\mathbf{x}, t) \geq 0$, $i = 1, \ldots, p$, *such that for every* $\boldsymbol{\theta} \in \Omega_\theta$ *and all* $\mathbf{x} \in \mathbb{R}^n$, $t \in \mathbb{R}_{\geq 0}$ *there exist* $\boldsymbol{\theta}_f \in \mathbb{R}^p$ *and* $\boldsymbol{\theta}_b \in \mathbb{R}^p$ *ensuring that the following inequality holds:*

$$|f_{2,i}(\mathbf{x}, \boldsymbol{\theta}, t)| \leq \theta_{f,i} \cdot \delta_{f,i}(\mathbf{x}, t) + \theta_{b,i}, \quad i = 1, \ldots, p. \tag{5.55}$$

According to Assumption 5.8 for every $\boldsymbol{\theta} \in \Omega_\theta$ there are $\Delta_f : \mathbb{R}^p \times \mathbb{R}_{\geq 0} \to \mathbb{R}^{p+1}$, $\Delta_f(\mathbf{x}, t) = \delta_f(\mathbf{x}, t) \oplus 1$, and $\boldsymbol{\eta} \in \mathbb{R}^{p+1}$ such that

$$\|\mathbf{f}_2(\mathbf{x}, \boldsymbol{\theta}, t)\| \leq \boldsymbol{\eta}^{\mathrm{T}} \Delta_f(\mathbf{x}, t). \tag{5.56}$$

Systems satisfying conditions (5.55) and (5.56) include a broad class of equations of which the right-hand side is locally bounded in $\boldsymbol{\theta}$. In this respect Assumption 5.8 is not at all restrictive. A somewhat restricting factor, however, is that the functions $\delta_f(\mathbf{x}, t)$ and $\Delta_f(\mathbf{x}, t)$ in (5.55) and (5.56) are supposed to be known. Let us now impose one more technical condition on $f(\mathbf{x}, \boldsymbol{\theta}, t)$.

Assumption 5.9 *There exist* $\mathbf{h}_\epsilon : \mathbb{R}^p \times \mathbb{R}^n \times \mathbb{R}_{\geq 0} \to \mathbb{R}^p$, $\mathbf{h}_\xi : \mathbb{R}^p \to \mathbb{R}^p$, *and* $\mathbf{h}_\epsilon, \mathbf{h}_\xi \in \mathcal{C}^1$ *such that*

$$\|\mathbf{h}_\epsilon(\boldsymbol{\xi}, \mathbf{x}, t) - \mathbf{h}_\epsilon(\mathbf{x}_2, \mathbf{x}, t)\| \geq |f(\mathbf{x}_1 \oplus \mathbf{h}_\xi(\boldsymbol{\xi}), \boldsymbol{\theta}, t) - f(\mathbf{x}_1 \oplus \mathbf{x}_2, \boldsymbol{\theta}, t)|,$$
$$\forall \boldsymbol{\theta} \in \Omega_\theta. \tag{5.57}$$

Assumption 5.9 holds, obviously, for the functions $f(\mathbf{x}, \boldsymbol{\theta}, t)$ which are Lipschitz in $\mathbf{x}$.

Consider

$$\dot{\boldsymbol{\xi}} = (\mathcal{H}(\boldsymbol{\xi}, \mathbf{x}, t)^{\mathrm{T}} \mathcal{H}(\boldsymbol{\xi}, \mathbf{x}, t) + \beta)(\mathbf{x}_2 - \boldsymbol{\xi}) + \mathbf{g}_2(\mathbf{x}, t)u + v, \ \beta > 0, \tag{5.58}$$

where

$$\mathcal{H}(\boldsymbol{\xi}, \mathbf{x}, t) = \int_0^1 \frac{\partial \mathbf{h}_\epsilon(\mathbf{s}(\lambda, \boldsymbol{\xi}, \mathbf{x}), \mathbf{x}, t)}{\partial \mathbf{s}} d\lambda, \ \mathbf{s}(\lambda, \boldsymbol{\xi}, \mathbf{x}) = \lambda \mathbf{x}_2 + (1 - \lambda)\boldsymbol{\xi},$$

and v is defined as

$$\begin{aligned} v &= -\hat{\boldsymbol{\eta}}^{\mathrm{T}} \Delta_f(\mathbf{x}, t) \cdot \operatorname{sign}(\boldsymbol{\xi} - \mathbf{x}_2), \\ \dot{\hat{\boldsymbol{\eta}}} &= \Gamma_\eta \Delta_f(\mathbf{x}, t) \cdot \operatorname{sign}(\boldsymbol{\xi} - \mathbf{x}_2)^{\mathrm{T}}(\boldsymbol{\xi} - \mathbf{x}_2), \ \Gamma_\eta > 0. \end{aligned} \tag{5.59}$$

Some asymptotic properties of (5.58) and (5.59) are formulated below.

Theorem 5.3 *Consider serial interconnection of (5.1) and (5.58) plus (5.59). Let system (5.1) satisfy Assumption 5.8, and let $\mathbf{h}_\epsilon(\cdot)$ in (5.58) satisfy Assumption 5.9 with $\mathbf{h}_\xi = \boldsymbol{\xi}$. Then*

(1) the system (5.58) and (5.59) is complete in the sense that $\mathbf{x}(t) \in L^n_\infty[t_0, T], \mathbf{x} \in \mathcal{C}^0 \Rightarrow \boldsymbol{\xi}(t) \in L^p_\infty[t_0, T]$;
(2) the following estimate holds for all $\mathbf{x}_0 \in \mathbb{R}^n$ and $\boldsymbol{\theta} \in \Omega_\theta$:

$$\begin{aligned} &\|f(\mathbf{x}_1 \oplus \mathbf{h}_\xi, \boldsymbol{\theta}, t) - f(\mathbf{x}_1 \oplus \mathbf{x}_2, \boldsymbol{\theta}, t)\|_{2,[t_0,T^*]} \\ &\quad \le \frac{1}{\sqrt{2}} \left(\|\boldsymbol{\xi}(t_0) - \mathbf{x}_2(t_0)\|^2 + \|\hat{\boldsymbol{\eta}}(t_0) - \boldsymbol{\eta}\|^2_{\Gamma_\eta^{-1}} \right)^{\frac{1}{2}}, \end{aligned} \tag{5.60}$$

where T^* is the maximal time of existence of the solution $\mathbf{x}(t, \mathbf{x}_0)$.

The theorem allows one to replace the problem of finding functions Ψ satisfying Assumption 5.5 or Assumption 5.7 with that of checking Assumptions 5.8 and 5.9. These assumptions are not linked with any integrability constraints. Indeed, if the function $\mathbf{f}_2(\mathbf{x}, \boldsymbol{\theta}, t)$ is locally bounded in $\boldsymbol{\theta}$ and $f(\mathbf{x}, \boldsymbol{\theta}, t)$ is Lipschitz in $\mathbf{x}_2$ then an extension (5.47) satisfying these new assumptions will always exist. Moreover, the theorem ensures that the system (5.58) and (5.59) can serve as an extension satisfying Assumption 5.2. Thus according to Theorem 5.2, Problem 5.1 can be solved for a broad range of systems with locally bounded right-hand side. Particular algorithms of adaptation follow from (5.52), (5.54), (5.58), and (5.59).

Notice, however, that the right-hand side of (5.58) is not guaranteed to be continuous. It could be made continuous for systems in which the vector fields governing the dynamics of partition $\mathbf{x}_2''$ of $\mathbf{x}_2$ are linearly parametrized (see Section 5.3). In general, however, finding extensions (5.47) with continuous right-hand side and at the same time satisfying Assumption 5.6 is a non-trivial task. Nonetheless, continuity and even differentiability could be needed in a number of applications. Adaptive regulation of cascaded systems is an example of such problems. Solving the problem of adaptive regulation for this class of systems usually involves calculation of partial derivatives (up to the nth order) of the vector fields on the right-hand side of the system (Krstić *et al.* 1992; Kolesnikov 1994). Thus an extension of the results to cascaded systems is needed. This extension is provided in the next section.

5.2.3 Direct adaptive control for systems with lower-triangular structure

Consider the following class of models:

$$\begin{aligned}\dot{x}_i &= f_i(x_1,\ldots,x_i,\boldsymbol{\theta}_i) + x_{i+1},\\ \dot{x}_n &= f_n(x_1,\ldots,x_n,\boldsymbol{\theta}_n) + u + \varepsilon(t),\\ &\varepsilon(t) \in L_2,\ \boldsymbol{\theta}_i \in \Omega_\theta,\ i = 1,\ldots,n-1, \boldsymbol{\theta}_n \in \Omega_\theta,\end{aligned} \tag{5.61}$$

where the functions $f_i(\cdot), i = 1,\ldots,n$, satisfy Assumptions 5.3 and 5.4, and the corresponding functions $\boldsymbol{\alpha}_i(\mathbf{x})$ are smooth (i.e. the function is differentiable infinitely many times). In addition, suppose that the following assumption holds.

Assumption 5.10 *There exist smooth functions $\bar{D}_i(\cdot) : \mathbb{R}^i \times \mathbb{R}^i \to \mathbb{R}$ such that*

$$(f_i(x_1,\ldots,x_i,\boldsymbol{\theta}_i) - f_i(x'_1,\ldots,x'_i,\boldsymbol{\theta}_i))^2 \le \bar{D}_i^2(\mathbf{x}_i,\mathbf{x}'_i)\|\mathbf{x}_i - \mathbf{x}'_i\|^2,\ \forall\,\boldsymbol{\theta}_i \in \Omega_\theta,$$

where $\mathbf{x}_i = (x_1,\ldots,x_i)^{\mathrm{T}}$ and $\mathbf{x}'_i = (x'_1,\ldots,x'_i)^{\mathrm{T}}$.

Let the control goal be to reach asymptotically the following manifold $\psi(x_1) = 0$ in the system's state space:

$$\lim_{t\to\infty} \psi(x_1(t)) = 0. \tag{5.62}$$

We start with the following lemma.

Lemma 5.1 *Consider*

$$\dot{x}_i = f_i(x_1,\ldots,x_i,\boldsymbol{\theta}_i) + \beta_i(\mathbf{x},t),\ \mathbf{x}_0 = \mathbf{x}(t_0), \tag{5.63}$$

$i = 1,\ldots,n$ and a smooth function $u(\mathbf{x},\mathbf{z},\boldsymbol{\theta}_0) : \mathbb{R}^n \times \mathbb{R}^m \times \mathbb{R}^d \to \mathbb{R}$. Let $[t_0, T]$, $T > t_0$, be an interval of existence of solutions of (5.63).

Let us suppose that $\boldsymbol{\theta}_0 \in \Omega_\theta^0$, where Ω_θ^0 is bounded. In addition let there exist functions $\bar{F}(\mathbf{x},\mathbf{x}',\mathbf{z})$ and $\bar{D}_i(\mathbf{x},\mathbf{x}')$, $i = 1,\ldots,n$ such that

(1) $(u(\mathbf{x},\mathbf{z},\boldsymbol{\theta}_0) - u(\mathbf{x}',\mathbf{z},\boldsymbol{\theta}_0))^2 \le \|\mathbf{x}-\mathbf{x}'\|^2,\ \bar{F}^2(\mathbf{x},\mathbf{x}',\mathbf{z}), \forall\,\boldsymbol{\theta}_0 \in \Omega_\theta^0$, with $\mathbf{x},\mathbf{x}' \in \mathbb{R}^n$,
(2) $(f_i(x_1,\ldots,x_i,\boldsymbol{\theta}_i) - f_i(x'_1,\ldots,x'_i,\boldsymbol{\theta}_i))^2 \le \|\tilde{\mathbf{x}}_i - \tilde{\mathbf{x}}'_i\|^2 \bar{D}_i^2(\mathbf{x},\mathbf{x}')\ \forall\,\boldsymbol{\theta}_i \in \Omega_\theta$, with $\tilde{\mathbf{x}}_i, \tilde{\mathbf{x}}'_i \in \mathbb{R}^n$,

$$\begin{aligned}\tilde{\mathbf{x}}_i &= (x_1,\ldots,x_i,0,\ldots,0)^{\mathrm{T}},\\ \tilde{\mathbf{x}}'_i &= (x'_1,\ldots,x'_i,0,\ldots,0)^{\mathrm{T}}.\end{aligned}$$

Finally, let there exist and be known functions $\boldsymbol{\alpha}_i(\mathbf{x})$ such that Assumptions 5.3 and 5.4 hold for the respective functions $f_i(x_1,\ldots,x_i,\boldsymbol{\theta}_i)$.

Then there exist $\boldsymbol{\xi}(t) : \mathbb{R} \to \mathbb{R}^n$, $\boldsymbol{\nu}(t) : \mathbb{R} \to \mathbb{R}^m$, *smooth functions* $\mathbf{f}_\xi(\cdot)$ *and* $\mathbf{f}_\nu(\cdot)$, *and the corresponding system*

$$\begin{aligned} \dot{\boldsymbol{\xi}} &= \mathbf{f}_\xi(\mathbf{x}, \boldsymbol{\xi}, \mathbf{z}, \boldsymbol{\nu}), \ \boldsymbol{\xi}_0 \in \mathbb{R}^n, \\ \dot{\boldsymbol{\nu}} &= \mathbf{f}_\nu(\mathbf{x}, \boldsymbol{\xi}, \mathbf{z}, \boldsymbol{\nu}), \ \boldsymbol{\nu}_0 \in \mathbb{R}^m, \end{aligned} \tag{5.64}$$

such that

(1) $u(\mathbf{x}, \mathbf{z}, \boldsymbol{\theta}_0) - u(\mathbf{q}_i, \mathbf{z}, \boldsymbol{\theta}_0) \in L_2[t_0, T]$, $i = 1, \ldots, n$,

$$\mathbf{q}_i = (\xi_1, \ldots, \xi_i, x_{i+1}, \ldots, x_n)^{\mathrm{T}};$$

(2) $f_i(x_1, \ldots, x_i, \boldsymbol{\theta}_i) - f_i(\xi_1, \ldots, \xi_{i-1}, x_i, \boldsymbol{\theta}_i) \in L_2[t_0, T]$, $i = 2, \ldots, n$;
(3) $\mathbf{x} \in L_\infty[t_0, T] \Rightarrow \boldsymbol{\xi}, \boldsymbol{\nu} \in L_\infty[t_0, T]$.

Moreover, the corresponding L_2*- and* L_∞*-norms can be bounded from above by constants whose values do not depend on* T.

Lemma 5.1 allows us to prove the following result.

Theorem 5.4 *Let system (5.61) and the goal functional* $\psi(x_1) = 0$ *be given. Suppose that the function* ψ *is smooth, and* $\psi(x_1) \in L_\infty[t_0, T] \Rightarrow x_1 \in L_\infty[t_0, T]$. *Let* $f_i(x_1, \ldots, x_i, \boldsymbol{\theta}_i)$ *in (5.61) be smooth and satisfy Assumptions 5.3 and 5.4 with smooth functions* $\boldsymbol{\alpha}_i(x_1, \ldots, x_i)$, *respectively. In addition, suppose that Assumption 5.10 holds.*

Then there exists an auxiliary system

$$\begin{aligned} &\dot{\boldsymbol{\xi}} = \mathbf{f}_\xi(\mathbf{x}, \boldsymbol{\xi}, \boldsymbol{\nu}), \ \dot{\boldsymbol{\nu}} = \mathbf{f}_\nu(\mathbf{x}, \boldsymbol{\xi}, \boldsymbol{\nu}), \\ &\boldsymbol{\xi}_0 \in \mathbb{R}^n, \ \ \boldsymbol{\nu}_0 \in \mathbb{R}^m, \end{aligned} \tag{5.65}$$

as well as smooth functions $\psi_i(x_i, t)$, $i = 1, \ldots, n$, $\hat{\boldsymbol{\theta}}_P(\mathbf{x}, \boldsymbol{\xi})$, *control input* $u(\mathbf{x}, \hat{\boldsymbol{\theta}}, \boldsymbol{\xi}, \boldsymbol{\nu})$, *and an adaptation algorithm*

$$\begin{aligned} \hat{\boldsymbol{\theta}}(\mathbf{x}, \boldsymbol{\xi}, \hat{\boldsymbol{\theta}}_I) &= \gamma(\hat{\boldsymbol{\theta}}_P(\mathbf{x}, \boldsymbol{\xi}) + \hat{\boldsymbol{\theta}}_I), \ \gamma > 0, \\ \dot{\hat{\boldsymbol{\theta}}}_I &= \mathbf{f}_{\hat{\theta}}(\mathbf{x}, \hat{\boldsymbol{\theta}}, \boldsymbol{\xi}, \boldsymbol{\nu}), \end{aligned} \tag{5.66}$$

such that

(1) $\psi_i(x_i, t), \psi \in L_2[t_0, \infty] \cap L_\infty[t_0, \infty]$, $\dot{\psi}, \dot{\psi}_i \in L_2[t_0, \infty]$, $i = 1, \ldots, n$;
(2) $\hat{\boldsymbol{\theta}} \in L_\infty[t_0, \infty]$ *and* $u(\mathbf{x}, \hat{\boldsymbol{\theta}}, \boldsymbol{\xi}, \boldsymbol{\nu}) - u(\mathbf{x}, \boldsymbol{\theta}_n, \boldsymbol{\xi}, \boldsymbol{\nu}) \in L_2[t_0, \infty]$;
(3) $\mathbf{x}, \boldsymbol{\xi}, \boldsymbol{\nu} \in L_\infty[t_0, \infty]$;
(4) if $\varepsilon(t) \in L_\infty[t_0, \infty]$, *then* $\dot{\psi}, \dot{\psi}_i \in L_\infty[t_0, \infty]$, *and the following holds:*

$$\lim_{t\to\infty} \psi(x_1(t)) = 0, \qquad \lim_{t\to\infty} \psi_i(x_i(t), t) = 0, \ i = 1, \ldots, n.$$

Similarly to Theorems 5.2 and 5.3, Theorem 5.4 establishes conditions (Assumption 5.10) allowing one to replace the integrability constraint (5.33) with the weaker requirement of finding an auxiliary system (extension) (5.47), (5.58), (5.59) or (5.65), satisfying Assumption 5.6. Such an auxiliary system, or extension, can be viewed as a functional observer $\mathcal{S}_o$ (see Section 4.3.2) for $\mathbf{x}$, ensuring the boundedness of the estimation error with respect to the norm (4.60). Thus, completeness and the boundedness of the system's state follow immediately from Theorem 4.6.

Our proof of the theorem is constructive (see the appendix to this chapter). This implies that it not only establishes the existence of an adaptive control law for (5.61), but also presents specific adaptation algorithms (e.g. (A5.44)). In order to illustrate how these algorithms can be constructed, two examples are considered below. In the first example we show how Theorems 5.2–5.4 can be used for designing adaptation algorithms for cascaded systems with linear parametrization. In the second example we demonstrate how the very same method can be applied to systems with nonlinear parametrization.

Example 5.2.1 Let us consider the following system:

$$\dot{x}_1 = x_1^2\theta_0 + x_2, \qquad \dot{x}_2 = x_1\theta_1 + x_2\theta_2 + u, \tag{5.67}$$

where the parameters θ_0, θ_1, and θ_2 are assumed to be unknown. The control goal is to steer the system towards the manifold $x_1 - 1 = 0$ in $\mathbb{R}^2$. To design adaptive algorithms in finite form for system (5.67), we follow the steps in the proof of Theorem 5.4.

(1) *Intermediate control design*. Derive the control function $u_1(x_1, \hat{\theta}_0)$, such that for the reduced system

$$\dot{x}_1 = x_1^2\theta_0 + u_1(x_1, \hat{\theta}_0) + \varepsilon_1(t), \ \varepsilon_1(t) \in L_2; \ \hat{\theta}_0 = \hat{\theta}_{0,P}(x_1) + \hat{\theta}_{0,I}(t),$$

attainment of the control goal is guaranteed: $\psi(x_1(t)) = x_1(t) - 1 \to 0$ as $t \to \infty$. Moreover, the function $u_1(x_1, \hat{\theta}_0(x_1, \hat{\theta}_{0,I}))$ should ensure that $\psi, \dot{\psi} \in L_2[t_0, \infty]$.

(2) *Embedding*. Extend the system dynamics with (or embed it into) the auxiliary system

$$\dot{\xi} = f_\xi(\mathbf{x}, \xi, \nu), \quad \dot{\nu} = f_\nu(\mathbf{x}, \xi, \nu) \tag{5.68}$$

in order to guarantee that

$$u(x_1, \hat{\theta}_0(x_1, \hat{\theta}_{0,I})) - u(\xi, \hat{\theta}_0(\xi, \hat{\theta}_{0,I})) \in L_2[t_0, T], \qquad x_1 - \xi \in L_2[t_0, T]. \tag{5.69}$$

These $L_2[t_0, T]$-norms are to be bounded by continuous functions of the initial conditions and, possible, parameters.

(3) *Control-function design.* Introduce a new goal function $\psi_2(x_2, t) = x_2 - u_1(\xi, \hat{\theta}_0(\xi, \hat{\theta}_{0,I}))$ and derive a control function $u(x_1, x_2, \xi, t)$ such that $\dot{\psi}_2 \in L_2[t_0, \infty]$, $\psi_2 \in L_2 \cap L_\infty[t_0, \infty]$. The last condition automatically implies that

$$\dot{x}_1 = x_1^2\theta_0 + x_2 = x_1^2\theta_0 + u_1(x_1, \hat{\theta}_0(x_1, \hat{\theta}_{0,I})) + \mu(t),$$

where

$$\begin{aligned}\mu(t) &= x_2 - u_1(x_1, \hat{\theta}_0(x_1, \hat{\theta}_{0,I}))\\ &= (x_2 - u_1(\xi, \hat{\theta}_0(\xi, \hat{\theta}_{0,I}))) + (u_1(\xi, \hat{\theta}_0(\xi, \hat{\theta}_{0,I}))\\ &\quad - u_1(x_1, \hat{\theta}_0(x_1, \hat{\theta}_{0,I}))) \in L_2[t_0, \infty].\end{aligned}$$

Therefore, according to the choice of the function $u_1(x_1, \hat{\theta}_0(x_1, \hat{\theta}_{0,I}))$, control $u(x_1, x_2, \xi, t)$ guarantees that $\psi(x_1(t)) \to 0$ as $t \to \infty$, with $\psi, \dot{\psi} \in L_2[t_0, \infty]$.

We begin by determining the function $u(x_1, \hat{\theta}_0(x_1, \hat{\theta}_{0,I}))$. Let $u_1(x_1, \hat{\theta}_0) = -C_1(x_1 - 1) - \hat{\theta}_0 x_1^2$, where $C_1 > 0$ is the design parameter and $\hat{\theta}_0$ satisfies the following differential equation (*virtual adaptation algorithm*):

$$\dot{\hat{\theta}}_0 = \gamma_0(C_1(x_1 - 1) + \dot{x}_1)x_1^2, \ \gamma_0 > 0. \tag{5.70}$$

It follows from Lemma A5.2 (see the proof of Lemma 5.1 in the appendix to this chapter) that the control function $u_1(x_1, \hat{\theta}_0)$ with algorithm (5.70) guarantees that $\psi, \dot{\psi} \in L_2[t_0, \infty]$, $\psi(x_1(t)) \to 0$ as $t \to \infty$. According to Theorem 5.1, the finite-form realization of (5.70) can be given as follows: $\hat{\theta}_0(x_1, \hat{\theta}_{0,I}(t)) = \gamma_0(\frac{1}{3}x_1^3 + \hat{\theta}_{0,I}(t))$, $\dot{\hat{\theta}}_{0,I} = C_1(x_1 - 1)x_1^2$. On substituting these into $u_1(x_1, \hat{\theta}_0)$ we get the following expression for $u_1(\cdot)$:

$$\begin{aligned}u_1(x_1, \hat{\theta}_0(x_1, \hat{\theta}_{0,I})) &= -C_1(x_1 - 1) - \gamma_0\left(\frac{1}{3}x_1^5 + x_1^2\hat{\theta}_{0,I}(t)\right),\\ \dot{\hat{\theta}}_{0,I} &= \psi(x_1)\alpha_1(x_1) = C_1(x_1 - 1)x_1^2.\end{aligned} \tag{5.71}$$

This completes step 1 of the synthesis procedure.

Let us design a system (5.68) that guarantees that (5.69) holds for the function (5.71). First, consider the difference

$$\begin{aligned}&u(x_1, \hat{\theta}_0(x_1, \hat{\theta}_{0,I})) - u(\xi, \hat{\theta}_0(\xi, \hat{\theta}_{0,I}))\\ &= -(x_1 - \xi)\left(C_1 + \gamma_0\left((x_1 + \xi)\hat{\theta}_{0,I} + \frac{1}{3}(x_1^4 + x_1^3\xi + x_1^2\xi^2 + x_1\xi^3 + \xi^4)\right)\right),\end{aligned} \tag{5.72}$$

and denote $F(x_1,\xi,\hat{\theta}_{I,0}) = C_1+\gamma_0((x_1+\xi)\hat{\theta}_{0,I}+\frac{1}{3}(x_1^4+x_1^3\xi+x_1^2\xi^2+x_1\xi^3+\xi^4))$. It follows from Lemma 5.1 that there exists a system (5.68) such that condition (5.69) holds. This system can be given by the equation

$$\dot{\xi} = (x_1 - \xi)(F^2(x_1,\xi,\hat{\theta}_{0,I}) + 1) + x_1^2\hat{\theta}_\xi + x_2, \tag{5.73}$$

where $\hat{\theta}_\xi$ satisfies the following differential equation: $\dot{\hat{\theta}}_\xi = (x_1 - \xi + \dot{x}_1 - \dot{\xi})x_1^2$. The finite-form realization of this algorithm[11] follows from Theorem 5.1, and it can be written as

$$\begin{aligned}
\hat{\theta}_\xi &= \frac{1}{3}x_1^3 + \hat{\theta}_{\xi,I},\\
\dot{\hat{\theta}}_{\xi,I} &= (x_1 - \xi)x_1^2 - x_1^2((x_1 - \xi)(F^2(x_1,\xi,\hat{\theta}_{0,I}) + 1) + x_1^2\hat{\theta}_\xi + x_2).
\end{aligned} \tag{5.74}$$

Taking into account (5.74) and (5.73), the system (5.68) which ensures (5.69) can be represented as follows:

$$\begin{aligned}
\dot{\xi} &= (x_1 - \xi)(F^2(x_1,\xi,\hat{\theta}_{0,I}) + 1) + \frac{1}{3}x_1^5 + \hat{\theta}_{\xi,I}(t)x_1^2 + x_2,\\
\dot{\hat{\theta}}_{\xi,I} &= (x_1 - \xi)x_1^2 - x_1^2((x_1 - \xi)(F^2(x_1,\xi,\hat{\theta}_{0,I}) + 1)\\
&\quad + \frac{1}{3}x_1^5 + \hat{\theta}_{\xi,I}(t)x_1^2 + x_2).
\end{aligned} \tag{5.75}$$

This completes the second step of the iteration.

To conclude the procedure we consider the new target set

$$x_2 - u_1(\xi,\hat{\theta}_0(\xi,\hat{\theta}_{0,I})) = 0$$

and goal function

$$\psi_2(x_2,t) = x_2 - u_1(\xi,\hat{\theta}_0(\xi,\hat{\theta}_{0,I})) = x_2 + C_1(\xi - 1) + \gamma_0\left(\frac{1}{3}\xi^5 + \hat{\theta}_{0,I}\xi^2\right).$$

[11] The introduction of algorithms (5.74) is not necessary because the original system is linearly parametrized and condition (5.69) can be satisfied even with conventional adaptation schemes. Nevertheless, we would like to keep the calculations consistent with the steps in the proof of Theorem 5.4 in order to show how the method operates, in preparation for cases in which the right-hand sides are nonlinearly parametrized.

Let us write the derivative with respect to time t of the function $\psi_2(\cdot)$:

$$\dot{\psi}_2 = \dot{x}_2 - \frac{\partial u_1(\xi, \hat{\theta}_0(\xi, \hat{\theta}_{0,I}))}{\partial \xi}\dot{\xi} - \frac{\partial u_1(\xi, \hat{\theta}_0(\xi, \hat{\theta}_{0,I}))}{\partial \hat{\theta}_{0,I}}\dot{\hat{\theta}}_{0,I}$$

$$= x_1\theta_1 + x_2\theta_2 + u + \gamma_0 C_1 \xi^2 (x_1 - 1)x_1^2 + \left(C_1 + \gamma_0\left(\frac{5}{3}\xi^4 + 2\xi\hat{\theta}_{0,I}\right)\right)$$

$$\times \left((x_1 - \xi)(F^2(x_1, \xi, \hat{\theta}_{0,I}) + 1) + \gamma_0\left(\frac{1}{3}x_1^5 + \hat{\theta}_{\xi,I}(t)x_1^2\right) + x_2\right).$$

Therefore, the control function

$$u = -\xi\hat{\theta}_1 - x_2\hat{\theta}_2 - \gamma_0 C_1 \xi^2 (x_1 - 1)x_1^2$$

$$- C_2\left(x_2 + C_1(\xi - 1) + \gamma_0\left(\frac{1}{3}\xi^5 + \hat{\theta}_{0,I}\xi^2\right)\right)$$

$$- \left(C_1 + \gamma_0\left(\frac{5}{3}\xi^4 + 2\xi\hat{\theta}_{0,I}\right)\right)$$

$$\times \left((x_1 - \xi)(F^2(x_1, \xi, \hat{\theta}_{0,I}) + 1) + \gamma_0\left(\frac{1}{3}x_1^5 + \hat{\theta}_{\xi,I}(t)x_1^2\right) + x_2\right), \quad (5.76)$$

where $C_2 > 0$ is a design parameter, ensures that $\dot{\psi}_2 = -C_2\psi_2(x_2, t) + x_1\theta_1 + x_2\theta_2 - \xi\hat{\theta}_1 - x_2\hat{\theta}_2$. Taking condition (5.69) into account, we can rewrite the derivative $\dot{\psi}_2$ as $\dot{\psi}_2 = -C_2\psi_2(x_2, t) + \xi\theta_1 + x_2\theta_2 - \xi\hat{\theta}_1 - x_2\hat{\theta}_2 + \varepsilon(t)$, where $\varepsilon(t) = (x_1 - \xi)\theta_1 \in L_2[t_0, T]$ (where T is the maximal interval of existence of the system's solution). It follows from Lemma A5.1 that the adaptation algorithm

$$\begin{aligned} \dot{\hat{\theta}}_1 &= \gamma_0(C_2\psi_2(x_2, t) + \dot{\psi}_2)\alpha_1(\xi), \\ \dot{\hat{\theta}}_2 &= \gamma_0(C_2\psi_2(x_2, t) + \dot{\psi}_2)\alpha_2(x_2), \ \alpha_1(\xi) = \xi, \ \alpha_2(x_2) = x_2, \end{aligned} \quad (5.77)$$

guarantees that $\psi_2 \in L_2 \cap L_\infty[t_0, T]$ and $\dot{\psi}_2 \in L_2[t_0, T]$. The realization of algorithms (5.77) can be obtained from Theorem 5.1:

$$\hat{\theta}_1(x_2, \xi, \hat{\theta}_{0,I}, t) = \gamma_0\left(\left(x_2 + C_1(\xi - 1) + \gamma_0\left(\frac{1}{3}\xi^5 + \hat{\theta}_{0,I}\xi^2\right)\right)\xi + \hat{\theta}_{1,I}(t)\right),$$

$$\dot{\hat{\theta}}_{1,I} = C_2\left(x_2 + C_1(\xi - 1) + \gamma_0\left(\frac{1}{3}\xi^5 + \hat{\theta}_{0,I}\xi^2\right)\right)\left(\xi - \frac{\dot{\xi}}{C_2}\right),$$

$$\hat{\theta}_2(x_2, \xi, \hat{\theta}_{0,I}, t) = \gamma_0\left(\frac{x_2^2}{2} + \hat{\theta}_{2,I}(t)\right), \quad (5.78)$$

$$\dot{\hat{\theta}}_{2,I} = C_2\left(x_2 + C_1(\xi - 1) + \gamma_0\left(\frac{1}{3}\xi^5 + \hat{\theta}_{0,I}\xi^2\right)\right)x_2 + \frac{\partial \Psi_2}{\partial \xi}\dot{\xi} + \frac{\partial \Psi_2}{\partial \hat{\theta}_{0,I}}\dot{\hat{\theta}}_{0,I},$$

where $\Psi_2(x_2, \xi, \hat{\theta}_{0,I}) = \int \psi_2(x_2, t)\partial\alpha_2(x_2)/\partial x_2 \; dx_2 = x_2^2/2 + (C_1(\xi - 1) + \gamma_0(\frac{1}{3}\xi^5 + \hat{\theta}_{0,I}\xi^2))x_2$. Given that the $L_2[t_0, T]$-norms of ε, ψ_2, and $\dot{\psi}_2$ are bounded by continuous functions of the initial conditions and parameters, we can conclude that $x_1(t)$ and $x_2(t)$ can be bounded by some continuous functions of the initial conditions and parameters. This implies that solutions of the combined system exist for all $t \geq t_0$ and are bounded, and that $\psi(x_1(t)) \to 0$ as $t \to \infty$.

We would also like to compare the performance of the proposed control scheme with that of adaptive backstepping control algorithms (Krstić *et al.* 1992; Kanellakopoulos *et al.* 1991). Adaptive backstepping design for system (5.67) according to Kanellakopoulos *et al.* (1991) results in the following control algorithm:

$$\begin{aligned}
u_1 &= -C_2(x_2 + C_1(x_1 - 1) + \hat{\theta}_3 x_1^2) - \gamma_0 x_1^4(x_1 - 1) - x_2(C_1 + 2x_1\hat{\theta}_3) \\
&\quad - (C_1 x_1^2 + 2\hat{\theta}_3 x_1^3)\hat{\theta} - x_1\hat{\theta}_1 - x_2\hat{\theta}_2, \\
\dot{\hat{\theta}} &= \gamma_0(x_2 + C_1(x_1 - 1) + \hat{\theta}_3 x_1^2)x_1^2(C_1 + 2\hat{\theta}_3 x_1), \\
\dot{\hat{\theta}}_1 &= \gamma_0(x_2 + C_1(x_1 - 1) + \hat{\theta}_3 x_1^2)x_1, \\
\dot{\hat{\theta}}_2 &= \gamma_0(x_2 + C_1(x_1 - 1) + \hat{\theta}_3 x_1^2)x_2, \\
\dot{\hat{\theta}}_3 &= \gamma_0(x_1 - 1)x_1^2,
\end{aligned} \tag{5.79}$$

where $C_1 > 0$, $C_2 > 0$, and $\gamma_0 > 0$ are parameters. As before, the parameters C_1 and C_2 stand for the feedback gains, and γ_0 is the adaptation gain. Adaptive back-stepping with tuning functions (Krstić *et al.* 1992) results in

$$\begin{aligned}
u_1 &= -C_2(x_2 + C_1(x_1 - 1) + x_1^2\hat{\theta}) - (x_1 - 1) \\
&\quad - (C_1 + 2x_1\hat{\theta})(x_2 + \hat{\theta}x_1^2) - x_1^2\tau - x_1\hat{\theta}_1 - x_2\hat{\theta}_2, \\
\dot{\hat{\theta}} &= \tau; \quad \tau = \gamma_0((x_1 - 1)x_1^2 + (x_2 + C_1(x_1 - 1) + x_1^2\hat{\theta})x_1^2(C_1 + 2x_1\hat{\theta})), \\
\dot{\hat{\theta}}_1 &= \gamma_0(x_2 + C_1(x_1 - 1) + x_1^2\hat{\theta})x_1, \\
\dot{\hat{\theta}}_2 &= \gamma_0(x_2 + C_1(x_1 - 1) + x_1^2\hat{\theta})x_2.
\end{aligned} \tag{5.80}$$

The meanings of the parameters C_1, C_2, and γ_0 in (5.80) are analogous to those in (5.79).

We simulated the adaptive system dynamics for the set of parameters $\theta_0 = \theta_1 = 1$, $\theta_2 = 0.5$, and $C_2 = C_1 = \gamma = 1$ and initial conditions $x_1(0) = 2$, $x_2(0) = 0.2$, $\hat{\theta}_3(0) = \hat{\theta}_0(0) = 3$, $\hat{\theta}_1(0) = \hat{\theta}_2(0) = -2$, $\xi_2(0) = 0$, and $\xi_1(0) = x_1(0)$. The initial conditions for $\hat{\theta}_{1,I}(0)$, $\hat{\theta}_{2,I}(0)$, and $\hat{\theta}_{3,I}(0)$ in (5.78) where chosen to satisfy $\hat{\theta}_1(0) = \hat{\theta}_2(0) = -2$ and $\hat{\theta}_3(0) = 3$. As an additional measure of performance, we introduced the variable $\Delta\boldsymbol{\theta}(t) = \|\boldsymbol{\theta} - \hat{\boldsymbol{\theta}}(t)\|$, which is the Euclidean distance

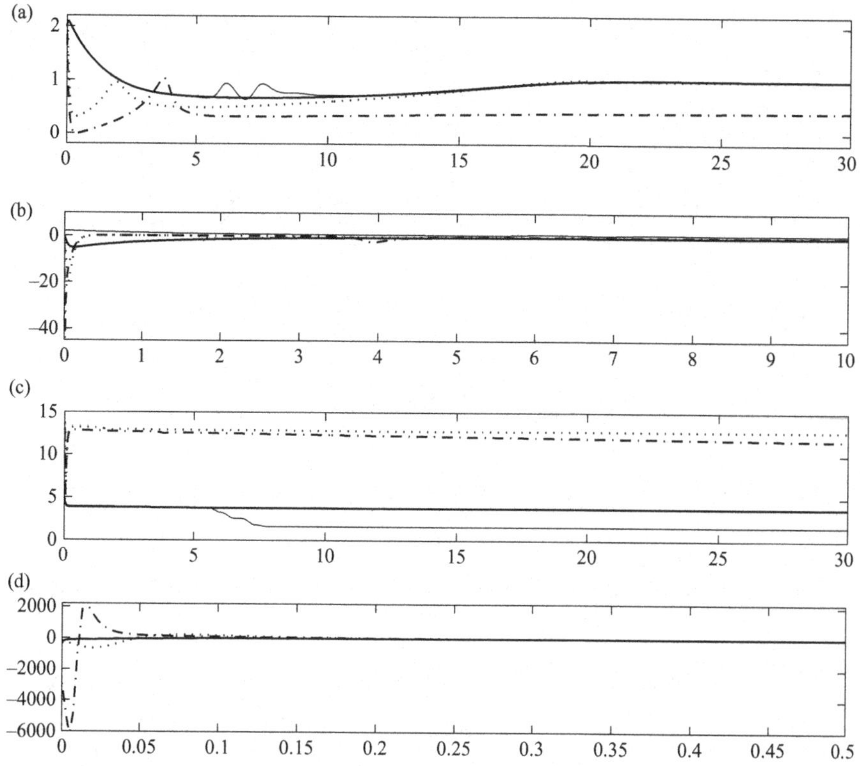

Figure 5.2 Plots of system (5.67) trajectories with control functions (5.76) and (5.78) (thick solid lines), (5.79) (dotted line), and (5.80) (dash–dotted line). In (a) x_1 is shown as a function of time, in (b) x_2 is depicted as a function of time, in (c) $\Delta\boldsymbol{\theta}$ is shown as a function of time, and in (d) u is shown as a function of time.

between estimates $\hat{\boldsymbol{\theta}}(t)$ and the real values of the parameters. Simulation results are presented in Figure 5.2. For the given set of initial conditions and controller parameters, the transient performance of the proposed adaptive algorithms is better than that of conventional algorithms. In addition, we calculated the integral $I = \int_0^T u_1^2(\tau)d\tau$, $T = 500$, for every controller for the system solutions. The values of the functional I indicate how much energy is spent to achieve the control goal. For the control function (5.76) and (5.78) $I = 627.10$, for the adaptive backstepping controller (5.79) $I = 13\,329.28$, and for the controller (5.80) $I = 263\,872.58$.

The picture remained the same when we varied the controller parameters C_1, C_2, and γ_0. In particular we let $C_1 = C_2 = c$ and varied the parameter c in the interval $[1, 5]$. The parameter γ_0 was chosen randomly in the interval $[0.1, 2]$.

For other values of the initial conditions $x_1(0)$ and $x_2(0)$ there was no definite winner. Yet, on average, the performance of the proposed scheme in terms of the L_2-norms calculated for the error function $x_1(t) - 1$ and control $u(t)$ was superior to the best of (5.79) and (5.80). The results are shown in Figure 5.3. Simulations also

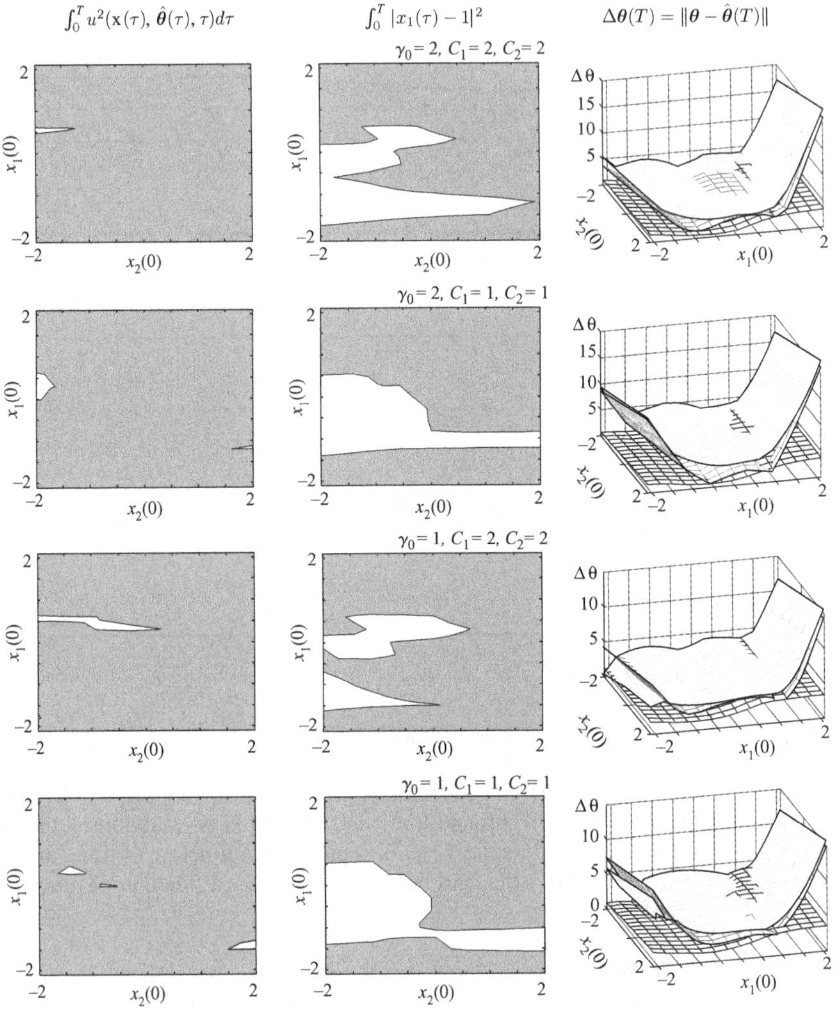

Figure 5.3 Performance diagrams comparing backstepping algorithms (5.79) and (5.80) with the currently proposed scheme (5.76) and (5.78) for the domain of initial conditions $x_1(0) \in [-2, 2]$, $x_2(0) \in [-2, 2]$ and parameter values $\theta_0 = 1$, $\theta_1 = 1$, and $\theta_2 = 1$. Each row contains the results for a single value of the controller parameters C_1, C_2, and γ_0. The left column contains the data on the L_2-norm of $u(t)$, the middle column stands for the L_2-norm of the signal $x_1(t) - 1$, and the right column represents the results for $\Delta\boldsymbol{\theta} = \|\boldsymbol{\theta} - \hat{\boldsymbol{\theta}}(T)\|$. The simulation time T was set to 30 s. White and shaded surfaces in the right column correspond to backstepping algorithms (5.79) and (5.80) respectively; the black "gridded" surface corresponds to adaptive feedback (5.76) and (5.78). White areas in the left and middle columns represent domains of initial conditions where either of the two backstepping algorithms (5.79) and (5.80) outperforms our proposed scheme with regard to the values of L_2-norms for signals $u(t)$ and $x_1(t) - 1$, respectively. The filled areas mark those domains where our scheme works better than both (5.79) and (5.80).

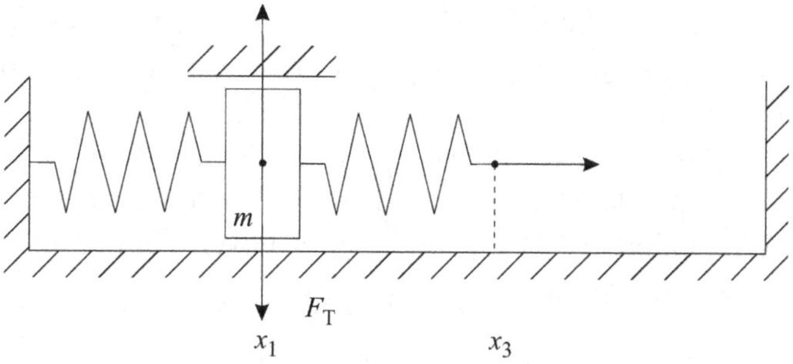

Figure 5.4 A spring–mass system.

indicate that, for the proposed algorithm (5.76) and (5.78), the uncertainty measure $\Delta\boldsymbol{\theta}(T)$ did not increase compared with $\Delta\boldsymbol{\theta}(0) = 4.609$. Standard backstepping algorithms, however, often led to a substantial increase in $\Delta\boldsymbol{\theta}(T)$.

The next example illustrates the application of the method to systems of which the models contain nonlinearly parametrized uncertainties.

Example 5.2.2 Let us consider a mass–spring system with unknown stick/slip friction (or, in general, nonlinear damping) and actuator dynamics. The system is schematically depicted in Figure 5.4. The equations which govern the system's dynamics are derived explicitly from Newton's laws:

$$\begin{aligned}
\dot{x}_1 &= x_2, \\
\dot{x}_2 &= -kx_1 + k(x_3 - x_1) - \tanh(S_f x_2)(C_1 + \theta_{1,2} e^{-\theta_{1,1} x_2^2}), \\
\dot{x}_3 &= \theta_2 x_3 + u.
\end{aligned} \tag{5.81}$$

The coefficients k in (5.81) denote elastic stiffness, and the term $\tanh(S_f x_2)(C_1 + \theta_{1,2} e^{-\theta_{1,1} x_2^2})$, $S_f = 50$ models friction forces acting on the connected bodies. The coefficient C_1 stands for Coulomb friction, and the parameters $\theta_{1,1}$ and $\theta_{1,2}$ parametrize Stribeck friction forces. The parameter θ_2 denotes the time constant of the actuator. For simplicity we assume that $k = C_1 = 1$. Equations (5.81) describe a practically relevant process representing, among others, systems of electromechanical valves, artificial/natural muscles (though with a different type of nonlinearly parametrized damping), and various haptic interface systems (Lawrence *et al.* 1998).

The lumped Coulomb static friction and elastic stiffness are defined mostly by the physical properties of the materials and can be estimated a priori. The Stribeck force, however, is more sensitive to the changes in operating conditions of the system (such as position) due to the principal spatial heterogeneity of the contact surfaces. Therefore, adaptation is needed. The adaptation should, however, be non-dominating in order to avoid overshooting in control and thus provide good transient behavior of the system.

Let the control goal be to steer the system to the manifold $x_1 = 1$. System (5.81) has lower-triangular structure and therefore the results of Theorem 5.4 are applicable if Assumptions 5.3, 5.4, and 5.10 hold. Given that the velocity x_2, is, in principle, bounded and

$$\theta_{1,2}e^{-\theta_{1,1}x_2^2} = e^{-\theta_{1,1}x_2^2+\log\theta_{1,2}},$$

we can conclude that Assumptions 5.3 and 5.4 are satisfied, at least locally, with $\boldsymbol{\alpha}(\mathbf{x}) = (x_2^3, -x_2)^{\mathrm{T}}$. Assumption 5.10 is also satisfied since the nonlinearities are locally Lipschitz in $\mathbf{x}$. Consider the first two equations, where x_3 is replaced with a virtual input u_1:

$$\begin{aligned}\dot{x}_1 &= x_2,\\ \dot{x}_2 &= -2x_1 - \tanh(S_{\mathrm{f}}x_2)(1+\theta_{1,2}e^{-\theta_{1,1}x_2^2}) + u_1.\end{aligned} \tag{5.82}$$

Let $\psi(x_1) = x_1 - 1$. There are no uncertainties in the first equation; we therefore select $\psi_1(x_1, x_2) = x_1 - 1 + x_2$ and

$$u_1(\mathbf{x}, \hat{\boldsymbol{\theta}}_1) = -\psi_1(x_1, x_2) + 2x_1 - x_2 + \tanh(S_{\mathrm{f}}x_2)(1 + e^{-\hat{\theta}_{1,1}x_2^2+\log\hat{\theta}_{1,2}}), \tag{5.83}$$

which ensures that the following holds:

$$\dot{\psi}_1 = -\psi_1(x_1, x_2) + \tanh(S_{\mathrm{f}}x_2)(e^{-\hat{\theta}_{1,1}x_2^2+\log\hat{\theta}_{1,2}} - e^{-\theta_{1,1}x_2^2+\log\theta_{1,2}}). \tag{5.84}$$

The adaptation algorithms for $\hat{\theta}_{1,1}$ and $\hat{\theta}_{1,2}$ are as follows:

$$\begin{aligned}\hat{\theta}_{1,1}(x_1, x_2, t) &= -\gamma(-\psi_1(x_1, x_2)x_2^3 - \Psi_{1,1}(x_1, x_2) + \hat{\theta}_{1,1,I}) + \hat{\theta}_{1,1}(0), \quad \gamma > 0,\\ \dot{\hat{\theta}}_{1,1,I} &= -\psi_1(x_1, x_2)x_2^3 + \frac{\partial\Psi_{1,1}(x_1, x_2)}{\partial x_1}x_2,\\ \Psi_{1,1}(x_1, x_2) &= -(x_1 - 1)x_2^3 - \frac{3}{4}x_2^4;\end{aligned} \tag{5.85}$$

$$\begin{aligned}\hat{\theta}_{1,2}(x_1, x_2, t) &= -\gamma(\psi_1(x_1, x_2)x_2 - \Psi_{1,2}(x_1, x_2) + \hat{\theta}_{1,2,I}) + \hat{\theta}_{1,2}(0), \quad \gamma > 0,\\ \dot{\hat{\theta}}_{1,2,I} &= \psi_1(x_1, x_2)x_2 + \frac{\partial\Psi_{1,2}(x_1, x_2)}{\partial x_1}x_2,\\ \Psi_{1,2}(x_1, x_2) &= (x_1 - 1)x_2 + \frac{x_2^2}{2}.\end{aligned} \tag{5.86}$$

The functions $\Psi_{1,1} = \int \psi_1(x_1, x_2) 3x_2^2 \, dx_2$ and $\Psi_{1,2} = \int \psi_1(x_1, x_2) dx_2$ are chosen according to Assumption 5.5 of Theorem 5.1 to ensure that $\hat{\theta}_{1,1}$ and $\hat{\theta}_{1,2}$ satisfy the following differential equations:

$$\dot{\hat{\theta}}_{1,1} = \gamma(\dot{\psi}_1 + \psi_1(x_1, x_2)) x_2^3,$$
$$\dot{\hat{\theta}}_{1,2} = -\gamma(\dot{\psi}_1 + \psi_1(x_1, x_2)) x_2. \tag{5.87}$$

According to Lemma A5.2 we can conclude that solutions of the system (5.82), (5.83), and (5.87) satisfy $\lim_{t\to\infty} \psi(x_1(t), x_2(t)) = 0$. Notice that the finite-form realizations (5.85) and (5.86) of algorithms (5.87) are derived without introducing an auxiliary system (5.64). This is because the derivative $\dot{x}_1$ does not depend on the vector $\boldsymbol{\theta}$ explicitly and hence we can compensate explicitly for $\partial \Psi_{1,1}(x_1, x_2)/\partial x_1 \dot{x}_1$ and $\partial \Psi_{1,2}(x_1, x_2)/\partial x_1 \dot{x}_1$ in $\dot{\hat{\theta}}_{1,1,I}$ and $\dot{\hat{\theta}}_{1,2,I}$ as well.

This completes the first step of our iterative synthesis procedure. Let us now consider the original system (5.81) and select $\psi_2(x_1, x_2, x_3, t) = u_1(x_1, x_2, t) - x_3$.[12] Unlike in the case considered previously, the derivative $\dot{\psi}_2$ depends not only on $f_3(\mathbf{x}, \theta_2) = x_2\theta_2 + u$ but also on $f_2(\mathbf{x}, \boldsymbol{\theta}_1)$. The sum of two monotonic functions, however, generally is known to be not a monotonic function with respect to $\boldsymbol{\theta}_1 \oplus \theta_2$. Therefore, in order to satisfy Assumptions 5.3 and 5.4 in this case we need to replace the variable x_2 in ψ_2 with a new variable, ξ. In other words, we shall embed the original system into one of higher order, ensuring that $u_1(x_1, x_2, t) - u_1(x_1, \xi, t) \in L_2[t_0, T]$ (see Lemmas 5.1 and A5.1 for details). Notice that according to the realizability requirements (the existence of finite-form realizations) and the requirements of Lemma A5.2 this embedding should also guarantee that $f_3(\mathbf{x}, \theta_2) - f_3(x_1 \oplus \xi \oplus x_3, \theta_2) \in L_2[t_0, T]$. In our example, however, the function f_3 does not depend on x_2. Hence, it suffices to show that $u_1(x_1, x_2, t) - u_1(x_1, \xi, t) \in L_2[t_0, T]$, and that the norm is a continuous function merely of the initial conditions and parameters, in order to guarantee both convergence of ψ to 0 and derivative-independent realization of the adaptation algorithm.

Let us consider the difference $u_1(x_1, x_2, t) - u_1(x_1, \xi, t)$:

$$\begin{aligned} u_1(x_1, x_2, t) - u_1(x_1, \xi, t) &= -2(x_2 - \xi) \\ &\quad + \tanh(S_f x_2) e^{-\frac{x_2^6}{4} + \hat{\theta}_{1,1,I}(t) x_2^2 - \frac{x_2^2}{2} + \hat{\theta}_{1,2,I}(t)} \\ &\quad - \tanh(S_f \xi) e^{-\frac{\xi^6}{4} + \hat{\theta}_{1,1,I}(t)\xi^2 - \frac{\xi^2}{2} + \hat{\theta}_{1,2,I}(t)}. \end{aligned}$$

[12] For compactness of notation we supposed here that the function u_1 depends on t explicitly rather than implicitly through $\hat{\theta}_{1,1,I}(t)$ and $\hat{\theta}_{1,2,I}(t)$.

On applying the mean-value theorem we get that

$$|u_1(x_1, x_2, t) - u_1(x_1, \xi, t)| \leq |x_2 - \xi|(2 + S_f F(t)),$$

where

$$F(t) = \max_{0 \leq \tau \leq t} F_0(\tau); \tag{5.88}$$

$$F_0(t) = \max_{\lambda \in [0,1]} \left\{ \left| \frac{\partial}{\partial s} e^{-\frac{s^6}{4} + \hat{\theta}_{1,1,I}(t)s^2 - \frac{s^2}{2} + \hat{\theta}_{1,2,I}(t)} \right| \right\},$$

$$s = \lambda x_2(t) + (1 - \lambda)\xi(t). \tag{5.89}$$

Let us introduce the variable ξ:

$$\dot{\xi} = (x_2 - \xi)(1 + F^2(t)) - 2x_1 + x_3 - \tanh(S_f x_2)(1 + e^{-\theta_{\xi,1} x_2^2 + \theta_{\xi,2}}) \tag{5.90}$$

and consider $\dot{\psi}_\xi(x_2, \xi) = x_2 - \xi$:

$$\dot{\psi}_\xi = -\psi_\xi(1 + F^2(t)) + \tanh(S_f x_2)(e^{-\theta_{\xi,1} x_2^2 + \theta_{\xi,2}} - e^{-\theta_{1,1} x_2^2 + \theta_{1,2}}).$$

According to Lemma A5.2, the function $(x_2 - \xi)F(t) \in L_2[t_0, T]$ (and, consequently, the difference $u_1(x_1, x_2, t) - u_1(x_1, \xi, t) \in L_2[t_0, T]$) if $\theta_{\xi,1}$ and $\theta_{\xi,2}$ satisfy the following differential equations:

$$\dot{\theta}_{\xi,1} = \gamma(\dot{\psi}_\xi + \psi_\xi(1 + F^2(t)))x_2^3,$$

$$\dot{\theta}_{\xi,2} = -\gamma(\dot{\psi}_\xi + \psi_\xi(1 + F^2(t)))x_2. \tag{5.91}$$

The integral-differential realization of (5.91) is

$$\theta_{\xi,1} = -\gamma(-(x_2 - \xi)x_2^3 - \Psi_{\xi,1}(x_2, \xi) + \hat{\theta}_{\xi,1,I}),$$

$$\dot{\theta}_{\xi,1,I} = -(x_2 - \xi)x_2^3 + \frac{\partial \Psi_{\xi,1}(x_2, \xi)}{\partial \xi}\dot{\xi}, \tag{5.92}$$

$$\Psi_{\xi,1}(x_2, \xi) = -\int 3(x_2 - \xi)x_2^2\, dx_2 = -\frac{3}{4}x_2^4 + \xi x_2^3;$$

$$\theta_{\xi,2} = -\gamma((x_2 - \xi)x_2 - \Psi_{\xi,2}(x_2, \xi) + \hat{\theta}_{\xi,2,I}),$$

$$\dot{\theta}_{\xi,2,I} = (x_2 - \xi)x_2 + \frac{\partial \Psi_{\xi,2}(x_2, \xi)}{\partial \xi}\dot{\xi}, \tag{5.93}$$

$$\Psi_{\xi,2}(x_2, \xi) = \int (x_2 - \xi)dx_2 = \frac{1}{2}x_2^2 - \xi x_2.$$

Equations (5.90), (5.92), and (5.93) form the desired extension of the original system.

Let us finally introduce $\psi_2(x_1, x_3, \xi, t) = u_1(x_1, \xi, t) - x_3 = 0$. Ensuring that $\psi_2 \in L_2 \cap L_\infty[t_0, T]$ and that the corresponding norm depends only on the initial conditions and parameters will assure that the trajectories of (5.82), (5.85), and (5.86) are bounded and $\psi(x_1) \to 0$ as $t \to \infty$. Then boundedness of x_3 will follow from boundedness of $x_2 - \xi$ and smoothness of $u_1(x_1, \xi, t)$. Consider $\dot{\psi}_2$:

$$\dot{\psi}_2 = \frac{\partial u_1(x_1, \xi, t)}{\partial x_1} x_2 + \frac{\partial u_1(x_1, \xi, t)}{\partial \xi} \dot{\xi} + \frac{\partial u_1(x_1, \xi, t)}{\partial t} - x_3 \theta_2 - u.$$

Letting

$$u = \frac{\partial u_1(x_1, \xi, t)}{\partial x_1} x_2 + \frac{\partial u_1(x_1, \xi, t)}{\partial \xi} \dot{\xi} + \frac{\partial u_1(x_1, \xi, t)}{\partial t} - x_3 \hat{\theta}_2 + \psi_2(x_1, \xi, x_3, t), \quad (5.94)$$

we obtain that

$$\dot{\psi}_2 = -\psi_2(x_1, \xi, x_3, t) + x_3 \hat{\theta}_2 - x_3 \theta_2. \quad (5.95)$$

Notice that because (5.95) is linearly parametrized we can now complete the design by applying conventional approaches. For the sake of consistency, however, we derive estimates $\hat{\theta}_2$ according to the proof of Theorem 5.4. The resulting equations for the estimates $\hat{\theta}_2$ can be given as follows:

$$\begin{aligned} \hat{\theta}_2(x_1, \xi, x_3, t) &= -\gamma(\psi_2(x_1, \xi, x_3) x_3 - \Psi_2(x_1, \xi, x_3) + \hat{\theta}_{2,I}), \\ \dot{\hat{\theta}}_{2,I} &= \psi_2(x_1, \xi, x_3) x_3 + \frac{\partial \Psi_2(x_1, \xi, x_3)}{\partial x_1} x_2 + \frac{\partial \Psi_2(x_1, \xi, x_3)}{\partial \xi} \dot{\xi}, \quad (5.96) \\ \Psi_2(x_1, \xi, x_3) &= \int (u_1(x_1, \xi) - x_3) dx_3 = u_1(x_1, \xi) x_3 - \frac{1}{2} x_3^2. \end{aligned}$$

According to Lemma A5.2 the algorithms (5.96) guarantee that $\psi_2 \in L_2 \cap L_\infty[t_0, T]$, and that the norm can be bounded by a continuous function of the initial conditions and parameters. This in turn implies that $x_1, x_2, x_3 \in L_\infty[t_0, \infty]$. Moreover, $\psi_1(x_1(t), x_2(t)) \to 0$ and $\psi(x_1(t)) \to 0$ as $t \to \infty$. We simulated system (5.81) with the control function (5.94) and (5.96) and the auxiliary system (5.90), (5.92), and (5.93) for the following set of initial conditions and parameter values: $x_1(0) = 2$, $x_2(0) = 1$, $x_3(0) = 1$, $\theta_{1,1} = 2$, $\theta_{1,2} = 3$, $\theta_2 = 2$, $\xi(0) = 2$, $\hat{\theta}_{1,1,I}(0) = 0$, $\hat{\theta}_{1,2,I}(0) = -3$, $\theta_{\xi,1,I}(0) = 0$, and $\theta_{\xi,1,I}(0) = -3$, $\hat{\theta}_{2,I}(0) = 0$, and $\gamma = 1$. Results of computer simulations (the trajectory $x_1(t)$) of the system with our adaptive control are shown in Figure 5.5. The system approaches the goal manifold $\psi(x_1) = 0$ asymptotically as required.

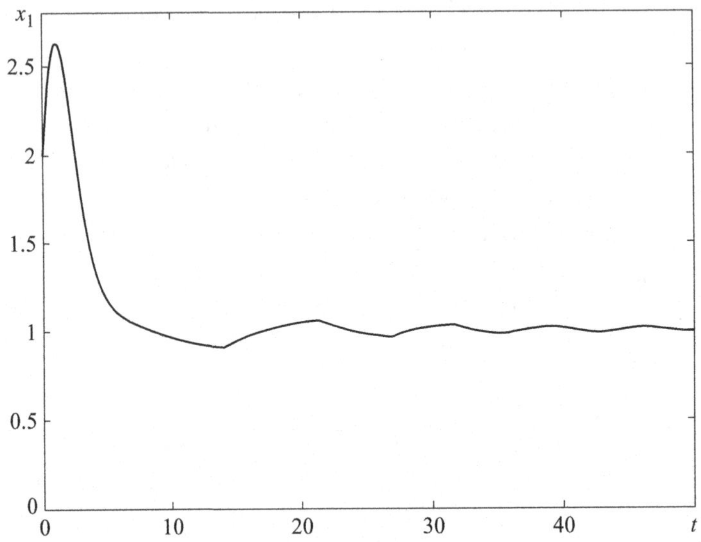

Figure 5.5 The trajectory $x_1(t)$ as a function of time.

5.3 Adaptive regulation to invariant sets

In the previous sections of the chapter we focused on determining algorithms of adaptation that could be viewed as solutions to Problem 5.1. Asymptotic characterizations of the resulting systems have been limited so far to

$$\lim_{t\to\infty} \psi(\mathbf{x}, t) = 0$$

(see e.g. Theorems 5.1 and 5.4). The question, however, is whether we can state anything stronger than the limiting relation above regarding the asymptotic properties of the system's solutions. It turns out that for a class of systems this possibility could exist. In particular, it is possible to show that the adaptation schemes developed here are capable of steering the system's state to invariant sets of which the location can be estimated a priori. One of the mechanisms enabling this possibility is requirement (5.21) in Problem 5.1 (i.e. the condition demanding that the influence of uncertainty on the target dynamics is compensated for up to a perturbation from $L_2[t_0, \infty]$).

5.3.1 Systems with parametric uncertainties and nonlinear parametrization

Consider a subclass of (5.1):

$$\begin{aligned} \dot{\mathbf{x}}_1 &= \mathbf{f}_1(\mathbf{x}, \boldsymbol{\zeta}) + \mathbf{g}_1(\mathbf{x}, \boldsymbol{\zeta})u, \\ \dot{\mathbf{x}}_2 &= \mathbf{f}_2(\mathbf{x}, \boldsymbol{\theta}, \boldsymbol{\zeta}) + \mathbf{g}_2(\mathbf{x}, \boldsymbol{\zeta})u, \\ \dot{\boldsymbol{\zeta}} &= S(\boldsymbol{\zeta}, \mathbf{x}), \end{aligned} \tag{5.97}$$

where the system

$$\dot{\boldsymbol{\zeta}} = S(\boldsymbol{\zeta}, \mathbf{x}), \; S : \mathbb{R}^{\zeta} \times \mathbb{R}^n \to \mathbb{R}^{\zeta} \tag{5.98}$$

is complete and admits an $L^n_\infty[t_0, \infty] \mapsto L^{\xi}_\infty[t_0, \infty]$ margin (bounded-input–bounded-state property). The following corollary from Theorem 5.1 holds for such systems.

Corollary 5.2 *Consider the system (5.11), (5.32), (5.34), and (5.97) with $\varepsilon(t) = 0$, and suppose that Assumptions 5.1, 5.2, 5.3, and 5.5*[13] *hold. In addition, let us suppose that the right-hand side of the combined system is locally Lipschitz, properties H1, H2, and H4 hold, the functions $\boldsymbol{\alpha}(\mathbf{x}, t)$ and $\partial\psi(\mathbf{x}, t)/\partial t$ are locally bounded in $\mathbf{x}$, and system (5.13) admits an $L^1_2[t_0, \infty] \mapsto L^1_p[t_0, \infty]$ margin.*

Then solutions of the system are bounded for all $\mathbf{x}(t_0)$ and $\boldsymbol{\zeta}(t_0)$, and converge to the maximal invariant set in

$$\{\mathbf{x} \in \mathbb{R}^n, \; \boldsymbol{\zeta} \in \mathbb{R}^{\zeta}, \; \hat{\boldsymbol{\theta}} \in \mathbb{R}^d | f(\mathbf{x}, \boldsymbol{\theta}, \boldsymbol{\zeta}) - f(\mathbf{x}, \hat{\boldsymbol{\theta}}, \boldsymbol{\zeta}) = 0\}. \tag{5.99}$$

Corollary 5.2 states that all limit motions in the controlled system will satisfy

$$\begin{aligned} \dot{\psi} &= -\varphi(\psi, \omega, \boldsymbol{\zeta}), \\ \dot{\boldsymbol{\zeta}} &= S(\mathbf{x}, \boldsymbol{\zeta}). \end{aligned} \tag{5.100}$$

In other words, the intersection of the invariant sets of (5.100) and the set of "compensated uncertainties" defined by (5.99) contains limit sets of the controlled system. This gives us the possibility of establishing adaptation mechanisms for systems of which the target motions are not fully known a priori, as functions of state or time, but can still be defined as a limit set of (5.100). An example illustrating this idea is discussed in the next section.

5.3.2 Systems with signal uncertainties and linear parametrization

Let us consider the following class of systems:

$$\begin{aligned} \dot{\mathbf{x}} &= \mathbf{f}(\mathbf{x}) + G_u(\boldsymbol{\phi}(\mathbf{x})\boldsymbol{\theta} + \mathbf{u}), \\ \dot{\boldsymbol{\theta}} &= S(\boldsymbol{\theta}), \; \boldsymbol{\theta}(t_0) \in \Omega_\theta \subset \mathbb{R}^d, \end{aligned} \tag{5.101}$$

where $\mathbf{f} : \mathbb{R}^n \to \mathbb{R}^n, \boldsymbol{\phi} : \mathbb{R}^n \to \mathbb{R}^{m \times d}$, are C^1-smooth vector fields, $G_u \in \mathbb{R}^{n \times m}, \boldsymbol{\theta}$ is the vector of variables of which the exact values are unknown, and $S : \mathbb{R}^d \to \mathbb{R}^d$, $S \in C^1$ is a known function. The initial conditions $\boldsymbol{\theta}(t_0) = \boldsymbol{\theta}_0 \in \Omega_\theta$ are supposed

[13] In the statement of the theorem the functions $\psi(\mathbf{x}, t)$, $\mathbf{f}(\mathbf{x}, \boldsymbol{\theta}, t)$, and $\mathbf{g}(\mathbf{x}, t)$ are allowed to depend on t explicitly. Here we replace this dependence with the implicit one, through the variable $\boldsymbol{\zeta}(t)$: $\psi(\mathbf{x}, \boldsymbol{\zeta}(t))$, $\mathbf{f}(\mathbf{x}, \boldsymbol{\theta}, \boldsymbol{\zeta}(t))$, $\mathbf{g}(\mathbf{x}, \boldsymbol{\zeta}(t))$.

to be unknown. Let $\Omega_S(\Omega_\theta) \subseteq \Omega_\theta$, and let Ω_θ be bounded. Our goal is to steer the system's state to the following *target set*:

$$\Omega^*(\mathbf{x}) \subset \mathbb{R}^n. \tag{5.102}$$

A very similar problem was stated and analyzed in detail in Panteley *et al.* (2002). Here we illustrate how this problem can be solved by employing methods developed in the previous sections.

An important feature of this problem, compared with the ones we considered earlier, is that the target set (5.102) is no longer a surface of which the functional (implicit) definition is known. Now it is a mere set in $\mathbb{R}^n$ satisfying the following assumption.

Assumption 5.11 *$\Omega^*(\mathbf{x})$ is a a bounded and closed subset of $\mathbb{R}^n$.*

With respect to the motions $\boldsymbol{\theta}(t, \boldsymbol{\theta}_0)$, we will consider only those which satisfy the following assumption.

Assumption 5.12 *There exists a positive definite matrix $H = H^{\mathrm{T}} \in \mathbb{R}^{d\times d}$, such that the following holds for the function $S : \mathbb{R}^d \to \mathbb{R}^d$ in (5.101):*

$$H\frac{\partial S(\boldsymbol{\theta})}{\partial \boldsymbol{\theta}} + \frac{\partial S(\boldsymbol{\theta})}{\partial \boldsymbol{\theta}}^{\mathrm{T}} H \leq 0, \quad \forall\, \boldsymbol{\theta} \in \mathbb{R}^d.$$

According to this assumption the vector $\boldsymbol{\theta}(t)$ not only represents the standard parametric uncertainty but also can model motions of which the solutions are stable in the sense of Lyapunov. With regard to the input $\mathbf{u}$ we suppose that the following holds.

Assumption 5.13 *For the given $\Omega^*(\mathbf{x})$ defined by (5.4) and system (5.101) there is a function $\mathbf{u}_0(\mathbf{x}), \mathbf{u}_0 \in C^1$ such that*

$$G_u\mathbf{u}_0(\mathbf{x}) + \mathbf{f}(\mathbf{x}) = \mathbf{f}_0(\mathbf{x}).$$

Moreover, $\Omega^*(\mathbf{x}) \supset \Omega_{\mathbf{f}_0}(\mathbf{x})$ for all $\mathbf{x}_0 \in \mathbb{R}^n$, where the flow $\mathbf{x}(t, \mathbf{x}_0, t_0)$ is generated by

$$\dot{\mathbf{x}} = \mathbf{f}_0(\mathbf{x}). \tag{5.103}$$

Finally, let us introduce two additional alternative technical hypotheses with regard to the target set (5.102). The first one is formulated in Assumption 5.14, and the second one is presented in Assumption 5.15.

Assumption 5.14 *There exist functions $\psi(\mathbf{x}) : \mathbb{R}^n \to \mathbb{R}$, $\varphi : \mathbb{R} \to \mathbb{R}$, such that*

(1) $\Omega^* \supseteq \Omega_{\mathbf{f}_0}(\Omega_\psi)$, $\Omega_\psi = \{\mathbf{x} \in \mathbb{R}^n | \, \mathbf{x} : \varphi(\psi(\mathbf{x})) = 0\}$, *that is* $\Omega^*(\mathbf{x})$ *contains invariant sets of system (5.103) restricted to* Ω_ψ*;*

(2) *there exists a function* $\beta(\mathbf{x}) : \mathbb{R}^n \to \mathbb{R}_{\geq 0}$ *such that* $\beta(\mathbf{x})$ *is separated away from zero and satisfies*

$$\psi \frac{\partial \psi(\mathbf{x})}{\partial \mathbf{x}} \mathbf{f}_0(\mathbf{x}) \leq -\beta(\mathbf{x})\varphi(\psi)\psi,$$
$$\int_0^\psi \varphi(\sigma) d\sigma \geq 0, \qquad \lim_{\psi \to \infty} \int_0^\psi \varphi(\sigma) d\sigma = \infty; \tag{5.104}$$

(3) *for the function* $\psi(\mathbf{x}) : \mathbb{R}^n \to \mathbb{R}$, $\psi(\mathbf{x}) \in C^1$, *with* $\partial\psi/\partial\mathbf{x}$ *being Lipschitz, Assumption 5.1 holds.*

Notice that $\psi(\mathbf{x})$ and $\varphi(\psi)$ are not required to be sign-definite.

Assumption 5.15 *Consider (5.103) with an additive perturbation* $\boldsymbol{\varepsilon}_0$*:*

$$\dot{\mathbf{x}} = \mathbf{f}_0(\mathbf{x}) + \boldsymbol{\varepsilon}_0(t), \; \boldsymbol{\varepsilon}_0 \in L_2^n[t_0, \infty] \cap C^1. \tag{5.105}$$

System (5.105) admits an $L_2^n[t_0, \infty] \to L_\infty^n[t_0, \infty]$ *margin, and* $\Omega^* \supseteq \Omega_{\mathbf{f}_0}$*.*

Let us now proceed with determining the existence conditions for a feedback

$$\mathbf{u}(\mathbf{x}, \boldsymbol{\xi}) = \mathbf{u}(\mathbf{x}, \boldsymbol{\xi}, \hat{\boldsymbol{\theta}}), \; \dot{\boldsymbol{\xi}} = \mathbf{f}_\xi(\mathbf{x}, \boldsymbol{\xi}, \hat{\boldsymbol{\theta}}), \; \boldsymbol{\xi} \in \mathbb{R}^k$$

steering state $\mathbf{x}(t)$ of system (5.101) to Ω^* as $t \to \infty$. In order to derive these conditions we employ the method of the virtual algorithm of adaptation. First, we specify a suitable class of feedbacks and look for a virtual adaptation algorithm (a system of differential equations of which the right-hand side depends on unmeasured quantities $\boldsymbol{\theta}$ explicitly) ensuring that $\mathbf{x}(t) \to \Omega^*$ as $t \to \infty$. Such algorithms and their properties are defined in Lemma A5.2. In order to obtain an adaptation scheme that does not require information about the values of $\boldsymbol{\theta}$ we rely on the embedding technique described in Section 5.2.2. The main results are provided in Theorems 5.5 and 5.6. For the sake of compactness these results are formulated in the existential form. Nevertheless, the proofs are constructive, offering specific procedures for finding control inputs that steer the system's state to Ω^* adaptively.

Theorem 5.5 *Let system (5.101) be given and suppose that Assumptions 5.11–5.14 hold. In addition, suppose that there is a* C^1*-smooth function* $\kappa(\mathbf{x})$ *such that*

$$\left\| \frac{\partial \psi(\mathbf{x})}{\partial \mathbf{x}} \right\| \leq |\kappa(\mathbf{x})|.$$

Then there exists a system

$$\begin{aligned}\dot{\boldsymbol{\xi}} &= \mathbf{f}_{\xi}(\mathbf{x}, \boldsymbol{\xi}, \boldsymbol{\nu}),\\ \dot{\boldsymbol{\nu}} &= \mathbf{f}_{\nu}(\mathbf{x}, \boldsymbol{\xi}, \boldsymbol{\nu}),\ \boldsymbol{\xi} \in \mathbb{R}^n,\ \boldsymbol{\nu} \in \mathbb{R}^d,\end{aligned} \tag{5.106}$$

a control function $\mathbf{u}(\mathbf{x}, \hat{\boldsymbol{\theta}}) = \mathbf{u}_0(\mathbf{x}) - \phi(\boldsymbol{\xi})\hat{\boldsymbol{\theta}}(t)$, *and an adaptation algorithm*

$$\begin{aligned}\hat{\boldsymbol{\theta}} &= (H^{-1}\Psi(\boldsymbol{\xi})\mathbf{x} + \hat{\boldsymbol{\theta}}_I(t)),\\ \Psi(\boldsymbol{\xi}) &= (\kappa^2(\boldsymbol{\xi}) + 1)(G_u\phi(\boldsymbol{\xi}))^{\mathrm{T}},\\ \dot{\hat{\boldsymbol{\theta}}}_I &= S(\hat{\boldsymbol{\theta}}) - H^{-1}\frac{\partial\Psi(\boldsymbol{\xi})}{\partial\boldsymbol{\xi}}\mathbf{f}_{\xi}(\mathbf{x}, \boldsymbol{\xi}, \boldsymbol{\nu})\mathbf{x} - H^{-1}\Psi(\boldsymbol{\xi})\mathbf{f}_0(\mathbf{x}),\end{aligned} \tag{5.107}$$

such that the following properties hold:

(1) $\hat{\boldsymbol{\theta}}(t), \mathbf{x}(t) \in L^n_{\infty}[t_0, \infty]$;
(2) $\mathbf{x}(t)$ *converges to* Ω^* *asymptotically as* $t \to \infty$;
(3) if the function $G_u\phi(\boldsymbol{\xi}(t))$ *is persistently exciting and* $S(\boldsymbol{\theta}) \equiv 0$, *then* $\hat{\boldsymbol{\theta}}(t, \hat{\boldsymbol{\theta}}_0, t_0)$ *converges to* $\boldsymbol{\theta}_0$ *as* $t \to \infty$.

Theorem 5.6 *Consider system (5.101), and let Assumptions 5.11–5.13 and 5.15 hold. Then there is a system (5.106), a control input* $\mathbf{u}(\mathbf{x}, \hat{\boldsymbol{\theta}}) = \mathbf{u}_0(\mathbf{x}) - \phi(\boldsymbol{\xi})\hat{\boldsymbol{\theta}}(t)$, *and an adaptation algorithm (5.107) with* $\kappa(\boldsymbol{\xi}) \equiv 0$ *such that statements (1)–(3) of Theorem 5.5 hold.*

5.4 Adaptive control of interconnected dynamical systems

In the previous sections of this chapter, when analyzing and designing algorithms of adaptive regulation and control, the spectrum of unmodeled dynamics and the perturbation was limited to functions from $L_2[t_0, \infty]$. No specific attention was paid to the nature of the perturbation dynamics either. Now we wish to consider this issue in greater detail. In particular, we consider the following class of models:

$$\begin{aligned}\dot{\mathbf{x}}_1 &= \mathbf{f}_1(\mathbf{x}) + \mathbf{g}_1(\mathbf{x})u,\\ \dot{\mathbf{x}}_2 &= \mathbf{f}_2(\mathbf{x}, \boldsymbol{\theta}) + \mathbf{z}(\mathbf{x}, \mathbf{q}(t), t) + \mathbf{g}_2(\mathbf{x})u,\ \ \mathbf{x}(t_0) = \mathbf{x}_0,\end{aligned} \tag{5.108}$$

where signals $\mathbf{q}(t)$ are generated by

$$\dot{\mathbf{q}} = \mathbf{f}_q(\mathbf{x}, \mathbf{q}, t),\ \ \mathbf{q}(t_0) = \mathbf{q}_0 \in \mathbb{R}^s. \tag{5.109}$$

System (5.109) is supposed to be complete for all $\mathbf{x}(t) \in L^n_{\infty}[t_0, \infty)$. Equations (5.109) can model unmeasured but modeled perturbations; interestingly, they can also stand for another compartment of a larger adapting system affecting the dynamics of (5.108) through coupling. These cases are considered and analyzed separately in the next sections.

5.4.1 Systems with unmodeled dynamics

As before we suppose that the goal of regulation is to ensure that

$$\lim_{t\to\infty} \psi(\mathbf{x}(t,\mathbf{x}_0),t) = 0,$$

where $\psi(\mathbf{x},t)$ is the function satisfying Assumption 5.1. Let the control input u be defined as in (5.10). Then

$$\dot{\psi} = -\varphi(\psi,\boldsymbol{\omega},t) + f(\mathbf{x},\boldsymbol{\theta},t) - f(\mathbf{x},\hat{\boldsymbol{\theta}},t) + z(\mathbf{x},\mathbf{q},t),\ z = L_{\mathbf{z}(\mathbf{x},\mathbf{q},t)}\psi. \quad (5.110)$$

We will assume that the function $z(\mathbf{x},\mathbf{q},t) \in \mathcal{C}^1$ satisfies

$$|z(\mathbf{x},\mathbf{q},t)| \leq |h_x(\mathbf{x},t)| + |h_q(\mathbf{q},t)|, \quad (5.111)$$

where $h_x : \mathbb{R}^n \times \mathbb{R}_{\geq 0} \to \mathbb{R}$, $h_q : \mathbb{R}^s \times \mathbb{R}_{\geq 0} \to \mathbb{R}$, $h_x(\cdot), h_q(\cdot) \in \mathcal{C}^0$. Our analysis of the system (5.108), (5.109), and (5.10) will be based on the small-gain argument. Therefore some quantitative knowledge about gain margins (5.14) is needed. In particular, we suppose that the following additional information is available.

Assumption 5.16 *There exist functions $\gamma_{h_x}, \gamma_{h_q}, \gamma_{\psi,2}, \gamma_{x,\infty} \in \mathcal{K}$ and constants $\beta_{h_x}, \beta_{h_q}, \beta_{\psi,2}$, and $\beta_{x,\infty}$ such that*

(1) system (5.108) satisfies

$$\|h_x(\mathbf{x}(t),t)\|_{2,[t_0,T]} \leq \gamma_{h_x}(\|\psi(\mathbf{x}(t),t)\|_{\infty,[t_0,T]}) + \beta_{h_x}; \quad (5.112)$$

(2) system (5.109) satisfies

$$\|h_q(\mathbf{q}(t),t)\|_{2,[t_0,T]} \leq \gamma_{h_q}(\|\mathbf{x}(t)\|_{\infty,[t_0,T]}) + \beta_{h_q}; \quad (5.113)$$

(3) the function $\psi(\mathbf{x},t)$ majorizes state $\mathbf{x}$ of (5.108),

$$\|\mathbf{x}(t)\|_{\infty,[t_0,T]} \leq \gamma_{x,\infty}(\|\psi(\mathbf{x}(t),t)\|_{\infty,[t_0,T]}) + \beta_{x,\infty}; \quad (5.114)$$

(4) system (5.13) admits $L_2[t_0,T] \mapsto L_2[t_0,T]$ margin,

$$\|\psi(t)\|_{\infty,[t_0,T]} \leq \gamma_{\psi,2}(\|\zeta(t)\|_{2,[t_0,T]}) + \beta_{\psi,2}. \quad (5.115)$$

Inequalities (5.114) and (5.115) in Assumption 5.16 constitute slightly stronger versions of Assumptions 5.1 and 5.2. Inequalities (5.112) and (5.113) characterize the interaction between the perturbation subsystem, (5.109), and (5.108). Now we are ready to formulate the following result.

Theorem 5.7 *Consider the combined system (5.10), (5.34), and (5.108)–(5.110). Suppose that Assumptions 5.3–5.5 and 5.16 hold, and that there exist functions*

$$\lambda_1(\cdot), \lambda_2(\cdot), \rho_i(\cdot) \in \mathcal{K}_\infty$$

such that

$$\begin{aligned}
&(Id + \lambda_2) \circ (Id - \gamma_{h_{q,q}})^{-1} \circ (\rho_9 + I_d) \circ \gamma_{h_{q,x}} \\
&\quad \circ (Id + \lambda_1) \circ (Id - \gamma_{h_{x,x}})^{-1} \circ (\rho_8 + I_d) \circ \gamma_{h_{x,q}}(s) \le s, \\
&(Id + \lambda_1) \circ (Id - \gamma_{h_{x,x}})^{-1} \circ (\rho_8 + I_d) \circ \gamma_{h_{x,q}} \\
&\quad \circ (Id + \lambda_2) \circ (Id - \gamma_{h_{q,q}})^{-1} \circ (\rho_9 + I_d) \circ \gamma_{h_{q,x}}(s); \le s,
\end{aligned}$$

for all $s \ge s_0$, *where*

$$\begin{aligned}
\gamma_{h_{x,x}}(s) &= \gamma_{h_x} \circ (\rho_4 + Id) \circ \gamma_{\psi,2} \circ (\rho_1 + Id) \circ (\rho_2 + Id)(C_D s), \\
\gamma_{h_{x,q}}(s) &= \gamma_{h_x} \circ (\rho_4 + Id) \circ \rho_4^{-1} \circ (\rho_5 + Id) \circ \gamma_{\psi,2} \circ (\rho_1 + Id) \\
&\quad \circ (\rho_2 + Id) \circ \rho_2^{-1}(C_D s), \\
\gamma_{h_{q,x}}(s) &= \gamma_{h_q} \circ (\rho_3 + Id) \circ \gamma_{x,\infty} \circ (\rho_6 + Id) \circ \gamma_{\psi,2} \\
&\quad \circ (\rho_1 + Id) \circ (\rho_2 + Id)(C_D s); \\
\gamma_{h_{q,q}}(s) &= \gamma_{h_q} \circ (\rho_3 + Id) \circ \gamma_{x,\infty} \circ (\rho_6 + Id) \circ \rho_6^{-1} \circ (\rho_7 + Id) \\
&\quad \circ \gamma_{\psi,2} \circ (\rho_1 + Id) \circ (\rho_2 + Id) \circ \rho_2^{-1}(C_D s), \quad C_D = 1 + \frac{D}{D_1}.
\end{aligned}$$

Then

(1) *solutions of the system exist for all* $[t_0, \infty)$, *and* $\mathbf{x}(t) \in L_\infty^n[t_0, \infty]$, $\hat{\boldsymbol{\theta}}(t) \in L_\infty^d[t_0, \infty]$; *moreover,* $\mathbf{q}(t) \in L_\infty^s[t_0, \infty]$, *provided that system (5.109) admits* $L_\infty \mapsto L_\infty$ *margin;*
(2) $f(\mathbf{x}(t), \boldsymbol{\theta}) - f(\mathbf{x}(t), \hat{\boldsymbol{\theta}}(t)) \in L_2^1[t_0, \infty]$; *in addition, if* $f(\cdot, \cdot) \in C^1$ *and* $z(\mathbf{x}, \mathbf{q}, t)$ *and* $\boldsymbol{\alpha}(\mathbf{x}, t)$ *are locally bounded uniformly in t, then*

$$\lim_{t \to \infty} f(\mathbf{x}(t), \boldsymbol{\theta}) - f(\mathbf{x}(t), \hat{\boldsymbol{\theta}}(t)) = 0. \tag{5.116}$$

Theorem 5.7 establishes sufficient conditions ensuring the existence of a successful adaptation scheme for systems (5.108) in the presence of unmodeled dynamics and nonlinear parametrization. The result is based on the nonlinear small-gain theorem, and thus the conditions are rather conservative. Despite this obvious limitation, an advantage of this approach is that knowledge of the Lyapunov functions describing motion in the individual subsystems is not required. Notice also that some specific properties of the uncertainty model (coefficient C_D) are inherent parts of the resulting small-gain condition.

A practically relevant feature of the adaptation scheme discussed above is that asymptotic compensation of the influence of uncertainty on the target dynamics, (5.116), is still guaranteed. This enables reconstruction of the true values of $\boldsymbol{\theta}$, provided that a nonlinear persistency-of-excitation condition holds (Cao *et al.* 2003; Tyukin and van Leeuwen 2005) (see also Theorem 5.9). In addition, if the mappings $\gamma_{h_x}, \gamma_{h_q}, \gamma_{\psi,2}$, and $\gamma_{x,\infty}$ in Assumption 5.16 are linear then the requirements on the functions $\gamma_{h_x}, \gamma_{h_q}, \gamma_{\psi,2}$, and $\gamma_{x,\infty}$ in Theorem 5.7 can be reduced to

$$\gamma_{h_{x,x}} = \gamma_{h_x} \cdot \gamma_{\psi,2} \cdot C_D < 1, \qquad \gamma_{h_{q,q}} = \gamma_{h_q} \cdot \gamma_{x,\infty} \cdot \gamma_{\psi,2} \cdot C_D < 1,$$

$$\frac{\gamma_{h_{q,x}}}{1 - \gamma_{h_{q,q}}} \frac{\gamma_{h_{x,q}}}{1 - \gamma_{h_{x,x}}} < 1,$$

$$\gamma_{h_{x,q}} = \gamma_{h_x} \cdot \gamma_{\psi,2} \cdot C_D, \qquad \gamma_{h_{q,x}} = \gamma_{h_q} \cdot \gamma_{x,\infty} \cdot \gamma_{\psi,2} \cdot C_D.$$

5.4.2 Decentralized adaptive control

So far we have considered adaptive control and regulation problems for systems consisting of the controlled subsystem (5.108) and perturbations governed by e.g. (5.109). Let us now consider the case of *interconnected adapting* systems $\mathcal{S}_x$ and $\mathcal{S}_y$. Regulatory, control, and adapting mechanisms in one of these subsystems do not necessarily account for the other. The question is what can we say about the overall behavior of such interconnection?

Questions of this kind are inherent to the domain of decentralized adaptive control. There is a large literature on this topic containing successful solutions to problems of adaptive stabilization (Gavel and Siljak 1989; Jain and Khorrami 1997), tracking (Ioannou 1986; Jain and Khorrami 1997; Shi and Singh 1992; Spooner and Passino 1996), and output regulation (Jiang 2000; Ye and Huang 2003) of linear and nonlinear systems. In most of these cases the problem of decentralized control is solved within the conventional framework of adaptive stabilization/tracking/regulation by a family of linearly parametrized controllers. While these results may be successfully implemented in a large variety of technical and artificial systems, there is room for further improvements, in particular when the target dynamics of the systems is not stable in the Lyapunov sense but intermittent, meta-stable, or multi-stable, or when the uncertainties are nonlinearly parametrized and no domination of the uncertainties by feedback is allowed.

Here we take advantage of the results from Sections 5.1 and 5.2 and show how these issues can be addressed simultaneously for a class of nonlinear dynamical systems. Our contribution is that we provide conditions ensuring forward-completeness, boundedness, and asymptotic reaching of the goal for a pair of interconnected systems with uncertain coupling and parameters. The method does not require either the availability of a Lyapunov function for the desired motions in each subsystem or linear parametrization of the controllers. Moreover, the results can straightforwardly be extended to interconnection of arbitrarily many (but still a finite number of) subsystems.

Consider interconnection of systems $\mathcal{S}_x$ and $\mathcal{S}_y$:

$$\mathcal{S}_x : \quad \begin{aligned} \dot{\mathbf{x}}_1 &= \mathbf{f}_1(\mathbf{x}) + \mathbf{g}_1(\mathbf{x})u_x, \\ \dot{\mathbf{x}}_2 &= \mathbf{f}_2(\mathbf{x}, \boldsymbol{\theta}_x) + \gamma_y(\mathbf{y}, t) + \mathbf{g}_2(\mathbf{x})u_x, \end{aligned} \tag{5.117}$$

$$\mathcal{S}_y : \quad \begin{aligned} \dot{\mathbf{y}}_1 &= \mathbf{q}_1(\mathbf{y}) + \mathbf{z}_1(\mathbf{y})u_y, \\ \dot{\mathbf{y}}_2 &= \mathbf{q}_2(\mathbf{y}, \boldsymbol{\theta}_y) + \gamma_x(\mathbf{x}, t) + \mathbf{z}_2(\mathbf{y})u_y, \end{aligned} \tag{5.118}$$

where $\mathbf{x} \in \mathbb{R}^{n_x}$ and $\mathbf{y} \in \mathbb{R}^{n_y}$ are the state vectors of systems $\mathcal{S}_x$ and $\mathcal{S}_y$, respectively, the vectors $\boldsymbol{\theta}_x \in \mathbb{R}^{n_{\theta x}}$ and $\boldsymbol{\theta}_y \in \mathbb{R}^{n_{\theta y}}$ are unknown parameters, and the functions $\mathbf{f} = \mathbf{f}_1(\mathbf{x}) \oplus \mathbf{f}_2(\mathbf{x}, \boldsymbol{\theta}_x) : \mathbb{R}^{n_x} \times \mathbb{R}^{n_{\theta x}} \to \mathbb{R}^{n_x}$, $\mathbf{q} = \mathbf{q}_1(\mathbf{y}) \oplus \mathbf{q}_2(\mathbf{y}, \boldsymbol{\theta}_y) : \mathbb{R}^{n_y} \times \mathbb{R}^{n_{\theta y}} \to \mathbb{R}^{n_y}$, $\mathbf{g} = \mathbf{g}_1(\mathbf{x}) \oplus \mathbf{g}_2(\mathbf{x}) : \mathbb{R}^{n_x} \to \mathbb{R}^{n_x}$, and $\mathbf{z} = \mathbf{z}_1(\mathbf{y}) \oplus \mathbf{z}_2(\mathbf{x}) : \mathbb{R}^{n_y} \to \mathbb{R}^{n_y}$ are continuous and locally bounded. The functions $\gamma_y : \mathbb{R}^{n_y} \times \mathbb{R}_{\geq 0} \to \mathbb{R}_n$ and $\gamma_x : \mathbb{R}^{n_x} \times \mathbb{R}_{\geq 0} \to \mathbb{R}^{n_y}$, stand for nonlinear, non-stationary, and, in general, unknown couplings between systems $\mathcal{S}_x$ and $\mathcal{S}_y$, and $u_x \in \mathbb{R}$ and $u_y \in \mathbb{R}$ are the control inputs.

Let $\psi_x : \mathbb{R}^{n_x} \times \mathbb{R}_{\geq 0} \to \mathbb{R}$ and $\psi_y : \mathbb{R}^{n_y} \times \mathbb{R}_{\geq 0} \to \mathbb{R}$ be the goal functions for systems $\mathcal{S}_x$ and $\mathcal{S}_y$, respectively. In other words, for some values $\varepsilon_x \in \mathbb{R}_{\geq 0}$ and $\varepsilon_y \in \mathbb{R}_{\geq 0}$ and time instant $t^* \in \mathbb{R}_{\geq 0}$, the inequalities

$$\|\psi_x(\mathbf{x}(t), t)\|_{\infty,[t^*,\infty]} \leq \varepsilon_x, \qquad \|\psi_y(\mathbf{y}(t), t)\|_{\infty,[t^*,\infty]} \leq \varepsilon_y \tag{5.119}$$

specify the desired state of the interconnection (5.117) and (5.118). Our goal is to specify the class of functions $u_x(\mathbf{x}, t)$ and $u_y(\mathbf{y}, t)$ constituting solutions to part 1 of Problem 5.3, in particular ensuring that for all $\boldsymbol{\theta}_x \in \mathbb{R}^{n_{\theta x}}$ and $\boldsymbol{\theta}_y \in \mathbb{R}^{n_{\theta y}}$

(1) the interconnection (5.117) and (5.118) is complete;
(2) the trajectories $\mathbf{x}(t)$ and $\mathbf{y}(t)$ are bounded;
(3) for given values of ε_x and ε_y, some $t^* \in \mathbb{R}_{\geq 0}$ exists such that inequalities (5.119) are satisfied or, possibly, the functions $\psi_x(\mathbf{x}(t), t)$ and $\psi_y(\mathbf{y}(t), t)$ converge to zero as $t \to \infty$.

The function $u_x(\cdot)$ should not depend explicitly on $\mathbf{y}$ and, symmetrically, the function $u_y(\cdot)$ should not depend explicitly on $\mathbf{x}$. The general structure of the desired configuration of the control scheme is provided in Figure 5.6.

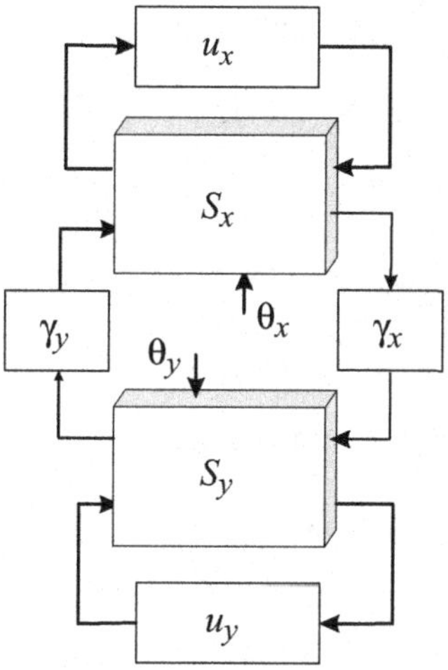

Figure 5.6 The general structure of the interconnection (5.117) and (5.118).

Consider the following functions:

$$u_x(\mathbf{x},\hat{\boldsymbol{\theta}}_x,\boldsymbol{\omega}_x,t) = (L_{\mathbf{g}(\mathbf{x})}\psi_x(\mathbf{x},t))^{-1} \times \left(-L_{\mathbf{f}(\mathbf{x},\hat{\boldsymbol{\theta}}_x)}\psi_x(\mathbf{x},t) - \varphi_x(\psi_x,\boldsymbol{\omega}_x,t) - \frac{\partial\psi_x(\mathbf{x},t)}{\partial t}\right), \quad (5.120)$$

$$\varphi_x : \mathbb{R}\times\mathbb{R}^w\times\mathbb{R}_{\geq 0}\to\mathbb{R},$$

$$u_y(\mathbf{y},\hat{\boldsymbol{\theta}}_y,\boldsymbol{\omega}_y,t) = (L_{\mathbf{z}(\mathbf{y})}\psi_y(\mathbf{y},t))^{-1} \times \left(-L_{\mathbf{q}(\mathbf{y},\hat{\boldsymbol{\theta}}_y)}\psi_y(\mathbf{y},t) - \varphi_y(\psi_y,\boldsymbol{\omega}_y,t) - \frac{\partial\psi_y(\mathbf{y},t)}{\partial t}\right), \quad (5.121)$$

$$\varphi_y : \mathbb{R}\times\mathbb{R}^w\times\mathbb{R}_{\geq 0}\to\mathbb{R}.$$

These functions transform the original equations (5.117) and (5.118) into the following form:

$$\begin{aligned}\dot{\psi}_x &= -\varphi_x(\psi_x,\boldsymbol{\omega}_x,t) + f_x(\mathbf{x},\boldsymbol{\theta}_x,t) - f_x(\mathbf{x},\hat{\boldsymbol{\theta}}_x,t) + h_y(\mathbf{x},\mathbf{y},t),\\ \dot{\psi}_y &= -\varphi_y(\psi_x,\boldsymbol{\omega}_y,t) + f_y(\mathbf{y},\boldsymbol{\theta}_y,t) - f_y(\mathbf{y},\hat{\boldsymbol{\theta}}_y,t) + h_x(\mathbf{x},\mathbf{y},t),\end{aligned} \quad (5.122)$$

where

$$h_x(\mathbf{x},\mathbf{y},t) = L_{\gamma_y(\mathbf{y},t)}\psi_x(\mathbf{x},t), \qquad h_y(\mathbf{x},\mathbf{y},t) = L_{\gamma_x(\mathbf{x},t)}\psi_y(\mathbf{y},t),$$

$$f_x(\mathbf{x},\boldsymbol{\theta}_x,t) = L_{\mathbf{f}(\mathbf{x},\boldsymbol{\theta}_x)}\psi_x(\mathbf{x},t), \qquad f_y(\mathbf{x},\boldsymbol{\theta}_y,t) = L_{\mathbf{q}(\mathbf{y},\boldsymbol{\theta}_y)}\psi_y(\mathbf{y},t).$$

Consider the following adaptation algorithms:

$$\begin{aligned}\hat{\boldsymbol{\theta}}_x(\mathbf{x},t) &= \Gamma_x(\hat{\boldsymbol{\theta}}_{P,x}(\mathbf{x},t)+\hat{\boldsymbol{\theta}}_{I,x}(t)),\ \Gamma_x \in \mathbb{R}^{d\times d},\ \Gamma_x > 0,\\ \hat{\boldsymbol{\theta}}_{P,x}(\mathbf{x},t) &= \psi_x(\mathbf{x},t)\boldsymbol{\alpha}_x(\mathbf{x},t) - \Psi_x(\mathbf{x},t),\\ \dot{\hat{\boldsymbol{\theta}}}_{I,x} &= \varphi_x(\psi_x(\mathbf{x},t),\boldsymbol{\omega}_x,t)\boldsymbol{\alpha}_x(\mathbf{x},t) + \mathcal{R}_x(\mathbf{x},\hat{\boldsymbol{\theta}}_x,u_x(\mathbf{x},\hat{\boldsymbol{\theta}}_x,t),t),\end{aligned} \tag{5.123}$$

$$\begin{aligned}\hat{\boldsymbol{\theta}}_y(\mathbf{x},t) &= \Gamma_y(\hat{\boldsymbol{\theta}}_{P,y}(\mathbf{y},t)+\hat{\boldsymbol{\theta}}_{I,y}(t)),\ \Gamma_y \in \mathbb{R}^{d\times d},\ \Gamma_y > 0,\\ \hat{\boldsymbol{\theta}}_{P,y}(\mathbf{y},t) &= \psi_y(\mathbf{y},t)\boldsymbol{\alpha}_y(\mathbf{y},t) - \Psi_y(\mathbf{y},t),\\ \dot{\hat{\boldsymbol{\theta}}}_{I,y} &= \varphi_y(\psi_y(\mathbf{y},t),\boldsymbol{\omega}_y,t)\boldsymbol{\alpha}_y(\mathbf{y},t) + \mathcal{R}_y(\mathbf{x},\hat{\boldsymbol{\theta}}_y,u_y(\mathbf{y},\hat{\boldsymbol{\theta}}_y,t),t),\end{aligned} \tag{5.124}$$

where $\mathcal{R}_x(\cdot)$ and $\mathcal{R}_y(\cdot)$ are defined as in (5.34), and the functions $\boldsymbol{\alpha}_x(\cdot)$, $\boldsymbol{\alpha}_y(\cdot)$, $\Psi_x(\cdot)$ and $\Psi_y(\cdot)$ will be specified later. Now we are ready to formulate the following result.

Theorem 5.8 *Let systems (5.117) and (5.118) be given. Furthermore, suppose that the following conditions hold:*

(1) the functions $\psi_x(\mathbf{x},t)$ and $\psi_y(\mathbf{y},t)$ satisfy Assumption 5.1 for systems (5.117) and (5.118), respectively;

(2) the systems

$$\dot{\psi}_x = -\varphi_x(\psi_x,\boldsymbol{\omega}_x,t)+\zeta_x(t), \qquad \dot{\psi}_y = -\varphi_y(\psi_y,\boldsymbol{\omega}_y,t)+\zeta_y(t) \tag{5.125}$$

satisfy Assumption 5.2 with corresponding margins

$$\gamma_{x_{\infty,2}}(\psi_{x0},\boldsymbol{\omega}_x,\|\zeta_x(t)\|_{2,[t_0,T]}), \qquad \gamma_{y_{\infty,2}}(\psi_{y0},\boldsymbol{\omega}_y,\|\zeta_y(t)\|_{2,[t_0,T]});$$

(3) systems (5.125) admit $L_2^1[t_0,\infty] \mapsto L_2^1[t_0,\infty]$ margins, that is

$$\begin{aligned}&\|\psi_x(\mathbf{x}(t),t)\|_{2,[t_0,T]} \le C_{\gamma_x} + \gamma_{x_{2,2}}(\|\zeta_x(t)\|_{2,[t_0,T]}),\\ &\|\psi_y(\mathbf{y}(t),t)\|_{2,[t_0,T]} \le C_{\gamma_y} + \gamma_{y_{2,2}}(\|\zeta_y(t)\|_{2,[t_0,T]}),\\ &C_{\gamma_x},\ C_{\gamma_y} \in \mathbb{R}_{\ge 0},\quad \gamma_{x_{2,2}},\quad \gamma_{y_{2,2}} \in \mathcal{K}_\infty;\end{aligned} \tag{5.126}$$

(4) the functions $f_x(\mathbf{x},\boldsymbol{\theta}_x,t)$ and $f_y(\mathbf{y},\boldsymbol{\theta}_y,t)$ satisfy Assumptions 5.3 and 5.4 with corresponding constants D_x, D_{x_1}, and D_y, D_{y_1} and functions $\boldsymbol{\alpha}_x(\mathbf{x},t)$ and $\boldsymbol{\alpha}_y(\mathbf{y},t)$ from (5.123) and (5.124);

(5) the functions $h_x(\mathbf{x},\mathbf{y},t)$ and $h_y(\mathbf{x},\mathbf{y},t)$ satisfy the following inequalities:

$$\begin{aligned}\|h_x(\mathbf{x},\mathbf{y},t)\| \le \beta_x\|\psi_x(\mathbf{x},t)\|, \qquad \|h_y(\mathbf{x},\mathbf{y},t)\| \le \beta_y\|\psi_y(\mathbf{y},t)\|,\\ \beta_x,\beta_y \in \mathbb{R}_{\ge 0}.\end{aligned} \tag{5.127}$$

Finally, let the functions $\Psi_x(\mathbf{x},t)$ and $\Psi_y(\mathbf{y},t)$ in (5.123) and (5.124) satisfy Assumption 5.5 with $\mathcal{B} = 0$ for systems (5.117) and (5.118), respectively, and *there exist functions* $\rho_1(\cdot)$, $\rho_2(\cdot)$, $\rho_3(\cdot) > Id(\cdot) \in \mathcal{K}_\infty$ *and constant* $\bar{\Delta} \in \mathbb{R}_{\geq 0}$ *such that the inequality*

$$\beta_y \circ \gamma_{y2,2} \circ \rho_1 \circ \left(\frac{D_y}{D_{y,1}} + 1\right) \circ \rho_3 \circ \beta_x \circ \gamma_{x2,2} \circ \rho_2 \circ \left(\frac{D_x}{D_{x,1}} + 1\right)(\Delta) < \Delta \quad (5.128)$$

holds for all $\Delta \geq \bar{\Delta}$.

Then

(C1) The interconnection (5.117) and (5.118) with controls (5.120) and (5.121) is forward-complete and the trajectories $\mathbf{x}(t)$ *and* $\mathbf{y}(t)$ *are bounded. Furthermore,*

(C2) if properties H1 and H4 hold for the functions $f_x(\mathbf{x}, \boldsymbol{\theta}_x, t)$, $f_y(\mathbf{y}, \boldsymbol{\theta}_y, t)$, $h_x(\mathbf{x}, \mathbf{y}, t)$, *and* $h_y(\mathbf{x}, \mathbf{y}, t)$, *and also for the functions* $\varphi_x(\psi_x, \boldsymbol{\omega}_x, t)$ *and* $\varphi_y(\psi_y, \boldsymbol{\omega}_y, t)$, *then*

$$\lim_{t\to\infty} \psi_x(\mathbf{x}(t), t) = 0, \qquad \lim_{t\to\infty} \psi_y(\mathbf{y}(t), t) = 0. \quad (5.129)$$

Moreover,

(C3) if property H2 holds for $f_x(\mathbf{x}, \boldsymbol{\theta}_x, t)$ *and* $f_y(\mathbf{y}, \boldsymbol{\theta}_y, t)$, *and the functions* $\boldsymbol{\alpha}_x(\mathbf{x}, t)$, $\partial\psi_x(\mathbf{x}, t)/\partial t$, $\boldsymbol{\alpha}_y(\mathbf{y}, t)$, *and* $\partial\psi_y(\mathbf{y}, t)/\partial t$ *are locally bounded with respect to* $\mathbf{x}$ *and* $\mathbf{y}$ *uniformly in* t, *then*

$$\begin{aligned} &\lim_{t\to\infty} f_x(\mathbf{x}(t), \boldsymbol{\theta}_x, t) - f_x(\mathbf{x}(t), \hat{\boldsymbol{\theta}}_x(t), t) = 0, \\ &\lim_{t\to\infty} f_y(\mathbf{y}(t), \boldsymbol{\theta}_y, t) - f_y(\mathbf{y}(t), \hat{\boldsymbol{\theta}}_y(t), t) = 0. \end{aligned} \quad (5.130)$$

Let us briefly comment on the conditions and assumptions of Theorem 5.8. Conditions (1) and (2) specify restrictions on the goal functionals, which restrictions are similar to those of Theorem 5.1. Condition (3) is analogous to requirement (3) in Theorem 5.1, whereas condition (5) specifies uncertainties in the coupling functions $h_x(\cdot)$ and $h_y(\cdot)$ in terms of their growth rates w.r.t. $\psi_x(\cdot)$ and $\psi_y(\cdot)$. We observe here that this property is needed in order to characterize the L_2-norms of the functions $h_x(\mathbf{x}(t), \mathbf{y}(t), t)$ and $h_y(\mathbf{x}(t), \mathbf{y}(t), t)$ in terms of the L_2-norms of the functions $\psi_x(\mathbf{x}(t), t)$ and $\psi_y(\mathbf{y}(t), t)$. Therefore, it is possible to replace requirement (5.127) with the following set of conditions:

$$\begin{aligned} \|h_x(\mathbf{x}(t), \mathbf{y}(t), t)\|_{2,[t_0,T]} &\leq \beta_x \|\psi_x(\mathbf{x}(t), t)\|_{2,[t_0,T]} + C_x, \\ \|h_y(\mathbf{x}(t), \mathbf{y}(t), t)\|_{2,[t_0,T]} &\leq \beta_y \|\psi_y(\mathbf{y}(t), t)\|_{2,[t_0,T]} + C_y. \end{aligned} \quad (5.131)$$

The replacement will allow us to extend the results of Theorem 5.8 to interconnections of systems where the coupling functions do not depend explicitly on $\psi_x(\mathbf{x}(t), t)$ and $\psi_y(\mathbf{y}(t), t)$. We illustrate this possibility later in Example 5.4.1.

Condition (5.128) is the small-gain condition with respect to the $L_2^1[t_0, T]$-norms for the interconnection (5.117) and (5.118) with control (5.120) and (5.121). When the mappings $\gamma_{x_{2,2}}(\cdot)$ and $\gamma_{y_{2,2}}(\cdot)$ in (5.125) are majorized by linear functions

$$\gamma_{x_{2,2}}(\Delta) \leq g_{x_{2,2}}\Delta, \qquad \gamma_{y_{2,2}}(\Delta) \leq g_{y_{2,2}}\Delta, \ \Delta \geq 0,$$

condition (5.128) reduces to the much simpler one

$$\beta_y \beta_x g_{x_{2,2}} g_{y_{2,2}} \left(\frac{D_y}{D_{y,1}} + 1\right)\left(\frac{D_x}{D_{x,1}} + 1\right) < 1.$$

Notice also that the mappings $\gamma_{x_{2,2}}(\cdot)$ and $\gamma_{y_{2,2}}(\cdot)$ are defined by the properties of the target dynamics (5.125), and, in principle, these can be made arbitrarily small. This, together with (5.128), eventually leads to the following conclusion: the smaller the L_2-gains of the target dynamics of systems $\mathcal{S}_x$ and $\mathcal{S}_y$, the wider the class of nonlinearities (bounds for β_x and β_y, domains of D_x, $D_{1,x}$, D_y, and $D_{1,y}$) which admit a solution to Problem 5.3 can be.

Example 5.4.1 Let us illustrate the application of Theorem 5.8 to the problem of decentralized control of two coupled oscillators with nonlinear damping. Consider the following interconnected systems:

$$\begin{cases} \dot{x}_1 = x_2, \\ \dot{x}_2 = f_x(x_1, \theta_x) + k_1 y_1 + u_x, \end{cases}$$
$$\begin{cases} \dot{y}_1 = y_2, \\ \dot{y}_2 = f_y(y_1, \theta_y) + k_2 x_1 + u_y, \end{cases} \tag{5.132}$$

where $k_1, k_2 \in \mathbb{R}$ are uncertain parameters of the coupling, the functions $f(x_1, \theta_x)$ and $f(y_1, \theta_y)$ stand for the nonlinear damping terms, and θ_x and θ_y are unknown parameters. For illustrative purpose we assume the following mathematical model for the functions $f_x(\cdot)$ and $f_y(\cdot)$ in (5.132):

$$\begin{aligned} f_x(x_1, \theta_x) &= \theta_x(x_1 - x_0) + 0.5\sin(\theta_x(x_1 - x_0)), \\ f_y(y_1, \theta_y) &= \theta_y(y_1 - y_0) + 0.6\sin(\theta_y(y_1 - y_0)), \end{aligned} \tag{5.133}$$

where x_0 and y_0 are known.

Let the control goal be to steer states $\mathbf{x}$ and $\mathbf{y}$ to the origin. Consider the following goal functions:

$$\psi_x(\mathbf{x}, t) = x_1 + x_2, \qquad \psi_y(\mathbf{y}, t) = y_1 + y_2. \tag{5.134}$$

Taking into account (5.132) and (5.134), we can derive that

$$\dot{x}_1 = -x_1 + \psi_x(\mathbf{x}, t), \qquad \dot{y}_1 = -y_1 + \psi_y(\mathbf{y}, t). \tag{5.135}$$

This automatically implies that

$$\|x_1(t)\|_{\infty,[t_0,T]} \leq \|x_1(t_0)\| + \|\psi_x(\mathbf{x}(t),t)\|_{\infty,[t_0,T]},$$
$$\|y_1(t)\|_{\infty,[t_0,T]} \leq \|y_1(t_0)\| + \|\psi_y(\mathbf{y}(t),t)\|_{\infty,[t_0,T]}.$$

Hence, Assumption 5.1 is satisfied for chosen goal functions $\psi_x(\cdot)$ and $\psi_y(\cdot)$. Notice also that the equalities (5.135) imply that

$$\begin{aligned} \|x_1(t)\|_{2,[t_0,T]} &\leq 2^{-1/2}\|x_1(t_0)\| + \|\psi_x(\mathbf{x},t)\|_{2,[t_0,T]}, \\ \|y_1(t)\|_{2,[t_0,T]} &\leq 2^{-1/2}\|y_1(t_0)\| + \|\psi_y(\mathbf{y},t)\|_{2,[t_0,T]}. \end{aligned} \tag{5.136}$$

Moreover, according to (5.135), the properties

$$\begin{aligned} \lim_{t\to\infty} \psi_x(\mathbf{x}(t),t) &= \lim_{t\to\infty} x_1(t) + x_2(t) = 0, \\ \lim_{t\to\infty} \psi_y(\mathbf{y}(t),t) &= \lim_{t\to\infty} y_1(t) + y_2(t) = 0 \end{aligned} \tag{5.137}$$

guarantee that

$$\lim_{t\to\infty} x_1(t) = 0, \qquad \lim_{t\to\infty} x_2(t) = 0, \qquad \lim_{t\to\infty} y_1(t) = 0, \qquad \lim_{t\to\infty} y_2(t) = 0.$$

Hence, (5.137) ensures that the control goal is attained.

According to (5.120) and (5.121) the control functions

$$\begin{aligned} u_x &= -\lambda_x\psi_x - x_2 - f_x(x_1,\hat{\theta}_x), \\ u_y &= -\lambda_y\psi_y - y_2 - f_y(y_1,\hat{\theta}_y), \ \lambda_x, \ \lambda_y > 0 \end{aligned} \tag{5.138}$$

transform system (5.132) into the following form:

$$\begin{aligned} \dot{\psi}_x &= -\lambda_x\psi_x + f_x(x_1,\theta_x) - f_x(x_1,\hat{\theta}_x) + k_1 y_1, \\ \dot{\psi}_x &= -\lambda_x\psi_x + f_x(x_1,\theta_x) - f_x(x_1,\hat{\theta}_x) + k_2 x_1. \end{aligned} \tag{5.139}$$

Notice that the systems

$$\dot{\psi}_x = -\lambda_x\psi_x + \xi_x(t), \qquad \dot{\psi}_y = -\lambda_y\psi_t + \xi_y(t)$$

satisfy Assumption 5.2 with

$$\gamma_{x_{2,2}} = \frac{1}{\lambda_x}\|\psi_x(\mathbf{x}(t),t)\|_{2,[t_0,T]}, \qquad \gamma_{y_{2,2}} = \frac{1}{\lambda_y}\|\psi_y(\mathbf{y}(t),t)\|_{2,[t_0,T]},$$

respectively, and the functions $f_x(\cdot)$ and $f_y(\cdot)$ satisfy Assumptions 5.3 and 5.4 with

$$D_x = 1.5, \qquad D_{x,1} = 0.5, \qquad \alpha_x(\mathbf{x},t) = x_1 - x_0,$$
$$D_y = 1.6, \qquad D_{y,1} = 0.4, \qquad \alpha_y(\mathbf{y},t) = y_1 - y_0.$$

Hence conditions (1)–(4) of Theorem 5.8 are satisfied. Furthermore, according to the remarks regarding condition (5) of the theorem, the requirements (5.127) can be replaced with the implicit constraints (5.131). These, however, according to (5.136), also hold with $\beta_x = k_1$ and $\beta_y = k_2$.

Given that $\alpha_x(\mathbf{x}, t) = x_1 - x_0$ and $\alpha_y(\mathbf{y}, t) = y_1 - y_0$, Assumption 5.5 will be satisfied for functions $\alpha_x(\mathbf{x}, t)$ and $\alpha_y(\mathbf{y}, t)$ with $\Psi_x(\cdot) = 0$ and $\Psi_y(\cdot) = 0$. Therefore, the adaptation algorithms (5.123) and (5.124) will have the following form:

$$\begin{aligned}
\hat{\theta}_x &= \Gamma_x((x_1 + x_2)(x_1 - x_0) + \hat{\theta}_{x,I}),\\
\dot{\hat{\theta}}_{x,I} &= \lambda_x(x_1 + x_2)(x_1 - x_0) - (x_1 + x_2)x_2,\\
\hat{\theta}_y &= \Gamma_y((y_1 + y_2)(y_1 - y_0) + \hat{\theta}_{y,I}),\\
\dot{\hat{\theta}}_{y,I} &= \lambda_y(y_1 + y_2)(y_1 - y_0) - (y_1 + y_2)y_2.
\end{aligned} \tag{5.140}$$

Hence, according to Theorem 5.8, boundedness of the solutions in the closed-loop system (5.139) and (5.140) is ensured upon satisfying the following condition

$$\frac{k_1 k_2}{\lambda_x \lambda_y}\left(1 + \frac{D_x}{D_{x,1}}\right)\left(1 + \frac{D_y}{D_{y,1}}\right) < 1 \Rightarrow k_1 k_2 < \frac{\lambda_x \lambda_y}{20}. \tag{5.141}$$

Moreover, given that properties H1, H2, and H4 hold for the chosen functions $\psi_x(\mathbf{x}, t)$ and $\psi_y(\mathbf{y}, t)$, condition (5.141) guarantees that the limiting relations (5.129) and (5.130) hold.

Trajectories of the closed-loop system (5.132), (5.138), and (5.140) with the parameter values $\Gamma_x = \Gamma_y = 1$, $\lambda_x = \lambda_y = 2$, $x_0 = y_0 = 1$, and $\theta_x = \theta_y = 1$ and initial conditions $x_1(0) = -1$, $x_2(0) = 0$, $y_1(0) = 1$, $y_2(0) = 0$, $\hat{\theta}_{x,I}(0) = -1$, and $\hat{\theta}_{y,I}(0) = -2$ are provided in Figure 5.7. Despite the fact that the values of the coupling varied substantially from one simulation to another, this variation had hardly any effect on the transient behavior of the system.

5.5 Non-dominating adaptive control for dynamical systems with nonlinear parametrization of a general kind

In the previous sections of this chapter the classes of nonlinearly parametrized uncertainties were limited to those satisfying Assumptions 5.3 and 5.4. Although this class of functions covers a broad range of models, there are models that do not fit these assumptions. Thus additional tools to account for these situations are needed.

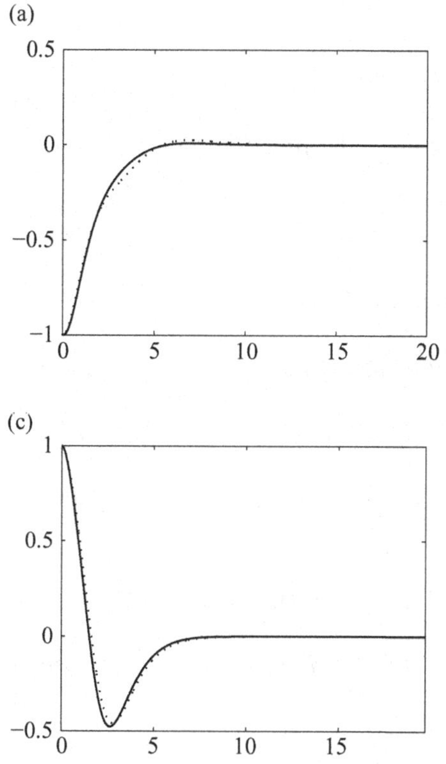

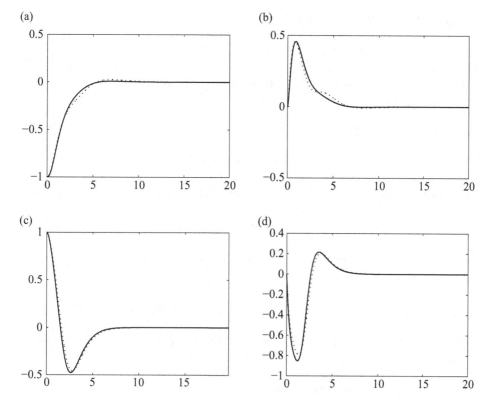

Figure 5.7 Plots of trajectories $x_1(t)$ (a), $x_2(t)$ (b), $y_1(t)$ (c), and $y_2(t)$ (d) as functions of t in the closed-loop system (5.132), (5.138), and (5.140). The dotted lines correspond to the case when $k_1 = k_2 = 0.4$, and the solid lines stand for solutions obtained with values of the coupling of $k_1 = 1$ and $k_2 = 0.1$.

Consider systems that can be transformed by means of static or dynamic feedback[14] into the following form:

$$\dot{\mathbf{x}} = \mathbf{f}_0(\mathbf{x}, t) + \mathbf{f}(\boldsymbol{\xi}(t), \boldsymbol{\theta}) - \mathbf{f}(\boldsymbol{\xi}(t), \hat{\boldsymbol{\theta}}) + \boldsymbol{\varepsilon}(t), \tag{5.142}$$

where

$$\boldsymbol{\varepsilon}(t) \in L_\infty^m[t_0, \infty], \qquad \|\boldsymbol{\varepsilon}(\tau)\|_{\infty,[t_0,t]} \leq \Delta_\varepsilon$$

is an external perturbation with known Δ_ε, and $\mathbf{x} \in \mathbb{R}^n$. The function $\boldsymbol{\xi} : \mathbb{R}_{\geq 0} \to \mathbb{R}^\xi$ is a function of time, which possibly includes available measurements of the state, $\boldsymbol{\theta}, \hat{\boldsymbol{\theta}} \in \Omega_\theta \subset \mathbb{R}^d$ are the unknown and estimated parameters of the function $\mathbf{f}(\cdot)$, respectively, and the set Ω_θ is bounded. We assume that the function $\mathbf{f}(\boldsymbol{\xi}(t), \boldsymbol{\theta})$ is

[14] Notice that conventional observers in control theory could be viewed as dynamic feedbacks.

locally bounded in $\boldsymbol{\theta}$ uniformly in $\boldsymbol{\xi}$:

$$\|\mathbf{f}(\boldsymbol{\xi}(t),\boldsymbol{\theta}) - \mathbf{f}(\boldsymbol{\xi}(t),\hat{\boldsymbol{\theta}})\| \leq D_f\|\boldsymbol{\theta} - \hat{\boldsymbol{\theta}}\| + \Delta_f,$$

and the values of $D_f \in \mathbb{R}_{\geq 0}$, Δ_f are available. The function $\mathbf{f}_0(\cdot)$ in (5.142) is assumed to satisfy the following condition.

Assumption 5.17 *The system*

$$\dot{\mathbf{x}} = \mathbf{f}_0(\mathbf{x},t) + \mathbf{u}(t) \tag{5.143}$$

is forward-complete. Furthermore, for all $\mathbf{u}(t)$ such that

$$\|\mathbf{u}(t)\|_{\infty,[t_0,t]} \leq \Delta_u + \|\mathbf{u}_0(\tau)\|_{\infty,[t_0,t]}, \ \Delta_u \in \mathbb{R}_{\geq 0}$$

there exists a bounded set $\mathcal{A}$, $c > 0$ and a function $\Delta : \mathbb{R}_{\geq 0} \to \mathbb{R}_{\geq 0}$ satisfying the following inequality:

$$\|\mathbf{x}(t)\|_{\mathcal{A}_{\Delta(\Delta_u)}} \leq \beta(t-t_0)\|\mathbf{x}(t_0)\|_{\mathcal{A}_{\Delta(\Delta_u)}} + c\|\mathbf{u}_0(\tau)\|_{\infty,[t_0,t]},$$

where $\beta(\cdot) : \mathbb{R}_{\geq 0} \to \mathbb{R}_{\geq 0}$, $\lim_{t\to\infty}\beta(t) = 0$ is a strictly decreasing continuous function.

Consider the following auxiliary system:

$$\dot{\boldsymbol{\lambda}} = S(\boldsymbol{\lambda}), \ \boldsymbol{\lambda}(t_0) = \boldsymbol{\lambda}_0 \in \Omega_\lambda \subset \mathbb{R}^\lambda, \tag{5.144}$$

where $\Omega_\lambda \subset \mathbb{R}^n$ is bounded, $\boldsymbol{\lambda}(t,\boldsymbol{\lambda}_0) \in \Omega_\lambda$ for all $t \in \mathbb{R}_{\geq 0}$, and $S(\lambda)$ is locally Lipschitz. Furthermore, suppose that the following assumption holds for system (5.144).

Assumption 5.18 *System (5.144) is Poisson stable in Ω_λ, that is*

$$\forall\,\boldsymbol{\lambda}' \in \Omega_\lambda,\ \delta \in \mathbb{R}_{\geq 0},\ t' \in \mathbb{R}_{\geq 0} \Rightarrow \exists t'' > t' + \delta : \ \|\boldsymbol{\lambda}(t'',\boldsymbol{\lambda}') - \boldsymbol{\lambda}'\| \leq \epsilon,$$

where ϵ is an arbitrarily small positive constant. Moreover, the trajectory $\lambda(t,\boldsymbol{\lambda}_0)$ is dense in Ω_λ:

$$\forall \boldsymbol{\lambda}' \in \Omega_\lambda,\ \epsilon \in \mathbb{R}_{>0} \Rightarrow \exists\, t \in \mathbb{R}_{\geq 0} : \ \|\boldsymbol{\lambda}' - \boldsymbol{\lambda}(t,\boldsymbol{\lambda}_0)\| < \epsilon.$$

Assumptions 5.17 and 5.18 allow us to formulate the following result.

Corollary 5.3 *Consider system (5.142) and suppose that the following conditions hold:*

(1) the vector field $\mathbf{f}_0(\mathbf{x},t)$ in (5.142) satisfies Assumption 5.17;

(2) there exists a (known) system (5.144) satisfying Assumption 5.18;
(3) there exists a locally Lipschitz $\boldsymbol{\eta}:\mathbb{R}^{\lambda}\rightarrow\mathbb{R}^{d}$*:*

$$\|\boldsymbol{\eta}(\boldsymbol{\lambda}')-\boldsymbol{\eta}(\boldsymbol{\lambda}'')\|\leq D_{\eta}\|\boldsymbol{\lambda}'-\boldsymbol{\lambda}''\|$$

such that the set $\boldsymbol{\eta}(\Omega_{\lambda})$ *is dense in* Ω_{θ}*;*
(4) system (5.142) has a steady-state characteristic with respect to the norm

$$\|\cdot\|_{\mathcal{A}_{\Delta(M)}},\ M=2\Delta_{f}+\Delta_{\varepsilon}+\delta$$

and input $\hat{\boldsymbol{\theta}}$*, where* δ *is some positive (arbitrarily small) constant.*

Consider the following interconnection of (5.142) and (5.144):

$$\begin{aligned}\dot{\mathbf{x}}&=\mathbf{f}_{0}(\mathbf{x},t)+\mathbf{f}(\boldsymbol{\xi}(t),\boldsymbol{\theta})-\mathbf{f}(\boldsymbol{\xi}(t),\hat{\boldsymbol{\theta}})+\boldsymbol{\varepsilon}(t),\\ \hat{\boldsymbol{\theta}}&=\boldsymbol{\eta}(\boldsymbol{\lambda}),\\ \dot{\boldsymbol{\lambda}}&=\gamma\|\mathbf{x}(t)\|_{\mathcal{A}_{\Delta(M)}}S(\boldsymbol{\lambda}),\end{aligned}\tag{5.145}$$

where $\gamma>0$ *satisfies the following inequality:*

$$\begin{aligned}\gamma&\leq\left(\beta_{t}^{-1}\left(\frac{d}{\kappa}\right)\right)^{-1}\frac{\kappa-1}{\kappa}\frac{1}{D_{\lambda}(\beta_{t}(0)(1+\kappa/(1-d))+1)}\\ D_{\lambda}&=c\cdot D_{f}\cdot D_{\eta}\cdot\max_{\boldsymbol{\lambda}\in\Omega\lambda}\|S(\boldsymbol{\lambda})\|\end{aligned}\tag{5.146}$$

for some $d\in(0,1)$*,* $\kappa\in(1,\infty)$*. Then, for* $\boldsymbol{\lambda}(t_0)=\boldsymbol{\lambda}_0$*, some* $\boldsymbol{\theta}'\in\Omega_{\theta}$ *and all* $\mathbf{x}(t_0)=\mathbf{x}_0\in\mathbb{R}^{n}$ *the following holds:*

$$\lim_{t\rightarrow\infty}\|\mathbf{x}(t)\|_{\mathcal{A}_{\Delta(M)}}=0,\qquad\lim_{t\rightarrow\infty}\hat{\boldsymbol{\theta}}(t)=\boldsymbol{\theta}'\in\Omega_{\theta}.\tag{5.147}$$

Notice that in the case in which the dynamics of (5.143) is exponentially stable with a rate of convergence equal to ρ and $\beta(0)=D_{\beta}$, condition (5.146) will have the following form:

$$\gamma\leq-\rho\left(\ln\left(\frac{d}{D_{\beta}\kappa}\right)\right)^{-1}\frac{\kappa-1}{\kappa}\frac{1}{D_{\lambda}(D_{\beta}(1+\kappa/(1-d)+1)}.$$

According to Corollary 5.3, for the rather general class of systems (5.142) it is possible to design an estimator $\hat{\boldsymbol{\theta}}(t)$ that guarantees not only that the "error" vector $\mathbf{x}(t)$ reaches a neighborhood of the origin, but also that the estimates $\hat{\boldsymbol{\theta}}(t)$ converge to some $\boldsymbol{\theta}'$ in Ω_{θ}. These two facts, together with additional nonlinear persistent excitation conditions,

$$\begin{aligned}&\exists T>0,\rho\in\mathcal{K}:\ \forall\mathcal{T}=[t,t+T],\ t\in\mathbb{R}_{\geq0}\Rightarrow\\ &\exists\tau\in\mathcal{T}:|\mathbf{f}(\boldsymbol{\xi}(\tau),\boldsymbol{\theta})-\mathbf{f}(\boldsymbol{\xi}(\tau),\boldsymbol{\theta}')|\geq\rho(\|\boldsymbol{\theta}-\boldsymbol{\theta}'\|),\end{aligned}$$

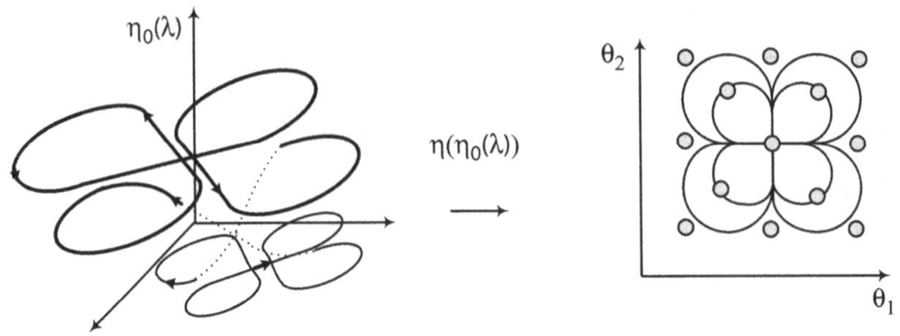

Figure 5.8 Trajectories $\boldsymbol{\eta}(\lambda)$ as projections of smooth closed curves. The gray circles correspond to the nodes of the searching grid. If we were to travel along the curve $\boldsymbol{\eta}(\lambda)$ periodically then we would visit the circle placed at the origin 16 times per period. Other circles would be visited less often, namely twice per period. Constructing such curves allowing us to take the occurrence of $\boldsymbol{\theta}$ in Ω_θ into account is not trivial. Nonetheless, this possibility can be exploited when optimizing the performance of adaptation.

in principle allow us to estimate the domain of convergence for $\hat{\boldsymbol{\theta}}(t)$.

Notice that if the frequencies (probabilities) of occurrence of $\boldsymbol{\theta}$ in Ω_θ are known then mappings $\boldsymbol{\eta}$ in (5.145) can be chosen to account for such information. This is illustrated in Figure 5.8.

In the next example we illustrate how Corollary 5.3 can be applied to solve the problem of adaptive regulation in the presence of nonlinear parametrization of uncertainty.

Example 5.5.1 Consider the following system

$$\dot{x} = -kx + \sin(x\theta + \theta) + u, \; k > 0, \; \theta \in [-a, a], \tag{5.148}$$

where θ is an unknown parameter and u is the control input. Without loss of generality we let $a = 1$ and $k = 1$. The problem is to estimate the parameter θ from measurements of x and steer the system to the origin. Clearly, the choice $u = -\sin(x\hat{\theta} + \hat{\theta})$ transforms (5.148) into

$$\dot{x} = -kx + \sin(x\theta + \theta) - \sin(x\hat{\theta} + \hat{\theta}), \tag{5.149}$$

which satisfies Assumption 5.17. Moreover, the system

$$\begin{aligned}
\dot{\lambda}_1 &= \lambda_1, \\
\dot{\lambda}_2 &= -\lambda_2, \; \lambda_1^2(t_0) + \lambda_2^2(t_0) = 1
\end{aligned}$$

with mapping $\eta = (1,\ 0)^{\mathrm{T}}\boldsymbol{\lambda}$ satisfies Assumption 5.18 and therefore

$$\begin{aligned} \dot{\lambda}_1 &= \gamma |x| \lambda_1, \\ \dot{\lambda}_2 &= -\gamma |x| \lambda_2,\ \lambda_1^2(t_0) + \lambda_2^2(t_0) = 1 \end{aligned} \tag{5.150}$$

would be a candidate for the control and parameter-estimation algorithm. According to Corollary 5.3, the goal will be attained if the parameter γ in (5.150) obeys the constraint

$$\begin{aligned} &\gamma \leq -\rho \left(\ln \left(\frac{d}{\kappa D_\beta} \right) \right)^{-1} \frac{\kappa - 1}{\kappa} \frac{1}{D_\lambda (D_\beta (1 + \kappa/(1-d)) + 1)}, \\ &\rho = k = 1,\ D_\beta = 1,\ D_\lambda = 1 \end{aligned}$$

for some $d \in (0, 1)$, $\kappa \in (1, \infty)$. Hence, on choosing, for example, $d = 0.5$ and $\kappa = 2$, we obtain that the choice

$$0 < \gamma < -\ln \left(\frac{0.5}{2} \right)^{-1} \frac{1}{2} \cdot \frac{1}{6} = 0.0601$$

suffices to ensure that

$$\lim_{t \to \infty} x(t) = 0, \qquad \lim_{t \to \infty} \hat{\theta}(t) = \theta.$$

We simulated the system (5.149) and (5.150) with $\theta = 0.3$ and $\gamma = 0.05$ and initial conditions $x(t_0)$ randomly distributed in the interval $[-1, 1]$. Results of the simulation are illustrated in Figure 5.9, where the phase plots of the system (5.149) and (5.150) as well as the trajectories of $\hat{\theta}(t)$ are given.

5.6 Parametric identification of dynamical systems with nonlinear parametrization

So far we have focused predominantly on control aspects of the problem of adaptation in dynamical systems. Let us now view this problem from a different angle: state and parameter inference. Two distinct cases will be considered below. We start with systems in which the uncertainties satisfy monotonicity restrictions (Assumptions 5.3 and 5.4). Then we continue by discussing a more general class of nonlinearly parametrized systems with merely Lipschitz nonlinearities.

5.6.1 Systems with monotone nonlinear parametrization

Let us formulate conditions ensuring convergence of the estimates $\hat{\boldsymbol{\theta}}(t)$ to $\boldsymbol{\theta}$ in the closed-loop system (5.1), (5.11), (5.32), and (5.34). When the mathematical

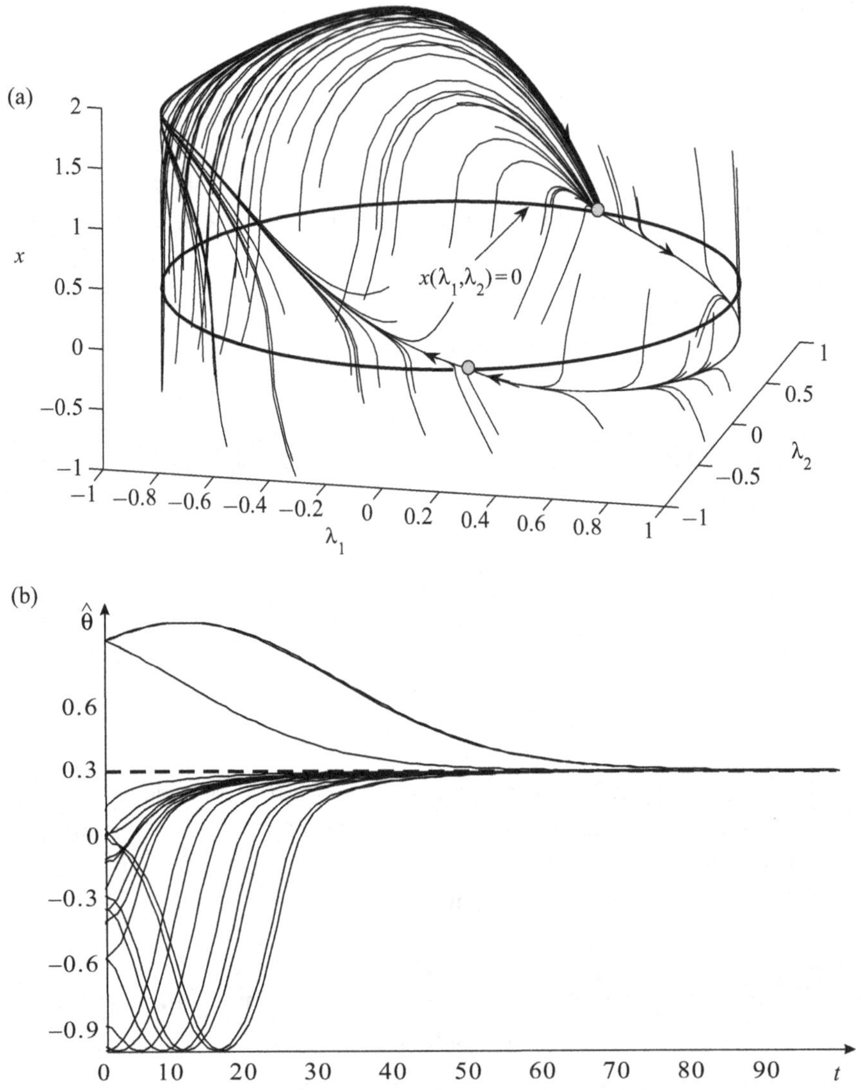

Figure 5.9 Trajectories of the system (5.149) and (5.150) (a) and the family of estimates $\hat{\theta}(t)$ of the parameter θ as functions of time t (b).

model of the uncertainties is linear in its parameters and does not depend on $\mathbf{x}$, i.e. $f(\mathbf{x},\boldsymbol{\theta},t) = \boldsymbol{\alpha}(t)^{\mathrm{T}}\boldsymbol{\theta}$, the usual requirement for convergence is that the signal $\boldsymbol{\alpha}(t)$ is *persistently exciting* (Sastry and Bodson 1989) (see also Definition 2.5.1). In particular, the function $\boldsymbol{\alpha}(t)$ is said to be persistently exciting iff there exist constants $\delta > 0$ and $L > 0$ such that for all $t \in \mathbb{R}_{\geq 0}$ the following holds:

$$\int_{t}^{t+L} \boldsymbol{\alpha}(\tau)\boldsymbol{\alpha}(\tau)^{\mathrm{T}}\, d\tau \geq \delta I. \tag{5.151}$$

Checking that condition (5.151) holds often necessitates knowledge of the signal $\boldsymbol{\alpha}(t)$ as a function of time. In the closed-loop system, however, relevant signals in the model of uncertainty $f(\mathbf{x}, \boldsymbol{\theta}, t)$ can depend on the state $\mathbf{x}$, initial conditions, uncertainties, parameters of the feedback, and initial time t_0. In order to take such dependences into account the notion of *uniform persistent excitation* has been suggested (Loria and Panteley 2002).

Definition 5.6.1 Let $\boldsymbol{\alpha} : \mathcal{D} \times \mathbb{R}_{\geq t_0} \rightarrow \mathbb{R}^m$, $\mathcal{D} \subset \mathbb{R}^d$ be a continuous function. We say that $\boldsymbol{\alpha}(\mathbf{r}, t)$ is uniformly persistently exciting in $\mathcal{D}$ if there exist $\delta \in \mathbb{R}_{>0}$ and $L \in \mathbb{R}_{>0}$ such that for each $\mathbf{r} \in \mathcal{D}$

$$\int_{t}^{t+L} \boldsymbol{\alpha}(\mathbf{r}, t)\boldsymbol{\alpha}(\mathbf{r}, t)^{\mathrm{T}}\, d\tau \geq \delta I \ \forall\, t \geq t_0. \tag{5.152}$$

In the case of models with linear parametrization of uncertainty, persistent excitation of the signal $\boldsymbol{\alpha}(t)$, i.e. inequality (5.151), implies that the following property holds:

$$\exists\, t' \in [t, t + L] : \ |\boldsymbol{\alpha}(t')^{\mathrm{T}}(\boldsymbol{\theta}_1 - \boldsymbol{\theta}_2)| \geq \delta \|\boldsymbol{\theta}_1 - \boldsymbol{\theta}_2\|. \tag{5.153}$$

In other words, the difference $|\boldsymbol{\alpha}(t)^{\mathrm{T}}(\boldsymbol{\theta}_1 - \boldsymbol{\theta}_2)|$ is *proportional* to the distance $\|\boldsymbol{\theta}_1 - \boldsymbol{\theta}_2\|$ in parameter space for some $t' \in [t, t + L]$. When dealing with nonlinear parametrization, it is useful to have a similar characterization that takes model nonlinearity into account. So it is natural to replace the linear term $\boldsymbol{\alpha}(t')^{\mathrm{T}}(\boldsymbol{\theta}_1 - \boldsymbol{\theta}_2)$, or the term $\boldsymbol{\alpha}(\mathbf{x}(t', \boldsymbol{\theta}, \mathbf{x}_0), t')^{\mathrm{T}}(\boldsymbol{\theta}_1 - \boldsymbol{\theta}_2)$ if the regressor is state-dependent, in (5.153) with its nonlinear substitute $f(\mathbf{x}(t'), \boldsymbol{\theta}_1, t') - f(\mathbf{x}(t'), \boldsymbol{\theta}_2, t')$, as has been done, for example, in Cao *et al.* (2003) for systems with convex/concave parametrization. It is also natural to replace the proportion $\delta\|\boldsymbol{\theta}_1 - \boldsymbol{\theta}_2\|$ in (5.153) with a nonlinear function. We, therefore, use the following notion of *nonlinear* persistent excitation.

Definition 5.6.2 A function $f(\mathbf{x}(t), \boldsymbol{\theta}, t) : \mathbb{R}^n \times \mathbb{R}^d \times \mathbb{R}_{\geq 0} \rightarrow \mathbb{R}$ is said to be nonlinearly persistently excited with respect to parameters $\boldsymbol{\theta} \in \Omega_\theta \subset \mathbb{R}^d$ iff there exist constant $L > 0$ and a function $\varrho : \mathbb{R}_{\geq 0} \rightarrow \mathbb{R}_{\geq 0}$, $\varrho \in \mathcal{K} \cap C^0$ such that for all $t \in \mathbb{R}_{\geq 0}$, $\boldsymbol{\theta}_1, \boldsymbol{\theta}_2 \in \Omega_\theta$ the following holds:

$$\exists\, t' \in [t, t + L] : \ |f(\mathbf{x}(t'), \boldsymbol{\theta}_1, t') - f(\mathbf{x}(t'), \boldsymbol{\theta}_2, t')| \geq \varrho(\|\boldsymbol{\theta}_1 - \boldsymbol{\theta}_2\|). \tag{5.154}$$

Properties (5.152) and (5.154) in Definitions 5.6.1 and 5.6.2 provide alternative characterizations of excitation in dynamical systems. While (5.151) accounts for properties of the signals in the uncertainty, (5.154) reflects the possibility of detecting parametric mismatches from the difference $f(\mathbf{x}(t), \boldsymbol{\theta}_1, t) - f(\mathbf{x}(t), \boldsymbol{\theta}_2, t)$. The following theorem presents corresponding alternatives for parameter convergence in the system (5.1), (5.12), (5.32), and (5.34).

Theorem 5.9 *Let the system (5.1), (5.12), (5.32), and (5.34) satisfy Assumptions 5.1–5.3. Let, in addition, Assumption 5.5 hold with $\mathcal{B}(\mathbf{x}, t) = 0$. Then $\mathbf{x}(t) \in L^n_\infty[t_0, \infty]$ and $\hat{\boldsymbol{\theta}}(t) \in L^d_\infty[t_0, \infty]$. Moreover, the limiting relation*

$$\lim_{t\to\infty} \hat{\boldsymbol{\theta}}(\mathbf{x}(t), t) = \boldsymbol{\theta}$$

is ensured if $\boldsymbol{\alpha}(\mathbf{x}, t)$ is locally bounded in $\mathbf{x}$ uniformly in t, and one of the following alternatives holds:

(1) function $\boldsymbol{\alpha}(\mathbf{x}(t), t)$ is persistently exciting, and hypothesis H3 holds;
(2) function $f(\mathbf{x}(t), \boldsymbol{\theta}, t)$ is nonlinearly persistently exciting, i.e. it satisfies condition (5.154); it satisfies hypotheses H1 and H2; the function $\varphi(\psi, \boldsymbol{\omega}, t)$ satisfies hypothesis H4; and the function $\partial\psi(\mathbf{x}, t)/\partial t$ is locally bounded in $\mathbf{x}$ uniformly in t.

If alternative (1) is satisfied, the estimates $\hat{\boldsymbol{\theta}}(\mathbf{x}(t), t)$ converge to $\boldsymbol{\theta}$ exponentially fast. If, in addition, $\boldsymbol{\alpha}(\mathbf{x}(t), t)$ is uniformly persistently exciting and Assumption 5.4 holds, then the convergence is uniform. The rate of convergence can be estimated as follows:

$$\|\hat{\boldsymbol{\theta}}(t) - \boldsymbol{\theta}\| \le e^{-\rho t}\|\hat{\boldsymbol{\theta}}(t_0) - \boldsymbol{\theta}\| D_\Gamma, \tag{5.155}$$

where

$$D_\Gamma = \left(\frac{\lambda_{\max}(\Gamma)}{\lambda_{\min}(\Gamma)}\right)^{\frac{1}{2}} e^{\rho L}, \quad \rho = \frac{\delta D_1 \lambda_{\min}(\Gamma)}{L(1 + \lambda_{\max}(\Gamma)\alpha_\infty^2 L^2)^2},$$
$$\alpha_\infty = \sup_{\|\mathbf{x}\| \le \|\mathbf{x}(t)\|_{\infty,[t_0,\infty]},\ t \ge t_0} \|\boldsymbol{\alpha}(\mathbf{x}, t)\|.$$

Theorem 5.9 considers error models (5.12) without a disturbance term $\varepsilon(t)$ but can straightforwardly be extended to ones with disturbance (5.11). As follows from alternative (1), the parameter-estimation subsystem is exponentially stable when $\boldsymbol{\alpha}(\mathbf{x}(t), t)$ is persistently exciting. This allows (sufficiently small) additive disturbances on the right-hand side of (5.12). When the excitation is uniform, convergence of the estimates $\hat{\boldsymbol{\theta}}(t)$ to a neighborhood of $\boldsymbol{\theta}$ is guaranteed for every $\varepsilon(t) \in L^1_\infty[t_0, \infty]$ by the inverse-Lyapunov-stability theorems (Khalil 2002). In the case of alternative (2), (5.154) guarantees convergence (5.24) without invoking Assumption 5.4 or hypothesis H3. However, the convergence might not be robust, which seems to be a natural tradeoff between generality of nonlinear parametrizations $f(\mathbf{x}, \boldsymbol{\theta}, t)$ and robustness with respect to unknown disturbances $\varepsilon(t)$.

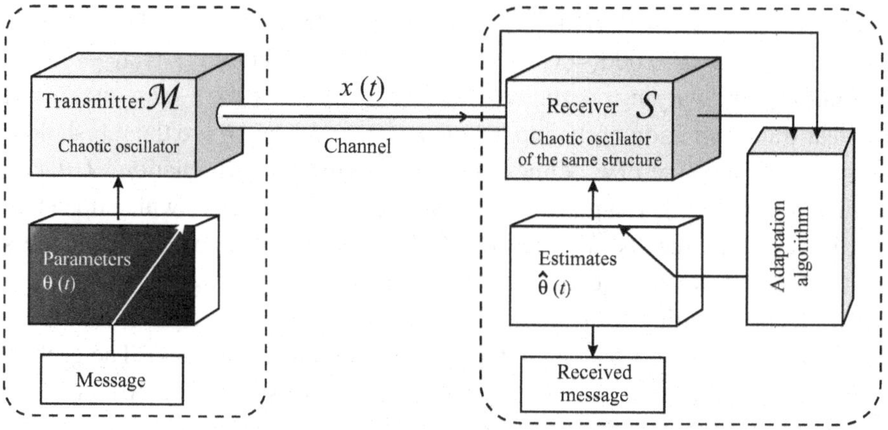

Figure 5.10 A simple scheme for secure communication based on chaotic nonlinear oscillators.

Let us illustrate the application of Theorem 5.9 with an example. The example will also demonstrate the applicability of our approach to problems in which the target dynamics is not necessarily stable.

Example 5.6.1 Consider a pair of coupled nonlinear forced oscillators $\mathcal{M}$ (master) and $\mathcal{S}$ (slave). Their inputs depend on the parameters $\theta \in \mathbb{R}$ in $\mathcal{M}$ and $\hat{\theta} \in \mathbb{R}$ in $\mathcal{S}$. The parameter θ can be understood as a "message" transmitted by the master. The parameter $\hat{\theta}$ in the slave subsystem $\mathcal{S}$ should track the message transmitted by $\mathcal{M}$. A review on problems of this kind is available in Yang (2004). Traditionally this problem is solved within observer-based approaches under the conditions that there is a stable synchrony between $\mathcal{M}$ and $\mathcal{S}$ at $\hat{\theta} = \theta$ and the oscillators are linearly parametrized in θ. The results formulated in Theorems 5.1 and 5.9 allow one to solve this problem without the need to demand existence of stable synchrony between subsystems at $\theta = \hat{\theta}$; also they enable extension of the approach to systems with nonlinear parametrization.

Let us consider a pair of coupled oscillators, for instance, the well-known Hindmarsh–Rose-model neurons (Hindmarsh and Rose 1984):

$$
\begin{aligned}
\mathcal{M} &: \begin{cases} \dot{x}_1 = f_1(x_1, x_2, x_3) + I(\theta, r(t)), \\ \dot{x}_2 = f_2(x_1, x_2, x_3), \\ \dot{x}_3 = f_3(x_1, x_2, x_3), \end{cases} \\
\mathcal{S} &: \begin{cases} \dot{y}_1 = f_1(y_1, y_2, y_3) + c(t)(y_1 - x_1) + I(\hat{\theta}, r(t)), \\ \dot{y}_2 = f_2(y_1, y_2, y_3), \\ \dot{y}_3 = f_3(y_1, y_2, y_3), \end{cases}
\end{aligned}
\tag{5.156}
$$

where $f_1(x_1,x_2,x_3) = -x_1^3 + 3x_1^2 + x_2 - x_3$, $f_2(x_1,x_2,x_3) = 1 - 5x_1^2 - x_2$, and $f_3(x_1,x_2,x_3) = 0.006(4(x_1 + 1.6) - x_3)$. The variable $c : \mathbb{R}_{\geq 0} \rightarrow \mathbb{R}_{\geq 0}$ is the coupling time-varying coefficient, $I(\theta, r(t)) = \kappa/(1 + \exp(-\theta - r(t)))$ is the nonlinear transformation of the input signal $r(t)$, and θ and $\hat{\theta}$ are the thresholds or "messages." The value of κ defines the upper bound for "stimulation" $I(\theta, r(t))$. The function $I(\theta, r(t))$ is a plausible model of the synaptic gates which open when the value of $r(t)$ exceeds the threshold θ. For the current choice of $f_i(\cdot)$ we set $\kappa = 4$, which allows chaotic bursting in (5.156).

A sufficient condition for synchronization between the $\mathcal{M}$ and $\mathcal{S}$ subsystems at $\hat{\theta} = \theta$ is $c(t) \geq c^* = 21.5$ (see Oud and Tyukin (2004) for details). The problem, however, is how to design $\hat{\theta}$ as a function of the state of system (5.156), input $r(t)$, and time t such that $\hat{\theta}$ tracks the values of $\boldsymbol{\theta}$ for $c(t) \geq 0$ below c^*.

First, we notice that $\mathbf{x}(t)$ and $\mathbf{y}(t)$ are bounded if $I(\theta, r(t))$ and $I(\hat{\theta}, r(t))$ are bounded: both $\mathcal{M}$ and $\mathcal{S}$ are semi-passive with a quadratic storage function, which is positive definite and radially unbounded outside a bounded domain in the state space of (5.156) (Oud and Tyukin 2004). For the error function ψ, we chose the difference $x_1 - y_1$. Because boundedness of the state for all $\hat{\theta}$, θ, $r(t)$ is already ensured by the strict semi-passivity of (5.156), checking Assumption 5.1 is not necessary for application of our method. Let us consider the dynamics of ψ:

$$\dot{\psi} = f_1(x_1,x_2,x_3) - f_1(y_1,y_2,y_3) - c(t)\psi + I(\theta, r(t)) - I(\hat{\theta}, r(t)). \quad (5.157)$$

Given that $\mathbf{x}(t)$, $\mathbf{y}(t)$, $I(\theta, r(t))$, and $I(\hat{\theta}, r(t))$ are bounded, Assumption 5.2 is satisfied for (5.157) with $c(t) \geq \delta$, where $\delta \in \mathbb{R}_{>0}$ is arbitrarily small. The function $\varphi(\psi, \boldsymbol{\omega}, t)$ in this case can be defined as $\varphi(\psi, \boldsymbol{\omega}, t) = -f_1(x_1(t), x_2(t), x_3(t)) + f_1(y_1(t), y_2(t), y_3(t)) + c(t)\psi$. Notice also that $I(\theta, r(t))$ is differentiable. Hence, according to Hadamard's lemma, the difference $I(\theta, r(t)) - I(\hat{\theta}, r(t))$ can be expressed as

$$I(\theta, r(t)) - I(\hat{\theta}, r(t)) = \int_0^1 \frac{\partial f(\theta\lambda + (1-\lambda)\hat{\theta}, r(t))}{\partial(\theta\lambda + (1-\lambda)\hat{\theta})} d\lambda (\theta - \hat{\theta}). \quad (5.158)$$

Given that $I(\theta, r(t))$ is strictly monotone in θ, the value of

$$\int_0^1 \frac{\partial f(\theta\lambda + (1-\lambda)\hat{\theta}, r(t))}{\partial(\theta\lambda + (1-\lambda)\hat{\theta})} d\lambda$$

is positive. Furthermore, $\partial I(\theta, r(t))/\partial\theta$ is ultimately bounded. Therefore, Assumption 5.3 is satisfied with $\alpha = \text{constant} = 1$. Because $\partial\alpha/\partial(\mathbf{x} \oplus \mathbf{y}) = 0$, Assumption 5.5 is satisfied with $\Psi(\mathbf{x} \oplus \mathbf{y}, t) = 0$.

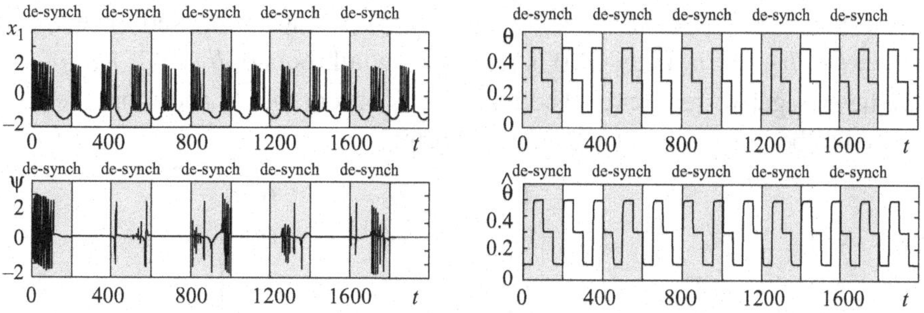

Figure 5.11 Trajectories of the system (5.156) and (5.159) as functions of time t. The input signal $r(t)$ was set as follows: $r(t) = \sin(0.001t)$. White areas mark the time instants at which the coupling variable $c(t)$ exceeds the critical value $c^* = 21.5$, $c(t) = 21.55$ ensuring synchronization between the $\mathcal{M}$ and $\mathcal{S}$ subsystems at $\hat{\theta} = \theta$. The shaded domains correspond to the time intervals in which conditions for synchronization are violated: $c(t) = 0.05$. Even though subsystems $\mathcal{M}$ and $\mathcal{S}$ fail to synchronize over these intervals, the message θ sent by $\mathcal{M}$ is successfully tracked by $\mathcal{S}$.

Using (5.32) and (5.34) we write the adaptation algorithm for $\hat{\theta}$ as

$$\begin{aligned} \hat{\theta} &= \gamma(x_1(t) - y_1(t) + \hat{\theta}_I(t)), \\ \dot{\hat{\theta}}_I &= -f_1(x_1, x_2, x_3) + f_1(y_1, y_2, y_3) + c(t)(x_1 - y_1), \ \gamma \in \mathbb{R}_{>0}. \end{aligned} \tag{5.159}$$

According to Theorem 5.1 and Corollary 5.1 the variable $\hat{\theta}$ is bounded and non-increasing. Hence, taking into account equality (5.158) and the fact that $\partial I(\theta, r(t))/\partial\theta$ is separated from zero for all bounded θ and $r(t)$, Assumption 5.4 holds for the nonlinearity $I(\theta, r(t))$ in the system (5.156) and (5.157) with bounded $r(t)$ and algorithm (5.159). Moreover, $\alpha(t) = 1$ is uniformly persistently exciting. Therefore, Theorem 5.9 applies and $\hat{\theta}$ converges to θ in (5.156) exponentially fast. In other words, the message θ as transmitted can be recovered exponentially fast using algorithm (5.159). This property is illustrated in Figure 5.11.

5.6.2 Observer-based state and parameter reconstruction for systems with Lipschitz parametrization

Let us now consider a somewhat more general class of single-input–single-output nonlinear systems:

$$\begin{aligned} \dot{\mathbf{x}} &= A\mathbf{x} + \mathbf{b}\boldsymbol{\varphi}(y, \boldsymbol{\lambda}, t)^{\mathrm{T}}\boldsymbol{\theta} + \mathbf{g}u(t) + \boldsymbol{\xi}(t), \\ y &= \mathbf{c}^{\mathrm{T}}\mathbf{x}, \end{aligned} \tag{5.160}$$

where $\mathbf{c}, \mathbf{b}, \mathbf{g} \in \mathbb{R}^n$, $A \in \mathbb{R}^{n\times n}$, and $\mathbf{c} = (1, 0, \dots, 0)^{\mathrm{T}}$ are supposed to be known, and $\mathbf{b} = (b_1, b_2, \dots, b_n)^{\mathrm{T}}, b_1 \neq 0$ is Hurwitz, i.e. all roots of $b_1 p^{n-1} + b_2 p^{n-2} + \cdots b_n$ have negative real parts. The pair $A, \mathbf{c}$ is observable,[15] and moreover A, is supposed to be in Brunovsky's canonical form:

$$A = \begin{pmatrix} a_1 & 1 & 0 & 0 & \dots & 0 \\ a_2 & 0 & 1 & 0 & \dots & 0 \\ \dots & \dots & \dots & \dots & \dots & \dots \\ a_{n-1} & 0 & 0 & \dots & 0 & 1 \\ a_n & 0 & 0 & \dots & 0 & 0 \end{pmatrix}.$$

In (5.160) $\mathbf{x} = \mathrm{col}(x_1, x_2, \dots, x_n) \in \mathbb{R}^n$ is the state vector, y is the measured output, $u : \mathbb{R} \to \mathbb{R}$, $u \in C^0$ is a known function, and $\boldsymbol{\xi} : \mathbb{R} \to \mathbb{R}^n$, $\boldsymbol{\xi} \in C^0$ is an unknown yet bounded continuous function,

$$\|\boldsymbol{\xi}(t)\| \leq \Delta_\xi, \ \Delta_\xi \in \mathbb{R}_{>0}, \ \forall\, t \geq t_0, \tag{5.161}$$

representing *unmodeled dynamics* (e.g. noise).[16] The system state $\mathbf{x}$ is not measured; only the values of the input $u(t)$ and the output $y(t) = x_1(t)$, $t \geq t_0$ in (5.160) are accessible over any time interval $[t_0, t]$ that belongs to the history of the system.

The function $\boldsymbol{\varphi} : \mathbb{R}\times\mathbb{R}^s\times\mathbb{R} \to \mathbb{R}^m, \boldsymbol{\varphi}(y, \boldsymbol{\lambda}, t) = \mathrm{col}(\varphi_1(y, \boldsymbol{\lambda}, t), \dots, \varphi_m(y, \boldsymbol{\lambda}, t))$ is assumed to be continuous and bounded in y, and t, and in addition is required to be Lipschitz in $\boldsymbol{\lambda}$ uniformly in y, and t:

$$\|\boldsymbol{\varphi}(y, \boldsymbol{\lambda}', t) - \boldsymbol{\varphi}(y, \boldsymbol{\lambda}'', t)\| \leq D\|\boldsymbol{\lambda}' - \boldsymbol{\lambda}''\|, \ D \in \mathbb{R}_{>0}. \tag{5.162}$$

The vectors $\boldsymbol{\lambda} = \mathrm{col}(\lambda_1, \dots, \lambda_s) \in \mathbb{R}^s$ and $\boldsymbol{\theta} = \mathrm{col}(\theta_1, \dots, \theta_m) \in \mathbb{R}^m$ are parameters of the system. The values of $\boldsymbol{\lambda}$ and $\boldsymbol{\theta}$ are supposed to be unknown, yet we assume that they belong to the hypercubes $\Omega_\lambda \subset \mathbb{R}^s$ and $\Omega_\theta \subset \mathbb{R}^m$ of which the bounds are known: $\theta_i \in [\theta_{i,\min}, \theta_{j,\max}]$ and $\lambda_j \in [\lambda_{j,\min}, \lambda_{j,\max}]$.

To reconstruct the unknown state and parameters of the system from the values of $y(t) = x_1(t), u(t)$, we should find an auxiliary system, i.e. an *adaptive observer*,

$$\dot{\mathbf{q}} = \mathbf{f}(\mathbf{q}, y, t), \ \mathbf{q}(t_0) = \mathbf{q}_0, \ \mathbf{q} \in \mathbb{R}^q, \tag{5.163}$$

such that for some given $\mathbf{h} : \mathbb{R}^q \to \mathbb{R}^{s+n+m}$, $\mathbf{q}_0, \delta \in \mathbb{R}_{>0}$, and $t_0 \in \mathbb{R}$ the following property holds:

$$\exists\, t' \geq t_0 : \|\mathbf{h}(\mathbf{q}(t, \mathbf{q}_0)) - \mathbf{w}(t)\| < \delta, \ \forall\, t \geq t', \tag{5.164}$$

[15] Let us recall that the pair $A, \mathbf{c}$ is called observable iff the rank of the $n \times n$ matrix $(\mathbf{c}, A^{\mathrm{T}}\mathbf{c}, (A^{\mathrm{T}})^2\mathbf{c}, \dots, (A^{\mathrm{T}})^{n-1}\mathbf{c})$ is n.

[16] For a technical reason we suppose here that the value of Δ_ξ is greater than zero. This requirement is not necessary in general (see the proof of Theorem 5.10 for details).

where $\mathbf{w}(t) = \mathrm{col}(\boldsymbol{\theta}, \boldsymbol{\lambda}, \mathbf{x}(t, \mathbf{x}_0))$.

The goal requirement for our adaptive observer is stated in the form of inequality (5.164), instead of the more usual requirement $\lim_{t\to\infty} \mathbf{h}(\mathbf{q}(t, \mathbf{q}_0)) - \mathbf{w}(t) = 0$. This is because we allow unmodeled dynamics, $\boldsymbol{\xi}(t)$, on the right-hand side of (5.160). We demonstrate below that asymptotic reconstruction of the state and parameters of (5.160) in the sense of (5.164) is achievable, subject to a set of standard persistency-of-excitation conditions on $\boldsymbol{\varphi}(y, \boldsymbol{\lambda}, t)$.

Traditionally, methods of observer design are based on the notion of Lyapunov stability of solutions. If a globally asymptotically Lyapunov-stable solution exists in the state space of the combined system (5.160) and (5.163) such that the requirements (5.164) are satisfied, then (5.163) is the observer needed. However, as we mentioned earlier, in systems with nonlinear parametrization and unmodeled perturbations distinct parameter values on the right-hand side of (5.160) may correspond to indistinguishable input–output behavior. Hence multiple distinct invariant sets in the extended state space will co-exist, and only local stability would generally be guaranteed.

Therefore, instead of the standard requirement that the dynamics of errors of state and parameter estimates is globally asymptotically stable in the sense of Lyapunov, we will require mere convergence of the estimates as specified by (5.164). This relaxation allows us to invoke a wider range of tools for accessing convergence such as in Chapter 4. We propose that an asymptotically converging observer for (5.160) consists of two coupled subsystems, $\mathcal{S}_{\mathrm{a}}$ and $\mathcal{S}_{\mathrm{w}}$. The role of subsystem $\mathcal{S}_{\mathrm{a}}$ is to provide estimates of the state and "linear" parameters $\boldsymbol{\theta}$ of (5.160), and the role of subsystem $\mathcal{S}_{\mathrm{w}}$ is to search for the values of "nonlinear" parameters $\boldsymbol{\lambda}$.

Let subsystem $\mathcal{S}_{\mathrm{a}}$ be defined as

$$\mathcal{S}_{\mathrm{a}}: \begin{cases} \dot{\hat{\mathbf{x}}} = A\hat{\mathbf{x}} + \boldsymbol{\ell}(\hat{y} - y) + \mathbf{b}\boldsymbol{\varphi}(y, \hat{\boldsymbol{\lambda}}, t)^{\mathrm{T}}\hat{\boldsymbol{\theta}} + \mathbf{g}u(t), \\ \dot{\hat{\boldsymbol{\theta}}} = -k(y - \hat{y})\boldsymbol{\varphi}(y, \hat{\boldsymbol{\lambda}}, t), \ k \in \mathbb{R}_{>0}, \\ \hat{y} = \mathbf{c}^{\mathrm{T}}\hat{\mathbf{x}}, \ \hat{\mathbf{x}}(t_0) \in \mathbb{R}^n, \ \hat{\boldsymbol{\theta}}(t_0) \in \mathbb{R}^m. \end{cases} \tag{5.165}$$

The vector $\hat{\boldsymbol{\theta}} = \mathrm{col}(\hat{\theta}_0, \hat{\theta}_1, \ldots, \hat{\theta}_n)$ in (5.165) is the vector of estimates of $\boldsymbol{\theta}$, and $\hat{\boldsymbol{\lambda}} = \mathrm{col}(\hat{\lambda}_1, \ldots, \hat{\lambda}_s)$ is the vector of estimates of $\boldsymbol{\lambda}$. The values of $\hat{\boldsymbol{\lambda}}$ will be specified later. The vector $\boldsymbol{\ell}$ in (5.165) is to be chosen such that the following condition holds:

$$\begin{aligned} &P(A + \boldsymbol{\ell}\,\mathbf{c}^{\mathrm{T}}) + (A + \boldsymbol{\ell}\,\mathbf{c}^{\mathrm{T}})^{\mathrm{T}}P \leq -Q, \\ &P\mathbf{b} = \mathbf{c}, \end{aligned} \tag{5.166}$$

where $P, Q > 0$ are some positive definite $n \times n$ matrices. This choice is always possible. Indeed, the pair $A, \mathbf{c}$ is observable and hence there exists an $\boldsymbol{\ell}$ such that the matrix $A + \boldsymbol{\ell}\mathbf{c}^{\mathrm{T}}$ is Hurwitz. Moreover, if $r_1, \ldots, r_n \in \mathbb{C}$ are arbitrary numbers in the left-half of the complex plane, then one can choose $\boldsymbol{\ell}$ such that $r_1, \ldots, r_n$ are zeros of the characteristic polynomial $\det(p - (A + \boldsymbol{\ell}\mathbf{c}^{\mathrm{T}}))$. Finally, notice that $H(p) = \mathbf{c}^{\mathrm{T}}(p - (A + \boldsymbol{\ell}\mathbf{c}^{\mathrm{T}}))^{-1}\mathbf{b} = (b_1 p^{n-1} + b_2 p^{n-2} + \cdots + b_n)/\det(p - (A + \boldsymbol{\ell}\mathbf{c}^{\mathrm{T}}))$. Thus there is an $\boldsymbol{\ell}$ such that $H(p) = 1/(p - r_n)$. The function $H(p)$ is strictly positive real.[17] Then, according to the Meyer–Kalman–Yakubovich lemma (please see the Appendix for details), there are positive definite symmetric matrices $P, Q > 0$ such that (5.166) holds. In addition, since A and $\boldsymbol{\ell}$ are known, we will assume that the matrices P and Q in (5.166) are also known.

If the values of $\boldsymbol{\lambda}$ are known then the substitution $\hat{\boldsymbol{\lambda}} = \boldsymbol{\lambda}$ reduces system (5.165) to a standard gradient-based adaptive observer for (5.160). The asymptotic properties of this observer have been studied widely in the literature (see e.g. Marino (1990), Bastin and Gevers (1988), and Kreisselmeier (1977)). The problem is that the values of $\boldsymbol{\lambda}$ are unknown and therefore these standard techniques cannot be applied directly. A solution would be to replace the value of $\boldsymbol{\lambda}$ with its estimate, $\hat{\boldsymbol{\lambda}}$, as is done in (5.165), where the estimate $\hat{\boldsymbol{\lambda}}$ is to be supplied by an additional subsystem. Two questions, however, arise with regard to this strategy: first, how to define this additional subsystem; and second, how to make sure that the estimates of $\hat{\boldsymbol{\theta}}$ and $\hat{\boldsymbol{\lambda}}$ will necessarily converge to the true values of $\boldsymbol{\theta}$ and $\boldsymbol{\lambda}$.

With respect to the first question, no particular assumptions are imposed on the function $\boldsymbol{\varphi}(y, \boldsymbol{\lambda}, t)$, except that it is bounded, Lipschitz in $\boldsymbol{\lambda}$, and continuous in y and t. Hence standard gradient-based estimation algorithms (Farza *et al.* 2009; Grip *et al.* 2010; Tyukin *et al.* 2007b) are not guaranteed to converge to a small neighborhood of $\boldsymbol{\lambda}$ even if the values of $\boldsymbol{\theta}$ are known. Therefore, searching-based exploration, which is partially discussed in the previous section, is a plausible option. The exploration can be realized, for example, by the following class of dynamical systems:

$$\dot{\mathbf{s}} = \mathbf{f}(\mathbf{s}), \ \mathbf{s}(t_0) = \mathbf{s}_0,$$
$$\hat{\boldsymbol{\lambda}} = \boldsymbol{\beta}(\mathbf{s}),$$

where $\mathbf{f} : \mathbb{R}^{n_s} \to \mathbb{R}^{n_s}$ and $\boldsymbol{\beta} : \mathbb{R}^{n_s} \to \mathbb{R}^{s}$ are Lipschitz, and $\mathbf{s}_0$ is such that the solution $\mathbf{s}(t, \mathbf{s}_0)$ is bounded. As before, let Ω_s be the set of all ω-limit points of $\mathbf{s}(t, \mathbf{s}_0)$, and suppose that this set is Poisson-stable. Finally, suppose that for every $\boldsymbol{\lambda} \in \Omega_\lambda$ there is an $\mathbf{s} \in \Omega_s$: $\boldsymbol{\beta}(\mathbf{s}) = \boldsymbol{\lambda}$. This implies that the corresponding trajectory $\boldsymbol{\beta}(\mathbf{s}(t, \mathbf{s}_0))$ is dense in Ω_λ:

$$\forall\, \boldsymbol{\lambda} \in \Omega_\lambda, \ \varepsilon \in \mathbb{R}_{>0}, t \geq t_0 \ \exists\, t' > t_0 : \ \|\boldsymbol{\lambda} - \boldsymbol{\beta}(\mathbf{s}(t', \mathbf{s}_0))\| < \varepsilon.$$

[17] A function $H : \mathbb{C} \to \mathbb{C}$ is said to be positive real if it is analytic in the open right half of the complex plane, and $\mathrm{Re}(H(p)) \geq 0 \forall p : \mathrm{Re}(p) \geq 0$. It is strictly positive real if the function $H(p)$ is not identically zero, and $H(p - \epsilon)$ is positive real for some $\epsilon > 0$.

Clearly, for any $\boldsymbol{\lambda} \in \Omega_\lambda$ and any $\varepsilon > 0$ there will exist a sequence of time instants $t_i : \lim_{i\to\infty} t_i = \infty$ such that $\|\hat{\boldsymbol{\lambda}}(t_i) - \boldsymbol{\lambda}\| < \varepsilon$. The issue, however, is how do we know that a reasonably small neighborhood of $\boldsymbol{\lambda}$ has been reached, and that the exploration should stop? This is closely related to the second, convergence, question.

In order to be able to stop in a small vicinity of $\boldsymbol{\lambda}$ an explorative subsystem must be supplied with a measure of error. In our case, the value of $y - \hat{y}$ is the most immediate candidate. Therefore, in what follows we propose that the evolution of the estimates $\hat{\boldsymbol{\lambda}} = \mathrm{col}(\hat{\lambda}_1, \ldots, \hat{\lambda}_s)$ in time is described by the following class of equations:

$$\begin{aligned} \dot{\mathbf{s}} &= \|y(t) - \hat{y}(t)\|_\varepsilon \mathbf{f}(\mathbf{s}), \ \varepsilon \in \mathbb{R}_{\geq 0}, \\ \boldsymbol{\lambda} &= \boldsymbol{\beta}(\mathbf{s}), \ \mathbf{s}(t_0) = \mathbf{s}_0. \end{aligned}$$

The number of systems satisfying these conditions is infinitely large. For simplicity, let us suppose that system $\mathcal{S}_\mathrm{w}$ is given by the following set of equations:

$$\mathcal{S}_\mathrm{w} : \begin{cases} \dot{s}_{2j-1} = \gamma \cdot \omega_j \cdot e \cdot \left(s_{2j-1} - s_{2j} - s_{2j-1}\left(s_{2j-1}^2 + s_{2j}^2\right)\right), \\ \dot{s}_{2j} = \gamma \cdot \omega_j \cdot e \cdot \left(s_{2j-1} + s_{2j} - s_{2j}\left(s_{2j-1}^2 + s_{2j}^2\right)\right), \\ \hat{\lambda}_j = \lambda_{j,\min} + \left((\lambda_{j,\max} - \lambda_{j,\min})/2\right)(s_{2j-1} + 1), \\ e = \sigma(\|y - \hat{y}\|_\varepsilon), \end{cases} \tag{5.167}$$

$$j = \{1, \ldots, s\}, \qquad s_{2j-1}^2(t_0) + s_{2j}^2(t_0) = 1, \tag{5.168}$$

where $\sigma(\cdot) : \mathbb{R} \to \mathbb{R}_{\geq 0}$ is a bounded and Lipschitz function: i.e. $\sigma(\upsilon) \leq S \in \mathbb{R}_{>0}$, and $|\sigma(\upsilon)| \leq |\upsilon|$ for all $\upsilon \in \mathbb{R}$. Moreover, we suppose that $\sigma(\upsilon) = 0$ iff $\upsilon = 0$. The parameters $\omega_j \in \mathbb{R}_{>0}$ in (5.167) are supposed to be *rationally independent*:

$$\sum \omega_j k_j \neq 0, \ \forall\, k_j \in \mathbb{Z}. \tag{5.169}$$

If $e = 1$ for all $t \geq t_0$ then the variables $(s_{2j-1}(t, \mathbf{s}_0), s_{2j}(t, \mathbf{s}_0))$ in (5.167) and (5.168) are

$$(s_{2j-1}(t, \mathbf{s}_0), s_{2j}(t, \mathbf{s}_0)) = \left(\cos\left(\gamma \int_{t_0}^t \omega_i \tau \, d\tau + a\right), \sin\left(\gamma \int_{t_0}^t \omega_i \tau d\tau + a\right)\right),$$

$a \in \mathbb{R}$.

They densely fill the corresponding invariant tori, and, since ω_j are rationally independent, trajectories $\hat{\lambda}_j(t, \mathbf{s}_0)$ are dense in Ω_λ (Arnold 1978).

It turns out that uniform persistency of excitation of $\varphi(y(t), \boldsymbol{\lambda}, t)$ in t and nonlinear persistency of excitation of the same function in $\boldsymbol{\lambda}$ suffice to ensure that the observer defined by (5.165)–(5.167) is capable of reconstructing the values of $\mathbf{x}$, $\boldsymbol{\theta}$, and $\boldsymbol{\lambda}$ asymptotically. The result is formally stated in the theorem below.

Theorem 5.10 *Consider system (5.160) together with an estimator defined by (5.165)–(5.167). Suppose that*

(1) the function $\boldsymbol{\varphi}(y, \boldsymbol{\lambda}, t)$ is uniformly persistently exciting in the sense of Definition 5.6.1 (condition (5.152) holds for the function $\boldsymbol{\varphi}(y(t, \boldsymbol{\theta}, t_0), \boldsymbol{\lambda}, t)$ in (5.160) with $\mathbf{r} = \mathrm{col}(\boldsymbol{\theta}, \boldsymbol{\lambda})$);
(2) the function $\boldsymbol{\theta}^{\mathrm{T}}\boldsymbol{\varphi}(y, \boldsymbol{\lambda}, t)$ is weakly nonlinearly persistently exciting in $\boldsymbol{\lambda}$:

$$\exists\, T,\ \beta \in \mathbb{R}_{>0} : \ \forall\, t \geq t_0, \boldsymbol{\lambda}, \boldsymbol{\lambda}' \in \Omega_\lambda\ \exists\, t' \in [t, t+T] :$$
$$\|\boldsymbol{\theta}^{\mathrm{T}}(\boldsymbol{\varphi}(y(t'), \boldsymbol{\lambda}, t') - \boldsymbol{\varphi}(y(t'), \boldsymbol{\lambda}', t'))\| \geq \beta \left\|\boldsymbol{\lambda}'\right\|_{\mathcal{E}(\boldsymbol{\lambda})}, \tag{5.170}$$

where $\mathcal{E}(\boldsymbol{\lambda}) = \{\boldsymbol{\lambda}' \in \Omega_\lambda | \boldsymbol{\varphi}(y, \boldsymbol{\lambda}', t) = \boldsymbol{\varphi}(y, \boldsymbol{\lambda}, t)\ \forall\, t \in \mathbb{R}\}$;
(3) the function $\boldsymbol{\varphi}(y, \boldsymbol{\lambda}, t)$ is Lipschitz in $\boldsymbol{\lambda}$, (5.162), and $\partial\boldsymbol{\varphi}(y, \boldsymbol{\lambda}, t)/\partial t$ is bounded.

Then there exist numbers $\varepsilon > 0$ and $\gamma^* > 0$, and functions $r_1, r_2 \in \mathcal{K}$ such that for all $\gamma \in (0, \gamma^*]$ the following hold along the trajectories of (5.165)–(5.167):

$$\lim_{t\to\infty} \|\hat{\mathbf{x}}(t) - \mathbf{x}(t)\|_\varepsilon = 0, \tag{5.171}$$

$$\limsup_{t\to\infty} \|\hat{\boldsymbol{\theta}}(t) - \boldsymbol{\theta}\| = r_1(\Delta_\xi), \tag{5.172}$$

$$\limsup_{t\to\infty} \left\|\hat{\boldsymbol{\lambda}}(t)\right\|_{\mathcal{E}(\boldsymbol{\lambda})} = r_2(\Delta_\xi). \tag{5.173}$$

The value of Δ_ξ is specified in (5.161).

Theorem 5.10 concludes this chapter. Here we have presented a range of adaptation algorithms particularly suitable for systems (models) with nonlinearly parametrized uncertainty and potentially unstable solutions in the target dynamics. In the next chapters we shall see how these algorithms can be applied to address several practical issues in the domain of mathematical modeling of neural systems and artificial intelligence. We will start with the problem of adaptive classification, and then proceed by discussing a prototype system for invariant template matching. We will conclude by showing how the set of techniques developed can be used in the problem of modeling evoked potentials in neural membranes.

Appendix to Chapter 5

A5.1 Proof of Theorem 5.1

Let us first show that property (1) holds. Consider solutions of the system (5.1), (5.11), and (5.32), and (5.34) passing through the point $\mathbf{x}(t_0)$, $\hat{\boldsymbol{\theta}}_I(t_0)$ for

$t \in [t_0, T^*]$.[18] Let us calculate formally the time-derivative of the function $\hat{\boldsymbol{\theta}}(\mathbf{x}, t)$:

$$\dot{\hat{\boldsymbol{\theta}}}(\mathbf{x}, t) = \Gamma(\dot{\hat{\boldsymbol{\theta}}}_P + \dot{\hat{\boldsymbol{\theta}}}_I) = \Gamma(\dot{\psi}\boldsymbol{\alpha}(\mathbf{x}, t) + \psi\dot{\boldsymbol{\alpha}}(\mathbf{x}, t) - \dot{\Psi}(\mathbf{x}, t) + \dot{\hat{\boldsymbol{\theta}}}_I).$$

Notice that

$$\begin{aligned}\psi\dot{\boldsymbol{\alpha}}(\mathbf{x}, t) - \dot{\Psi}(\mathbf{x}, t) + \dot{\hat{\boldsymbol{\theta}}}_I = {} & \psi(\mathbf{x}, t)\frac{\partial\boldsymbol{\alpha}(\mathbf{x}, t)}{\partial\mathbf{x}_1}\dot{\mathbf{x}}_1 + \psi(\mathbf{x}, t)\frac{\partial\boldsymbol{\alpha}(\mathbf{x})}{\partial\mathbf{x}_2}\dot{\mathbf{x}}_2 \\ & + \psi(\mathbf{x}, t)\frac{\partial\boldsymbol{\alpha}(\mathbf{x}, t)}{\partial t} - \frac{\partial\Psi(\mathbf{x}, t)}{\partial\mathbf{x}_1}\dot{\mathbf{x}}_1 \\ & - \frac{\partial\Psi(\mathbf{x}, t)}{\partial\mathbf{x}_2}\dot{\mathbf{x}}_2 - \frac{\partial\Psi(\mathbf{x}, t)}{\partial t} + \dot{\hat{\boldsymbol{\theta}}}_I.\end{aligned} \tag{A5.1}$$

According to Assumption 5.5 $\partial\Psi(\mathbf{x}, t)/\partial\mathbf{x}_2 = \psi(\mathbf{x}, t)\partial\boldsymbol{\alpha}(\mathbf{x}, t)/\partial\mathbf{x}_2 + \mathcal{B}(\mathbf{x}, t)$. Then, taking (A5.1) into account, we can obtain

$$\begin{aligned}\psi\dot{\boldsymbol{\alpha}}(\mathbf{x}, t) - \dot{\Psi}(\mathbf{x}, t) + \dot{\hat{\boldsymbol{\theta}}}_I = {} & \left(\psi(\mathbf{x}, t)\frac{\partial\boldsymbol{\alpha}(\mathbf{x}, t)}{\partial\mathbf{x}_1} - \frac{\partial\Psi}{\partial\mathbf{x}_1}\right)\dot{\mathbf{x}}_1 \\ & + \psi(\mathbf{x}, t)\frac{\partial\boldsymbol{\alpha}(\mathbf{x}, t)}{\partial t} - \frac{\Psi(\mathbf{x}, t)}{\partial t} \\ & - \mathcal{B}(\mathbf{x}, t)(\mathbf{f}_2(\mathbf{x}, \boldsymbol{\theta}, t) + \mathbf{g}_2(\mathbf{x}, t)u) + \dot{\hat{\boldsymbol{\theta}}}_I.\end{aligned} \tag{A5.2}$$

Notice that, according to the proposed notation, we can rewrite the term

$$\left(\psi(\mathbf{x}, t)\frac{\partial\boldsymbol{\alpha}(\mathbf{x}, t)}{\partial\mathbf{x}_1} - \frac{\partial\Psi}{\partial\mathbf{x}_1}\right)\dot{\mathbf{x}}_1$$

in (A5.2) in the following way:

$$\big(\psi(\mathbf{x}, t)L_{\mathbf{f}_1}\boldsymbol{\alpha}(\mathbf{x}, t) - L_{\mathbf{f}_1}\Psi(\mathbf{x}, t)\big) + \big(\psi(\mathbf{x}, t)L_{\mathbf{g}_1}\boldsymbol{\alpha}(\mathbf{x}, t) - L_{\mathbf{g}_1}\Psi(\mathbf{x}, t)\big)\, u(\mathbf{x}, \hat{\boldsymbol{\theta}}, t).$$

Hence it follows from (5.32) and (A5.2) that $\psi\dot{\boldsymbol{\alpha}}(\mathbf{x}, t) - \dot{\Psi}(\mathbf{x}, t) + \dot{\hat{\boldsymbol{\theta}}}_I = \varphi(\psi, \boldsymbol{\omega}, t)\boldsymbol{\alpha}(\mathbf{x}, t) + \mathcal{B}(\mathbf{x}, t)(\mathbf{f}_2(\mathbf{x}, \hat{\boldsymbol{\theta}}, t) - \mathbf{f}_2(\mathbf{x}, \boldsymbol{\theta}, t))$. The derivative $\dot{\hat{\boldsymbol{\theta}}}(\mathbf{x}, t)$ can therefore be written in the following manner:

$$\dot{\hat{\boldsymbol{\theta}}} = \Gamma((\dot{\psi} + \varphi(\psi, \boldsymbol{\omega}, t))\boldsymbol{\alpha}(\mathbf{x}, t) + \mathcal{B}(\mathbf{x}, t)(\mathbf{f}_2(\mathbf{x}, \hat{\boldsymbol{\theta}}, t) - \mathbf{f}_2(\mathbf{x}, \boldsymbol{\theta}, t))). \tag{A5.3}$$

Consider the following positive definite function:

$$V_{\hat{\theta}}(\hat{\boldsymbol{\theta}}, \boldsymbol{\theta}, t) = \frac{1}{2}\|\hat{\boldsymbol{\theta}} - \boldsymbol{\theta}\|^2_{\Gamma^{-1}} + \frac{D}{4D_1^2}\int_t^{\infty} \varepsilon^2(\tau)d\tau. \tag{A5.4}$$

[18] In accordance with the formulation of the theorem, the interval $[t_0, T^*]$ is the interval of existence of the solutions.

Its time-derivative can be derived according to (A5.3) as follows:

$$\dot{V}_{\hat{\theta}}(\hat{\boldsymbol{\theta}},\boldsymbol{\theta},t) = (\varphi(\psi,\boldsymbol{\omega},t)+\dot{\psi})(\hat{\boldsymbol{\theta}}-\boldsymbol{\theta})^{\mathrm{T}}\boldsymbol{\alpha}(\mathbf{x},t) + (\hat{\boldsymbol{\theta}}-\boldsymbol{\theta})^{\mathrm{T}}\mathcal{B}(\mathbf{x},t)(\mathbf{f}_2(\mathbf{x},\hat{\boldsymbol{\theta}},t)-\mathbf{f}_2(\mathbf{x},\boldsymbol{\theta},t)) - \frac{D}{4D_1^2}\varepsilon^2(t). \quad \text{(A5.5)}$$

Let $\mathcal{B}(\mathbf{x},t) \neq 0$, then consider the difference $\mathbf{f}_2(\mathbf{x},\hat{\boldsymbol{\theta}},t)-\mathbf{f}_2(\mathbf{x},\boldsymbol{\theta},t)$. Applying Hadamard's lemma, we represent this difference in the following way:

$$\mathbf{f}_2(\mathbf{x},\hat{\boldsymbol{\theta}},t)-\mathbf{f}_2(\mathbf{x},\boldsymbol{\theta},t) = \int_0^1 \frac{\partial \mathbf{f}_2(\mathbf{x},\mathbf{s}(\lambda),t)}{\partial \mathbf{s}}d\lambda(\hat{\boldsymbol{\theta}}-\boldsymbol{\theta}),\quad \mathbf{s}(\lambda)=\hat{\boldsymbol{\theta}}\lambda+\boldsymbol{\theta}(1-\lambda).$$

According to Assumption 5.5, term $(\hat{\boldsymbol{\theta}}-\boldsymbol{\theta})^{\mathrm{T}}\mathcal{B}(\mathbf{x},t)(\mathbf{f}_2(\mathbf{x},\hat{\boldsymbol{\theta}},t)-\mathbf{f}_2(\mathbf{x},\boldsymbol{\theta},t))$ is negative semi-definite; hence, using Assumptions 5.3 and 5.4 and equality (5.11), we can estimate the derivative $\dot{V}_{\hat{\theta}}$ as

$$\begin{aligned}\dot{V}_{\hat{\theta}}(\hat{\boldsymbol{\theta}},\boldsymbol{\theta},t) &\leq -(f(\mathbf{x},\hat{\boldsymbol{\theta}},t)-f(\mathbf{x},\boldsymbol{\theta},t)+\varepsilon(t))(\hat{\boldsymbol{\theta}}-\boldsymbol{\theta})^{\mathrm{T}}\boldsymbol{\alpha}(\mathbf{x},t) - \frac{D}{4D_1^2}\varepsilon^2(t)\\ &\leq -\frac{1}{D}(f(\mathbf{x},\hat{\boldsymbol{\theta}},t)-f(\mathbf{x},\boldsymbol{\theta},t))^2\\ &\quad + \frac{1}{D_1}|\varepsilon(t)||f(\mathbf{x},\hat{\boldsymbol{\theta}},t)-f(\mathbf{x},\boldsymbol{\theta},t)| - \frac{D}{4D_1^2}\varepsilon^2(t)\\ &\leq -\frac{1}{D}\left(|f(\mathbf{x},\hat{\boldsymbol{\theta}},t)-f(\mathbf{x},\boldsymbol{\theta},t)| - \frac{D}{2D_1}\varepsilon(t)\right)^2 \leq 0. \end{aligned} \quad \text{(A5.6)}$$

It follows immediately from (A5.6) and (A5.4) that

$$\|\hat{\boldsymbol{\theta}}(t)-\boldsymbol{\theta}\|^2_{\Gamma^{-1}} \leq \|\hat{\boldsymbol{\theta}}(t_0)-\boldsymbol{\theta}\|^2_{\Gamma^{-1}} + \frac{D}{2D_1^2}\|\varepsilon(t)\|^2_{2,[t_0,\infty]}. \quad \text{(A5.7)}$$

Given that inequality (A5.4) implies $\|\hat{\boldsymbol{\theta}}(t)-\boldsymbol{\theta}\|^2_{\Gamma^{-1}} \leq \|\hat{\boldsymbol{\theta}}(t_0)-\boldsymbol{\theta}\|^2_{\Gamma^{-1}} + (D/(2D_1^2))\|\varepsilon(t)\|^2_{2,[t_0,T^*]}$, we can conclude that $\hat{\boldsymbol{\theta}}(t) \in L^2_\infty[t_0,T^*]$. Moreover,

$$|f(\mathbf{x}(t),\hat{\boldsymbol{\theta}}(t),t)-f(\mathbf{x}(t),\boldsymbol{\theta},t)| - \frac{D}{2D_1}\varepsilon(t) \in L^1_2[t_0,T^*].$$

In particular,

$$\left\| |f(\mathbf{x}(t),\hat{\boldsymbol{\theta}}(t),t)-f(\mathbf{x}(t),\boldsymbol{\theta},t)| - \frac{D}{2D_1}\varepsilon(t)\right\|^2_{2,[t_0,T^*]} \leq \frac{D}{2}\|\boldsymbol{\theta}-\hat{\boldsymbol{\theta}}(t_0)\|^2_{\Gamma^{-1}} + \frac{D^2}{4D_1^2}\|\varepsilon(t)\|^2_{2,[t_0,T^*]}. \quad \text{(A5.8)}$$

Hence $f(\mathbf{x}(t),\hat{\boldsymbol{\theta}}(t),t) - f(\mathbf{x}(t),\boldsymbol{\theta},t) \in L_2^1[t_0,T^*]$ as a sum of two functions from $L_2^1[t_0,T^*]$. In order to estimate the upper bound of the norm $\|f(\mathbf{x}(t),\hat{\boldsymbol{\theta}}(t),t) - f(\mathbf{x}(t),\boldsymbol{\theta},t)\|_{2,[t_0,T^*]}$ from (A5.8), we use the Minkowski inequality

$$\left\| f(\mathbf{x}(t),\hat{\boldsymbol{\theta}}(t),t) - f(\mathbf{x}(t),\boldsymbol{\theta},t)| - \frac{D}{2D_1}\varepsilon(t)\right\|_{2,[t_0,T^*]}$$
$$\leq \left(\frac{D}{2}\|\boldsymbol{\theta} - \hat{\boldsymbol{\theta}}(t_0)\|^2_{\Gamma^{-1}}\right)^{\frac{1}{2}} + \frac{D}{2D_1}\|\varepsilon(t)\|_{2,[t_0,T^*]},$$

and then apply the triangle inequality to the functions from $L_2^1[t_0,T^*]$:

$$\|f(\mathbf{x}(t),\hat{\boldsymbol{\theta}}(t),t) - f(\mathbf{x}(t),\boldsymbol{\theta},t)\|_{2,[t_0,T^*]}$$
$$\leq \left\| f(\mathbf{x}(t),\hat{\boldsymbol{\theta}}(t),t) - f(\mathbf{x}(t),\boldsymbol{\theta},t) - \frac{D}{2D_1}\varepsilon(t)\right\|_{2,[t_0,T^*]} + \frac{D}{2D_1}\|\varepsilon(t)\|_{2,[t_0,T^*]}$$
$$\leq \left(\frac{D}{2}\|\boldsymbol{\theta} - \hat{\boldsymbol{\theta}}(t_0)\|^2_{\Gamma^{-1}}\right)^{\frac{1}{2}} + \frac{D}{D_1}\|\varepsilon(t)\|_{2,[t_0,T^*]}. \tag{A5.9}$$

Therefore, property (1) is proven.

Let us prove property (2). In order to do this we have to check first whether the solutions of the closed-loop system are defined for all $t \in \mathbb{R}_{\geq 0}$, i.e. they do not reach infinity in finite time. We prove this by a contradiction argument. Indeed, let there exist a time instant t_s such that $\|\mathbf{x}(t_s)\| = \infty$. It follows from (1), however, that $f(\mathbf{x}(t),\hat{\boldsymbol{\theta}}(t),t) - f(\mathbf{x}(t),\boldsymbol{\theta},t) \in L_2^1[t_0,t_s]$. Moreover, according to (A5.9) the norm

$$\|f(\mathbf{x}(t),\hat{\boldsymbol{\theta}}(t),t) - f(\mathbf{x}(t),\boldsymbol{\theta},t)\|_{2,[t_0,t_s]}$$

can be bounded from above by a continuous function of $\boldsymbol{\theta}$, $\hat{\boldsymbol{\theta}}(t_0)$, Γ, and $\|\varepsilon(t)\|_{2,[t_0,\infty]}$. Let us denote this bound by D_f. Notice that D_f does not depend on t_s. Consider the system (5.11) for $t \in [t_0,t_s]$: $\dot{\psi} = f(\mathbf{x},\boldsymbol{\theta},t) - f(\mathbf{x},\hat{\boldsymbol{\theta}},t) - \varphi(\psi,\boldsymbol{\omega},t) + \varepsilon(t)$. Given that $f(\mathbf{x}(t),\boldsymbol{\theta},t) - f(\mathbf{x}(t),\hat{\boldsymbol{\theta}}(t),t), \varepsilon(t) \in L_2^1[t_0,t_s]$ and taking Assumption 5.2 into account, we automatically obtain that $\psi(\mathbf{x}(t),t) \in L^1_\infty[t_0,t_s]$. In particular, using the triangle inequality and the fact that the function $\gamma_{\infty,2}(\psi(\mathbf{x}_0,t_0),\boldsymbol{\omega},M)$ in Assumption 5.2 is non-decreasing in M, we can estimate the norm $\|\psi(\mathbf{x}(t),t)\|_{\infty,[t_0,t_s]}$ as follows:

$$\|\psi(\mathbf{x}(t),t)\|_{\infty,[t_0,t_s]} \leq \gamma_{\infty,2}\left(\psi(\mathbf{x}_0,t_0),\boldsymbol{\omega},D_f + \|\varepsilon(t)\|^2_{2,[t_0,\infty]}\right). \tag{A5.10}$$

According to Assumption 5.1 the following inequality holds:

$$\|\mathbf{x}(t)\|_{\infty,[t_0,t_s]} \leq \tilde{\gamma}\left(\mathbf{x}_0,\boldsymbol{\theta},\|\psi(\mathbf{x}(t),t)\|_{\infty,[t_0,t_s]}\right).$$

Hence

$$\|\mathbf{x}(t)\|_{\infty,[t_0,t_s]} \leq \tilde{\gamma}\left(\mathbf{x}_0, \boldsymbol{\theta}, \gamma_{\infty,2}\left(\psi(\mathbf{x}_0, t_0), \boldsymbol{\omega}, D_f + \|\varepsilon(t)\|^2_{2,[t_0,\infty]}\right)\right). \quad \text{(A5.11)}$$

Because a superposition of locally bounded functions is locally bounded, we conclude that $\|\mathbf{x}(t)\|_{\infty[t_0,t_s]}$ is bounded. This, however, contradicts the previous claim that $\|\mathbf{x}(t_s)\| = \infty$. On taking (A5.7) into account as well as the fact that $\|\varepsilon(t)\|_{2,[t_0,T^*]}$ in (A5.7) is bounded from above by $\|\varepsilon(t)\|_{2,[t_0,\infty]}$, we can derive that both $\hat{\boldsymbol{\theta}}(\mathbf{x}(t), t)$ and $\hat{\boldsymbol{\theta}}_I(t)$ are bounded for every $t \in \mathbb{R}_{\geq 0}$. Moreover, according to (A5.10), (A5.11), and (A5.7) these bounds are (locally bounded) functions of initial conditions and parameters. Therefore, $\mathbf{x}(t) \in L^n_\infty[t_0, \infty]$ and $\hat{\boldsymbol{\theta}}(\mathbf{x}(t), t) \in L^d_\infty[t_0, \infty]$. Inequality (5.39) follows immediately from (A5.9) and (5.14), and the triangle inequality. Property (2) is proven.

Let us show that property (3) holds. It is assumed that the system (5.13) admits $L^1_2[t_0, \infty] \mapsto L^1_p[t_0, \infty]$, $p > 1$ margin. In addition, we have just shown that $f(\mathbf{x}(t), \boldsymbol{\theta}, t) - f(\mathbf{x}(t), \hat{\boldsymbol{\theta}}(t), t), \varepsilon(t) \in L_2[t_0, \infty]$. Hence, taking into account (5.11), we conclude that $\psi(\mathbf{x}(t), t) \in L^1_p[t_0, \infty]$, $p > 1$. On the other hand, given that $f(\mathbf{x}, \hat{\boldsymbol{\theta}}, t)$ and $\varphi(\psi, \boldsymbol{\omega}, t)$ are locally bounded with respect to their first two arguments uniformly in t and that $\mathbf{x}(t) \in L^n_\infty[t_0, \infty]$, $\psi(\mathbf{x}(t), t) \in L^1_\infty[t_0, \infty]$, $\hat{\boldsymbol{\theta}}(t) \in L^d_\infty[t_0, \infty]$, and $\boldsymbol{\theta} \in \Omega_\theta$, the signal $\varphi(\psi(\mathbf{x}(t), t), \boldsymbol{\omega}, t) + f(\mathbf{x}(t), \boldsymbol{\theta}, t) - f(\mathbf{x}(t), \hat{\boldsymbol{\theta}}(t), t)$ is bounded. Thus $\varepsilon(t) \in L^1_\infty[t_0, \infty]$ implies that $\dot{\psi}$ is bounded as well and (3) is guaranteed by Barbalat's lemma.

To complete the proof of the theorem (property (4)), consider the time-derivative of the function $f(\mathbf{x}, \hat{\boldsymbol{\theta}}, t)$

$$\begin{aligned}\frac{d}{dt} f(\mathbf{x}, \hat{\boldsymbol{\theta}}, t) &= L_{[\mathbf{f}(\mathbf{x},\boldsymbol{\theta},t)+\mathbf{g}(\mathbf{x},t)u(\mathbf{x},\hat{\boldsymbol{\theta}},t)]} f(\mathbf{x}, \hat{\boldsymbol{\theta}}, t) \\ &\quad + \frac{\partial f(\mathbf{x}, \hat{\boldsymbol{\theta}}, t)}{\partial \hat{\boldsymbol{\theta}}} \Gamma(\varphi(\psi, \boldsymbol{\omega}, t) + \dot{\psi})\boldsymbol{\alpha}(\mathbf{x}, t) + \frac{\partial f(\mathbf{x}, \hat{\boldsymbol{\theta}}, t)}{\partial t}.\end{aligned}$$

Taking into account that $\mathbf{f}(\mathbf{x}, \boldsymbol{\theta}, t)$ and $\mathbf{g}(\mathbf{x}, t)$ are locally bounded with respect to $\mathbf{x}, \boldsymbol{\theta}$ uniformly in t, the function $f(\mathbf{x}, \boldsymbol{\theta}, t)$ is continuously differentiable in $\mathbf{x}$ and $\boldsymbol{\theta}$, the derivative $\partial f(\mathbf{x}, \boldsymbol{\theta}, t)/\partial t$ is locally bounded with respect to $\mathbf{x}$ and $\boldsymbol{\theta}$ uniformly in t, and the functions $\boldsymbol{\alpha}(\mathbf{x}, t)$ and $\partial\psi(\mathbf{x}, t)/\partial t$ are locally bounded with respect to $\mathbf{x}$ uniformly in t, then $d/dt(f(\mathbf{x}, \boldsymbol{\theta}, t) - f(\mathbf{x}, \hat{\boldsymbol{\theta}}, t))$ is bounded. Given that $f(\mathbf{x}(t), \boldsymbol{\theta}, t) - f(\mathbf{x}(t), \hat{\boldsymbol{\theta}}(t), t) \in L^1_2[t_0, \infty]$, we conclude by applying Barbalat's lemma that $f(\mathbf{x}(t), \boldsymbol{\theta}, t) - f(\mathbf{x}(t), \hat{\boldsymbol{\theta}}(t), t) \to 0$ as $t \to \infty$. □

A5.2 Proof of Corollary 5.1

Let $\varepsilon(t) \equiv 0$. By choosing the function $V_{\hat{\theta}}(\boldsymbol{\theta}, \hat{\boldsymbol{\theta}}, t)$ as in (A5.4), using (A5.5), and invoking Assumption 5.3 we obtain that

$$\begin{aligned} \dot{V}_{\hat{\theta}(\hat{\theta},\theta,t)} &\leq -(f(\mathbf{x}, \boldsymbol{\theta}, t) - f(\mathbf{x}, \hat{\boldsymbol{\theta}}, t))\boldsymbol{\alpha}(\mathbf{x}, t)^{\mathrm{T}}(\boldsymbol{\theta} - \hat{\boldsymbol{\theta}}) \\ &\leq -\frac{1}{D}(f(\mathbf{x}, \boldsymbol{\theta}, t) - f(\mathbf{x}, \hat{\boldsymbol{\theta}}, t))^2. \end{aligned} \quad \text{(A5.12)}$$

The equality (A5.12) and the fact that $\varepsilon(t) \equiv 0$ in (A5.4) imply that the norm $\|\hat{\boldsymbol{\theta}} - \boldsymbol{\theta}\|^2_{\Gamma^{-1}}$ is non-increasing. Furthermore, (A5.12) implies that

$$\|f(\mathbf{x}(t), \boldsymbol{\theta}, t) - f(\mathbf{x}(t), \hat{\boldsymbol{\theta}}(t), t)\|_{2,[t_0,T^*]} \leq \left(\frac{D}{2}\|\hat{\boldsymbol{\theta}}(t_0) - \boldsymbol{\theta}\|^2_{\Gamma^{-1}}\right)^{\frac{1}{2}}. \quad \text{(A5.13)}$$

This proves property (1). Taking into account (A5.13) and given that Assumptions 5.1 and 5.2 are satisfied, we can conclude that $\mathbf{x}(t) \in L^n_\infty[t_0, \infty]$ and $\psi(\mathbf{x}(t), t) \in L^1_\infty[t_0, \infty]$ and that the following estimate holds:

$$\|\psi(\mathbf{x}(t), t)\|_{\infty,[t_0,\infty]} \leq \gamma_{\infty,2}\left(\psi(\mathbf{x}_0, t_0), \boldsymbol{\omega}, \left(\frac{D}{2}\|\hat{\boldsymbol{\theta}}(t_0) - \boldsymbol{\theta}\|^2_{\Gamma^{-1}}\right)^{\frac{1}{2}}\right). \quad \text{(A5.14)}$$

Hence (2) is also proven. Properties (3) and (4) follow by virtue of the same arguments as in the proof of Theorem 5.1. Therefore, (5) is proven. □

A5.3 Proof of Theorem 5.2

According to the conditions of the theorem, system (5.47) is complete. In addition, the differentiability of $\psi(\cdot)$, $\boldsymbol{\alpha}(\cdot)$, and $\Psi(\cdot)$ implies that the derivative $\dot{\hat{\boldsymbol{\theta}}}(\mathbf{x}, t)$ is defined in a small vicinity t_0. Moreover, according to Assumption 5.7 this derivative satisfies

$$\dot{\hat{\boldsymbol{\theta}}} = (\dot{\psi} + \varphi(\psi, \boldsymbol{\omega}, t))\boldsymbol{\alpha}(\mathbf{x}_1 \oplus \mathbf{x}'_2 \oplus \mathbf{h}_\xi, t). \quad \text{(A5.15)}$$

Finally,

$$\dot{\psi} = -\varphi(\psi, \boldsymbol{\omega}, t) + f(\mathbf{x}_1 \oplus \mathbf{x}'_2 \oplus \mathbf{h}_\xi, \boldsymbol{\theta}, t) - f(\mathbf{x}_1 \oplus \mathbf{x}'_2 \oplus \mathbf{h}_\xi, \hat{\boldsymbol{\theta}}, t) + \varepsilon_\xi(t), \quad \text{(A5.16)}$$

where $\varepsilon_\xi(t) \in L_2[t_0, T]$ and $\|\varepsilon(t)\|_{2,[t_0,T]} \leq \Delta_\xi(\boldsymbol{\theta}, \mathbf{x}_0)$. Thus property (1) follows explicitly from the proof of Theorem 5.1. If the combined system is complete then the remaining properties also follow from the proof of Theorem 5.1. Completeness of the combined system, however, follows from Theorem 4.6 presented in Chapter 4. □

A5.4 Proof of Theorem 5.3

The proof of this theorem is based on the ideas presented in Loria *et al.* (2001). Consider serial interconnection of system (5.1) and the system (5.58) and (5.59):

$$\begin{aligned} \dot{\mathbf{x}}_2 &= \mathbf{f}_2(\mathbf{x}, \boldsymbol{\theta}, t) + \mathbf{g}_2(\mathbf{x}, t)u, \\ \dot{\boldsymbol{\xi}} &= \left(\mathcal{H}(\boldsymbol{\xi}, \mathbf{x}, t)^{\mathrm{T}}\mathcal{H}(\boldsymbol{\xi}, \mathbf{x}, t) + \beta\right)(\mathbf{x}_2 - \boldsymbol{\xi}) + \mathbf{g}_2(\mathbf{x}, t)u + \boldsymbol{v}. \end{aligned} \tag{A5.17}$$

Using (A5.17) and invoking Hadamard's lemma,

$$\mathcal{H}(\boldsymbol{\xi}, \mathbf{x}, t)(\mathbf{x}_2 - \boldsymbol{\xi}) = \mathbf{h}_\epsilon(\mathbf{x}_2, \mathbf{x}, t) - \mathbf{h}_\epsilon(\boldsymbol{\xi}, \mathbf{x}, t), \tag{A5.18}$$

we can conclude that the differential equation for $\boldsymbol{\epsilon} = \boldsymbol{\xi} - \mathbf{x}_2$ can be written as

$$\begin{aligned} \dot{\boldsymbol{\epsilon}} &= -(\mathcal{H}(\boldsymbol{\xi}, \mathbf{x}, t)^{\mathrm{T}}\mathcal{H}(\boldsymbol{\xi}, \mathbf{x}, t) + \beta)\boldsymbol{\epsilon} - \mathbf{f}_2(\mathbf{x}, \boldsymbol{\theta}, t) - \hat{\boldsymbol{\eta}}^{\mathrm{T}}\Delta_f(\mathbf{x}, t) \cdot \mathrm{sign}(\boldsymbol{\epsilon}), \\ \dot{\hat{\boldsymbol{\eta}}} &= \Gamma_\eta \Delta_f(\mathbf{x}, t) \cdot \mathrm{sign}(\boldsymbol{\epsilon})^{\mathrm{T}} \cdot \boldsymbol{\epsilon}, \ \Gamma_\eta > 0. \end{aligned} \tag{A5.19}$$

Let us introduce the following positive definite function:

$$V_\epsilon(\boldsymbol{\epsilon}, \hat{\boldsymbol{\eta}}) = \frac{1}{2}\left(\boldsymbol{\epsilon}^{\mathrm{T}}\boldsymbol{\epsilon} + \|\hat{\boldsymbol{\eta}} - \boldsymbol{\eta}\|^2_{\Gamma_\eta^{-1}}\right), \tag{A5.20}$$

where the vector $\boldsymbol{\eta}$ satisfies (5.56). The derivative $\dot{V}_\epsilon$ can be expressed as

$$\dot{V}_\epsilon = -\|\mathcal{H}(\boldsymbol{\xi}, \mathbf{x}, t)\boldsymbol{\epsilon}\|^2 - \beta\|\boldsymbol{\epsilon}\|^2 - \boldsymbol{\epsilon}^{\mathrm{T}}\mathbf{f}_2(\mathbf{x}, \boldsymbol{\theta}, t) - \boldsymbol{\eta}^{\mathrm{T}}\Delta_f(\mathbf{x}, t) \cdot \mathrm{sign}(\boldsymbol{\epsilon})^{\mathrm{T}} \cdot \boldsymbol{\epsilon}. \tag{A5.21}$$

Taking into account (5.56), we can estimate $\dot{V}_\epsilon$ in (A5.21) as follows:

$$\dot{V}_\epsilon \leq -\|\mathcal{H}(\boldsymbol{\xi}, \mathbf{x}, t)\boldsymbol{\epsilon}\|^2 - \beta\|\boldsymbol{\epsilon}\|^2 - \boldsymbol{\eta}^{\mathrm{T}}\Delta_f(\mathbf{x}, t)(\|\boldsymbol{\epsilon}\|_{\mathcal{F}} - \|\boldsymbol{\epsilon}\|), \tag{A5.22}$$

where

$$\|\boldsymbol{\epsilon}\|_{\mathcal{F}} = \sum_{i=1}^{p} |\epsilon_i|.$$

Given that

$$\|\boldsymbol{\epsilon}\|_{\mathcal{F}} \geq \|\boldsymbol{\epsilon}\|,$$

and taking (A5.22) into account, we can see that the following estimate holds:

$$\dot{V}_\epsilon \leq -\|\mathcal{H}(\boldsymbol{\xi}, \mathbf{x}, t)\boldsymbol{\epsilon}\|^2 - \beta\|\boldsymbol{\epsilon}\|^2 \leq 0. \tag{A5.23}$$

Thus

$$\mathbf{x}(t) \in L^n_\infty[t_0, T], \mathbf{x} \in \mathcal{C}^0 \Rightarrow \boldsymbol{\xi}(t) \in L^p_\infty[t_0, T],$$

i.e. the system (5.58) and (5.59) is complete. Moreover, (A5.23) implies that

$$\int_{t_0}^{T^*} \|\mathcal{H}(\boldsymbol{\xi}, \mathbf{x}, t)\boldsymbol{\epsilon}\|^2 \leq V_\epsilon(\boldsymbol{\epsilon}(t_0), \boldsymbol{\eta}(t_0)) - V_\epsilon(\boldsymbol{\epsilon}(t), \boldsymbol{\eta}(t))$$
$$\leq \frac{1}{2}\left(\|\boldsymbol{\xi}(t_0) - \mathbf{x}_2(t_0)\|^2 + \|\hat{\boldsymbol{\eta}}(t_0) - \boldsymbol{\eta}\|^2_{\Gamma_\eta^{-1}}\right). \quad \text{(A5.24)}$$

Hence, on taking into account (A5.18), (5.57), and (A5.24) we obtain

$$\|f(\mathbf{x}_1(t) \oplus \mathbf{h}_\xi(\boldsymbol{\xi}(t)), \boldsymbol{\theta}, t) - f(\mathbf{x}_1(t) \oplus \mathbf{x}_2(t), \boldsymbol{\theta}, t)\|_{2,[t_0,T^*]}$$
$$\leq \|\mathbf{h}_\varepsilon(\boldsymbol{\xi}, \mathbf{x}, t) - \mathbf{h}_\varepsilon(\mathbf{x}_2, \mathbf{x}, t)\|_{2,[t_0,T^*]}$$
$$\leq \frac{1}{\sqrt{2}}\left(\|\boldsymbol{\xi}(t_0) - \mathbf{x}_2(t_0)\|^2 + \|\hat{\boldsymbol{\eta}}(t_0) - \boldsymbol{\eta}\|^2_{\Gamma_\eta^{-1}}\right)^{\frac{1}{2}},$$

where $[t_0, T^*]$ is the maximal interval of existence of $\mathbf{x}(t, \mathbf{x}_0, t_0)$ in forward time. □

A5.5 Proof of Lemma 5.1

For the sake of compactness we introduce the following notational agreements:

$$f_i(x_1, \ldots, x_i, \boldsymbol{\theta}_i) = f_i(\mathbf{x}, \boldsymbol{\theta}_i), \qquad \psi_{\xi_i} = x_i - \xi_i,$$
$$\varepsilon_i(t) = f_i(\mathbf{q}_{i-2}(t), \boldsymbol{\theta}_i) - f_i(\mathbf{q}_{i-1}(t), \boldsymbol{\theta}_i), \quad \text{(A5.25)}$$

$i = 2, \ldots, n$. In addition, we will need the following technical result.

Lemma A5.2 *Consider the system*

$$\dot{\mathbf{x}} = \mathbf{f}(\mathbf{x}, \boldsymbol{\theta}, u), \ \mathbf{x}_0 = \mathbf{x}(t_0), \quad \text{(A5.26)}$$

and suppose that $[t_0, T]$, $T > 0$ is an interval of existence of its solutions. In addition suppose that the following holds along the solutions of (A5.26):

$$\dot{\psi} = -\varphi(\psi)(1 + F(t)) + z(\mathbf{x}, \boldsymbol{\theta}, t) - z(\mathbf{x}, \hat{\boldsymbol{\theta}}, t) + \varepsilon(t). \quad \text{(A5.27)}$$

In (A5.27) the function $F : \mathbb{R}_{\geq 0} \to \mathbb{R}_{\geq 0}$, $F(t) \in C^0$, and the function $\psi(\mathbf{x}, t)$: $|\psi(\mathbf{x}, t)| \leq \delta \Rightarrow \|\mathbf{x}\| \leq \varepsilon(\delta)$, $\delta > 0$, $\varepsilon : \mathbb{R}_{\geq 0} \to \mathbb{R}_{\geq 0}$; $\varepsilon(t) \in C^0$, $\varepsilon(t) \in L_2$, $\varphi(\psi) \in C^0$, $\varphi(\psi)\psi > 0 \ \forall \psi \neq 0$, $\lim_{\psi \to \infty} \int_0^\psi \varphi(\varsigma)d\varsigma = \infty$. Let, in addition, Assumptions 5.3 and 5.4 hold for $z(\mathbf{x}, \hat{\boldsymbol{\theta}}, t)$ with $F(t) \equiv 0$, $\varepsilon(t) \equiv 0$, and let $\hat{\boldsymbol{\theta}}$ evolve according to

$$\dot{\hat{\boldsymbol{\theta}}} = \Gamma(\dot{\psi} + \varphi(\psi)(1 + F(t)))\boldsymbol{\alpha}(\mathbf{x}, t), \ \Gamma > 0. \quad \text{(A5.28)}$$

Then

(1) $\psi(\mathbf{x},t) \in L_\infty[t_0,T]$, $\varphi(\psi(\mathbf{x},t)) \in L_2[t_0,T] \cap L_\infty[t_0,T]$, $\sqrt{F(t)}\varphi(\psi(\mathbf{x},t)) \in L_2[t_0,T]$; $\hat{\boldsymbol{\theta}} \in L_\infty[t_0,T]$;

(2) $z(\mathbf{x},\boldsymbol{\theta},t) - z(\mathbf{x},\hat{\boldsymbol{\theta}},t) \in L_2[t_0,T]$;

(3) $F(t) \in L_\infty[t_0,T] \Rightarrow \dot{\psi} \in L_2[t_0,T]$.

The corresponding L_2- and L_∞-norms can be bounded from above by constants whose values do not depend on T. Furthermore, if $\varepsilon(t) \in L_\infty$, $\mathbf{f}(\cdot)$ in (A5.26) is continuous in $\mathbf{x}$ uniformly in u, u is bounded, and $z(\mathbf{x},\hat{\boldsymbol{\theta}},t)$ is locally bounded in $\mathbf{x}$ and $\hat{\boldsymbol{\theta}}$ uniformly in t then

(4) *solutions of (A5.26) exist for all $t \geq t_0$ and $\psi(\mathbf{x}(t),t) \to 0$ as $t \to \infty$.*

Proof of Lemma A5.2. Notice that, since the function $\mathbf{f}(\mathbf{x},\boldsymbol{\theta},u(\mathbf{x},\hat{\boldsymbol{\theta}},t))$ is continuous in $\mathbf{x}$, $\hat{\boldsymbol{\theta}}$, and t, there is a non-empty interval $[t_0,T]$, $T > t_0$ such that solution of the combined system is defined on $[t_0,T]$. Since $\varepsilon(t) \in L_2$ implies that $\int_t^\infty \varepsilon^2(\tau)d\tau < \infty$ the following function is defined for all $t \in [t_0,T]$:

$$V_{\hat{\theta}}(\hat{\boldsymbol{\theta}},\boldsymbol{\theta},t) = \frac{D}{4D_1^2}\int_t^\infty \varepsilon^2(\tau)d\tau + \frac{1}{2}\|\hat{\boldsymbol{\theta}} - \boldsymbol{\theta}\|^2_{\Gamma^{-1}}.$$

Let

$$\begin{aligned}\dot{V}_{\hat{\theta}} &= -\frac{D}{4D_1^2}\varepsilon^2(t) + (\varphi(\psi)(1+F(t)) + \dot{\psi})(\hat{\boldsymbol{\theta}} - \boldsymbol{\theta})^{\mathrm{T}}\boldsymbol{\alpha}(\mathbf{x},t)\\ &= -\frac{D}{4D_1^2}\varepsilon^2(t) + (z(\mathbf{x},\boldsymbol{\theta},t) - z(\mathbf{x},\hat{\boldsymbol{\theta}},t) + \varepsilon(t))(\hat{\boldsymbol{\theta}} - \boldsymbol{\theta})^{\mathrm{T}}\boldsymbol{\alpha}(\mathbf{x},t).\end{aligned}$$

According to Assumptions 5.3 and 5.4, the derivative $\dot{V}_{\hat{\theta}}$ can be estimated as follows:

$$\dot{V}_{\hat{\theta}} \leq -\frac{1}{D}\left(|z(\mathbf{x},\boldsymbol{\theta},t) - z(\mathbf{x},\hat{\boldsymbol{\theta}},t)| - \frac{D}{2D_1}|\varepsilon(t)|\right)^2 \qquad \text{(A5.29)}$$

Equation (A5.29) implies that $z(\mathbf{x},\hat{\boldsymbol{\theta}},t) - z(\mathbf{x},\boldsymbol{\theta},t) \in L_2[t_0,T]$, and that $\|z(\mathbf{x},\hat{\boldsymbol{\theta}},t) - z(\mathbf{x},\boldsymbol{\theta},t)\|_{2,[t_0,T]}$ can be bounded from above by a constant whose value does not depend on T. Let us denote $\mu(t) = \varepsilon(t) + z(\mathbf{x}(t),\hat{\boldsymbol{\theta}}(t),t) - z(\mathbf{x}(t),\hat{\boldsymbol{\theta}}^*,t)$. Thus (A5.27) can be rewritten as

$$\dot{\psi} = -(1+F(t))\varphi(\psi) + \mu(t),$$

where the function $\mu(t) \in L_2[t_0,T]$. Indeed, $\mu(t)$ is a sum of two functions from $L_2[t_0,T]$, and $\|\mu(t)\|_{2,[t_0,T]}$ can again be bounded from above by a constant whose value does not depend on T. Consider now the following non-negative function $V_1(\psi, t)$:

$$V_1(\psi, t) = \int_0^{\psi} \varphi(\xi) d\xi + \frac{1}{4} \int_t^T \mu^2(\tau) d\tau.$$

Its derivative is

$$\begin{aligned}\dot{V}_1 &= -\varphi(\psi)(1 + F(t))\varphi(\psi) + \varphi(\psi)\mu(t) - \frac{1}{4}\mu^2(t) \\ &\leq -F(t)\varphi^2(\psi) - \left(\varphi(\psi) - \frac{1}{2}\mu(t)\right)^2. \end{aligned} \tag{A5.30}$$

Inequality (A5.30) implies that $\psi(\mathbf{x}, t)$, $\varphi(\psi(\mathbf{x}(t), t)) \in L_\infty[t_0, T]$, and that $\|\psi(\mathbf{x}(t), t)\|_{\infty,[t_0,T]}$ and $\|\varphi(\psi(\mathbf{x}(t), t))\|_{\infty,[t_0,T]}$ can be bounded from above by a constant whose value does not depend on T. Moreover,

$$\sqrt{F(t)}\varphi(\psi(\mathbf{x}(t), t)) \in L_2[t_0, T], \qquad \varphi(\psi(\mathbf{x}(t), t)) - \mu(t)/2 \in L_2[t_0, T],$$

with

$$\begin{aligned}\|\sqrt{F(t)}\varphi(\psi(\mathbf{x}(t), t))\|_{2,[t_0,T]} &\leq V_1(\psi(\mathbf{x}(t_0), t_0))^{1/2}, \\ \|\varphi(\psi(\mathbf{x}(t), t)) - \mu(t)/2\|_{2,t_0,T} &\leq V_1(\psi(\mathbf{x}(t_0), t_0))^{1/2}.\end{aligned}$$

Given that $\mu(t) \in L_2[t_0, T]$ we obtain that $\varphi(\psi(\mathbf{x}(t), t)) \in L_2[t_0, T]$, and the corresponding L_2-norms can be bounded from above by constants independent of T. Thus properties (1) and (2) are proven.

Let $F(t) \in L_\infty[t_0, T]$. Then $(1 + F(t))\varphi(\psi(\mathbf{x}(t), t)) \in L_2[t_0, T]$, and (A5.27) implies that $\dot{\psi} \in L_2[t_0, T]$. This proves property (3). In order to show that (4) holds, we notice that (A5.29) implies that $\|\hat{\boldsymbol{\theta}}(t)\|_{\infty,[t_0,T]}$ can be bounded from above by a constant whose value does not depend on T. Since boundedness of $\psi(\mathbf{x}, t)$ implies that $\mathbf{x}$ is bounded, we can conclude that $\|\mathbf{x}(t)\|_{\infty,[t_0,T]}$ can be bounded from above by a constant independent of T. This, in turn, implies that solutions of (A5.26) are defined for all $t \geq t_0$. Finally, notice that $\dot{\psi}$ is bounded, provided that $\varepsilon(t)$ is bounded and $\mathbf{z}(\mathbf{x}, \boldsymbol{\theta}, t)$ is locally bounded uniformly in t. Hence, invoking Barbalat's lemma, we can conclude that $\psi(\mathbf{x}(t), t) \to 0$ as $t \to \infty$. □

Consider now the following equations:

$$\dot{\xi}_i = \left(\left(\bar{F}_i^2(\mathbf{q}_{i-1}, \mathbf{q}_i, \mathbf{z}) + \sum_{j=i}^{k} \bar{D}_{j+1}^2(\mathbf{q}_{i-1}, \mathbf{q}_i)\right) + 1\right)(x_i - \xi_i)$$

$$+ f_i(\mathbf{q}_{i-1}, \hat{\boldsymbol{\theta}}_{\xi_i}) + \beta_i(\mathbf{x}, t),$$

$$\dot{\hat{\boldsymbol{\theta}}}_{\xi_i} = \gamma_{\xi_i}\left(\psi_{\xi_i}\left(\left(\bar{F}_i^2(\mathbf{q}_{i-1}, \mathbf{q}_i, \mathbf{z}) + \sum_{j=i}^{k} \bar{D}_{j+1}^2(\mathbf{q}_{i-1}, \mathbf{q}_i)\right) + 1\right) + \dot{\psi}_{\xi_i}\right)\boldsymbol{\alpha}_i(\mathbf{q}_{i-1}), \tag{A5.31}$$

where $i = 1, \ldots, k$, $\gamma_{\xi_i} > 0$, and $\bar{F}_i(\mathbf{q}_{i-1}, \mathbf{q}_i, \mathbf{z}) = \bar{F}(\mathbf{q}_{i-1}, \mathbf{q}_i, \mathbf{z})$. Taking into account (5.63) and (A5.31), we can write

$$\dot{\psi}_{\xi_i} = -\left(\left(\bar{F}_i^2(\mathbf{q}_{i-1}, \mathbf{q}_i, \mathbf{z}) + \sum_{j=i}^{k} \bar{D}_{j+1}^2(\mathbf{q}_{i-1}, \mathbf{q}_i)\right) + 1\right)\psi_{\xi_i}$$

$$- f_i(\mathbf{q}_{i-1}, \hat{\boldsymbol{\theta}}_{\xi_i}) + f_i(\mathbf{x}, \boldsymbol{\theta}_i). \tag{A5.32}$$

Let $k = 1$ in (A5.31). Then according to Lemma A5.1 there exists $C_{1,u} \in \mathbb{R}_{>0}$ ($C_{1,u} \in \mathbb{R}_{>0}$ does not depend on T) such that $\|u(\mathbf{x}, \mathbf{z}, \boldsymbol{\theta}_0) - u(\mathbf{q}_1, \mathbf{z}, \boldsymbol{\theta}_0)\|_{2,[t_0,T]} \leq C_{1,u}$, along solutions of (A5.31).

Let $k = 2$, then taking the equations for $\dot{\psi}_{\xi_1}$ and $\dot{\hat{\boldsymbol{\theta}}}_{\xi_1}$ into account and invoking Lemma A5.1, we obtain that

$$(\bar{F}_1^2(\mathbf{q}_0, \mathbf{q}_1, \mathbf{z}) + \bar{D}_2^2(\mathbf{q}_0, \mathbf{q}_1))^{1/2}(x_1 - \xi_1) \in L_2, \quad x_1 - \xi_1 \in L_\infty[t_0, T],$$
$$\hat{\boldsymbol{\theta}}_{\xi_1} \in L_\infty[t_0, T].$$

Hence

$$\bar{F}_1(\mathbf{q}_0, \mathbf{q}_1, \mathbf{z})(x_1 - \xi_1) \in L_2[t_0, T], \qquad \bar{D}_2(\mathbf{q}_0, \mathbf{q}_1)(x_1 - \xi_1) \in L_2[t_0, T],$$

provided that

$$\sqrt{\bar{F}_1^2(\mathbf{q}_0, \mathbf{q}_1, \mathbf{z}) + \bar{D}_2^2(\mathbf{q}_0, \mathbf{q}_1)} \geq |\bar{F}_1(\mathbf{q}_0, \mathbf{q}_1, \mathbf{z})|$$

and

$$\sqrt{\bar{F}_1^2(\mathbf{q}_0, \mathbf{q}_1, \mathbf{z}) + \bar{D}_2^2(\mathbf{q}_0, \mathbf{q}_1)} \geq |\bar{D}_2(\mathbf{q}_0, \mathbf{q}_1)|.$$

Therefore we can conclude that there exist $C_{2,u}$ and $C_{2,f} \in \mathbb{R}_{>0}$, independent of T, such that

$$\|u(\mathbf{q}_0, \mathbf{z}, \boldsymbol{\theta}_0) - u(\mathbf{q}_1, \mathbf{z}, \boldsymbol{\theta}_0)\|_{2,[t_0,T]} \leq C_{2,u},$$
$$\|f_2(\mathbf{q}_0, \boldsymbol{\theta}_2) - f_2(\mathbf{q}_1, \boldsymbol{\theta}_2)\|_{2,[t_0,T]} \leq C_{2,f},$$

provided that

$$|u(\mathbf{q}_0,\mathbf{z},\boldsymbol{\theta}_0) - u(\mathbf{q}_1,\mathbf{z},\boldsymbol{\theta}_0)| \leq |\bar{F}_1(\mathbf{q}_0,\mathbf{q}_1,\mathbf{z})(x_1 - \xi_1)|,$$
$$|f_2(\mathbf{q}_0,\boldsymbol{\theta}_2) - f_2(\mathbf{q}_1,\boldsymbol{\theta}_2)| \leq |\bar{D}_2(\mathbf{q}_0,\mathbf{q}_1)(x_1 - \xi_1)|.$$

Let us denote

$$\varepsilon_2(t) = f_2(\mathbf{q}_0,\boldsymbol{\theta}_2) - f_2(\mathbf{q}_1,\boldsymbol{\theta}_2) = f_2(\mathbf{x},\boldsymbol{\theta}_2) - f_2(\mathbf{q}_1,\boldsymbol{\theta}_2).$$

Thus, according to (A5.31) and (A5.32), the following holds:

$$\begin{aligned}
\dot{\psi}_{\xi_2} &= -(\bar{F}_2^2(\mathbf{q}_1,\mathbf{q}_2,\mathbf{z}) + 1)\psi_{\xi_2} - f_2(\mathbf{q}_1,\hat{\boldsymbol{\theta}}_{\xi_2}) + f_2(\mathbf{q}_1,\boldsymbol{\theta}_2) + \varepsilon_2(t),\\
\dot{\hat{\boldsymbol{\theta}}}_{\xi_2} &= \gamma_{\xi_2}(\psi_{\xi_2}(\bar{F}_2^2(\mathbf{q}_1,\mathbf{q}_2,\mathbf{z}) + 1) + \dot{\psi}_{\xi_2})\boldsymbol{\alpha}_2(\mathbf{q}_1),\ \gamma_{\xi_2} > 0,\\
&\varepsilon_2(t) \in L_2[t_0,T],\ \|\varepsilon_2(t)\|_{2,[t_0,T]} \leq C_{2,f}.
\end{aligned} \tag{A5.33}$$

Applying Lemma A5.1 to (A5.32) and (A5.33), and using the inequality

$$|u(\mathbf{q}_1,\mathbf{z},\boldsymbol{\theta}_0) - u(\mathbf{q}_2,\mathbf{z},\boldsymbol{\theta}_0)| \leq |\bar{F}_2(\mathbf{q}_1,\mathbf{q}_2,\mathbf{z})(x_2 - \xi_2)|,$$

we can write

$$\begin{aligned}
&\|u(\mathbf{q}_0,\mathbf{z},\boldsymbol{\theta}_0) - u(\mathbf{q}_1,\mathbf{z},\boldsymbol{\theta}_0)\|_{2,[t_0,T]} \leq C_{1,u},\ \|x_1 - \xi_1\|_{\infty,[t_0,T]} \leq C_{1,x},\\
&\|u(\mathbf{q}_1,\mathbf{z},\boldsymbol{\theta}_0) - u(\mathbf{q}_2,\mathbf{z},\boldsymbol{\theta}_0)\|_{2,[t_0,T]} \leq C_{2,u},\ \|x_2 - \xi_2\|_{\infty,[t_0,T]} \leq C_{2,x},\\
&\|f_2(\mathbf{q}_0,\boldsymbol{\theta}_2) - f_2(\mathbf{q}_1,\boldsymbol{\theta}_2)\|_{2,[t_0,T]} \leq C_{2,f},\ \hat{\boldsymbol{\theta}}_{\xi_1} \in L_\infty[t_0,T],\ \hat{\boldsymbol{\theta}}_{\xi_2} \in L_\infty[t_0,T],
\end{aligned}$$

where the constants $C_{1,u}, C_{2,u}, C_{1,x}, C_{2,x}$, and $C_{2,f}$ do not depend on T. Moreover,

$$\|u(\mathbf{q}_0,\mathbf{z},\boldsymbol{\theta}_0) - u(\mathbf{q}_2,\mathbf{z},\boldsymbol{\theta}_0)\| \leq C_{1,u} + C_{2,u},$$

$$\dot{\hat{\boldsymbol{\theta}}}_{\xi_1} = \gamma_{\xi_1}\left(\psi_{\xi_1}\left(\left(\bar{F}_1^2(\mathbf{q}_0,\mathbf{q}_1,\mathbf{z}) + \sum_{j=1}^{k}\bar{D}_{j+1}^2(\mathbf{q}_0,\mathbf{q}_1)\right) + 1\right) + \dot{\psi}_{\xi_1}\right)\boldsymbol{\alpha}_1(\mathbf{q}_0).$$

According to Lemma A5.2 the following property holds:

$$\left(\bar{F}_1^2(\mathbf{q}_0,\mathbf{q}_1,\mathbf{z}) + \sum_{j=1}^{k}\bar{D}_{j+1}^2(\mathbf{q}_0,\mathbf{q}_1)\right)^{\frac{1}{2}}(x_1 - \xi_1) \in L_2[t_0,T],$$

and the corresponding L_2-norm can be bounded from above by a constant that is independent of T. Hence there exist constants $C_{1,u}$ and $C_{i,f}$, which are independent of T, such that

$$\|u(\mathbf{x},\mathbf{z},\boldsymbol{\theta}_0) - u(\mathbf{q}_1,\mathbf{z},\boldsymbol{\theta}_0)\|_{2,[t_0,T]} \leq C_{1,u},$$
$$\|f_i(\mathbf{x},\boldsymbol{\theta}_i) - f_i(\mathbf{q}_1,\boldsymbol{\theta}_i)\|_{2,[t_0,T]} \leq C_{i,f},\quad i = 2,\ldots,k.$$

Thus the equation for ψ_{ξ_2} in (A5.32) can be written

$$\dot{\psi}_{\xi_2} = -\left(\left(\bar{F}_2^2(\mathbf{q}_1, \mathbf{q}_2, \mathbf{z}) + \sum_{j=3}^{k} \bar{D}_{j+1}^2(\mathbf{q}_1, \mathbf{q}_2)\right) + 1\right)\psi_{\xi_2}$$
$$- f_2(\mathbf{q}_1, \hat{\boldsymbol{\theta}}_{\xi_2}) + f_2(\mathbf{q}_1, \boldsymbol{\theta}_2) + \varepsilon_2(t),$$

where $\varepsilon_2(t) \in L_2[t_0, T]$, and $\|\varepsilon_2(t)\|_{2,[t_0,T]}$ can be bounded from above by a constant whose value does not depend on T.

According to Lemma A5.2 there are constants $C_{i,f,2}$, which are independent of T, such that the trajectories of (A5.31) satisfy

$$\|u(\mathbf{q}_1, \mathbf{z}, \boldsymbol{\theta}_0) - u(\mathbf{q}_2, \mathbf{z}, \boldsymbol{\theta}_0)\|_{2,[t_0,T]} \leq C_{2,u}$$
$$\|f_i(\mathbf{q}_1, \boldsymbol{\theta}_i) - f_i(\mathbf{q}_2, \boldsymbol{\theta}_i)\| \leq C_{i,f,2}, \quad i = 3, \ldots, k.$$

This implies that

$$\|f_3(\mathbf{x}, \boldsymbol{\theta}_3) - f_3(\mathbf{q}_2, \boldsymbol{\theta}_3)\|_{2,[t_0,T]} \leq C_{3,f} + C_{3,f,2}.$$

By following the same logic we can show that the error model for (A5.32), $2 \leq k \leq n$, can be expressed as

$$\dot{\psi}_{\xi_i} = -\left(\left(\bar{F}_i^2(\mathbf{q}_{i-1}, \mathbf{q}_i, \mathbf{z}) + \sum_{j=i+1}^{k} \bar{D}_{j+1}^2(\mathbf{q}_{i-1}, \mathbf{q}_i)\right) + 1\right)\psi_{\xi_i}$$
$$- f_i(\mathbf{q}_{i-1}, \hat{\boldsymbol{\theta}}_{\xi_i}) + f_i(\mathbf{q}_{i-1}, \boldsymbol{\theta}_i) + \varepsilon_i(t),$$

where $\varepsilon_i(t) \in L_2[t_0, T]$, and $\|\varepsilon_i(t)\|_{2,[t_0,T]}$ can be bounded from above by constants that are independent of T. Thus, invoking Lemma A5.2, one can conclude that there exist constants $C_{j,u}$, $C_{i,f,j}$, and $C_{i,x}$, which are independent of T, such that

$$\|u(\mathbf{q}_{j-1}, \mathbf{z}, \boldsymbol{\theta}_0) - u(\mathbf{q}_j, \mathbf{z}, \boldsymbol{\theta}_0)\|_{2,[t_0,T]} \leq C_{j,u},$$
$$\|f_i(\mathbf{q}_{j-1}, \boldsymbol{\theta}_i) - f_i(\mathbf{q}_j, \boldsymbol{\theta}_i)\|_{2,[t_0,T]} \leq C_{i,f,j},$$
$$\|x_i - \xi_i\|_{\infty,[t_0,T]} \leq C_{i,x},$$
$$\hat{\boldsymbol{\theta}}_{\xi_i} \in L_\infty[t_0, T], \quad i = j, \ldots, n.$$

This, in turn, implies that $\|u(\mathbf{x}, \mathbf{z}, \boldsymbol{\theta}_0) - u(\mathbf{q}_i, \mathbf{z}, \boldsymbol{\theta}_0)\|_{2,[t_0,T]} \leq \sum_{j=1}^{i} C_{j,u}$. The proof will be completed if we show that system (A5.31) can be realized in a form that does not involve knowledge or measurement of $\boldsymbol{\theta}$. In particular, we should seek a realization of

$$\dot{\hat{\boldsymbol{\theta}}}_{\xi_i} = \gamma_{\xi_i}\left(\psi_{\xi_i}\left(\left(\bar{F}_i^2(\mathbf{q}_{i-1},\mathbf{q}_i,\mathbf{z})\sum_{j=i}^{k}\bar{D}_{j+1}^2(\mathbf{q}_{i-1},\mathbf{q}_i)\right)+1\right)\dot{\psi}_{\xi_i}\right)\boldsymbol{\alpha}_i(\mathbf{q}_{i-1}), \tag{A5.34}$$

where $\gamma_{\xi_i} > 0$, $i = 1,\ldots,k$, that does not require knowledge of unknown $\boldsymbol{\theta}_i$.

Notice that the functions $f_i(\mathbf{x},\boldsymbol{\theta}_i)$, $i < n$ do not depend on x_j, $j = i+1,\ldots,n$. Therefore $\partial\alpha_i(\mathbf{x})/\partial x_j = 0$. Hence there exist finite-form realizations of (A5.34):[19]

$$\begin{aligned}
\hat{\boldsymbol{\theta}}_{\xi_i}(\mathbf{q}_{i-1},\xi_i,t) &= \gamma_{\xi_i}(\hat{\boldsymbol{\theta}}_{\xi_i,P}(\mathbf{q}_{i-1},\xi_i)+\hat{\boldsymbol{\theta}}_{\xi_i,I}(t)),\\
\hat{\boldsymbol{\theta}}_{\xi_i,P}(\mathbf{q}_{i-1},\xi_i) &= \psi_{\xi_i}(x_i,\xi_i)\boldsymbol{\alpha}_i(\mathbf{q}_{i-1}) - \Psi_{\xi_i}(\mathbf{q}_{i-1},\xi_i), \qquad \text{(A5.35)}\\
\dot{\hat{\boldsymbol{\theta}}}_{\xi_i,I} &= \left(\left(\bar{F}_i^2(\mathbf{q}_{i-1},\mathbf{q}_i,\mathbf{z})+\sum_{j=i}^{k}\bar{D}_{j+1}^2(\mathbf{q}_{i-1},\mathbf{q}_i)\right)+1\right)\psi_{\xi_i}(x_i,\xi_i)\boldsymbol{\alpha}_i(\mathbf{q}_{i-1})\\
&\quad+\sum_{j=1}^{i}\frac{\partial\Psi_{\xi_i}(\mathbf{q}_{i-1},\xi_i)}{\partial\xi_j}\dot{\xi}_j - \sum_{j=1}^{i-1}\psi_{\xi_i}(x_i,\xi_i)\frac{\partial\boldsymbol{\alpha}_i(\mathbf{q}_{i-1})}{\partial\xi_j}\dot{\xi}_j,\\
\Psi_{\xi_i}(\mathbf{q}_{i-1},\xi_i) &= \int_{x_i(0)}^{x_i(t)}\psi_{\xi_i}(x_i,\xi_i)\frac{\partial\boldsymbol{\alpha}_i(\mathbf{q}_{i-1})}{\partial x_i}dx_i,
\end{aligned}$$

where $\gamma_{\xi_i} > 0$. Notice also that $\mathbf{x} \in L_\infty[t_0,T]$ implies $\boldsymbol{\xi} \in L_\infty[t_0,T]$, and hence $\hat{\boldsymbol{\theta}}_{\xi_i,P}(\mathbf{q}_{i-1},\xi_i) \in L_\infty[t_0,T]$, provided that $\hat{\boldsymbol{\theta}}_{\xi_i,P}(\mathbf{q}_{i-1},\xi_i)$ is smooth (locally bounded). Given that $\hat{\boldsymbol{\theta}}_{\xi_i} = \gamma_{\xi_i}(\hat{\boldsymbol{\theta}}_{\xi_i,P}(\mathbf{q}_{i-1},\xi_i)+\hat{\boldsymbol{\theta}}_{\xi_i,I})$ and $\hat{\boldsymbol{\theta}}_{\xi_i}$, $\hat{\boldsymbol{\theta}}_{\xi_i,P}(\mathbf{q}_{i-1},\xi_i) \in L_\infty[t_0,T]$, we obtain $\hat{\boldsymbol{\theta}}_{\xi_i,I} \in L_\infty[t_0,T]$ for $\mathbf{x} \in L_\infty[t_0,T]$. Denoting $\boldsymbol{\nu} = \hat{\boldsymbol{\theta}}_{\xi_i,I}$, one can easily see that the system (A5.31) and (A5.35) can straightforwardly be transformed into system (5.64) satisfying conditions (1)–(3) of the lemma. □

A5.6 Proof of Theorem 5.4

The proof will carried out by induction w.r.t. the number of equations in (5.61).

Basis of induction. Let the original system be described by an ordinary differential equation of first order:

$$\dot{x}_1 = f_1(x_1,\boldsymbol{\theta}_1) + u + \varepsilon_1(t),\ \varepsilon_1(t) \in L_2[t_0,\infty].$$

[19] The key factor here is that the function $\boldsymbol{\alpha}_i(\mathbf{q}_{i-1})$ does not depend on any components of the state vector, except probably x_i. All other components $x_1,\ldots,x_{i-1}$ are replaced with ξ_j, $j = 1,\ldots,i-1$, respectively. Their derivatives do not depend on $\boldsymbol{\theta}_1,\ldots,\boldsymbol{\theta}_i$.

According to the assumptions of the theorem the functions $f_1(x_1,\boldsymbol{\theta}_1)$ and $\boldsymbol{\alpha}_1(x_1)$ are smooth. Let the function u be chosen such that

$$\begin{aligned} u(x_1,\hat{\theta}_1) &= -f_1(x_1,\hat{\boldsymbol{\theta}}_1) - \varphi_1(\psi(x_1)), \\ \hat{\boldsymbol{\theta}}_1 &= \gamma_1(\hat{\boldsymbol{\theta}}_{1,P}(x_1) + \hat{\boldsymbol{\theta}}_{1,I}(t)),\ \gamma_1 > 0, \\ \hat{\boldsymbol{\theta}}_{1,P}(x_1) &= \psi(x_1)\boldsymbol{\alpha}_1(x_1) - \Psi(x_1), \\ \Psi(x_1) &= \int_{x_1(0)}^{x_1(t)} \psi(x_1)\frac{\partial \boldsymbol{\alpha}_1(x_1)}{\partial x_1}dx_1, \\ \dot{\hat{\boldsymbol{\theta}}}_{1,I} &= \varphi_1(\psi(x_1))\boldsymbol{\alpha}_1(x_1), \end{aligned} \tag{A5.36}$$

where the function φ satisfies the assumptions of Lemma A5.2. The function u is smooth. Thus solutions of the system are defined on some non-empty interval $[t_0,T]$, $T > t_0$. Let us denote $\psi_1(x_1,t) = \psi(x_1)$. According to Lemma A5.2 the properties $\psi_1(x_1(t),t) \in L_2[t_0,T] \cap L_\infty[t_0,T]$ and $\dot{\psi}_1 \in L_2[t_0,T]$ hold[20] along the solutions of

$$\dot{x}_1 = f_1(x_1,\boldsymbol{\theta}_1) + u(x_1,\hat{\boldsymbol{\theta}}_1) + \varepsilon_1(t),\ \varepsilon_1(t) \in L_2[t_0,\infty]. \tag{A5.37}$$

Given that the function u is smooth, and that $\|x_1\|_{\infty,[t_0,T]}$ and $\|\hat{\boldsymbol{\theta}}_1\|_{\infty,[t_0,T]}$ can be bounded from above by constants that do not depend on T, we can conclude that the solutions of (A5.37) and (A5.36) are defined for all $t \geq t_0$, and properties (1)–(4) of the theorem hold. This proves the basis of induction.

Inductive hypothesis. Let the theorem hold for systems in which $\dim(\mathbf{x}) = i > 1$. This implies that there exist $u_i(\mathbf{x}_i,\hat{\boldsymbol{\theta}}_i,\boldsymbol{\xi}_i,\boldsymbol{\nu}_i)$, $\mathbf{x}_i,\boldsymbol{\xi}_i \in \mathbb{R}^i$, $\mathbf{x}_i = (x_1,\ldots,x_i)^{\mathrm{T}}$, $\boldsymbol{\xi}_i = (\xi_1,\ldots,\xi_i)^{\mathrm{T}}$, and functions $\psi_j(x_j(t),t)$, $j = 1,\ldots,i$ such that $\psi_j(x_j(t),t) \in L_2 \cap L_\infty$ and $\dot{\psi}_j \in L_2$ for (5.61) with $\dim(\mathbf{x}) \leq i$:

$$\begin{aligned} \dot{x}_j &= f_j(x_1,\ldots,x_j,\boldsymbol{\theta}_j) + x_{j+1},\ j \in \{1,\ldots,i-1\}, \\ \dot{x}_i &= f_i(x_1,\ldots,x_i,\boldsymbol{\theta}_i) + u_i + \varepsilon_i(t),\ \varepsilon_i(t) \in L_2. \end{aligned} \tag{A5.38}$$

The theorem will be proven if we can show that the same properties hold for systems (5.61) with $\dim(\mathbf{x}) = i+1$, provided that the theorem holds for systems (A5.38).

Inductive step. According to the inductive hypothesis the function u_i is smooth. Therefore, Hadamard's lemma assures the existence of $F(\mathbf{x}_i,\mathbf{x}'_i,\hat{\boldsymbol{\theta}}_i,\boldsymbol{\xi}_i,\boldsymbol{\nu}_i)$ such that $u_i(\mathbf{x}_i,\hat{\boldsymbol{\theta}}_i,\boldsymbol{\xi}_i,\boldsymbol{\nu}_i) - u_i(\mathbf{x}'_i,\hat{\boldsymbol{\theta}}_i,\boldsymbol{\xi}_i,\boldsymbol{\nu}_i) = F(\mathbf{x}_i,\mathbf{x}'_i,\hat{\boldsymbol{\theta}}_i,\boldsymbol{\xi}_i,\boldsymbol{\nu}_i)(\mathbf{x}_i - \mathbf{x}'_i)$. Let us denote $\bar{F}^2_{i+1}(\mathbf{x}_i,\mathbf{x}'_i,\hat{\boldsymbol{\theta}}_i,\boldsymbol{\xi}_i,\boldsymbol{\nu}_i) = \|F(\mathbf{x}_i,\mathbf{x}'_i,\hat{\boldsymbol{\theta}}_i,\boldsymbol{\xi}_i,\boldsymbol{\nu}_i)\|^2$. According to the conditions of the theorem the functions $f_j(\mathbf{x}_j,\boldsymbol{\theta}_j)$, $j = 1,\ldots,i+1$ satisfy the following inequalities:

$$(f_j(\mathbf{x}_j,\boldsymbol{\theta}_j) - f_j(\mathbf{x}'_j,\boldsymbol{\theta}_j))^2 \leq \|\mathbf{x}_j - \mathbf{x}'_j\|^2 \bar{D}^2_j(\mathbf{x}_j,\mathbf{x}'_j)\ \forall\, \boldsymbol{\theta}_j \in \Omega_\theta.$$

20 In order to see this one needs to derive $\dot{\hat{\theta}}_1$ and apply Lemma A5.2 to the resulting system (A5.37).

Thus Lemma 5.1 guarantees the existence of the following system of differential equations:

$$\dot{\boldsymbol{\xi}}_{i+1} = \mathbf{f}_{\xi_{i+1}}(\boldsymbol{\xi}_{i+1}, \mathbf{x}, \mathbf{z}, \boldsymbol{\nu}_{i+1}), \quad \boldsymbol{\xi} \in \mathbb{R}^i,$$
$$\dot{\boldsymbol{\nu}}_{i+1} = \mathbf{f}_{\nu_{i+1}}(\boldsymbol{\xi}_{i+1}, \mathbf{x}, \mathbf{z}), \quad \mathbf{z} = \hat{\boldsymbol{\theta}}_i \oplus \boldsymbol{\xi}_i \oplus \boldsymbol{\nu}_i,$$

such that

$$u_i(\mathbf{x}_i, \hat{\boldsymbol{\theta}}_i(\mathbf{x}_i, \boldsymbol{\xi}_i, \hat{\boldsymbol{\theta}}_{I,i}), \boldsymbol{\xi}_i, \boldsymbol{\nu}_i) - u_i(\boldsymbol{\xi}_{i+1}, \hat{\boldsymbol{\theta}}_i(\boldsymbol{\xi}_{i+1}, \boldsymbol{\xi}_i, \hat{\boldsymbol{\theta}}_{I,i}), \boldsymbol{\xi}_i, \boldsymbol{\nu}_i) \in L_2[t_0, T],$$
$$f_{i+1}(\mathbf{x}_{i+1}, \boldsymbol{\theta}_{i+1}) - f_{i+1}(\boldsymbol{\xi}_{i+1} \oplus x_{i+1}, \boldsymbol{\theta}_{i+1}) \in L_2[t_0, T],$$

where $[t_0, T]$, $T \geq t_0$ is an interval of existence of the solution $\mathbf{x}(t)$. Notice that the L_2-norms can be bounded from above by a constant whose value does not depend on T.

Let us introduce the new goal function

$$\psi_{i+1}(x_{i+1}, \boldsymbol{\xi}_{i+1}, \boldsymbol{\xi}_i, \boldsymbol{\nu}_i, \hat{\boldsymbol{\theta}}_{I,i}) = x_{i+1} - u_i(\boldsymbol{\xi}_{i+1}, \hat{\boldsymbol{\theta}}_i(\boldsymbol{\xi}_{i+1}, \boldsymbol{\xi}_i, \hat{\boldsymbol{\theta}}_{I,i}), \boldsymbol{\xi}_i, \boldsymbol{\nu}_i)$$

and consider its derivative $\dot{\psi}_{i+1}$:

$$\begin{aligned}\dot{\psi}_{i+1} = {} & f_{i+1}(\mathbf{x}_{i+1}, \boldsymbol{\theta}_{i+1}) + u_{i+1} \\ & - L_{\mathbf{f}_{\xi_i}} u_i(\boldsymbol{\xi}_{i+1}, \hat{\boldsymbol{\theta}}_i(\boldsymbol{\xi}_{i+1}, \boldsymbol{\xi}_i, \hat{\boldsymbol{\theta}}_{I,i}), \boldsymbol{\xi}_i, \boldsymbol{\nu}_i) \\ & - L_{\mathbf{f}_{\nu_i}} u_i(\boldsymbol{\xi}_{i+1}, \hat{\boldsymbol{\theta}}_i(\boldsymbol{\xi}_{i+1}, \boldsymbol{\xi}_i, \hat{\boldsymbol{\theta}}_{I,i}), \boldsymbol{\xi}_i, \boldsymbol{\nu}_i) \\ & - L_{\mathbf{f}_{\xi_{i+1}}} u_i(\boldsymbol{\xi}_{i+1}, \hat{\boldsymbol{\theta}}_i(\boldsymbol{\xi}_{i+1}, \boldsymbol{\xi}_i, \hat{\boldsymbol{\theta}}_{I,i}), \boldsymbol{\xi}_i, \boldsymbol{\nu}_i) \\ & - L_{\mathbf{f}_{\hat{\theta}_i}} u_i(\boldsymbol{\xi}_{i+1}, \hat{\boldsymbol{\theta}}_i(\boldsymbol{\xi}_{i+1}, \boldsymbol{\xi}_i, \hat{\boldsymbol{\theta}}_{I,i}), \boldsymbol{\xi}_i, \boldsymbol{\nu}_i).\end{aligned} \tag{A5.39}$$

Denoting $\varepsilon_{i+1}(t) = f_{i+1}(\mathbf{x}_{i+1}, \boldsymbol{\theta}_{i+1}) - f_{i+1}(\boldsymbol{\xi}_{i+1} \oplus x_{i+1}, \boldsymbol{\theta}_{i+1})$, we can rewrite (A5.39) as

$$\begin{aligned}\dot{\psi}_{i+1} = {} & \varepsilon_{i+1}(t) + f_{i+1}(\boldsymbol{\xi}_{i+1} \oplus x_{i+1}, \boldsymbol{\theta}_{i+1}) \\ & + u_{i+1} - L_{\mathbf{f}_{\xi_i}} u_i(\boldsymbol{\xi}_{i+1}, \hat{\boldsymbol{\theta}}_i(\boldsymbol{\xi}_{i+1}, \boldsymbol{\xi}_i, \hat{\boldsymbol{\theta}}_{I,i}), \boldsymbol{\xi}_i, \boldsymbol{\nu}_i) \\ & - L_{\mathbf{f}_{\nu_i}} u_i(\boldsymbol{\xi}_{i+1}, \hat{\boldsymbol{\theta}}_i(\boldsymbol{\xi}_{i+1}, \boldsymbol{\xi}_i, \hat{\boldsymbol{\theta}}_{I,i}), \boldsymbol{\xi}_i, \boldsymbol{\nu}_i) \\ & - L_{\mathbf{f}_{\xi_{i+1}}} u_i(\boldsymbol{\xi}_{i+1}, \hat{\boldsymbol{\theta}}_i(\boldsymbol{\xi}_{i+1}, \boldsymbol{\xi}_i, \hat{\boldsymbol{\theta}}_{I,i}), \boldsymbol{\xi}_i, \boldsymbol{\nu}_i) \\ & - L_{\mathbf{f}_{\hat{\theta}_i}} u_i(\boldsymbol{\xi}_{i+1}, \hat{\boldsymbol{\theta}}_i(\boldsymbol{\xi}_{i+1}, \boldsymbol{\xi}_i, \hat{\boldsymbol{\theta}}_{I,i}), \boldsymbol{\xi}_i, \boldsymbol{\nu}_i).\end{aligned} \tag{A5.40}$$

We set

$$\begin{aligned} u_{i+1} = &-\varphi_{i+1}(\psi_{i+1}(x_{i+1}, \boldsymbol{\xi}_{i+1}, \boldsymbol{\xi}_i, \boldsymbol{\nu}_i, \hat{\boldsymbol{\theta}}_{I,i})) \\ &+ L_{\mathbf{f}_{\xi_i}} u_i(\boldsymbol{\xi}_{i+1}, \hat{\boldsymbol{\theta}}_i(\boldsymbol{\xi}_{i+1}, \boldsymbol{\xi}_i, \hat{\boldsymbol{\theta}}_{I,i}), \boldsymbol{\xi}_i, \boldsymbol{\nu}_i) \\ &+ L_{\mathbf{f}_{\nu_i}} u_i(\boldsymbol{\xi}_{i+1}, \hat{\boldsymbol{\theta}}_i(\boldsymbol{\xi}_{i+1}, \boldsymbol{\xi}_i, \hat{\boldsymbol{\theta}}_{I,i}), \boldsymbol{\xi}_i, \boldsymbol{\nu}_i) \\ &+ L_{\mathbf{f}_{\xi_{i+1}}} u_i(\boldsymbol{\xi}_{i+1}, \hat{\boldsymbol{\theta}}_i(\boldsymbol{\xi}_{i+1}, \boldsymbol{\xi}_i, \hat{\boldsymbol{\theta}}_{I,i}), \boldsymbol{\xi}_i, \boldsymbol{\nu}_i) \\ &+ L_{\mathbf{f}_{\hat{\theta}_i}} u_i(\boldsymbol{\xi}_{i+1}, \hat{\boldsymbol{\theta}}_i(\boldsymbol{\xi}_{i+1}, \boldsymbol{\xi}_i, \hat{\boldsymbol{\theta}}_{I,i}), \boldsymbol{\xi}_i, \boldsymbol{\nu}_i) \\ &- f_{i+1}(\boldsymbol{\xi}_{i+1} \oplus x_{i+1}, \hat{\boldsymbol{\theta}}_{i+1}). \end{aligned} \tag{A5.41}$$

On letting, for the sake of notational compactness,

$$\psi_{i+1}(x_{i+1}, \boldsymbol{\xi}_{i+1}(t), \boldsymbol{\xi}_i(t), \boldsymbol{\nu}_i(t), \hat{\boldsymbol{\theta}}_{I,i}(t)) = \psi_{i+1}(x_{i+1}, t)$$

(where $\boldsymbol{\xi}_{i+1}(t), \boldsymbol{\xi}_i(t), \boldsymbol{\nu}_i(t)$, and $\hat{\boldsymbol{\theta}}_{I,i}(t)$ are treated as functions of t) and substituting (A5.41) and (A5.40) into (A5.39) we obtain

$$\begin{aligned} \dot{\psi}_{i+1} = &-\varphi_{i+1}(\psi_{i+1}(x_{i+1}, t)) + f_{i+1}(\boldsymbol{\xi}_{i+1} \oplus x_{i+1}, \boldsymbol{\theta}_{i+1}) \\ &- f_{i+1}(\boldsymbol{\xi}_{i+1} \oplus x_{i+1}, \hat{\boldsymbol{\theta}}_{i+1}) + \varepsilon_{i+1}(t). \end{aligned} \tag{A5.42}$$

It follows from the conditions of the theorem that there is a function $\boldsymbol{\alpha}_{i+1}(\boldsymbol{\xi}_{i+1} \oplus x_{i+1})$ satisfying Assumptions 5.3 and 5.4 for the function $f_{i+1}(\boldsymbol{\xi}_{i+1} \oplus x_{i+1}, \boldsymbol{\theta}_{i+1})$.

Finally, consider the following equations for $\hat{\boldsymbol{\theta}}_{i+1}$:

$$\dot{\hat{\boldsymbol{\theta}}}_{i+1} = \gamma_{i+1}(\dot{\psi}_{i+1} + \varphi_{i+1}(\psi_{i+1}(x_{i+1}, t)))\boldsymbol{\alpha}_{i+1}(\boldsymbol{\xi}_{i+1} \oplus x_{i+1}), \quad \gamma_{i+1} \in \mathbb{R}_{>0}. \tag{A5.43}$$

As has already been demonstrated by (A5.35), algorithm (A5.43) can be realized in an integro-differential form that does not require knowledge/measurement of $\dot{\psi}_{i+1}$:

$$\begin{aligned} \hat{\boldsymbol{\theta}}_{i+1}(\boldsymbol{\xi}_{i+1} \oplus x_{i+1}, t) &= \gamma_{i+1}(\hat{\boldsymbol{\theta}}_{i+1,P}(\boldsymbol{\xi}_{i+1} \oplus x_{i+1}, t) + \hat{\boldsymbol{\theta}}_{i+1,I}(t)), \ \gamma_{i+1} > 0, \\ \hat{\boldsymbol{\theta}}_{i+1,P}(\boldsymbol{\xi}_{i+1} \oplus x_{i+1}, t) &= \psi_{i+1}(x_{i+1}, t)\boldsymbol{\alpha}_{i+1}(\boldsymbol{\xi}_{i+1} \oplus x_{i+1}) - \Psi_{i+1}(\boldsymbol{\xi}_{i+1} \oplus x_{i+1}, t); \\ \dot{\hat{\boldsymbol{\theta}}}_{i+1,I} &= \varphi_{i+1}(\psi(x_{i+1}, t))\boldsymbol{\alpha}_{i+1}(\boldsymbol{\xi}_{i+1} \oplus x_{i+1}) \\ &\quad - L_{\mathbf{f}_\xi}\boldsymbol{\alpha}_{i+1}(\boldsymbol{\xi}_{i+1} \oplus x_{i+1}) \\ &\quad + L_{\mathbf{f}_\xi}\Psi_{i+1}(\boldsymbol{\xi}_{i+1} \oplus x_{i+1}, t) + \partial\Psi_{i+1}(\boldsymbol{\xi}_{i+1} \oplus x_{i+1}, t)/\partial t, \end{aligned} \tag{A5.44}$$

where

$$\Psi_{i+1}(\boldsymbol{\xi}_{i+1} \oplus x_{i+1}, t) = \int_{x_{i+1}(0)}^{x_{i+1}(t)} \psi_{i+1}(x_{i+1}, t)\frac{\partial\boldsymbol{\alpha}_{i+1}(\boldsymbol{\xi}_{i+1} \oplus x_{i+1})}{\partial x_{i+1}} dx_{i+1}.$$

According to Lemma A5.2 the following properties hold for (A5.42), (A5.43), and (A5.44): $\hat{\boldsymbol{\theta}}_{i+1}(\boldsymbol{\xi}_{i+1} \oplus x_{i+1}, t) \in L_\infty[t_0, T]$, $\dot{\psi}_{i+1} \in L_2[t_0, T]$, and $\varphi_{i+1}(\psi_{i+1}(x_{i+1}, t)) \in L_2[t_0, T] \cap L_\infty[t_0, T]$, where the corresponding norms can be bounded from above by constants that are independent of T. Moreover, $\varepsilon_{i+1}(t) \in L_2[t_0, T]$ implies that

$$
\begin{aligned}
&u_{i+1}(\mathbf{x}_i, \hat{\boldsymbol{\theta}}_{i+1}, \boldsymbol{\xi}_{i+1}, \boldsymbol{\xi}_i, \boldsymbol{v}_{i+1}, \boldsymbol{v}_i, \hat{\boldsymbol{\theta}}_{I,i}) \\
&\quad - u_{i+1}(\mathbf{x}_i, \boldsymbol{\theta}_{i+1}, \boldsymbol{\xi}_{i+1}, \boldsymbol{\xi}_i, \boldsymbol{v}_{i+1}, \boldsymbol{v}_i, \hat{\boldsymbol{\theta}}_{I,i}) \in L_2[t_0, T].
\end{aligned}
$$

Finally, let us denote

$$
u_{i+1}(\mathbf{x}_i, \hat{\boldsymbol{\theta}}_{i+1}, \boldsymbol{\xi}_{i+1}, \tilde{\boldsymbol{v}}_{i+1}) = u_{i+1}(\mathbf{x}_i, \hat{\boldsymbol{\theta}}_{i+1}, \boldsymbol{\xi}_{i+1}, \boldsymbol{\xi}_i, \boldsymbol{v}_{i+1}, \boldsymbol{v}_i, \hat{\boldsymbol{\theta}}_{I,i}),
$$

with $\tilde{\boldsymbol{v}}_{i+1} = \boldsymbol{\xi}_i \oplus \boldsymbol{v}_{i+1} \oplus \boldsymbol{v}_i \oplus \hat{\boldsymbol{\theta}}_{I,i}$, and choose

$$
\varphi_{i+1}(\cdot): \ |\varphi_{i+1}(\cdot)| \geq k_{k+1}|\psi_{i+1}|, \ k_{i+1} > 0.
$$

This guarantees that $\psi_{i+1}(x_{i+1}, t) \in L_2[t_0, T] \cap L_\infty[t_0, T]$. Hence, according to the inductive hypothesis, $\psi_k(x_k(t), t) \in L_2[t_0, T] \cap L_\infty[t_0, T]$ and $\dot{\psi}_k \in L_2[t_0, T]$ for all $k = 1, \ldots, i$, $\psi \in L_2[t_0, T] \cap L_\infty[t_0, T]$, and $\dot{\psi} \in L_2[t_0, T]$. Given that (1) the corresponding norms can be bounded from above by constants whose values do not depend on T, and (2) the right-hand side of the resulting system is continuous and locally bounded, one can conclude that solutions of the overall system are defined for all $t \geq t_0$, provided that $\|x_{i+1}\|_{\infty,[t_0,T]}$ has an upper bound that does not depend on T.

In order to see that this is indeed the case consider

$$
\varepsilon_i(t) = u_i(\mathbf{x}_i, \hat{\boldsymbol{\theta}}_i(\mathbf{x}_i, \boldsymbol{\xi}_i, \hat{\boldsymbol{\theta}}_{I,i}), \boldsymbol{\xi}_i, \boldsymbol{v}_i) - u_i(\boldsymbol{\xi}_{i+1}, \hat{\boldsymbol{\theta}}_i(\boldsymbol{\xi}_{i+1}, \boldsymbol{\xi}_i, \hat{\boldsymbol{\theta}}_{I,i}), \boldsymbol{\xi}_i, \boldsymbol{v}_i).
$$

Given that the function u_i is smooth, the sup-norm of this difference admits an upper bound that does not depend on T. On the other hand, we have just shown that the sup-norm for

$$
\psi_{i+1} = x_{i+1} - u_i(\boldsymbol{\xi}_{i+1}, \hat{\boldsymbol{\theta}}_i(\boldsymbol{\xi}_{i+1}, \boldsymbol{\xi}_i, \hat{\boldsymbol{\theta}}_{I,i}), \boldsymbol{\xi}_i, \boldsymbol{v}_i)
$$

can be bounded from above by a constant that does not depend on T. Hence $\|x_{i+1}\|_{\infty,[t_0,T]}$ has an upper bound that does not depend on T, and thus statements (1), (2), and (3) of the theorem hold.

The derivatives $\dot{\psi}_j$, $j = 1, \ldots, i$ are bounded because $\varepsilon_i(t)$ are bounded (according to the inductive hypothesis the theorem holds for $j = 1, \ldots, i$). If the function $\varepsilon(t)$ is bounded then $\dot{\psi}_{i+1}$ is bounded as well. Moreover $u_{i+1}(\cdot)$ and $f_{i+1}(\cdot)$ are smooth, and $\mathbf{x}_{i+1}, \boldsymbol{\xi}_{i+1}, \boldsymbol{v}_{i+1}$, and $\hat{\boldsymbol{\theta}}_{i+1}$ are bounded. Then, as follows from Lemma A5.2, $\psi_{i+1} \to 0$ as $t \to \infty$. Hence statement (4) is proven. □

A5.7 Proof of Corollary 5.2

First we notice that Theorem 5.1 and the bounded-input–bounded-output property of (5.98) guarantee that the state of the combined system is bounded. Thus, according to statements (3) and (4) of Theorem 5.1 the following property holds:

$$\lim_{t\to\infty} f(\mathbf{x}(t),\boldsymbol{\theta},\boldsymbol{\zeta}(t)) - f(\mathbf{x}(t),\hat{\boldsymbol{\theta}}(t),\boldsymbol{\zeta}(t)) = 0. \tag{A5.45}$$

In this case Lemma 4.1 applies, and we can conclude that an ω-limit set of the system exists. Continuity of $f(\cdot)$ and property (A5.45) taken together imply that every limit point $p = \mathbf{x}^p \oplus \hat{\boldsymbol{\theta}}^p \oplus \boldsymbol{\zeta}^p$ from this ω-limit set satisfies

$$f(\mathbf{x}^p,\boldsymbol{\theta},\boldsymbol{\zeta}^p) - f(\mathbf{x}^p,\hat{\boldsymbol{\theta}}^p,\boldsymbol{\zeta}^p) = 0.$$

Therefore, the ω-limit set of the combined system is contained in (5.99). According to Lemma 4.1, solutions of the system converge asymptotically to the ω-limit set. The latter set is invariant, and it is contained in (5.99). Hence, solutions of the combined system must necessarily converge to the maximal invariant set in (5.99). □

A5.8 Proof of Theorem 5.5

Consider the following dynamic feedback:

$$\mathbf{u}(\mathbf{x},\hat{\boldsymbol{\theta}}) = \mathbf{u}_0(\mathbf{x}) - \boldsymbol{\phi}(\boldsymbol{\xi})\hat{\boldsymbol{\theta}}(t).$$

This feedback transforms (5.101) into

$$\dot{\mathbf{x}} = \mathbf{f}_0(\mathbf{x}) + G_u\boldsymbol{\phi}(\boldsymbol{\xi})(\boldsymbol{\theta} - \hat{\boldsymbol{\theta}}(t)) + G_u(\boldsymbol{\phi}(\mathbf{x}) - \boldsymbol{\phi}(\boldsymbol{\xi}))\boldsymbol{\theta}. \tag{A5.46}$$

Let us denote $G_u\boldsymbol{\phi}(\mathbf{x}) = \boldsymbol{\alpha}(\mathbf{x})$, and consider

$$\begin{aligned}
\dot{\mathbf{x}} &= \mathbf{f}_0(\mathbf{x}) + \boldsymbol{\alpha}(\boldsymbol{\xi})(\boldsymbol{\theta} - \hat{\boldsymbol{\theta}}) + \boldsymbol{\varepsilon}(t),\\
\dot{\boldsymbol{\theta}} &= S(\boldsymbol{\theta}),\\
\dot{\hat{\boldsymbol{\theta}}} &= S(\hat{\boldsymbol{\theta}}) + H^{-1}(\kappa^2(\boldsymbol{\xi}) + 1)\boldsymbol{\alpha}(\boldsymbol{\xi})^{\mathrm{T}}(\boldsymbol{\alpha}(\boldsymbol{\xi})(\boldsymbol{\theta} - \hat{\boldsymbol{\theta}}) + \boldsymbol{\varepsilon}(t)),\\
&\kappa(\boldsymbol{\xi}) : \mathbb{R}^n \to \mathbb{R},\ \kappa \in \mathcal{C}^1,\\
&\boldsymbol{\xi},\boldsymbol{\varepsilon} : [t_0,T] \to \mathbb{R}^n,\ \boldsymbol{\xi} \in \mathcal{C}^1,\ T > t_0.
\end{aligned} \tag{A5.47}$$

Lemma A5.2 *Consider system (A5.47) and suppose that Assumptions 5.11–5.14 hold. In addition, let $\|\kappa(\boldsymbol{\xi}(t))\boldsymbol{\varepsilon}(t)\|_{2,[t_0,T]}$ and $\|\boldsymbol{\varepsilon}(t)\|_{2,[t_0,T]}$ be bounded from above by constants that do not depend on T.*

Then

(1) *there exists* C_θ, *which is independent of* T, *such that* $\|\hat{\boldsymbol{\theta}}(t)\|_{\infty,[t_0,T]} \leq C_\theta$, *provided that* $\boldsymbol{\theta}(t_0) \in \Omega_\theta$ *and* $\hat{\boldsymbol{\theta}}(t_0) \in \mathbb{R}^d$;

(2) *there exist* C_κ *and* C_α, *which are independent of* T, *such that*

$$\|\kappa(\boldsymbol{\xi})\boldsymbol{\alpha}(\boldsymbol{\xi})(\hat{\boldsymbol{\theta}}(t) - \boldsymbol{\theta}(t))\|_{2,[t_0,T]} \leq C_\kappa,$$

$$\|\boldsymbol{\alpha}(\boldsymbol{\xi})(\hat{\boldsymbol{\theta}}(t) - \boldsymbol{\theta}(t))\|_{2,[t_0,T]} \leq C_\alpha;$$

(3) *if in addition* $\|\partial\psi(\mathbf{x})/\partial\mathbf{x}\| \leq |\kappa(\mathbf{x})|$, *and there exists* C_x, *which is independent of* T, *such that* $\|\mathbf{x}(t) - \boldsymbol{\xi}(t)\|_{\infty,[t_0,T]} \leq C_x$, *then* $\mathbf{x} \in L_\infty[t_0, T]$; *the* L_∞*-norm of* $\mathbf{x}(t)$ *can be bounded from above by a constant that is independent of* T;

(4) *if, independently of condition (3),* $\boldsymbol{\varepsilon}(t) \equiv 0$, *the function* $\boldsymbol{\xi}$ *is defined for all* $t \geq t_0$ *and bounded,* $S(\boldsymbol{\theta}) \equiv 0$, *and* $\alpha(\boldsymbol{\xi})$ *is persistently exciting,*

$$\exists\, L, \delta \geq 0 : \int_t^{t+L} \alpha(\boldsymbol{\xi}(\tau))^{\mathrm{T}}\alpha(\boldsymbol{\xi}(\tau)) \geq \delta I_d \;\forall\, t \geq t_0,$$

then trajectories $\hat{\boldsymbol{\theta}}(t)$ *converge to* $\boldsymbol{\theta}_0$ *exponentially fast.*

Proof of Lemma A5.2. Let us start by showing that statements (1) and (2) hold. Consider the following function:

$$V_\theta(\boldsymbol{\theta}, \hat{\boldsymbol{\theta}}, t) = \|\boldsymbol{\theta} - \hat{\boldsymbol{\theta}}\|_H^2 + \epsilon = (\boldsymbol{\theta} - \hat{\boldsymbol{\theta}})^{\mathrm{T}} H (\boldsymbol{\theta} - \hat{\boldsymbol{\theta}}) + \epsilon,$$

where $\epsilon(t) = \frac{1}{2}\int_t^{\mathrm{T}} (\kappa^2(\boldsymbol{\xi}(\tau)) + 1)\boldsymbol{\varepsilon}^{\mathrm{T}}(\tau)\boldsymbol{\varepsilon}(\tau)d\tau \geq 0$. According to the assumptions of the lemma, the norm $\|\kappa(\boldsymbol{\xi}(t))\boldsymbol{\varepsilon}(t)\|_{2,[t_0,T]}$ can be bounded from above by a constant that is independent of T. This implies that there is a constant C_ϵ, which is independent of T, such that $\epsilon(t) < C_\epsilon$ for all $t \in [t_0, T]$. Consider the derivative $\dot{V}_\theta$:

$$\begin{aligned} \dot{V}_\theta &= (\boldsymbol{\theta} - \hat{\boldsymbol{\theta}})^{\mathrm{T}} H (S(\boldsymbol{\theta}) - S(\hat{\boldsymbol{\theta}})) + (S(\boldsymbol{\theta}) - S(\hat{\boldsymbol{\theta}}))^{\mathrm{T}} H (\boldsymbol{\theta} - \hat{\boldsymbol{\theta}}) \\ &\quad - 2(\kappa^2(\boldsymbol{\xi}) + 1)\Big((\boldsymbol{\theta} - \hat{\boldsymbol{\theta}})^{\mathrm{T}} \boldsymbol{\alpha}^{\mathrm{T}}(\boldsymbol{\xi})\boldsymbol{\alpha}(\boldsymbol{\xi})(\boldsymbol{\theta} - \hat{\boldsymbol{\theta}}) \\ &\qquad + (\boldsymbol{\theta} - \hat{\boldsymbol{\theta}})^{\mathrm{T}} \boldsymbol{\alpha}^{\mathrm{T}}(\boldsymbol{\xi})\boldsymbol{\varepsilon}(t) + \frac{\|\boldsymbol{\varepsilon}(t)\|^2}{4} \Big) \\ &= (\boldsymbol{\theta} - \hat{\boldsymbol{\theta}})^{\mathrm{T}} H (S(\boldsymbol{\theta}) - S(\hat{\boldsymbol{\theta}})) + (S(\boldsymbol{\theta}) - S(\hat{\boldsymbol{\theta}}))^{\mathrm{T}} H (\boldsymbol{\theta} - \hat{\boldsymbol{\theta}}) \\ &\quad - 2(\kappa^2(\boldsymbol{\xi}) + 1)\|(\boldsymbol{\theta} - \hat{\boldsymbol{\theta}})^{\mathrm{T}} \boldsymbol{\alpha}^{\mathrm{T}}(\boldsymbol{\xi}) + 0.5\boldsymbol{\varepsilon}(t)\|^2. \end{aligned} \tag{A5.48}$$

The function $S(\cdot) \in C^1$, hence Hadamard's lemma applies: $S(\boldsymbol{\theta}) - S(\hat{\boldsymbol{\theta}}) = \int_0^1 [\partial S(\mathbf{z}(\lambda))/\partial\mathbf{z}(\lambda)]d\lambda(\boldsymbol{\theta} - \hat{\boldsymbol{\theta}})$, $\mathbf{z}(\lambda) = \boldsymbol{\theta}\lambda + \hat{\boldsymbol{\theta}}(1 - \lambda)$. Using the mean-value theorem, we therefore obtain that

$$S(\boldsymbol{\theta}) - S(\hat{\boldsymbol{\theta}}) = \frac{\partial S(\mathbf{z})}{\partial \mathbf{z}}(\boldsymbol{\theta} - \hat{\boldsymbol{\theta}})$$

for some $\mathbf{z} \in \mathbb{R}^d$. The last inequality implies that

$$\begin{aligned}\dot{V}_\theta &= (\boldsymbol{\theta} - \hat{\boldsymbol{\theta}})^{\mathrm{T}} \left(\frac{\partial S(\mathbf{z})}{\partial \mathbf{z}}^{\mathrm{T}} H + H \frac{\partial S(\mathbf{z})}{\partial \mathbf{z}} \right) (\boldsymbol{\theta} - \hat{\boldsymbol{\theta}}) \\ &\quad - 2(\kappa^2(\boldsymbol{\xi}) + 1) \|(\boldsymbol{\theta} - \hat{\boldsymbol{\theta}})^{\mathrm{T}} \boldsymbol{\alpha}^{\mathrm{T}}(\boldsymbol{\xi}) + 0.5\boldsymbol{\varepsilon}(t)\|^2 \\ &\leq -2(\kappa^2(\boldsymbol{\xi}) + 1) \|(\boldsymbol{\theta} - \hat{\boldsymbol{\theta}})^{\mathrm{T}} \boldsymbol{\alpha}^{\mathrm{T}}(\boldsymbol{\xi}) + 0.5\boldsymbol{\varepsilon}(t)\|^2 \leq 0. \end{aligned} \tag{A5.49}$$

According to (A5.49),

$$\|\hat{\boldsymbol{\theta}}(t) - \boldsymbol{\theta}(t)\|_H^2 \leq \|\hat{\boldsymbol{\theta}}(t_0) - \boldsymbol{\theta}(t_0)\|_H^2 + C_\epsilon.$$

Hence

$$\|\hat{\boldsymbol{\theta}}(t) - \boldsymbol{\theta}(t)\| = (\lambda_{\max}(H)/\lambda_{\min}(H)) \|\hat{\boldsymbol{\theta}}(t_0) - \boldsymbol{\theta}(t_0)\|^2 + C_\epsilon/\lambda_{\min}(H).$$

Given that $(\lambda_{\max}(H)/\lambda_{\min}(H)) \|\hat{\boldsymbol{\theta}}(t_0) - \boldsymbol{\theta}(t_0)\|^2 + C_\epsilon/\lambda_{\min}(H)$ is independent of T and that $\boldsymbol{\theta}(t, \boldsymbol{\theta}_0, t_0) \subset \Omega(\Omega_\theta) \subseteq \Omega_\theta$, where Ω_θ is bounded, we can conclude that statement (1) holds.

In order to see that statement (2) holds we notice that the function $V(\boldsymbol{\theta}, \hat{\boldsymbol{\theta}}, t)$ is non-increasing and is bounded from below. Therefore

$$\begin{aligned}&\int_{t_0}^{t} (\kappa^2(\boldsymbol{\xi}(\tau)) + 1) \|(\boldsymbol{\theta}(\tau) - \hat{\boldsymbol{\theta}}(\tau))^{\mathrm{T}} \boldsymbol{\alpha}^{\mathrm{T}}(\boldsymbol{\xi}(\tau)) + 0.5\boldsymbol{\varepsilon}(\tau)\|^2 d\tau \\ &\quad \leq 0.5 V(\boldsymbol{\theta}(t_0), \hat{\boldsymbol{\theta}}(t_0), t_0), \ t \in [t_0, T].\end{aligned}$$

Hence the norms $\|\kappa(\boldsymbol{\xi})(\boldsymbol{\theta} - \hat{\boldsymbol{\theta}})^{\mathrm{T}} \boldsymbol{\alpha}^{\mathrm{T}}(\boldsymbol{\xi}) + 0.5\kappa(\boldsymbol{\xi})\boldsymbol{\varepsilon}\|_{2,[t_0,T]}$ and $\|(\boldsymbol{\theta} - \hat{\boldsymbol{\theta}})^{\mathrm{T}} \boldsymbol{\alpha}^{\mathrm{T}}(\boldsymbol{\xi}) + 0.5\boldsymbol{\varepsilon}\|_{2,[t_0,T]}$ can be bounded from above by a constant that is independent of T. This automatically implies (invoking the triangle inequality) that the same holds for $\|\kappa(\boldsymbol{\xi})(\boldsymbol{\theta} - \hat{\boldsymbol{\theta}})^{\mathrm{T}} \boldsymbol{\alpha}^{\mathrm{T}}(\boldsymbol{\xi})\|_{2,[t_0,T]}$ and $\|(\boldsymbol{\theta} - \hat{\boldsymbol{\theta}})^{\mathrm{T}} \boldsymbol{\alpha}^{\mathrm{T}}(\boldsymbol{\xi})\|_{2,[t_0,T]}$, provided that $\|\kappa(\boldsymbol{\xi})\boldsymbol{\varepsilon}\|_{2,[t_0,T]}$ and $\|\boldsymbol{\varepsilon}\|_{2,[t_0,T]}$ can be bounded from above by constants that are independent of T. Thus statement (2) of the lemma holds.

In order to show that statement (3) holds, consider

$$\begin{aligned}\dot{\psi} &= \frac{\partial \psi}{\partial \mathbf{x}}\mathbf{f}_0(\mathbf{x}) + \frac{\partial \psi(\mathbf{x})}{\partial \mathbf{x}}\boldsymbol{\alpha}(\boldsymbol{\xi})(\boldsymbol{\theta} - \hat{\boldsymbol{\theta}}) + \frac{\partial \psi}{\partial \mathbf{x}}\boldsymbol{\varepsilon}(t) \\ &= \frac{\partial \psi}{\partial \mathbf{x}}\mathbf{f}_0(\mathbf{x}) + \left(\frac{\partial \psi(\mathbf{x})}{\partial \mathbf{x}} - \frac{\partial \psi(\boldsymbol{\xi})}{\partial \boldsymbol{\xi}}\right) \\ &\quad \times \left(\boldsymbol{\alpha}(\boldsymbol{\xi})(\boldsymbol{\theta} - \hat{\boldsymbol{\theta}}) + \boldsymbol{\varepsilon}(t)\right) + \frac{\partial \psi(\boldsymbol{\xi})}{\partial \boldsymbol{\xi}}\left(\boldsymbol{\alpha}(\boldsymbol{\xi})(\boldsymbol{\theta} - \hat{\boldsymbol{\theta}}) + \boldsymbol{\varepsilon}(t)\right). \end{aligned} \tag{A5.50}$$

Given that $\psi \in C^1$ and $\partial \psi / \partial \mathbf{x}$ is Lipschitz, the property $\|\mathbf{x} - \boldsymbol{\xi}\|_{\infty,[t_0,T]} \leq C_x$ implies that: $\|\partial \psi(\mathbf{x})/\partial \mathbf{x} - \partial \psi(\boldsymbol{\xi})/\partial \boldsymbol{\xi}\|$ is bounded, and a bound can be found such that it does not depend on T. Moreover, $\|\partial \psi(\boldsymbol{\xi})/\partial \boldsymbol{\xi}\| \leq \kappa(\boldsymbol{\xi})$. Hence we can rewrite (A5.50) as

$$\dot{\psi} = \frac{\partial \psi(\mathbf{x})}{\partial \mathbf{x}}\mathbf{f}_0(\mathbf{x}) + \mu(t), \ \|\mu(t)\|_{2,[t_0,T]} \leq C_\mu, \tag{A5.51}$$

where C_μ does not depend on T. Notice that the function $\beta(\mathbf{x})$ is separated away from zero: $\exists \delta > 0 : \ \beta(\mathbf{x}) > 2\delta \ \forall \mathbf{x} \in \mathbb{R}^n$. Consider the following function:

$$V_\psi = \int_0^\psi \varphi(\sigma) d\sigma + \frac{1}{4\delta}\int_t^{\mathrm{T}} \mu^2(\tau) d\tau. \tag{A5.52}$$

Taking Assumption 5.14 and (A5.51) into account, one can derive that $\dot{V}_\psi \leq -\beta(\mathbf{x})\varphi^2(\psi) + \varphi(\psi)\mu(t) - (1/(4\delta))\mu^2(t) \leq -2\delta\varphi^2(\psi) + \varphi(\psi)\mu(t) - (1/(4\delta))\mu^2(t) = -\delta\varphi^2(\psi) - \delta(\varphi(\psi) - (1/(2\delta))\mu(t))^2 \leq 0$. Thus, according to Assumption 5.14 $\|\psi\|_{\infty,[t_0,T]}$ can be bounded from above by a constant that is independent of T, and, consequently, Assumption 5.1 implies that the same applies to $\|\mathbf{x}\|_{\infty,[t_0,T]}$. This proves statement (3).

Let us now show that $\hat{\boldsymbol{\theta}}(t)$ converges to $\hat{\boldsymbol{\theta}}(t_0)$ exponentially fast, provided that conditions of statement (4) hold. First notice that in this case solutions $\hat{\boldsymbol{\theta}}(t)$ exist for all $t \geq t_0$. Let us denote $\tilde{\boldsymbol{\theta}} = \boldsymbol{\theta} - \hat{\boldsymbol{\theta}}$. Then

$$\begin{aligned}\dot{\tilde{\boldsymbol{\theta}}} &= S(\boldsymbol{\theta}) - S(\hat{\boldsymbol{\theta}}) - H^{-1}(\kappa^2(\boldsymbol{\xi}) + 1)\boldsymbol{\alpha}(\boldsymbol{\xi})^{\mathrm{T}}\boldsymbol{\alpha}(\boldsymbol{\xi})\tilde{\boldsymbol{\theta}} \\ &= -H^{-1}(\kappa^2(\boldsymbol{\xi}) + 1)\boldsymbol{\alpha}(\boldsymbol{\xi})^{\mathrm{T}}\boldsymbol{\alpha}(\boldsymbol{\xi})\tilde{\boldsymbol{\theta}}. \end{aligned} \tag{A5.53}$$

According to Lemma 2.5 there exist $\lambda, D_0 \in \mathbb{R}_{>0}$ such that: $\|\tilde{\boldsymbol{\theta}}(t)\| \leq D_0 e^{-\lambda(t-t_0)}\|\tilde{\boldsymbol{\theta}}(t_0)\|$. □

Now consider systems (5.101) with locally Lipschitz $\phi_i(\mathbf{x})$:

$$\phi(\mathbf{x}) : \mathbb{R}^n \to \mathbb{R}^{d\times m},$$

$$\phi(\mathbf{x}) = \begin{pmatrix} \phi_{1,1}(\mathbf{x}), & \dots, & \phi_{1,d}(\mathbf{x}) \\ \dots, & \dots, & \dots \\ \phi_{m,1}(\mathbf{x}), & \dots, & \phi_{m,d}(\mathbf{x}) \end{pmatrix},$$

$$\phi_i(\mathbf{x}) = (\phi_{i,1}(\mathbf{x}), \dots, \phi_{i,d}(\mathbf{x})).$$

We ask whether there is a C^1-smooth function $\boldsymbol{\xi}(t)$ such that $\|(\boldsymbol{\alpha}(\mathbf{x}) - \boldsymbol{\alpha}(\boldsymbol{\xi}))\boldsymbol{\theta}(t)\|_{2,[t_0,T]}$ and $\|\kappa(\boldsymbol{\xi})(\boldsymbol{\alpha}(\mathbf{x}) - \boldsymbol{\alpha}(\boldsymbol{\xi}))\boldsymbol{\theta}(t)\|_{2,[t_0,T]}$ can be bounded from above by constants that are independent of T, $T \geq t_0$.

Lemma A5.3 *Consider system (5.101), and let* $[t_0, T]$, $T \geq t_0$ *be an interval of the system's solution. Let the functions* $\phi_i(\mathbf{x})$ *defined in (A5.54) be locally Lipschitz:*

$$\|\phi_i(\mathbf{x}) - \phi_i(\boldsymbol{\xi})\| \leq \lambda_i(\mathbf{x}, \boldsymbol{\xi})\|\mathbf{x} - \boldsymbol{\xi}\|,$$

where $\lambda_i(\mathbf{x}, \boldsymbol{\xi}) : \mathbb{R}^n \times \mathbb{R}^n \to \mathbb{R}_{\geq 0}$, $\lambda(\mathbf{x}, \boldsymbol{\xi})$ *are locally bounded with respect to* $\mathbf{x}$ *and* $\boldsymbol{\xi}$. *Furthermore, suppose that Assumption 5.12 holds. Then there exists a system*

$$\dot{\boldsymbol{\xi}} = \mathbf{f}(\mathbf{x}) + G_u\mathbf{u} + \lambda(\mathbf{x}, \boldsymbol{\xi})(\mathbf{x} - \boldsymbol{\xi}) + G_u\phi(\mathbf{x})\boldsymbol{\nu},$$

$$\dot{\boldsymbol{\nu}} = S(\boldsymbol{\nu}) + H^{-1}(G_u\phi(\mathbf{x}))^{\mathrm{T}}(\mathbf{x} - \boldsymbol{\xi})^{\mathrm{T}},$$

$$\lambda(\mathbf{x}, \boldsymbol{\xi}) = 1 + \sum_{i=1}^{m} \lambda_i^2(\mathbf{x}, \boldsymbol{\xi})(1 + \kappa^2(\boldsymbol{\xi})), \tag{A5.54}$$

such that

(1) *there are constants* D_α *and* D_κ, *which are independent of* T: $\|(\boldsymbol{\alpha}(\mathbf{x}) - \boldsymbol{\alpha}(\boldsymbol{\xi}))\boldsymbol{\theta}\|_{2,[t_0,T]} \leq D_\alpha$ *and* $\|\kappa(\boldsymbol{\xi})(\boldsymbol{\alpha}(\mathbf{x}) - \boldsymbol{\alpha}(\boldsymbol{\xi}))\boldsymbol{\theta}\|_{2,[t_0,T]} \leq D_\kappa$, *provided that* $\boldsymbol{\theta}$ *is bounded;*

(2) $\mathbf{x} \in L_\infty \Rightarrow \boldsymbol{\xi} \in L_\infty$; *if, in addition,* $\mathbf{x}(t)$ *is uniformly continuous then* $\lim_{t\to\infty} \mathbf{x}(t) - \boldsymbol{\xi}(t) = 0$.

Proof of Lemma A5.3. Consider the following function V_ξ: $V_\xi = 0.5\|\mathbf{x} - \boldsymbol{\xi}\|^2 + 0.5\|\boldsymbol{\theta} - \boldsymbol{\nu}\|_H^2$. Its derivative can be estimated as

$$\dot{V}_\xi \leq -\lambda(\mathbf{x}, \boldsymbol{\xi})\|\mathbf{x} - \boldsymbol{\xi}\|^2 + (\mathbf{x} - \boldsymbol{\xi})^{\mathrm{T}}G_u\phi(\mathbf{x})(\boldsymbol{\theta} - \boldsymbol{\nu})$$
$$+ (\boldsymbol{\theta} - \boldsymbol{\nu})^{\mathrm{T}}(G_u\phi(\mathbf{x}))^{\mathrm{T}}(\mathbf{x} - \boldsymbol{\xi}) \leq -\lambda(\mathbf{x}, \boldsymbol{\xi})\|\mathbf{x} - \boldsymbol{\xi}\|^2.$$

The last inequality implies that

$$\|\lambda_i(\mathbf{x},\boldsymbol{\xi})(\mathbf{x}-\boldsymbol{\xi})\|^2_{2,[t_0,T]} \leq V_{\xi}(\mathbf{x}(t_0),\boldsymbol{\xi}(t_0),\boldsymbol{\theta}(t_0),\boldsymbol{\nu}(t_0)),$$
$$\|\kappa(\boldsymbol{\xi})\lambda_i(\mathbf{x},\boldsymbol{\xi})(\mathbf{x}-\boldsymbol{\xi})\|^2_{2,[t_0,T]} \leq V_{\xi}(\mathbf{x}(t_0),\boldsymbol{\xi}(t_0),\boldsymbol{\theta}(t_0),\boldsymbol{\nu}(t_0)).$$

Given that $\|\phi_i(\mathbf{x}) - \phi_i(\boldsymbol{\xi})\| \leq \lambda_i(\mathbf{x},\boldsymbol{\xi})\|\mathbf{x}-\boldsymbol{\xi}\|$, we can conclude that $\|\phi_i(\mathbf{x}) - \phi_i(\boldsymbol{\xi})\|^2_{2,[t_0,T]} \leq V_{\xi}(\mathbf{x}(t_0),\boldsymbol{\xi}(t_0),\boldsymbol{\theta}(t_0),\boldsymbol{\nu}(t_0))$ and $\|\kappa(\boldsymbol{\xi})\|\phi_i(\mathbf{x}) - \phi_i(\boldsymbol{\xi})\|_{2,[t_0,T]} \leq V_{\xi}(\mathbf{x}(t_0),\boldsymbol{\xi}(t_0),\boldsymbol{\theta}(t_0),\boldsymbol{\nu}(t_0))$. Since $\boldsymbol{\theta}(t)$ is bounded for all $t \geq t_0$ and G_u is bounded as well, we can conclude that there exist D_α and D_κ, which are independent of T, such that $\|G_u(\phi_i(\mathbf{x}) - \phi_i(\boldsymbol{\xi}))\boldsymbol{\theta}(t)\|_{2,[t_0,T]} \leq D_\alpha$ and $\|\kappa(\boldsymbol{\xi})G_u(\phi_i(\mathbf{x}) - \phi_i(\boldsymbol{\xi}))\boldsymbol{\theta}(t)\|_{2,[t_0,T]} \leq D_\kappa$.

Finally, notice that the function V_ξ is non-increasing with t and is radially unbounded. This assures that $\boldsymbol{\xi}(t)$ is bounded as long as $\mathbf{x}(t)$ exists and is bounded. Moreover, $\lambda(\mathbf{x},\boldsymbol{\xi}) > 1$ implies that

$$\|\mathbf{x}-\boldsymbol{\xi}\|^2_{2,[t_0,T]} \leq V_{\xi}(\mathbf{x}(t_0),\boldsymbol{\xi}(t_0),\boldsymbol{\theta}(t_0),\boldsymbol{\nu}(t_0)).$$

Let $\mathbf{x}(t)$ be defined for all $t \geq t_0$ and $\mathbf{x}(t) \in L_\infty$. Then the right-hand side of the system is locally bounded. This ensures that $\|\mathbf{x}(t)-\boldsymbol{\xi}(t)\|^2$ is uniformly continuous, and hence $\lim_{t\to\infty} \mathbf{x}(t) - \boldsymbol{\xi}(t) = 0$. □

Let us now continue with the proof of the theorem. Consider trajectories $\boldsymbol{\theta}(t)$ and $\mathbf{x}(t)$ generated by $\boldsymbol{\theta}(t)$:

$$\dot{\mathbf{x}} = \mathbf{f}(\mathbf{x}) + G_u(\phi(\mathbf{x})\boldsymbol{\theta} + \mathbf{u});\ \dot{\boldsymbol{\theta}} = S(\boldsymbol{\theta}),\ \boldsymbol{\theta}(t_0) \in \Omega_\theta. \tag{A5.55}$$

Let these trajectories be defined for all $t \in [t_0, T]$, $T \geq t_0$. According to Lemma A5.3 there is a system of type (A5.54):

$$\begin{aligned}
\dot{\boldsymbol{\xi}} &= \mathbf{f}(\mathbf{x}) + G_u\mathbf{u} + \lambda(\mathbf{x},\boldsymbol{\xi})(\mathbf{x}-\boldsymbol{\xi}) + G_u\phi(\mathbf{x})\boldsymbol{\nu},\\
\dot{\boldsymbol{\nu}} &= S(\boldsymbol{\nu}) + H^{-1}(G_u\phi(\mathbf{x}))^{\mathrm{T}}(\mathbf{x}-\boldsymbol{\xi})^{\mathrm{T}},\\
\lambda(\mathbf{x},\boldsymbol{\xi}) &= 1 + \sum_{i=1}^{m}\lambda_i^2(\mathbf{x},\boldsymbol{\xi})(1+\kappa^2(\boldsymbol{\xi})),
\end{aligned} \tag{A5.56}$$

such that $\|G_u(\phi(\mathbf{x}) - \phi(\boldsymbol{\xi}))\boldsymbol{\theta}\|_{2,[t_0,T]}$ and $\|\kappa(\boldsymbol{\xi})G_u(\phi(\mathbf{x}) - \phi(\boldsymbol{\xi}))\boldsymbol{\theta}\|_{2,[t_0,T]}$ can be bounded from above by constants that are independent of T. On denoting $\boldsymbol{\alpha}(\boldsymbol{\xi}) = G_u\phi(\boldsymbol{\xi})$ and $\boldsymbol{\varepsilon}(t) = (\boldsymbol{\alpha}(\mathbf{x}) - \boldsymbol{\alpha}(\boldsymbol{\xi}))\boldsymbol{\theta}(t)$, and taking into account that $\mathbf{u}(\mathbf{x},\hat{\boldsymbol{\theta}}) = \mathbf{u}_0(\mathbf{x}) - \phi(\boldsymbol{\xi})\hat{\boldsymbol{\theta}}(t)$, we can rewrite equations (A5.55) as follows:

$$\begin{aligned}
\dot{\mathbf{x}} &= \mathbf{f}_0(\mathbf{x}) + \boldsymbol{\alpha}(\boldsymbol{\xi})(\boldsymbol{\theta} - \hat{\boldsymbol{\theta}}(t)) + \boldsymbol{\varepsilon}(t),\\
\dot{\boldsymbol{\theta}} &= S(\boldsymbol{\theta}).
\end{aligned} \tag{A5.57}$$

Moreover, (A5.57) and (5.107) give rise to the following equation for $\dot{\hat{\theta}}$:

$$\dot{\hat{\theta}} = S(\hat{\theta}) + H^{-1}(\kappa^2(\xi) + 1)\alpha^{\mathrm{T}}(\xi)(\alpha(\xi)(\theta - \hat{\theta}) + \varepsilon(t)). \tag{A5.58}$$

Notice that the right-hand sides of the combined system are continuous. Therefore solutions are defined at least locally in an interval $[t_0, T]$, $T \geq t_0$. According to Lemma A5.2 we have that $\|\mathbf{x}(t)\|_{\infty,[t_0,T]}$, $\|\theta(t)\|_{\infty,[t_0,T]}$, and $\|\hat{\theta}(t)\|_{\infty,[t_0,T]}$ can be bounded from above by constants that are independent of T. Moreover, Lemma A5.3 states that the same holds for $\|\xi(t)\|_{\infty,[t_0,T]}$. Thus solutions of the combined system are defined for all $t \geq t_0$, and are bounded. This proves statement (1) of the theorem.

According to Lemma A5.3, $\mathbf{x}(t) - \xi(t) \to 0$ as $t \to \infty$. This fact together with the uniform asymptotic stability of (A5.58) at $\varepsilon(t) \equiv 0$ (see also Lemma 2.5) guarantees that $\hat{\theta}(t, \hat{\theta}_0, t_0) \to \theta_0$ as $t \to \infty$. This proves statement (3).

Let us show that $\mathbf{x}(t) \to \Omega^*$ as $t \to \infty$. Consider the extended system

$$\begin{aligned}
\dot{\mathbf{x}} &= \mathbf{f}_0(\mathbf{x}) + \alpha(\xi)(\theta - \hat{\theta}) + [(\alpha(\mathbf{x}) - \alpha(\xi))\theta],\\
\dot{\theta} &= S(\theta),\\
\dot{\hat{\theta}} &= S(\hat{\theta}) + H^{-1}(\kappa^2(\xi) + 1)\alpha(\xi)^{\mathrm{T}}(\alpha(\xi)(\theta - \hat{\theta}) + [(\alpha(\mathbf{x}) - \alpha(\xi))\theta]),\\
\dot{\xi} &= \mathbf{f}(\mathbf{x}) + G_u\mathbf{u} + \lambda(\mathbf{x}, \xi)(\mathbf{x} - \xi) + G_u\phi(\mathbf{x})\nu,\\
\dot{\nu} &= S(\nu) + H^{-1}(G_u\phi(\mathbf{x}))^{\mathrm{T}}(\mathbf{x} - \xi)^{\mathrm{T}},\\
\dot{\epsilon}_0 &= -\left\|\frac{\partial\psi(\mathbf{x})}{\partial\mathbf{x}}(\alpha(\xi)(\theta - \hat{\theta}) + [(\alpha(\mathbf{x}) - \alpha(\xi))\theta])\right\|^2,\\
\dot{\epsilon}_1 &= -\|(\alpha(\mathbf{x}) - \alpha(\xi))\theta\|^2,\\
\dot{\epsilon}_2 &= -\|\alpha(\xi)(\theta - \hat{\theta})\|^2.
\end{aligned} \tag{A5.59}$$

We have shown already that all variables in (A5.59) are bounded except, probably, for the component ϵ_0. Boundedness of $\epsilon_0(t)$, however, follows immediately from the facts that $\partial\psi(\mathbf{x})/\partial\mathbf{x}$ is locally bounded and that the norms $\|(\alpha(\mathbf{x}) - \alpha(\xi))\theta\|_{2,[t_0,T]}$ and $\|\alpha(\xi)(\theta - \hat{\theta})\|_{2,[t_0,T]}$ can be bounded from above by constants that are independent of T.

Consider the function $V = \int_0^{\psi} \varphi(\sigma)d\sigma + \|\theta - \hat{\theta}\|_H^2 + (1/(4\delta))\epsilon_0(t) + \frac{1}{4}\epsilon_1(t) + \epsilon_2(t)$. Its derivative is $\dot{V} \leq -\delta\varphi^2(\psi) - \|\alpha(\xi)(\theta - \hat{\theta}) + 0.5\varepsilon(t)\|^2 - \|\alpha(\xi)(\theta - \hat{\theta}) + \varepsilon(t)\|^2 \leq -\delta\varphi^2(\psi) - \|\alpha(\xi)(\theta - \hat{\theta}) + \varepsilon(t)\|^2$, where $\varepsilon(t) = (\alpha(\mathbf{x}(t)) - \alpha(\xi(t)))\theta(t)$. According to La Salle's invariance principle (La Salle 1976), trajectories $(\mathbf{x}(t), \hat{\theta}(t))$ converge (as $t \to \infty$) to the maximal invariant set in $\Omega_\psi \times \Omega_\theta^1$, where $\Omega_\psi = \{\mathbf{x} \in \mathbb{R}^n | \mathbf{x} : \varphi(\psi(\mathbf{x})) = 0\}$, and $\Omega_\theta^1 : \{\hat{\theta} \in \mathbb{R}^d, \xi \in \mathbb{R}^n,\ \mathbf{x} \in \mathbb{R}^n | \alpha(\xi)(\theta - \hat{\theta}) + [(\alpha(\mathbf{x}) -$

$\boldsymbol{\alpha}(\boldsymbol{\xi}))\boldsymbol{\theta}] = 0\}$. This set is clearly contained in the direct sum of the maximal invariant set of

$$\dot{\mathbf{x}} = \mathbf{f}_0(\mathbf{x}) \tag{A5.60}$$

and Ω_θ^1. According to Assumption 5.14, Ω^* includes invariant sets of (A5.60) restricted to Ω_ψ. Thus $\mathbf{x}(t, \mathbf{x}_0) \to \Omega^*$ as $t \to \infty$. □

A5.9 Proof of Theorem 5.6

Consider system (A5.56). Since the right-hand side of the combined system is continuous there is a non-empty interval $[t_0, T]$, $T > t_0$ in which solutions of the system are defined. Lemma A5.3 and Assumption 5.12 taken together guarantee that the norm $\|G_u(\phi(\mathbf{x}) - \phi(\boldsymbol{\xi}))\boldsymbol{\theta}\|_{2,[t_0,T]}$ can be bounded from above by a constant that is independent of T. In addition, there is a constant C_θ, which is independent of T, such that $\|\hat{\boldsymbol{\theta}}(t)\|_{\infty,[t_0,T]} \leq C_\theta$ (this follows straightforwardly from the proof of Theorem 5.5 if we let $\kappa(\boldsymbol{\xi}) \equiv 0$ in (A5.49)). Moreover, Lemma A5.2 ensures that the norm $\|G_u\phi(\boldsymbol{\xi})(\boldsymbol{\theta} - \hat{\boldsymbol{\theta}})\|_{2,[t_0,T]}$ can be bounded from above by a constant that is independent of T. Let us denote $\boldsymbol{\varepsilon}_0(t) = G_u\phi(\boldsymbol{\xi})(\boldsymbol{\theta} - \hat{\boldsymbol{\theta}}) + G_u(\phi(\mathbf{x}) - \phi(\boldsymbol{\xi}))\boldsymbol{\theta}$. Then solutions $\mathbf{x}(t)$ of (5.101) should satisfy

$$\dot{\mathbf{x}} = \mathbf{f}_0(\mathbf{x}) + \boldsymbol{\varepsilon}_0(t), \tag{A5.61}$$

where $\boldsymbol{\varepsilon}_0(t)$ is continuous, and $\|\boldsymbol{\varepsilon}_0(t)\|_{2,[t_0,T]}$ can be bounded from above by a constant that is independent of T. System (A5.61), by virtue of the assumptions, admits $L_2 \to L_\infty$ margin. Thus we can conclude that $\mathbf{x}(t)$ is defined for all $t \geq t_0$ and is bounded. This proves statement (1) of the theorem. Statement (3) follows from Lemma A5.2. Let us show that $\mathbf{x}(t) \to \Omega^*$ as $t \to \infty$. To this end, consider system (A5.59) without the variable ϵ_0. As has been shown earlier, solutions of (A5.59) are bounded. Let $V = \|\boldsymbol{\theta} - \hat{\boldsymbol{\theta}}\|_H^2 + \frac{1}{4}\epsilon_1(t) + \epsilon_2(t)$. Its derivative can be estimated as $\dot{V} \leq -\|\boldsymbol{\alpha}(\boldsymbol{\xi})(\boldsymbol{\theta} - \hat{\boldsymbol{\theta}}) + \boldsymbol{\varepsilon}(t)\|^2 = -\|G_u\phi(\boldsymbol{\xi})(\boldsymbol{\theta} - \hat{\boldsymbol{\theta}}) + G_u(\phi(\mathbf{x}) - \phi(\boldsymbol{\xi}))\boldsymbol{\theta}\|^2$. Thus, according to La Salle's invariance principle, $\mathbf{x}(t) \to \Omega^*$ as $t \to \infty$. □

A5.10 Proof of Theorem 5.7

The right-hand sides of (5.108), (5.109), (5.32), and (5.34) are continuous and differentiable. Hence solutions of the combined system are defined on some non-empty interval $[t_0, T^*]$, $T^* > t_0$. According to (5.32), (5.34), Assumption 5.5, and (5.108), the derivative of $\hat{\boldsymbol{\theta}}$ can be written as

$$\dot{\hat{\boldsymbol{\theta}}} = \Gamma(\dot{\psi} + \varphi(\psi, \boldsymbol{\omega}, t))\boldsymbol{\alpha}(\mathbf{x}, t).$$

Consider

$$V_\theta(\boldsymbol{\theta}, \hat{\boldsymbol{\theta}}) = \frac{1}{2}\|\boldsymbol{\theta} - \hat{\boldsymbol{\theta}}\|^2_{\Gamma^{-1}}. \tag{A5.62}$$

Taking Assumptions 5.3 and 5.4 into account, we can write

$$\dot{V} \leq -\frac{1}{D}(f(\mathbf{x}, \boldsymbol{\theta}) - f(\mathbf{x}, \hat{\boldsymbol{\theta}}))^2 + \frac{1}{D_1}|f(\mathbf{x}, \boldsymbol{\theta}) - f(\mathbf{x}, \hat{\boldsymbol{\theta}})||z(\mathbf{x}, \mathbf{q}, t)|. \tag{A5.63}$$

Notice that (A5.63) can be transformed into

$$\dot{V} \leq -\frac{1}{D}\left(f(\mathbf{x}, \boldsymbol{\theta}) - f(\mathbf{x}, \hat{\boldsymbol{\theta}}) - \frac{D|z(\mathbf{x}, \mathbf{q}, t)|}{2D_1}\right)^2 + \frac{D}{4D_1^2}z^2(\mathbf{x}, \mathbf{q}, t). \tag{A5.64}$$

Using (A5.63) and (A5.64) we can derive the following estimate:

$$\begin{aligned} &\left\| f(\mathbf{x}(t), \boldsymbol{\theta}) - f(\mathbf{x}(t), \hat{\boldsymbol{\theta}}(t)) - \frac{D|z(\mathbf{x}(t), \mathbf{q}(t), t)|}{2D_1}\right\|^2_{2,[t_0,T^*]} \\ &\leq \frac{D}{2}\|\boldsymbol{\theta} - \hat{\boldsymbol{\theta}}(t_0)\|^2_{\Gamma^{-1}} + \frac{D^2}{4D_1^2}\|z(\mathbf{x}(t), \mathbf{q}(t), t)\|^2_{2,[t_0,T^*]}. \end{aligned}$$

Using the triangle inequality and that $\|\boldsymbol{\theta} - \hat{\boldsymbol{\theta}}\|^2_{\Gamma^{-1}} \leq \lambda_{\max}(\Gamma^{-1})\|\boldsymbol{\theta} - \hat{\boldsymbol{\theta}}\|^2 = 1/\lambda_{\min}(\Gamma)\|\boldsymbol{\theta} - \hat{\boldsymbol{\theta}}\|^2$, we arrive at

$$\begin{aligned} &\| f(\mathbf{x}(t), \boldsymbol{\theta}) - f(\mathbf{x}(t), \hat{\boldsymbol{\theta}}(t))\|_{2,[t_0,T^*]} \\ &\leq \left(\frac{D}{2\lambda_{\min}(\Gamma)}\right)^{\frac{1}{2}}\|\boldsymbol{\theta} - \hat{\boldsymbol{\theta}}(t_0)\| + \frac{D}{D_1}\|z(\mathbf{x}(t), \mathbf{q}(t), t)\|_{2,[t_0,T^*]}. \end{aligned}$$

Let us denote $\beta_\theta = (D/(2\lambda_{\min}(\Gamma)))^{1/2}\|\boldsymbol{\theta} - \hat{\boldsymbol{\theta}}(t_0)\|$ and $C_D = 1 + D/D_1$. Then (5.111) and

$$\begin{aligned} &\left(\int_{t_0}^{T^*}(|h_x(\mathbf{x}(t), t)| + |h_q(\mathbf{q}(t), t)|)^2 dt\right)^{\frac{1}{2}} \\ &\leq \left(\int_{t_0}^{T^*} h_x^2(\mathbf{x}(t), t)dt\right)^{\frac{1}{2}} + \left(\int_{t_0}^{T^*} h_q^2(\mathbf{q}(t), t)dt\right)^{\frac{1}{2}} \end{aligned} \tag{A5.65}$$

result in the following estimate:

$$\begin{aligned} &\| f(\mathbf{x}(t), \boldsymbol{\theta}) - f(\mathbf{x}(t), \hat{\boldsymbol{\theta}}(t))\|_{2,[t_0,T^*]} \\ &\leq \beta_\theta + (C_D - 1)(\|h_x(\mathbf{x}(t), t)\|_{2,[t_0,T^*]} + \|h_q(\mathbf{q}(t), t)\|_{2,[t_0,T^*]}). \end{aligned} \tag{A5.66}$$

Thus $\|f(\mathbf{x}(t),\boldsymbol{\theta}) - f(\mathbf{x}(t),\hat{\boldsymbol{\theta}}(t))\|_{2,[t_0,T^*]}$ can be bounded from above by the weighted sum of $\|h_x(\mathbf{x}(t),t)\|_{2,[t_0,T^*]}$, $\|h_q(\mathbf{q}(t),t)\|_{2,[t_0,T^*]}$, β_θ, and C_D. Now, (5.115) and (A5.66) allow us to estimate $\|\psi(\mathbf{x}(t),t)\|_{p,[t_0,T^*]}$ as

$$\begin{aligned}\|\psi(\mathbf{x}(t),t)\|_{\infty,[t_0,T^*]} \leq \gamma_{\psi,2}(\beta_\theta + C_D\|h_x(\mathbf{x}(t),t)\|_{2,[t_0,T^*]} \\ + C_D\|h_q(\mathbf{q}(t),t)\|_{2,[t_0,T^*]}) + \beta_{\psi,2}.\end{aligned} \tag{A5.67}$$

Let us recall one property of functions from $\mathcal{K}$ (Jiang *et al.* 1994).

Proposition A5.1 *Let $\gamma \in \mathcal{K}$, $\rho \in \mathcal{K}_\infty$, and $a, b \in \mathbb{R}_{\geq 0}$. Then*

$$\gamma(a+b) \leq \gamma((\rho + Id)(a)) + \gamma((\rho + Id) \circ \rho^{-1}(b)). \tag{A5.68}$$

Proof of Proposition A5.1. The proposition will be proven if we show that

$$\begin{aligned}\gamma(a+b) &\leq \gamma(a + r(a)) + \gamma(b + r^{-1}(b)) \\ &= \gamma(a + r(a)) + \gamma(r(r^{-1}(b)) + r^{-1}(b))\end{aligned}$$

holds for all $r \in \mathcal{K}_\infty$. Let us pick an arbitrary function r from $\mathcal{K}_\infty$. Consider two cases: $r(a) \geq b$ and $r(a) < b$. In the first case we get that $a + r(a) \geq a + b$. Given that γ is non-decreasing and non-negative, we obtain $\gamma(a+b) \leq \gamma(a + r(a)) + \gamma(b + r^{-1}(b))$ for all non-negative a and b. In the second case we can see that $a < r^{-1}(b)$, and hence $\gamma(a+b) \leq \gamma(r^{-1}(b) + b)$. □

Taking (A5.68) into account we can rewrite (A5.67) as

$$\begin{aligned}\|\psi(\mathbf{x}(t),t)\|_{\infty,[t_0,T^*]} \leq \gamma_{\psi,2}((\rho_1 + Id)(C_D\|h_x(\mathbf{x}(t),t)\|_{2,[t_0,T^*]} \\ + C_D\|h_q(\mathbf{q}(t),t)\|_{2,[t_0,T^*]})) \\ + \gamma_{\psi,2}((\rho_1 + Id) \circ \rho_1^{-1}(\beta_\theta)) + \beta_{\psi,2}, \ \rho_1 \in \mathcal{K}_\infty.\end{aligned}$$

According to (A5.68), we can now expand $\gamma \circ (\rho_1 + Id)(a+b) \in \mathcal{K}$ as

$$\begin{aligned}&\gamma \circ (\rho_1 + Id)(C_D\|h_x(\mathbf{x}(t),t)\|_{2,[t_0,T^*]} + C_D\|h_q(\mathbf{q}(t),t)\|_{2,[t_0,T^*]}) \\ &\quad \leq \gamma \circ (\rho_1 + Id) \circ (\rho_2 + Id)(C_D\|h_x(\mathbf{x}(t),t)\|_{2,[t_0,T^*]}) \\ &\qquad + \gamma \circ (\rho_1 + Id) \circ (\rho_2 + Id) \circ \rho_2^{-1}(C_D\|h_q(\mathbf{q}(t),t)\|_{2,[t_0,T^*]}), \ \rho_2 \in \mathcal{K}_\infty.\end{aligned}$$

Replacing $\gamma \circ (\rho_1 + Id)$ with $\gamma_{\psi,2} \circ (\rho_1 + Id)$ results in

$$\begin{aligned}\|\psi(\mathbf{x}(t),t)\|_{\infty,[t_0,T^*]} \leq \gamma_{\psi,2} \circ (\rho_1 + Id) \circ (\rho_2 + Id)(C_D\|h_x(\mathbf{x}(t),t)\|_{2,[t_0,T^*]}) \\ + \gamma_{\psi,2} \circ (\rho_1 + Id) \circ (\rho_2 + Id) \\ \circ \rho_2^{-1}(C_D\|h_q(\mathbf{q}(t),t)\|_{2,[t_0,T^*]}) \\ + \gamma_{\psi,2}((\rho_1 + Id) \circ \rho_1^{-1}(\beta_\theta)) + \beta_{\psi,2}.\end{aligned} \tag{A5.69}$$

On the other hand, the following holds

$$\begin{aligned}
\|h_q(\mathbf{q}(t),t)\|_{2,[t_0,T^*]} &\leq \gamma_{h_q} \circ (\rho_3 + Id) \circ \gamma_{x,\infty}(\|\psi(\mathbf{x}(t),t)\|_{\infty,[t_0,T^*]}) \\
&\quad + \gamma_{h_q} \circ (\rho_3 + Id) \circ \rho_3^{-1}(\beta_{x,p}) + \beta_{h_q}, \ \rho_3 \in \mathcal{K}_\infty, \\
\|h_x(\mathbf{x}(t),t)\|_{2,[t_0,T^*]} &\leq \gamma_{h_x}(\|\psi(\mathbf{x}(t),t)\|_{\infty,[t_0,T^*]}) + \beta_{h_x}.
\end{aligned} \tag{A5.70}$$

Hence, taking (A5.69) and (A5.70) into account, we obtain

$$\begin{aligned}
\|h_x(\mathbf{x}(t),t)\|_{2,[t_0,T^*]} &\leq \gamma_{h_x} \circ (\rho_4 + Id) \circ \gamma_{\psi,2} \circ (\rho_1 + Id), \\
&\quad \circ (\rho_2 + Id)(C_D\|h_x(\mathbf{x}(t),t)\|_{2,[t_0,T^*]}) \\
&\quad + \gamma_{h_x} \circ (\rho_4 + Id) \circ \rho_4^{-1} \circ (\rho_5 + Id) \circ \gamma_{\psi,2} \circ (\rho_1 + Id) \\
&\quad \circ (\rho_2 + Id) \circ -\rho_2^{-1}(C_D\|h_q(\mathbf{q}(t),t)\|_{2,[t_0,T^*]}) + \beta_1,
\end{aligned} \tag{A5.71}$$

$$\begin{aligned}
\|h_q(\mathbf{q}(t),t)\|_{2,[t_0,T^*]} &\leq \gamma_{h_q} \circ (\rho_3 + Id) \circ \gamma_{x,\infty} \circ (\rho_6 + Id) \circ \cdots \\
&\cdots \gamma_{\psi,2} \circ (\rho_1 + Id) \circ (\rho_2 + Id)(C_D\|h_x(\mathbf{x}(t),t)\|_{2,[t_0,T^*]}) \\
&+ \gamma_{h_q} \circ (\rho_3 + Id) \circ \gamma_{x,\infty} \circ (\rho_6 + Id) \circ \rho_6^{-1} \circ (\rho_7 + Id) \circ \cdots \\
&\cdots \gamma_{\psi,2} \circ (\rho_1 + Id) \circ (\rho_2 + Id) \circ \rho_2^{-1}(C_D\|h_q(\mathbf{q}(t),t)\|_{2,[t_0,T^*]}) + \beta_2,
\end{aligned} \tag{A5.72}$$

where $\beta_1, \beta_2 > 0$ and $\rho_4, \rho_5, \rho_6, \rho_7 \in \mathcal{K}_\infty$.

Let us denote

$$\begin{aligned}
\gamma_{h_{x,x}}(s) &= \gamma_{h_x} \circ (\rho_4 + Id) \circ \gamma_{\psi,2} \circ (\rho_1 + Id) \circ (\rho_2 + Id)(C_D s), \\
\gamma_{h_{x,q}}(s) &= \gamma_{h_x} \circ (\rho_4 + Id) \circ \rho_4^{-1} \circ (\rho_5 + Id) \circ \gamma_{\psi,2} \circ (\rho_1 + Id) \\
&\quad \circ (\rho_2 + Id) \circ \rho_2^{-1}(C_D s), \\
\gamma_{h_{q,x}}(s) &= \gamma_{h_q} \circ (\rho_3 + Id) \circ \gamma_{x,\infty} \circ (\rho_6 + Id) \circ \gamma_{\psi,2} \\
&\quad \circ (\rho_1 + Id) \circ (\rho_2 + Id)(C_D s), \\
\gamma_{h_{q,q}}(s) &= \gamma_{h_q} \circ (\rho_3 + Id) \circ \gamma_{x,\infty} \circ (\rho_6 + Id) \circ \rho_6^{-1} \circ (\rho_7 + Id) \\
&\quad \circ \gamma_{\psi,2} \circ (\rho_1 + Id) \circ (\rho_2 + Id) \circ \rho_2^{-1}(C_D s).
\end{aligned} \tag{A5.73}$$

Taking (A5.73) into account, we can now rewrite (A5.71) and (A5.72) as

$$\begin{aligned}\|h_x(\mathbf{x}(t),t)\|_{2,[t_0,T^*]} &\leq \gamma_{h_{x,x}}(\|h_x(\mathbf{x}(t),t)\|_{2,[t_0,T^*]}) \\ &\quad + \gamma_{h_{x,q}}(\|h_q(\mathbf{q}(t),t)\|_{2,[t_0,T^*]}) + \beta_1, \\ \|h_q(\mathbf{q}(t),t)\|_{2,[t_0,T^*]} &\leq \gamma_{h_{q,x}}(\|h_x(\mathbf{x}(t),t)\|_{2,[t_0,T^*]}) \\ &\quad + \gamma_{h_{q,q}}(\|h_q(\mathbf{q}(t),t)\|_{2,[t_0,T^*]}) + \beta_2.\end{aligned} \tag{A5.74}$$

According to the assumptions of the theorem, the following hold:

$$(Id - \gamma_{h_{x,x}}) \in \mathcal{K}_\infty, \qquad (Id - \gamma_{h_{q,q}}) \in \mathcal{K}_\infty.$$

Therefore (A5.74) allows us to obtain the following estimates:

$$\begin{aligned}\|h_x(\mathbf{x}(t),t)\|_{2,[t_0,T^*]} &\leq (Id - \gamma_{h_{x,x}})^{-1} \circ (\rho_8 + I_d) \\ &\quad \circ \gamma_{h_{x,q}}(\|h_q(\mathbf{q}(t),t)\|_{2,[t_0,T^*]}) + \beta_3, \\ \|h_q(\mathbf{q}(t),t)\|_{2,[t_0,T^*]} &\leq (Id - \gamma_{h_{q,q}})^{-1} \circ (\rho_9 + I_d) \\ &\quad \circ \gamma_{h_{q,x}}(\|h_x(\mathbf{x}(t),t)\|_{2,[t_0,T^*]}) + \beta_4,\end{aligned} \tag{A5.75}$$

where $\rho_8, \rho_9 \in \mathcal{K}_\infty$ and $\beta_3, \beta_4 \in \mathbb{R}_{\geq 0}$. According to Jiang *et al.* (1994), proof of Theorem 2.1, page 110, we can see that the right-hand side of (A5.74) is uniformly bounded with respect to T^*, provided that

$$\begin{aligned}&(Id + \lambda_2) \circ (Id - \gamma_{h_{q,q}})^{-1} \circ (\rho_9 + I_d) \circ \gamma_{h_{q,x}} \circ (Id + \lambda_1) \\ &\quad \circ (Id - \gamma_{h_{x,x}})^{-1} \circ (\rho_8 + I_d) \circ \gamma_{h_{x,q}}(s) \leq s, \\ &(Id + \lambda_1) \circ (Id - \gamma_{h_{x,x}})^{-1} \circ (\rho_8 + I_d) \circ \gamma_{h_{x,q}} \circ (Id + \lambda_2) \\ &\quad \circ (Id - \gamma_{h_{q,q}})^{-1} \circ (\rho_9 + I_d) \circ \gamma_{h_{q,x}}(s) \leq s,\end{aligned}$$

for all $s \geq s_0$ and some $\lambda_1, \lambda_2 \in \mathcal{K}_\infty$. This and (A5.65) imply that

$$\|z(\mathbf{x}(t),\mathbf{q}(t),t)\|_{2,[t_0,T^*]} \leq \|h_x(\mathbf{x}(t),t)\|_{2,[t_0,T^*]} + \|h_q(\mathbf{q}(t),t)\|_{2,[t_0,T^*]} \leq \beta_5, \tag{A5.76}$$

where $\beta_5 \in \mathbb{R}_{\geq 0}$ is a bounded function of the inital conditions and parameters of the combined system. Then, invoking the contradiction argument, using Assumptions 5.1 and 5.2, and taking (A5.63) into account, we conclude that $\mathbf{x}(t)$ and $\hat{\boldsymbol{\theta}}(t)$ generated by (5.108) are defined for all $t \geq t_0$. Furthermore, they are bounded. Boundedness of $\mathbf{q}(t)$ is ensured by the existence of the corresponding $L_\infty \mapsto L_\infty$ margin.

Statement (2) now follows from (A5.63), (A5.76), and Barbalat's lemma (see the proof of Theorem 5.1). □

A5.11 Proof of Theorem 5.8

Let us denote

$$\Delta f_x[t_0, T] = \|f_x(\mathbf{x}, \boldsymbol{\theta}_x, t) - f_x(\mathbf{x}, \hat{\boldsymbol{\theta}}_x, t)\|_{2,[t_0,T]},$$

$$\Delta f_y[t_0, T] = \|f_x(\mathbf{y}, \boldsymbol{\theta}_y, t) - f_y(\mathbf{y}, \hat{\boldsymbol{\theta}}_y, t)\|_{2,[t_0,T]}.$$

According to Theorem 5.1, the following estimates hold:

$$\Delta f_x[t_0, T] \leq C_x + \frac{D_x}{D_{1,x}}\|h_y(\mathbf{x}(t), \mathbf{y}(t), t)\|_{2,[t_0,T]}, \tag{A5.77}$$

$$\Delta f_y[t_0, T] \leq C_y + \frac{D_y}{D_{1,y}}\|h_x(\mathbf{x}(t), \mathbf{y}(t), t)\|_{2,[t_0,T]}, \tag{A5.78}$$

where C_x and C_y are constants that do not depend on T. Taking (A5.77) and (A5.78) into account, we can derive

$$\begin{aligned} &\Delta f_x[t_0, T] + \|h_y(\mathbf{x}(t), \mathbf{y}(t), t)\|_{2,[t_0,T]} \\ &\quad\leq C_x + \left(\frac{D_x}{D_{1,x}} + 1\right)\|h_y(\mathbf{x}(t), \mathbf{y}(t), t)\|_{2,[t_0,T]}, \end{aligned} \tag{A5.79}$$

$$\begin{aligned} &\Delta f_y[t_0, T] + \|h_x(\mathbf{x}(t), \mathbf{y}(t), t)\|_{2,[t_0,T]} \\ &\quad\leq C_y + \left(\frac{D_y}{D_{1,y}} + 1\right)\|h_x(\mathbf{x}(t), \mathbf{y}(t), t)\|_{2,[t_0,T]}. \end{aligned} \tag{A5.80}$$

The theorem then will be proven if we show that the $L_2^1[t_0, T]$-norms of the signals $h_x(\mathbf{x}(t), \mathbf{y}(t), t)$ and $h_y(\mathbf{x}(t), \mathbf{y}(t), t)$ are globally bounded with respect to T. Consider inequality (A5.68):

$$\gamma(a + b) \leq \gamma((\rho + Id)(a)) + \gamma((\rho + Id) \circ \rho^{-1}(b)), \ a, b \in \mathbb{R}_{\geq 0}, \ \gamma, \rho \in \mathcal{K}_\infty.$$

Applying this inequality to (A5.79) and (A5.80) and using (5.127) results in

$$\begin{aligned} &\|h_y(\mathbf{x}(t), \mathbf{y}(t), t)\|_{2,[t_0,T]} \\ &\quad\leq \beta_y \circ \gamma_{y2,2} \circ \rho_1\left(\left(\frac{D_y}{D_{1,y}} + 1\right)\|h_x(\mathbf{x}(t), \mathbf{y}(t), t)\|_{2,[t_0,T]}\right) + C_{y,1}, \\ &\|h_x(\mathbf{x}(t), \mathbf{y}(t), t)\|_{2,[t_0,T]} \\ &\quad\leq \beta_x \circ \gamma_{x2,2} \circ \rho_2\left(\left(\frac{D_x}{D_{1,x}} + 1\right)\|h_y(\mathbf{x}(t), \mathbf{y}(t), t)\|_{2,[t_0,T]}\right) + C_{x,1}, \end{aligned} \tag{A5.81}$$

where $\rho_1(\cdot), \rho_2(\cdot) \in \mathcal{K}_\infty$ and $\rho_1(\cdot), \rho_2(\cdot) > Id(\cdot)$. According to (A5.81), the fact that a function $\rho_3(\cdot) \in \mathcal{K}_\infty \geq Id(\cdot)$ exists such that the inequalities

$$\beta_y \circ \gamma_{y2,2} \circ \rho_1 \circ \left(\frac{D_y}{D_{y,1}} + 1\right) \circ \rho_3 \circ \beta_x \circ \gamma_{x2,2} \circ \rho_2 \circ \left(\frac{D_x}{D_{x,1}} + 1\right)(\Delta) < \Delta \,\forall\, \Delta \geq \bar{\Delta}$$

hold for some $\bar{\Delta} \in \mathbb{R}_{\geq 0}$ ensures that norms

$$\|h_y(\mathbf{x}(t), \mathbf{y}(t), t)\|_{2,[t_0,T]}, \ \|h_x(\mathbf{x}(t), \mathbf{y}(t), t)\|_{2,[t_0,T]}$$

can be bounded from above by constants that are independent of T. The remaining statements of the theorem follow explicitly from Theorem 5.1. □

A5.12 Proof of Corollary 5.3

Let $\boldsymbol{\lambda}(\tau, \lambda_0)$ be a solution of system (5.144). Consider it as a function of the variable τ. Let us pick some monotone, strictly increasing function σ such that the following holds:

$$\tau = \sigma(t), \ \sigma : \mathbb{R}_{\geq 0} \to \mathbb{R}_{\geq 0}.$$

Given that $\eta(\Omega_\lambda)$ is dense in Ω_θ, for any $\boldsymbol{\theta} \in \Omega_\theta$ there always exists a vector $\boldsymbol{\lambda}_\theta \in \Omega_\lambda$ such that $\eta(\boldsymbol{\lambda}_\theta) = \boldsymbol{\theta} + \epsilon_\theta$, where $\|\epsilon_\theta\|$ is arbitrarily small. Furthermore, $\boldsymbol{\lambda}(\tau)$ is dense in Ω_λ, hence there is a point $\boldsymbol{\lambda}^* = \boldsymbol{\lambda}(\tau^*, \boldsymbol{\lambda}_0)$, which is arbitrarily close to $\boldsymbol{\lambda}_\theta$. Consider the following difference:

$$\begin{aligned}\mathbf{f}(\boldsymbol{\xi}(t), \boldsymbol{\theta}) - \mathbf{f}(\boldsymbol{\xi}(t), \hat{\boldsymbol{\theta}}) = \mathbf{f}(\boldsymbol{\xi}(t), \boldsymbol{\theta}) - f(\boldsymbol{\xi}(t), \eta(\boldsymbol{\lambda}^*)) + \mathbf{f}(\boldsymbol{\xi}, \eta(\boldsymbol{\lambda}^*)) \\ - \mathbf{f}(\boldsymbol{\xi}, \eta(\boldsymbol{\lambda}(\sigma(t)))).\end{aligned}$$

The function $\mathbf{f}(\cdot)$ is locally bounded and $\eta(\cdot)$ is Lipschitz, so

$$\|\mathbf{f}(\boldsymbol{\xi}, \boldsymbol{\theta}) - \mathbf{f}(\boldsymbol{\xi}, \eta(\boldsymbol{\lambda}^*))\| \leq D_f \|\epsilon_\theta\| + \Delta_f = \Delta_\theta + \Delta_f,$$

where Δ_θ is arbitrarily small. Hence

$$\begin{aligned}\|\mathbf{f}(\boldsymbol{\xi}, \eta(\boldsymbol{\lambda}^*)) - \mathbf{f}(\boldsymbol{\xi}, \eta(\boldsymbol{\lambda}(\sigma(t))))\| &\leq D_f \|\eta(\boldsymbol{\lambda}^*) - \eta(\boldsymbol{\lambda}(\sigma(t)))\| \\ &\quad + \Delta_f + \Delta_\theta \leq D_f \cdot D_\eta \|\boldsymbol{\lambda}^* - \boldsymbol{\lambda}(\sigma(t))\| \\ &\quad + \Delta_f + \Delta_\theta. \end{aligned} \tag{A5.82}$$

Noticing that $\boldsymbol{\lambda}^* = \boldsymbol{\lambda}(\tau^*, \boldsymbol{\lambda}_0) = \boldsymbol{\lambda}(\sigma(t^*), \boldsymbol{\lambda}_0)$ and taking into account the Poisson stability of (5.144), we can always choose $\boldsymbol{\lambda}^*(\sigma^*, \boldsymbol{\lambda}_0)$ such that $\sigma^* > \sigma(t_0) = \tau_0$ for any $\tau_0 \in \mathbb{R}_{\geq 0}$. Hence, according to (A5.82), the following estimate holds:

$$\begin{aligned}&\|\mathbf{f}(\boldsymbol{\xi}, \eta(\boldsymbol{\lambda}^*)) - \mathbf{f}(\boldsymbol{\xi}, \eta(\boldsymbol{\lambda}(\sigma(t))))\| \\ &\leq D_f \cdot D_\eta \| \int_{\sigma(t)}^{\sigma^*} S(\boldsymbol{\lambda}(\sigma(\tau))) d\tau \| + \Delta_f + \Delta_\theta \\ &\leq D_f \cdot D_\eta \cdot \max_{\boldsymbol{\lambda} \in \Omega_\lambda} \|S(\boldsymbol{\lambda})\| |\sigma^* - \sigma(t)| = \mathcal{D} \cdot |\sigma^* - \sigma(t)| + \Delta_f + \Delta_\theta, \\ &\mathcal{D} = D_f \cdot D_\eta \cdot \max_{\boldsymbol{\lambda} \in \Omega_\lambda} \|S(\boldsymbol{\lambda})\|. \end{aligned} \tag{A5.83}$$

Denoting $\mathbf{u}(t) = \mathbf{f}(\boldsymbol{\xi}(t), \boldsymbol{\theta}) - \mathbf{f}(\boldsymbol{\xi}(t), \hat{\boldsymbol{\theta}}) + \boldsymbol{\varepsilon}(t)$, we can now conclude that

$$\begin{aligned}\|\mathbf{u}(t)\| &\leq \Delta_\epsilon + \Delta_f + \|\mathbf{f}(\boldsymbol{\xi}(t), \boldsymbol{\theta}) - f(\boldsymbol{\xi}(t), \boldsymbol{\eta}(\boldsymbol{\lambda}^*))\| + \mathcal{D} \cdot |\sigma^* - \sigma(t)| \\ &\leq \Delta_\epsilon + 2\Delta_f + \Delta_\theta + D_f\|\boldsymbol{\theta} - \boldsymbol{\eta}(\boldsymbol{\lambda}^*)\| + \mathcal{D} \cdot |\sigma^* - \sigma(t)|. \end{aligned} \tag{A5.84}$$

Notice that due to the denseness of $\boldsymbol{\lambda}(t, \boldsymbol{\lambda}_0)$ in Ω_λ it is always possible to choose $\boldsymbol{\lambda}^*$ such that

$$D_f\|\boldsymbol{\theta} - \boldsymbol{\eta}(\boldsymbol{\lambda}^*)\| = D_f\|\boldsymbol{\eta}(\boldsymbol{\lambda}_\theta) - \boldsymbol{\eta}(\boldsymbol{\lambda}^*)\| \leq D_f D_\eta \|\boldsymbol{\lambda}_\theta - \boldsymbol{\eta}(\boldsymbol{\lambda}^*)\| \leq \Delta_\lambda.$$

Hence, according to (A5.84), we have

$$\|\mathbf{u}(t)\|_{\infty,[t_0,t]} \leq 2\Delta_f + \Delta_\varepsilon + \delta + \mathcal{D} \cdot \|\sigma^* - \sigma(t)\|_{\infty,[t_0,t]},$$

where the term $\delta > \Delta_\theta + \Delta_\lambda$ can be made arbitrarily small.

Therefore Assumption 5.17 implies that the following inequality holds:

$$\|\mathbf{x}(t)\|_{\mathcal{A}_{\Delta(M)}} \leq \beta(t - t_0)\|\mathbf{x}(t_0)\|_{\mathcal{A}_{\Delta(M)}} + c \cdot \mathcal{D} \cdot \|\sigma^* - \sigma(t)\|_{\infty,[t_0,t]}. \tag{A5.85}$$

Let us now define $\sigma(t)$ as follows:

$$\sigma(t) = \int_{t_0}^{t} \gamma\|\psi(\mathbf{x}(\tau))\|_{\mathcal{A}_{\Delta(M)}}\, d\tau. \tag{A5.86}$$

Moreover, let us introduce the following notation:

$$h(t) = \sigma^* - \sigma(t) = \sigma^* - \int_{t_0}^{t} \gamma\, \|\psi(\mathbf{x}(\tau))\|_{\mathcal{A}_{\Delta(M)}}\, d\tau,$$

then for all $t',\ t \geq t_0, t \geq t'$ we have that

$$h(t') - h(t) = \int_{t'}^{t} \gamma\|\psi(\mathbf{x}(\tau))\|_{\mathcal{A}_{\Delta(M)}}\, d\tau.$$

Taking into account (A5.82) and (A5.83), the equality

$$\frac{\partial\boldsymbol{\lambda}(\sigma(t), \boldsymbol{\lambda}_0)}{dt} = \frac{\partial\sigma(t)}{dt} S(\boldsymbol{\lambda}(\sigma(t), \boldsymbol{\lambda}_0)) = \gamma\|\psi(\mathbf{x}(\tau))\|_{\mathcal{A}_{\Delta(M)}} S(\boldsymbol{\lambda}(\sigma(t), \boldsymbol{\lambda}_0)),$$

and (A5.85), and denoting $D_\lambda = c\mathcal{D}$, we can conclude that the following holds along the trajectories of (5.145):

$$\begin{aligned} \|\mathbf{x}(t)\|_{\mathcal{A}_{\Delta(M)}} &\leq \beta(t - t_0)\, \|\mathbf{x}(t_0)\|_{\mathcal{A}_{\Delta(M)}} + D_\lambda\|h(\tau)\|_{\infty,[t_0,t]}, \\ h(t_0) - h(t) &= \int_{t_0}^{t} \gamma\, \|\psi(\mathbf{x}(\tau))\|_{\mathcal{A}_{\Delta(M)}}\, d\tau. \end{aligned} \tag{A5.87}$$

Hence, according to Corollary 4.2, the limit relation (5.147) holds for all $|h(t_0)|$ and $\|\mathbf{x}(t_0)\|_{\mathcal{A}_{\Delta(M)}}$ that belong to the domain

$$\Omega_\gamma : \gamma \leq \left(\beta_t^{-1}\left(\frac{d}{\kappa}\right)\right)^{-1}\frac{\kappa-1}{\kappa} \times \frac{h(t_0)}{\beta_t(0)\|\mathbf{x}(t_0)\|_{\mathcal{A}_{\Delta+\delta}} + \beta_t(0)\cdot D_\lambda \cdot |h(t_0)|(1+\kappa/(1-d)) + D_\lambda|h(t_0)|}$$

for some $d < 1$, $\kappa > 1$. Notice, however, that $\|\mathbf{x}(t)\|_{\mathcal{A}_{\Delta+\delta}}$ is always bounded since $\mathbf{f}(\cdot)$ is Lipschitz in θ and both $\boldsymbol{\theta}$ and $\hat{\boldsymbol{\theta}}$ are bounded ($\boldsymbol{\eta}(\cdot)$ is Lipschitz and $\boldsymbol{\lambda}(t, \boldsymbol{\lambda}_0)$ is bounded according to the assumptions of the corollary). Moreover, due to the Poisson stability of (5.144) it is always possible to choose a point $\boldsymbol{\lambda}^*$ such that $h(t_0) = \sigma^*$ is arbitrarily large. Hence the choice of γ in (A5.87) as (5.146) suffices to ensure that $h(t)$ is bounded. Moreover, it follows that $h(t)$ converges to a limit as $t \to \infty$. This implies that $\gamma \int_{t_0}^{t} \|\mathbf{x}(\tau)\|_{\mathcal{A}_{\Delta(M)}}$ also converges as $t \to \infty$, and, consequently, $\boldsymbol{\lambda}(t, \boldsymbol{\lambda}_0)$ converges to some $\boldsymbol{\lambda}' \in \Omega_\lambda$. Hence

$$\lim_{t\to\infty} \boldsymbol{\theta}(\hat{t}) = \boldsymbol{\theta}'$$

holds for some $\boldsymbol{\theta}' \in \Omega_\theta$. According to the corollary conditions, system (5.143) has steady-state characteristics with respect to $\hat{\boldsymbol{\theta}}$. Then, in the same way as in the proof of Lemma 4.2, we can show that (5.147) holds. □

A5.13 Proof of Theorem 5.9

According to the theorem formulation, Assumptions 5.1, 5.2, 5.3, and 5.5 hold. Hence, on applying Corollary 5.1 we can conclude that the $\hat{\boldsymbol{\theta}}(t) \in L_\infty^d[t_0, \infty]$ and $\mathbf{x}(t) \in L_\infty^n[t_0, \infty]$. Let us show that the limiting relation (5.24) holds when alternative (1) is satisfied. To this end, consider the derivative $\dot{\hat{\boldsymbol{\theta}}}$:

$$\dot{\hat{\boldsymbol{\theta}}} = \Gamma(\dot{\psi} + \varphi(\psi))\boldsymbol{\alpha}(\mathbf{x}, t) = \Gamma(f(\mathbf{x}, \boldsymbol{\theta}, t) - f(\mathbf{x}, \hat{\boldsymbol{\theta}}, t))\boldsymbol{\alpha}(\mathbf{x}, t). \tag{A5.88}$$

Given that $\hat{\boldsymbol{\theta}}(t) \in L_\infty^d[t_0, \infty]$ and $\mathbf{x}(t) \in L_\infty^n[t_0, \infty]$, and that hypothesis H3 holds, the function $f_\psi(\mathbf{x}, \boldsymbol{\theta}, t)$ satisfies the following inequality for some D, $D_1 \in \mathbb{R}_{>0}$:

$$D_1|\boldsymbol{\alpha}(\mathbf{x}, t)^{\mathrm{T}}(\hat{\boldsymbol{\theta}} - \boldsymbol{\theta})| \leq |f(\mathbf{x}, \boldsymbol{\theta}, t) - f(\mathbf{x}, \hat{\boldsymbol{\theta}}, t)| \leq D|\boldsymbol{\alpha}(\mathbf{x}, t)^{\mathrm{T}}(\hat{\boldsymbol{\theta}} - \boldsymbol{\theta})|,$$
$$\boldsymbol{\alpha}(\mathbf{x}, t)^{\mathrm{T}}(\hat{\boldsymbol{\theta}} - \boldsymbol{\theta})(f(\mathbf{x}, \hat{\boldsymbol{\theta}}, t) - f(\mathbf{x}, \boldsymbol{\theta}, t)) \geq 0.$$

Therefore, there exists a function $\kappa : \mathbb{R}_{\geq 0} \to \mathbb{R}_{\geq 0}$, $D_1 \leq \kappa^2(t) \leq D$ such that

$$\begin{aligned} \dot{\hat{\boldsymbol{\theta}}} &= -\kappa^2(t)\Gamma\boldsymbol{\alpha}(\mathbf{x}, t)^{\mathrm{T}}(\hat{\boldsymbol{\theta}} - \boldsymbol{\theta})\boldsymbol{\alpha}(\mathbf{x}, t) \\ &= -\kappa^2(t)\Gamma\boldsymbol{\alpha}(\mathbf{x}, t)\boldsymbol{\alpha}(\mathbf{x}, t)^{\mathrm{T}}(\hat{\boldsymbol{\theta}} - \boldsymbol{\theta}). \end{aligned} \tag{A5.89}$$

Notice that $\Gamma = \Gamma^{\mathrm{T}}$ and $\Gamma > 0$. Then, according to Lemma 2.5, we have that

$$\|\hat{\boldsymbol{\theta}}(t) - \boldsymbol{\theta}\| \leq D_{\Gamma} e^{-\rho(t-t_0)} \|\hat{\boldsymbol{\theta}}(t_0) - \boldsymbol{\theta}\|,$$

where

$$D_{\Gamma} = \left(\frac{\lambda_{\max}(\Gamma)}{\lambda_{\min}(\Gamma)}\right)^{\frac{1}{2}} e^{\rho L}, \ \rho = \frac{\delta D_1 \lambda_{\min}(\Gamma)}{L(1 + \lambda_{\max}(\Gamma)\alpha_{\infty}^2 L^2)^2},$$

$$\alpha_{\infty} = \sup_{\|\mathbf{x}\| \leq \|\mathbf{x}(t)\|_{\infty,[t_0,\infty]},\ t \geq t_0} \|\boldsymbol{\alpha}(\mathbf{x}, t)\|.$$

This proves alternative (1) of the theorem.

Let us prove alternative (2). It follows immediately from Corollary 5.1 of Theorem 5.1 that

$$\lim_{t\to\infty} f(\mathbf{x}(t), \boldsymbol{\theta}, t) - f(\mathbf{x}(t), \hat{\boldsymbol{\theta}}(t), t) = 0. \tag{A5.90}$$

Furthermore, given that $\dot{\hat{\boldsymbol{\theta}}} = \Gamma(f(\mathbf{x}(t), \boldsymbol{\theta}, t) - f(\mathbf{x}(t), \hat{\boldsymbol{\theta}}(t), t))\boldsymbol{\alpha}(\mathbf{x}, t)$, $\mathbf{x}(t) \in L_{\infty}^n[t_0, \infty]$ and $\boldsymbol{\alpha}(\mathbf{x}, t)$ is locally bounded in $\mathbf{x}$ uniformly in t, we can conclude that $\dot{\hat{\boldsymbol{\theta}}} \to 0$ as $t \to \infty$. Let us divide the $\mathbb{R}_{\geq 0}$ into the following union of subintervals: $\mathbb{R}_{\geq 0} = \bigcup_{i=1}^{\infty} \Delta_i$, $\Delta_i = [t_i, t_i + T]$, $t_0 = 0$, $t_{i+1} = t_i + T$, $i \in \mathbb{N}$. The value of T is chosen to satisfy $T \geq L$, where L is the constant from Definition 5.6.2. The fact that $\dot{\hat{\boldsymbol{\theta}}} \to 0$ as $t \to \infty$ ensures that

$$\lim_{i\to\infty} \|\hat{\boldsymbol{\theta}}(s_i) - \hat{\boldsymbol{\theta}}(\tau_i)\| = 0, \ \forall\, s_i, \tau_i \in \Delta_i. \tag{A5.91}$$

In order to show this, let us integrate (A5.88):

$$\|\hat{\boldsymbol{\theta}}(s_i) - \hat{\boldsymbol{\theta}}(\tau_i)\| = \|\Gamma \int_{s_i}^{\tau_i} (f(\mathbf{x}(\tau), \boldsymbol{\theta}, \tau) - f(\mathbf{x}(\tau), \hat{\boldsymbol{\theta}}(\tau), \tau))\alpha(\mathbf{x}(\tau), \tau) d\tau\|. \tag{A5.92}$$

By applying the Cauchy–Schwartz inequality to (A5.92) and subsequently using the mean-value theorem we can obtain the following estimate:

$$\begin{aligned} &\|\hat{\boldsymbol{\theta}}(s_i) - \hat{\boldsymbol{\theta}}(\tau_i)\| \\ &\leq \int_{t_i}^{t_i+T} \|\Gamma\| \cdot |f(\mathbf{x}(\tau), \boldsymbol{\theta}, \tau) - f(\mathbf{x}(\tau), \hat{\boldsymbol{\theta}}(\tau), \tau)| \cdot \|\alpha(\mathbf{x}(\tau), \tau)\| d\tau \\ &= \|\Gamma\| \cdot T \cdot |f(\mathbf{x}(\tau_i'), \boldsymbol{\theta}, \tau_i') - f(\mathbf{x}(\tau_i'), \hat{\boldsymbol{\theta}}(\tau_i'), \tau_i')| \cdot \|\boldsymbol{\alpha}(\mathbf{x}(\tau_i'), \tau_i')\|, \\ &\tau_i' \in \Delta_i. \end{aligned} \tag{A5.93}$$

Given that the limiting relation (A5.90) holds, $\mathbf{x}(t) \in L^n_\infty[t_0, \infty]$, and $\boldsymbol{\alpha}(\mathbf{x}, t)$ is locally bounded uniformly in t, we can conclude from (A5.93) that the limiting relation (A5.91) holds.

Let us choose a sequence of points from $\mathbb{R}_{\geq 0}$: $\{\tau_i\}_{i=1}^{\infty}$ such that $\tau_i \in \Delta_i$, $i \in \mathbb{N}$. As follows from the nonlinear persistent excitation condition, (5.154), for every $\hat{\boldsymbol{\theta}}(\tau_i)$, $\tau_i \in \Delta_i$ there exists a point $t_i' \in \Delta_i$ such that the following inequality holds:

$$\|f(\mathbf{x}(t_i'), \boldsymbol{\theta}, t_i') - f(\mathbf{x}(t_i'), \hat{\boldsymbol{\theta}}(\tau_i), t_i')\| \geq \varrho(\|\boldsymbol{\theta} - \hat{\boldsymbol{\theta}}(\tau_i)\|) \geq 0. \tag{A5.94}$$

Let us consider the differences $f(\mathbf{x}(t_i'), \hat{\boldsymbol{\theta}}(\tau_i), t_i') - f(\mathbf{x}(t_i'), \hat{\boldsymbol{\theta}}(t_i'), t_i')$, $\tau_i, t_i' \in \Delta_i$. It follows immediately from H1, H2, and (A5.91) that

$$\lim_{i\to\infty} f(\mathbf{x}(t_i'), \hat{\boldsymbol{\theta}}(\tau_i), t_i') - f(\mathbf{x}(t_i'), \hat{\boldsymbol{\theta}}(t_i'), t_i') = 0, \quad \tau_i, t_i' \in \Delta_i. \tag{A5.95}$$

Taking into account (A5.95) and (A5.90), we can derive that

$$\begin{aligned}
&\lim_{i\to\infty} f(\mathbf{x}(t_i'), \boldsymbol{\theta}, t_i') - f(\mathbf{x}(t_i'), \hat{\boldsymbol{\theta}}(\tau_i), t_i') \\
&\quad = \lim_{i\to\infty} (f(\mathbf{x}(t_i'), \boldsymbol{\theta}, t_i') - f(\mathbf{x}(t_i'), \hat{\boldsymbol{\theta}}(t_i'), t_i')) \\
&\qquad + \lim_{i\to\infty} f(\mathbf{x}(t_i'), \hat{\boldsymbol{\theta}}(t_i'), t_i') - f(\mathbf{x}(t_i'), \hat{\boldsymbol{\theta}}(\tau_i), t_i') \\
&\quad = 0.
\end{aligned} \tag{A5.96}$$

According to (A5.96) and (A5.94), the sequence $\{\varrho(\|\boldsymbol{\theta} - \hat{\boldsymbol{\theta}}(\tau_i)\|)\}_{i=1}^{\infty}$ is bounded from above and below by two sequences converging to zero. Hence

$$\lim_{i\to\infty} \varrho(\|\boldsymbol{\theta} - \hat{\boldsymbol{\theta}}(\tau_i)\|) = 0.$$

Notice that $\varrho(\cdot) \in \mathcal{K} \cap \mathcal{C}^0$, which implies that

$$\lim_{i\to\infty} \|\boldsymbol{\theta} - \hat{\boldsymbol{\theta}}(\tau_i)\| = 0. \tag{A5.97}$$

In order to show that $\lim_{t\to\infty}(\boldsymbol{\theta} - \hat{\boldsymbol{\theta}}(t)) = 0$, notice that $\|\boldsymbol{\theta} - \hat{\boldsymbol{\theta}}(t)\| \leq \|\boldsymbol{\theta} - \hat{\boldsymbol{\theta}}(s_i)\|$, $s_i = \arg\max_{s\in\Delta_i} \|\boldsymbol{\theta} - \hat{\boldsymbol{\theta}}(s)\| \;\; \forall\, t \in \Delta_i$. Hence, by applying the triangle inequality $\|\boldsymbol{\theta} - \hat{\boldsymbol{\theta}}(s_i)\| \leq \|\boldsymbol{\theta} - \hat{\boldsymbol{\theta}}(\tau_i)\| + \|\hat{\boldsymbol{\theta}}(\tau_i) - \hat{\boldsymbol{\theta}}(s_i)\|$ and using (A5.91) and (A5.97), we can conclude that $\|\boldsymbol{\theta} - \hat{\boldsymbol{\theta}}(t)\|$ is bounded from above and below by two functions converging to zero. Hence, $\|\boldsymbol{\theta} - \hat{\boldsymbol{\theta}}(t)\| \to 0$ as $t \to \infty$ and (5.24) holds. □

A5.14 Proof of Theorem 5.10

Since the right-hand sides of (5.160) and (5.165)–(5.167) are continuous, there is at least one solution of the closed-loop system passing through $(t_0, \mathbf{s}_0, \hat{\boldsymbol{\theta}}(t_0), \hat{\mathbf{x}}(t_0),$

$\mathbf{x}(t_0)$), and this solution is defined at least locally. Let $[t_0, t_1), t_1 > g_0$ be an interval of existence of this solution. Given that

$$\frac{d}{dt}\sum_{j=1}^{s}\left(s_{2j-1}^2 + s_{2j}^2\right) = 0,$$

we can conclude that trajectories $\mathbf{s}(t, \mathbf{s}_0)$ are globally bounded for all $t \in [t_0, t_1)$: $\|\mathbf{s}(t, \mathbf{s}_0)\| \leq D_1$, $D_1 \in \mathbb{R}_{>0}$. Moreover, the value of D_1 does not depend on $\hat{\boldsymbol{\theta}}(t_0), \hat{\mathbf{x}}(t_0), \mathbf{x}(t_0)$ and parameters γ, ω_i, $\boldsymbol{\lambda}$, and $\boldsymbol{\theta}$. This automatically implies that the right-hand sides of (5.165) and (5.167) are Lipschitz in $\mathbf{x}$, $\hat{\mathbf{x}}$, $\hat{\boldsymbol{\theta}}$, and that the corresponding Lipschitz constants can be chosen such that they do not depend on $\mathbf{x}$, $\hat{\mathbf{x}}$, $\hat{\boldsymbol{\theta}}$, and $\mathbf{s}$. Thus, invoking the standard Gronwall lemma-based argument (see e.g. Khalil (2002)), one can conclude that solutions of the closed-loop system are defined for all $t \geq t_0$.

Let us denote

$$\mathbf{q}_1 = \hat{\mathbf{x}} - \mathbf{x}, \qquad \mathbf{q}_2 = \hat{\boldsymbol{\theta}} - \boldsymbol{\theta}, \qquad \upsilon(\hat{\boldsymbol{\lambda}}, \boldsymbol{\lambda}, t) = \boldsymbol{\theta}^{\mathrm{T}}(\boldsymbol{\varphi}(y, \hat{\boldsymbol{\lambda}}, t) - \boldsymbol{\varphi}(y, \boldsymbol{\lambda}, t)).$$

Then, according to (5.165) and (5.160), we can write that

$$\begin{pmatrix} \dot{\mathbf{q}}_1 \\ \dot{\mathbf{q}}_2 \end{pmatrix} = \begin{pmatrix} A + \boldsymbol{\ell}\mathbf{c}^{\mathrm{T}} & \mathbf{b}\boldsymbol{\varphi}(y, \hat{\boldsymbol{\lambda}}, t)^{\mathrm{T}} \\ -k\boldsymbol{\varphi}(y, \hat{\boldsymbol{\lambda}}, t)\mathbf{c}^{\mathrm{T}} & 0 \end{pmatrix} \begin{pmatrix} \mathbf{q}_1 \\ \mathbf{q}_2 \end{pmatrix}$$
$$+ \begin{pmatrix} \mathbf{b} \\ 0 \end{pmatrix} \upsilon(\hat{\boldsymbol{\lambda}}, \boldsymbol{\lambda}, t) - \boldsymbol{\xi}_0(t),\ \boldsymbol{\xi}_0(t) = \mathrm{col}(\boldsymbol{\xi}(t), 0). \tag{A5.98}$$

Let $\Phi(t, t_0)$, $\Phi(t_0, t_0) = I$ be the fundamental system of solutions of the homogeneous part of (A5.98). Then the solution of (A5.98) is defined as

$$\mathbf{q}(t) = \Phi(t, t_0)\mathbf{q}(t_0) + \int_{t_0}^{t} \Phi(t, \tau)(\mathbf{b}_0\upsilon(\hat{\boldsymbol{\lambda}}(\tau), \boldsymbol{\lambda}, \tau) - \boldsymbol{\xi}_0(\tau))d\tau,$$
$$t \geq t_0,\ \mathbf{b}_0 = \mathrm{col}(\mathbf{b}, 0). \tag{A5.99}$$

We are going to show that there exists $\bar{\gamma} \in \mathbb{R}_{>0}$ such that for all $\gamma \in [0, \bar{\gamma})$ solutions of (A5.98) and (5.167) are bounded. First we notice that the right-hand side of (5.167) is locally Lipschitz in s_i. Hence the following estimate holds:

$$\|\dot{\hat{\boldsymbol{\lambda}}}(t)\| \leq \gamma^* M,\ M \in \mathbb{R}_{>0}, \quad \gamma^* = \gamma \max_j\{\omega_j\}. \tag{A5.100}$$

As follows from the assumptions of the theorem, the function $\boldsymbol{\varphi}(y, \boldsymbol{\lambda}, t)$ is uniformly persistently exciting. This implies the existence of $L, \mu \in \mathbb{R}_{>0}$ such that

$$J(\boldsymbol{\lambda}, t) = \int_t^{t+L} \boldsymbol{\varphi}(y(\tau), \boldsymbol{\lambda}, \tau)\boldsymbol{\varphi}^{\mathrm{T}}(y(\tau), \boldsymbol{\lambda}, \tau)d\tau \geq \mu I \;\forall\, t \geq t_0,\; \boldsymbol{\lambda} \in \Omega_\lambda. \tag{A5.101}$$

Consider the following matrix:

$$\begin{aligned} J(\hat{\boldsymbol{\lambda}}(t), t) &- \int_t^{t+L} \boldsymbol{\varphi}(y(\tau), \hat{\boldsymbol{\lambda}}(\tau), \tau)\boldsymbol{\varphi}^{\mathrm{T}}(y(\tau), \hat{\boldsymbol{\lambda}}(\tau), \tau)d\tau \\ &= \int_t^{t+L} (\boldsymbol{\varphi}(y(\tau), \hat{\boldsymbol{\lambda}}(t), \tau) - \boldsymbol{\varphi}(y(\tau), \hat{\boldsymbol{\lambda}}(\tau), \tau))\boldsymbol{\varphi}^{\mathrm{T}}(y(\tau), \hat{\boldsymbol{\lambda}}(t), \tau)d\tau \\ &\quad + \int_t^{t+L} \boldsymbol{\varphi}(y(\tau), \hat{\boldsymbol{\lambda}}(\tau), \tau)(\boldsymbol{\varphi}(y(\tau), \hat{\boldsymbol{\lambda}}(t), \tau) - \boldsymbol{\varphi}(y(\tau), \hat{\boldsymbol{\lambda}}(\tau), \tau))^{\mathrm{T}}\, d\tau \\ &= J_1(\hat{\boldsymbol{\lambda}}(t), t) + J_2(\hat{\boldsymbol{\lambda}}(t), t). \end{aligned} \tag{A5.102}$$

Using the inequality

$$\|H\mathbf{z}\| \leq \max_{k,l} |h_{k,l}|\|\mathbf{z}\|, \; H \in \mathbb{R}^{m\times n}, \; \mathbf{z} \in \mathbb{R}^n$$

and given that $\|\boldsymbol{\varphi}(y(t), \boldsymbol{\lambda}, t)\| \leq B$ for all $t \geq t_0, \boldsymbol{\lambda} \in \Omega_\lambda$, we can conclude that the matrix (A5.102) satisfies

$$\begin{aligned} &\left|\mathbf{z}^{\mathrm{T}}(J_1(\hat{\boldsymbol{\lambda}}(t), t) + J_2(\hat{\boldsymbol{\lambda}}(t), t))\mathbf{z}\right| \\ &\quad \leq 2BD\|\hat{\boldsymbol{\lambda}}(t) - \hat{\boldsymbol{\lambda}}(\tau)\|_{\infty,[t,t+L]}\|\mathbf{z}\|^2, \; D \in \mathbb{R}_{>0}. \end{aligned}$$

Hence

$$\begin{aligned} &\int_t^{t+L} \boldsymbol{\varphi}(y(\tau), \hat{\boldsymbol{\lambda}}(\tau), \tau)\boldsymbol{\varphi}^{\mathrm{T}}(y(\tau), \hat{\boldsymbol{\lambda}}(\tau), \tau)d\tau \\ &\quad \geq J(\hat{\boldsymbol{\lambda}}(t), t) - 2BD\|\hat{\boldsymbol{\lambda}}(t) - \hat{\boldsymbol{\lambda}}(\tau)\|_{\infty,[t,t+L]}I \\ &\quad \geq (\mu - 2BD\|\hat{\boldsymbol{\lambda}}(t) - \hat{\boldsymbol{\lambda}}(\tau)\|_{\infty,[t,t+L]})I. \end{aligned} \tag{A5.103}$$

Taking (A5.100) and (A5.103) into account, we can conclude that

$$\int_t^{t+L} \boldsymbol{\varphi}(y(\tau), \hat{\boldsymbol{\lambda}}(\tau), \tau)\boldsymbol{\varphi}^{\mathrm{T}}(y(\tau), \hat{\boldsymbol{\lambda}}(\tau), \tau)d\tau \geq (\mu - 2BDLM\gamma^*)I. \tag{A5.104}$$

Then, by choosing γ for instance such that

$$\gamma^* = \frac{\mu}{4BDLM}, \tag{A5.105}$$

we can ensure that

$$\int_t^{t+L} \boldsymbol{\varphi}(y(\tau), \hat{\boldsymbol{\lambda}}(\tau), \tau)\boldsymbol{\varphi}^{\mathrm{T}}(y(\tau), \hat{\boldsymbol{\lambda}}(\tau), \tau)d\tau \geq \frac{\mu}{2} I \;\forall\, t \geq t_0.$$

Then according to Theorem 2.2, there exist $\rho,\ D_\rho \in \mathbb{R}_{>0}$ such that the following inequality holds:

$$\|\Phi(t, t_0)\mathbf{q}(t_0)\| \leq e^{-\rho(t-t_0)}\|\mathbf{q}(t_0)\| D_\rho. \tag{A5.106}$$

Therefore, taking (A5.99) and (A5.106) into account, we can conclude that, for any bounded and continuous function $\upsilon(t)$, the solutions of (A5.98) satisfy the following estimate:

$$\begin{aligned}\|\mathbf{q}(t)\| &= \left\| \Phi(t, t_0)\mathbf{q}(t_0) + \int_{t_0}^{t} \Phi(t, \tau)\mathbf{b}\upsilon(\tau)d\tau \right\| \\ &\leq e^{-\rho(t-t_0)} D_\rho \|\mathbf{q}(t_0)\| \\ &\quad + D_\rho \int_{t_0}^{t} e^{-\rho(t-\tau)}(\|\mathbf{b}_0\| |\upsilon(\hat{\boldsymbol{\lambda}}(\tau), \boldsymbol{\lambda}, \tau)| + \|\boldsymbol{\xi}_0(\tau)\|)d\tau.\end{aligned} \tag{A5.107}$$

Notice that

$$|\upsilon(\hat{\boldsymbol{\lambda}}(t), \boldsymbol{\lambda}, t)| \leq \|\boldsymbol{\theta}\| D \|\hat{\boldsymbol{\lambda}}(t) - \boldsymbol{\lambda}\|, \qquad \|\boldsymbol{\xi}_0(t)\| \leq \Delta_\xi.$$

Hence

$$\begin{aligned}\|\mathbf{q}(t)\| &\leq e^{-\rho(t-t_0)} D_\rho \|\mathbf{q}(t_0)\| + \|\boldsymbol{\theta}\| \|\mathbf{b}\| D D_\rho \int_{t_0}^{t} e^{-\rho(t-\tau)} \|\hat{\boldsymbol{\lambda}}(\tau) - \boldsymbol{\lambda}\| d\tau \\ &\quad + D_\rho \int_{t_0}^{t} e^{-\rho(t-\tau)} \Delta_\xi \, d\tau.\end{aligned}$$

Therefore, for all $t \geq t_0 \geq t'$ the following estimate holds:

$$\|\mathbf{q}(t)\| \leq e^{-\rho(t-t_0)} D_\rho \|\mathbf{q}(t_0)\| + \frac{\|\boldsymbol{\theta}\| \|\mathbf{b}\| D D_\rho}{\rho} \|\hat{\boldsymbol{\lambda}}(\tau) - \boldsymbol{\lambda}\|_{\infty,[t_0,t]} + \frac{\Delta_\xi D_\rho}{\rho}. \tag{A5.108}$$

Now consider solutions of the system

$$\begin{aligned}\dot{s}_{2j-1} &= \gamma \cdot \omega_j \left(s_{2j-1} - s_{2j} - s_{2j-1}\left(s_{2j-1}^2 + s_{2j}^2\right)\right), \\ \dot{s}_{2j} &= \gamma \cdot \omega_j \left(s_{2j-1} + s_{2j} - s_{2j}\left(s_{2j-1}^2 + s_{2j}^2\right)\right), \\ \hat{\lambda}_j &= \lambda_{j,\min} + ((\lambda_{j,\max} - \lambda_{j,\min})/2)(s_{2j-1} + 1),\end{aligned} \tag{A5.109}$$

with initial conditions (5.168). They are forward-invariant on $s_{2j-1}^2(t_0)+s_{2j}^2(t_0)=1$ and can be expressed as

$$(s_{2j-1}(t,\mathbf{s}_0), s_{2j}(t,\mathbf{s}_0)) = \left(\cos\left(\gamma \int_{t_0}^{t} \omega_i \tau d\tau + \beta\right), \right.$$
$$\left. \sin\left(\gamma \int_{t_0}^{t} \omega_i \tau d\tau + \beta\right)\right), \ \beta \in \mathbb{R}.$$

Taking into account that ω_j are rationally independent (condition (5.169) in the definition of the observer), we can conclude that the trajectories $\bar{s}_{2j-1}(t)$ densely fill an invariant n-dimensional torus (Arnold 1978), and that system (A5.109) with initial condition $s_{2j-1}^2(t_0) + s_{2j}^2(t_0) = 1$ is Poisson-stable in $\Omega_s = \{s_{2j-1} \in \mathbb{R} | s_{2j-1} \in [-1,1]\}$. This means that for any $\boldsymbol{\lambda} \in \Omega_\lambda$, (arbitrarily large) constant $T \in \mathbb{R}_{\geq 0}$, (arbitrarily small) constant $\Delta_\lambda \in \mathbb{R}_{\geq 0}$, and initial conditions on the torus there will exist $\bar{\boldsymbol{\lambda}}(t') \in \Omega_\lambda$, $t' - t_0 \geq T$, such that $\|\boldsymbol{\lambda} - \bar{\boldsymbol{\lambda}}(t')\| \leq \Delta_\lambda$, $\Delta_\lambda \in \mathbb{R}_{\geq 0}$. Expressing $\hat{\boldsymbol{\lambda}}(t)$ generated by (5.167) in terms of the function $\bar{\boldsymbol{\lambda}}(t)$ yields

$$\hat{\boldsymbol{\lambda}}(t) = \bar{\boldsymbol{\lambda}}\left(t_0 + \gamma \int_{t_0}^{t} \sigma(\|y(\tau) - \hat{y}(\tau)\|_\varepsilon) d\tau\right), \ t \geq t_0.$$

Denoting

$$h(t) = t' - \gamma \int_{t_0}^{t} \sigma(\|y(\tau) - \hat{y}(\tau)\|_\varepsilon) d\tau,$$

we obtain

$$\begin{aligned} \|\hat{\boldsymbol{\lambda}}(t) - \boldsymbol{\lambda}\| &\leq \Delta_\lambda + \|\hat{\boldsymbol{\lambda}}(t) - \bar{\boldsymbol{\lambda}}(t')\| \leq \Delta_\lambda + D_\lambda |h(t)|, \\ h(t) &= h(t_0) - \gamma \int_{t_0}^{t} \sigma(\|y(\tau) - \hat{y}(\tau)\|_\varepsilon) d\tau, \end{aligned} \tag{A5.110}$$

where D_λ is a positive constant and $h(t_0) = t'$. Hence the following holds along trajectories of the observer:

$$\begin{aligned} \|\mathbf{q}(t)\| &\leq e^{-\rho(t-t_0)} D_\rho \|\mathbf{q}(t_0)\| + c D_\lambda \|h(t)\|_{\infty,[t_0,t]} \\ &\quad + \frac{\Delta_\xi D_\rho}{\rho} + c\Delta_\lambda, \ c = \frac{\|\boldsymbol{\theta}\| \|\mathbf{b}\| D D_\rho}{\rho}, \\ h(t) &= h(t_0) - \gamma \int_{t_0}^{t} \sigma(\|y(\tau) - \hat{y}(\tau)\|_\varepsilon) d\tau \Rightarrow \\ h(t_0) &- \gamma \int_{t_0}^{t} \|\mathbf{q}(\tau)\|_\varepsilon \, d\tau \leq h(t) \leq h(t_0), \end{aligned} \tag{A5.111}$$

where Δ_λ is an arbitrarily small constant. Now, invoking Corollary 4.4, we can conclude that there exist γ^* and ε^* such that

$$\lim_{t\to\infty} \hat{\boldsymbol{\lambda}}(t) = \boldsymbol{\lambda}^*, \ \boldsymbol{\lambda}^* \in \Omega_\lambda. \tag{A5.112}$$

for all $\gamma \in (0, \gamma^*)$, $\varepsilon \geq \varepsilon^*$. Moreover

$$\lim_{t\to\infty} \|y(t) - \hat{y}(t)\|_\varepsilon = 0. \tag{A5.113}$$

The value of γ^*, as follows from Corollary 4.4, (A5.105), and the fact that the value of $h(t_0)$ can be chosen arbitrarily large, can be determined as

$$\gamma^* = \min\left\{\frac{\mu}{4BDLM}, \mathcal{G}\right\},$$

$$\mathcal{G} = \sup_{d\in(0,1),\ \psi\in(1,\infty)} \frac{\psi-1}{\psi}\rho^2\left[\ln\left(\frac{\psi D_\rho}{d}\right)\right]^{-1} \frac{1}{\|\boldsymbol{\theta}\| D D_\rho D_\lambda}\left(1+\frac{D_\rho \psi}{1-d}\right)^{-1}.$$

The value of ε^*, according to (4.112) and (A5.111), can be chosen as

$$\varepsilon^* = \left(\frac{\Delta_\xi D_\rho}{\rho} + c\Delta_\lambda\right)(D_\rho/K+1),$$

where K is a positive constant.

To proceed further we will need the following lemma.

Lemma A5.4 *Consider the system*

$$\begin{aligned} \dot{\mathbf{x}} &= A(t)\mathbf{x} + \mathbf{b}(t)u(t) + \mathbf{d}(t), \\ y &= \mathbf{c}^T\mathbf{x}, \quad \mathbf{c} \in \mathbb{R}^n, \quad \mathbf{x} \in \mathbb{R}^n, \end{aligned} \tag{A5.114}$$

where

$$\begin{aligned} &u : \mathbb{R} \to \mathbb{R},\ u \in \mathcal{C}^1,\ \mathbf{b} : \mathbb{R} \to \mathbb{R}^n,\ A(t) : \mathbb{R} \to \mathbb{R}^{n\times n}, \\ &\mathbf{d} : \mathbb{R} \to \mathbb{R},\ \mathbf{d}, \mathbf{b}, A \in \mathcal{C}^0, \end{aligned} \tag{A5.115}$$

and each entry of $A(t)$ is a globally bounded function of t. Suppose that the following properties hold for (A5.114):

(1) the origin of system (A5.114) is asymptotically stable at $u(t) = 0$, $\mathbf{d}(t) = 0$;
(2) the term $\|\mathbf{c}^T\Phi(t,\tau)\mathbf{d}(\tau)\| \leq \delta$, $\delta \in \mathbb{R}_{\geq 0}$ for all $t \in \mathbb{R}$, $\tau \leq t$, where $\Phi(t,\tau)$, $\Phi(t,t) = I$, is the fundamental system of solutions of the homogeneous part of (A5.114);
(3) the term $\mathbf{c}^T\mathbf{b}(t)$ is separated away from zero:

$$\exists\, \beta \in \mathbb{R}_{>0} : \ |\mathbf{c}^T\mathbf{b}(t)| \geq \beta\ \forall\, t \in \mathbb{R};$$

(4) the time derivative of $u(t)$ is bounded uniformly in t:

$$\exists\, \partial U \in \mathbb{R}_{>0}: \quad |\dot{u}(t)| \leq \partial U \ \forall\, t \in \mathbb{R}. \tag{A5.116}$$

Then

$$\|y(t)\|_{\infty,[t_0,\infty]} < \varepsilon, \ \varepsilon \in \mathbb{R}_{>0} \tag{A5.117}$$

implies the existence of $t'(\varepsilon) \geq t_0$ such that

$$2\sqrt{\frac{6\varepsilon\partial U}{\beta}} + \frac{2\delta}{\beta} \geq |u(t)|, \ \forall\, t \geq t'(\varepsilon),$$

provided that ε is sufficiently small.

Proof of Lemma A5.4. First, notice that closed-form solutions of (A5.114) are given by

$$\mathbf{x}(t) = \Phi(t,t_0)\mathbf{x}_0 + \int_{t_0}^{t} \Phi(t,\tau)[\mathbf{b}(\tau)u(\tau) + \mathbf{d}(\tau)]d\tau, \tag{A5.118}$$

where $\Phi(t,t_0)$ is the fundamental system of solutions of the homogeneous part of (A5.114). According to (A5.118) we have

$$y(t) = \mathbf{c}^{\mathrm{T}}\Phi(t,t_0)\mathbf{x}_0 + \int_{t_0}^{t} \mathbf{c}^{\mathrm{T}}\Phi(t,\tau)[\mathbf{b}(\tau)u(\tau) + \mathbf{d}(\tau)]d\tau.$$

Consider $y(t) - y(t-T)$, $T \in \mathbb{R}_{>0}$:

$$\begin{aligned} y(t) - y(t-T) &= \mathbf{c}^{\mathrm{T}}\left[\Phi(t,t_0) - \Phi(t-T,t_0)\right]\mathbf{x}_0 \\ &\quad + \int_{t-T}^{t} \mathbf{c}^{\mathrm{T}}\Phi(t,\tau)[\mathbf{b}(\tau)u(\tau) + \mathbf{d}(\tau)]d\tau. \end{aligned} \tag{A5.119}$$

Denoting

$$v(t,T) = \mathbf{c}^{\mathrm{T}}\left[\Phi(t,t_0) - \Phi(t-T,t_0)\right]\mathbf{x}_0$$

and applying the mean-value theorem to (A5.119), we obtain

$$y(t) - y(t-T) = v(t,T) + T\mathbf{c}^{\mathrm{T}}\Phi(t,\tau')[\mathbf{b}(\tau')u(\tau') + \mathbf{d}(\tau')], \ \tau' \in [t, t-T].$$

Hence

$$\begin{aligned} |y(t) - y(t-T)| &\geq T|\mathbf{c}^{\mathrm{T}}\Phi(t,\tau')\mathbf{b}(\tau')||u(\tau')| \\ &\quad - T|\mathbf{c}^{\mathrm{T}}\Phi(t,\tau')\mathbf{d}(\tau')| - |v(t,T)|, \ \tau' \in [t, t-T]. \end{aligned}$$

Because (A5.114) is asymptotically stable, the term $|v(t,T)| \to 0$ as $t \to \infty$, and there exists $t_1 \geq \mathbb{R}$ such that

$$|v(t,T)| < \varepsilon \ \forall\, t \geq t_1.$$

On the other hand, condition (A5.117) assures the existence of t_0 such that

$$|y(t) - y(t-T)| \leq 2\varepsilon \ \forall\, t \geq t_0, \ \forall\, T > 0.$$

Therefore there exists $t' = \max(t_1, t_0)$ such that for all $T \geq 0$ and $t \geq t'$ the following holds:

$$3\varepsilon \geq T|\mathbf{c}^{\mathrm{T}}\Phi(t,\tau')\mathbf{b}(\tau')||u(\tau')| - |\mathbf{c}^{\mathrm{T}}\Phi(t,\tau')\mathbf{d}(\tau')|, \ \tau' \in [t-T, t]. \quad \text{(A5.120)}$$

According to (A5.116) and (A5.115), inequality (A5.120) implies

$$3\varepsilon \geq (T|\mathbf{c}^{\mathrm{T}}\Phi(t,\tau')\mathbf{b}(\tau')||u(t)| - \delta) - T^2|\mathbf{c}^{\mathrm{T}}\Phi(t,\tau')\mathbf{b}(\tau')|\,\partial U, \ \tau' \in [t-T, t].$$

Because $\Phi(t,t) = I \ \forall\, t \in \mathbb{R}$, $\Phi(t,\tau')$ is continuous in τ', and $|\mathbf{c}^{\mathrm{T}}\Phi(t,t)\mathbf{b}(t)| \geq \beta$, we can conclude that there exists $T' \in \mathbb{R}_{>0}$ such that

$$|\mathbf{c}^{\mathrm{T}}\Phi(t,\tau')\mathbf{b}(\tau')| \geq M = \frac{\beta}{2} \ \forall\, \tau' \in [t-T', t]. \quad \text{(A5.121)}$$

On letting $0 < T \leq T'$ we obtain

$$\frac{3\varepsilon}{MT} + T\,\partial U + \frac{\delta}{M} \geq |u(t)| \ \forall\, t \geq t'. \quad \text{(A5.122)}$$

Optimizing the left-hand side of (A5.122) for T yields

$$2\sqrt{\frac{3\varepsilon\,\partial U}{M}} + \frac{\delta}{M} = 2\sqrt{\frac{6\varepsilon\,\partial U}{\beta}} + \frac{2\delta}{\beta} \geq |u(t)| \ \forall\, t \geq t',$$

provided that

$$\sqrt{\frac{6\varepsilon\,\partial U}{\beta}} \leq T', \quad \text{(A5.123)}$$

i.e. provided that ε is sufficiently small. □

Let us now go back to the proof of the theorem. As follows from (A5.113), there exists a time instant t' such that

$$\|y(t) - \hat{y}(t)\| \leq 2\varepsilon \ \forall\, t \geq t'.$$

In addition, the vectors $\mathbf{c}_0 = \mathrm{col}(\mathbf{c}, 0)$ and $\mathbf{b}_0$ in (A5.98) satisfy $\mathbf{c}_0^{\mathrm{T}}\mathbf{b}_0 = b_1 > 0$; the term $\mathbf{c}_0^{\mathrm{T}}\Phi(t,\tau)\boldsymbol{\xi}_0(\tau)$, $\tau < t$, where $\Phi(t,\tau)$, $\Phi(t,t) = I$ is the fundamental system of solutions of the homogeneous part of (A5.98), can be bounded from above as $\|\mathbf{c}_0^{\mathrm{T}}\Phi(t,\tau)\boldsymbol{\xi}_0(\tau)\| \leq v\Delta_{\xi}$, where v is some positive constant. Finally, according to the assumptions of the theorem, there exists a constant $\partial U \in \mathbb{R}_{>0}$ such that

$$\left|\frac{d}{dt}v_0(\hat{\boldsymbol{\lambda}}, \boldsymbol{\lambda}, t)\right| \leq \partial U.$$

Hence the conditions of Lemma A5.4 are satisfied for the system (A5.98) and (5.167) for all $t \geq t_0$.

Therefore, according to Lemma A5.4 there exists $t_1 \geq t_0$ such that

$$|\boldsymbol{\theta}^{\mathrm{T}}(\boldsymbol{\varphi}_0(\mathbf{x}_0(t), \boldsymbol{\lambda}, t) - \boldsymbol{\varphi}_0(\mathbf{x}_0(t), \hat{\boldsymbol{\lambda}}(t), t))| \leq 2\left(\sqrt{\frac{6\varepsilon\partial U}{b_1}} + \frac{\upsilon\Delta_\xi}{b_1}\right) \ \forall\, t \geq t_1 \geq t_0, \tag{A5.124}$$

provided that ε is sufficiently small. Hence, it follows from (A5.106) and (A5.107) that

$$\limsup_{t\to\infty} \|\hat{\boldsymbol{\theta}}(t) - \boldsymbol{\theta}\| \leq \frac{2D_\rho\|\mathbf{b}_0\|}{\rho}\left(\sqrt{\frac{6\varepsilon\partial U}{b_1}} + \frac{\upsilon\Delta_\xi}{b_1}\right) + \frac{\Delta_\xi D_\rho}{\rho}.$$

Notice that Δ_λ can be chosen arbitrarily small. In particular, we can choose it to be proportional to Δ_ξ. This proves statement (1) of the theorem.

To prove that statement (2) of the theorem holds, we notice that, according to the definition of weak nonlinear persistency of excitation, constants L and β must exist such that for all $t \in \mathbb{R}$ there exists $t' \in [t - L, t]$:

$$|\boldsymbol{\theta}^{\mathrm{T}}(\boldsymbol{\varphi}_0(\mathbf{x}_0(t'), \boldsymbol{\lambda}, t') - \boldsymbol{\varphi}_0(\mathbf{x}_0(t'), \hat{\boldsymbol{\lambda}}(t'), t'))| \geq \beta \cdot \mathrm{dist}(\hat{\boldsymbol{\lambda}}(t'), \mathcal{E}).$$

Combining this with (A5.124) yields

$$\forall\, t \geq t_1 \ \exists\, t' \in [t - L, t] : \mathrm{dist}(\hat{\boldsymbol{\lambda}}(t'), \mathcal{E}(\boldsymbol{\lambda})) \leq \frac{2}{\beta}\left(\sqrt{\frac{6\varepsilon\,\partial U}{b_1}} + \frac{\upsilon\Delta_\xi}{b_1}\right). \tag{A5.125}$$

Hence there is a sequence of t_i', $i = 1, 2, \ldots$ such that $3L \geq t_i' - t_{i-1}' \geq L$, and

$$\lim_{i\to\infty} \mathrm{dist}(\hat{\boldsymbol{\lambda}}(t_i'), \mathcal{E}(\boldsymbol{\lambda})) \leq \frac{2}{\beta}\left(\sqrt{\frac{6\varepsilon\partial U}{b_1}} + \frac{\upsilon\Delta_\xi}{b_1}\right).$$

Given that $\hat{\boldsymbol{\lambda}}(t_i')$ are bounded, we can choose t_j' such that

$$\lim_{t_j\to\infty} \hat{\boldsymbol{\lambda}}(t_j) = \bar{\boldsymbol{\lambda}} \in \Omega_\lambda.$$

On the other hand, according to (A5.112), $\hat{\boldsymbol{\lambda}}(t)$ converges to a point in Ω_λ as $t \to \infty$: $\lim_{t\to\infty} \hat{\boldsymbol{\lambda}}(t) = \boldsymbol{\lambda}^*$. Therefore $\boldsymbol{\lambda}^* = \bar{\boldsymbol{\lambda}}$ and

$$\mathrm{dist}(\boldsymbol{\lambda}^*, \mathcal{E}(\boldsymbol{\lambda})) \leq \frac{2}{\beta}\left(\sqrt{\frac{6\varepsilon\partial U}{b_1}} + \frac{\upsilon\Delta_\xi}{b_1}\right).$$

The latter inequality assures that

$$\limsup_{t\to\infty} \operatorname{dist}(\hat{\boldsymbol{\lambda}}(t), \mathcal{E}(\boldsymbol{\lambda})) \leq \frac{2}{\beta}\left(\sqrt{\frac{6\varepsilon \partial U}{b_1}} + \frac{\upsilon \Delta_\xi}{b_1}\right).$$

Hence, taking into account that the value of Δ_λ can be set proportional to Δ_ξ, statement (2) of the theorem follows. □

Part III

Applications

6

Adaptive behavior in recurrent neural networks with fixed weights

Recurrent neural networks (RNNs) with fixed weights are known to solve problems of adaptive classification, recognition, and control (Prokhorov *et al.* 2002a; Feldkamp *et al.* 1996; Feldkamp and Puskorius 1997; Younger *et al.* 1999; Lo 2001). When the objects to be classified are static, e.g. still images or vectors in $\mathbb{R}^n$, solutions to these problems are usually characterized in terms of convergence of the RNN state to an *attractor* (Hopfield 1982; Fuchs and Haken 1988). Each attractor corresponds to a specific class of objects and its basin determines which objects belong to the class. Conditions specifying convergence to an attractor are widely available in this case (Cohen and Grossberg 1983; Michel *et al.* 1989; Yang and Dillon 1994; Chen and Amari 2001; Lu and Chen 2003).

When the objects to be classified are dynamic, for instance nonlinearly parametrized functions of time of which the parameters are unknown a priori, no adequate theory exists that explains why the fixed-weight RNN approach is successful. At present, theoretical results are available to demonstrate that a single fixed-weight RNN of a certain type can *approximate* the solutions of multiple dynamical systems (Back and Chen 2002). Hence in principle, a fixed-weight RNN can behave adaptively with respect to changes of its input signals. These theoretical results, however, are restricted to the class of parameter-replacement networks (Chen and Chen 1995). The structure of these networks differs from that of the more commonly used recurrent multilayered perceptrons.

The question is whether adaptive behavior is inherent to other types of RNN. Several authors have given plausibility arguments that RNNs with conventional multilayer architecture should also have this ability (Feldkamp and Puskorius 1997; Prokhorov *et al.* 2002a). Here we will show how theoretical results presented in Chapters 4 and 5 can be used to derive a formal proof of adaptive behavior in fixed-weight nets (Tyukin *et al.* 2008b).

We consider adaptive behavior in fixed-weight RNNs with a view to their ability to *classify temporal signals* adaptively. We provide a formal proof that *continuous-time* recurrent neural networks with fixed weights can successfully classify and recognize nonlinear functions of time, which are allowed to be nonlinearly parametrized with unknown parameter values. The main idea behind our results consists of presenting a prototype dynamical system that solves the recognition problem. We then present a proof that an RNN with fixed weights can realize this system. We construct such a system using the concept of *weakly attracting sets* presented in Milnor (1985) and Gorban (2004) reviewed in Chapter 2 as well as the tests for convergence to such sets obtained in Chapter 4 (see also Tyukin *et al.* (2008a)). To show that our system can indeed be realized by an RNN with fixed weights, we employ classical results on function approximation by feed-forward networks (Cybenko 1989).

6.1 Signals to be classified

We consider the following set of signals:

$$\begin{aligned} &\mathcal{F} = \{f_i(\xi(t),\theta_i)\},\ i \in \{1,\ldots,N_f\},\\ &f_i : \mathbb{R}\times\mathbb{R} \to \mathbb{R},\ f_i(\cdot,\cdot) \in \mathcal{C}^0,\\ &\xi : \mathbb{R}_{\geq 0} \to \mathbb{R},\ \xi(\cdot) \in \mathcal{C}^1 \cap L_\infty[0,\infty], \end{aligned} \tag{6.1}$$

where $\theta_i \in \Omega_\theta \subset \mathbb{R}$ are parameters of which the values are unknown a priori, $\Omega_\theta = [\theta_{\min}, \theta_{\max}]$ is a bounded interval, and $\xi(t)$ is a known and bounded function. Signals $f_i(\xi(t),\theta_i)$ constitute the set of variables chosen to represent the state of an object.

For the given functions $f_i(\xi(t),\theta_i)$ and $\xi(t)$ we say that θ_i is *equivalent* to θ_i' iff

$$f_i(\xi(t),\theta_i) = f_i(\xi(t),\theta_i')\ \forall\, t \in \mathbb{R}_{\geq 0}. \tag{6.2}$$

Hence, an equivalence class for $\theta_i \in \Omega_\theta$ can be defined as

$$E_i(\theta_i) = \{\theta_i' \in \mathbb{R} \,|\, f_i(\xi(t),\theta_i) = f_i(\xi(t),\theta_i')\ \forall\, t \in \mathbb{R}_{\geq 0}\}. \tag{6.3}$$

Equivalence classes (6.3) determine sets of indistinguishable parametrizations of the ith signal. It is natural, therefore, to restrict ourselves to the problem of recognizing signals (6.1) up to their equivalence classes.

With respect to the equivalence classes $E_i(\theta_i)$, we further assume that there is at least one point $\theta_0 \in \mathbb{R}$ such that

$$\|\theta_0\|_{E_i(\theta_i)} \geq \Delta_\theta \in \mathbb{R}_{>0}\ \forall\, \theta_i \in \Omega_\theta. \tag{6.4}$$

Requirement (6.4) is a technical assumption. It is satisfied in a wide range of practically relevant situations in which the union of $E_i(\theta_i)$ for all i and θ_i belongs to an interval of $\mathbb{R}$. The requirement allows us to exclude from consideration pathological cases in which almost all points in Ω_θ are indistinguishable in the sense of condition (6.2).

Consider the case in which the values of $f_i(\xi(t), \theta_i)$ might not be available for direct observation. Assume that instead of functions $f_i(\xi(t), \theta_i)$ we access variables $s_i(t, s_{i,0}, \theta_i, \eta_i(t))$, which are solutions to the following ordinary differential equation:

$$\begin{aligned} \dot{s}_i &= -\varphi_i(s_i) + f_i(\xi(t), \theta_i) + \eta_i(t), \\ s_i(t_0) &= s_{i,0}, \ s_{i,0} \in \Omega_s \subset \mathbb{R}. \end{aligned} \tag{6.5}$$

In (6.5) the function $\eta_i : \mathbb{R}_{>0} \to \mathbb{R}$,

$$\eta_i(t) \in L_\infty[0, \infty], \ \|\eta_i(t)\|_{\infty,[0,\infty]} \leq \Delta_\eta \in \mathbb{R}_{\geq 0}, \tag{6.6}$$

corresponds to measurement noise. The value of Δ_η in (6.5) is supposed to be known, while the values of initial conditions $s_i(t_0)$ and functions $\varphi_i : \mathbb{R} \to \mathbb{R}$, $\varphi(\cdot) \in \mathcal{C}^1$ in (6.5) are assumed to be uncertain. We do, however, require that $\Omega_s = [s_{\min}, s_{\max}]$ is an interval and that the functions $\varphi_i(s_i)$ satisfy the following constraint:

$$\forall\, s_i \in \mathbb{R} \Rightarrow \varphi_{\min} \leq \frac{\partial \varphi_i(s_i)}{\partial s_i} \leq \varphi_{\max}, \ \varphi_{\min}, \varphi_{\max} \in \mathbb{R}_{>0}. \tag{6.7}$$

Condition (6.7) ensures that the dynamics of each variable $s_i(t, s_{i,0}, \theta_i, \eta_i(t))$ at $t \to \infty$ is uniquely determined in the absence of noise by $f_i(\xi(t), \theta_i)$, and the effects of initial conditions $s_{i,0}$ vanish with time asymptotically. In other words, solutions $s_i(t, s_{i,0}, \theta_i, 0)$ will asymptotically approach a function of time of which the shape is uniquely determined by $f_i(\xi(t), \theta_i)$.

The reason why we consider signals (6.5)–(6.7) instead of $f_i(\xi(t), \theta_i)$ is that in many systems, artificial or natural, measured physical quantities, represented here by signals $f_i(\xi(t), \theta_i)$, are often unavailable. In the domain of neural computation and modeling of neural systems this is the case when information about the stimuli is transferred from one node to another by dynamic synapses (Tsodyks *et al.* 1998). In robotics and motor control the intrinsic dynamics of sensors and moving mechanical parts often distort stimulus information $f_i(\xi(t), \theta_i)$. Notice also that solutions to the differential equation

$$\begin{aligned} \dot{s}_i &= -\tau(s_i - f_i(\xi(t), \theta_i)) + \eta_i(t), \ \tau \in \mathbb{R}_{>0}, \\ \eta_i(t) &= \frac{d}{dt}(f_i(\xi(t), \theta_i)) \end{aligned}$$

with initial conditions $s_i(t_0) = f_i(\xi(t_0), \theta_i)$ coincide with $f_i(\xi(t), \theta_i)$ for all $t \geq t_0$.[1] Therefore, in cases in which the derivative of $\xi(t)$ is uniformly bounded in t, it is always possible to consider signals $f_i(\xi(t), \theta_i)$ as if they were generated by (6.5)–(6.7) with appropriately chosen parameters and terms $\tau f_i(\xi(t), \theta_i)$ instead of $f_i(\xi(t), \theta_i)$. Hence, even in the absence of actual dynamics, we can still use the representation (6.5)–(6.7) for a broad class of signals (6.1), in which the time-derivative of $\xi(t)$ is uniformly bounded in t.

6.2 The class of recurrent neural networks

The following set of differential equations defines a recurrent neural network:

$$\dot{x}_j = \sum_{m=1}^{N} c_{j,m}\sigma(\mathbf{w}_{j,m}^{\mathrm{T}}\mathbf{u}(s(t), \xi(t), \mathbf{x}) + b_{j,m}), \ j \in \{1, \ldots, N_x\},$$
$$\mathbf{u}(s(t), \xi(t), \mathbf{x}(t)) = s(t) \oplus \xi(t) \oplus \mathbf{x},$$
$$\mathbf{x} = \mathrm{col}(x_1, \ldots, x_{N_x}), \ \mathbf{x}(t_0) = \mathbf{x}_0, \tag{6.8}$$

where functions $\sigma : \mathbb{R} \to \mathbb{R}$, $\sigma \in \mathcal{C}^0$ are sigmoid.[2] The vectors $\mathbf{c}_j = \mathrm{col}(c_{j,1}, \ldots, c_{j,N})$ and $\mathbf{b}_j = \mathrm{col}(b_{j,1}, \ldots, b_{j,N})$ and matrices $\boldsymbol{W}_j = (\mathbf{w}_{j,1}, \ldots, \mathbf{w}_{j,N})$ are parameters of the RNN. Functions $\xi(t)$, $s(t) : \mathbb{R}_{\geq 0} \to \mathbb{R}$, $s(t), \xi(t) \in \mathcal{C}^0$ are inputs, $\mathbf{x}$ is the state vector, and $\mathbf{x}_0$ is a vector of initial conditions.

Figure 6.1 shows the general structure of the network; it also illustrates how the network receives information about the original signals $f_i(\xi(t), \theta_i)$. According to the notation (6.8) the network maps two functions of time, $s(t)$ and $\xi(t)$, into the functions $x_1(t, \mathbf{x}_0)$, ..., $x_{N_x}(t, \mathbf{x}_0)$, which are the solutions of (6.8). In what follows we will consider the variables $s(t)$ and $\xi(t)$ as inputs to the network. The function of the first input, $s(t)$, is to communicate information about the signal to be classified. The function of the second input, $\xi(t)$, is to provide the network with enough information to ensure that classification is successful. Our choice of inputs is motivated by known results from systems and control theory; if a system adapts to a class of external signals it must contain a subsystem, or internal model, that is capable of generating all of the input signals (Conant and Ashby 1970; Sontag 2003). In our case this corresponds to the case in which the network is to model

[1] This can easily be verified by taking the time-derivative of $e_i(t) = s_i(t, s_{i,0}, \theta_i, \eta_i(t)) - f_i(\xi(t), \theta_i))$. It will satisfy the equation $\dot{e}_i = -\tau e_i$, yielding the target equality $e_i(t) = e^{-(t-t_0)} e_i(t_0) = 0 \ \forall \, t \geq t_0$.

[2] A function $\sigma : \mathbb{R} \to \mathbb{R}$ is called sigmoid if $\lim_{z\to\infty} \sigma(z) = 1$ and $\lim_{z\to-\infty} \sigma(z) = 0$. Choosing the function σ in (6.8) in the class of sigmoid functions is not critical for our analysis. In our proof we will require only that the sums $\sum_{i=1}^{N} c_i \sigma(\mathbf{w}_i^{\mathrm{T}}\mathbf{u} + b_i)$, c_i, $b_i \in \mathbb{R}$, $\mathbf{w}_i \in \mathbb{R}^{N_x+2}$ are dense in $\mathcal{C}^0([0, 1]^{N_x+2})$. Hence any continuous function σ satisfying this requirement can be used in (6.8). Detailed discussion and specification of such functions can be found in Chen and Chen (1995).

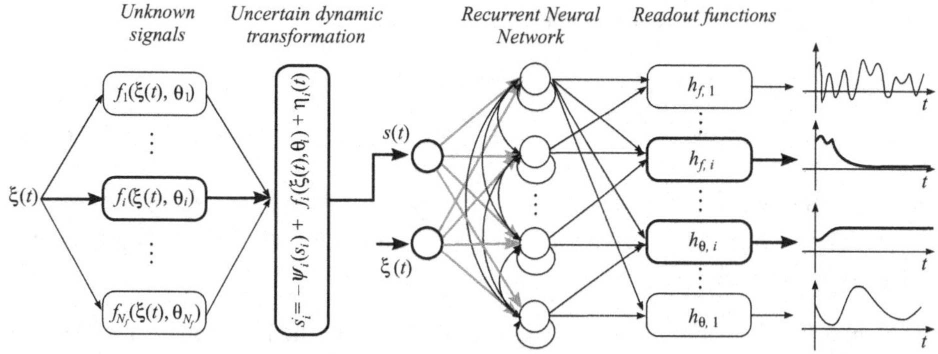

Figure 6.1 The structure of the network and routing of the signals. The network receives signals $s(t)$ and $\xi(t)$ on its inputs, and maps these functions into the trajectories of its state. Classification decisions are communicated through the state readout functions. The latter can in principle be realized by a feedforward component. Here we restrict our consideration to the mere existence of these functions (see Problem 6.1).

the functions $f_i(\xi(t), \theta_i)$. Hence in general knowledge of $\xi(t)$ is necessary for successful classification.[3]

6.3 Assumptions and statement of the problem

While the variable $\xi(t)$ is known a priori, the variable $s(t)$ is allowed to vary within the set of functions $s_i(t, s_{i,0}, \theta_i, \eta_i(t))$, which are the solutions of (6.5). In particular, we assume that the following condition is satisfied.

Assumption 6.1 *There exist $i \in N_f$, $\theta_i \in \Omega_\theta$, $s_{i,0} \in \Omega_s$, and $\eta_i(t)$ specified by (6.6) such that*

$$s(t) = s_i(t, s_{i,0}, \theta_i, \eta_i(t)) \; \forall\, t \geq 0. \tag{6.9}$$

We aim to determine whether there is a network of type (6.8) that is able to recover uncertain parameters i and θ_i from the input $s(t)$,[4] $t \geq t_0 \in \mathbb{R}_{\geq 0}$ within a finite interval of time for all $t_0 \in \mathbb{R}_{\geq 0}$. Informally, this means that there exist two

[3] In principle we could have refrained from using input $\xi(t)$ in our system under the assumption that such a signal can be generated by a fraction of the RNN's internal states. This, however, would unnecessarily complicate the analysis and increase the number of technical assumptions. Therefore we decided to use inputs $s(t)$ and $\xi(t)$ instead of just $s(t)$. Nevertheless, one should keep in mind such a possibility when deemed appropriate.

[4] Because the filters (6.5) are convergent, the effect of uncertainty in the parameter $s_{i,0}$ vanishes with time exponentially. Hence the only effective uncertainties are i and θ_i.

sets of functions of the network state $\mathbf{x}$ and input $s(t)$:

$$\begin{aligned}&\{h_{f,j}(\mathbf{x}(t),s(t))\},\ \{h_{\theta,j}(\mathbf{x}(t),s(t))\},\\&h_{f,j}:\mathbb{R}^{N_x}\times\mathbb{R}\to\mathbb{R},\ h_{\theta,j}:\mathbb{R}^{N_x}\times\mathbb{R}\to\mathbb{R},\ j\in\{1,\ldots,N_f\},\end{aligned}\tag{6.10}$$

such that the values of i and θ_i can be inferred from $\{h_{f,j}(\mathbf{x}(t),s(t))\}$ and $\{h_{\theta,j}(\mathbf{x}(t),s(t))\}$, respectively, within a given finite interval of time. Formally we can state this as follows.

Problem 6.1 Consider class $\mathcal{F}$ of signals (6.1), where the function $\xi(t)$ is known, and the values of the parameters θ_i are unknown a priori. Determine a recurrent neural network (6.8) such that the following properties hold:

(1) there is a set of initial conditions Ω_x such that $\mathbf{x}(t,\mathbf{x}_0)$ is bounded for all $\mathbf{x}_0\in\Omega_x$ and $t\geq t_0\in\mathbb{R}_{\geq 0}$; the volume of Ω_x is non-zero;
(2) there exists a set of output functions (6.10) such that, for all $\theta_i\in\Omega_\theta$, $s_{i,0}\in\Omega_s$, $t_0\in\mathbb{R}_{\geq 0}$, $\mathbf{x}_0\in\Omega_x$, and functions $\eta_i(t)$ given by (6.6), condition (6.9) implies the existence of a constant $\mathcal{T}\in\mathbb{R}_{>0}$, time instant $t'\in(t_0,t_0+\mathcal{T})$, (arbitrarily large) $T^*\in\mathbb{R}_{>0}$, and (arbitrarily small) $\varepsilon\in\mathbb{R}_{>0}$ and $\mathcal{D}\in\mathcal{K}_\infty$ such that

$$\begin{aligned}\|h_{f,i}(\mathbf{x}(t),s(t))\|_{\infty,[t',t'+T^*]}&<\varepsilon+\mathcal{D}(\Delta_\eta),\\ \inf_{\theta_i'\in E(\theta_i)}\|h_{\theta,i}(\mathbf{x}(t),s(t))-\theta_i'\|_{\infty,[t',t'+T^*]}&<\varepsilon+\mathcal{D}(\Delta_\eta).\end{aligned}\tag{6.11}$$

The functions $h_{f,i}(\cdot,\cdot)$ and $h_{\theta,i}(\cdot,\cdot)$ in Problem 6.1 are defined by (6.10). The existence of these functions implies a simple readout mechanism for recovering variables i and θ_i. When input $s(t)$ is generated by a signal from the ith class then $h_{f,i}(\mathbf{x}(t),s(t))$ must be in the neighborhood of zero for a sufficiently long time T^*, and the function $h_{f,i}(\mathbf{x}(t),s(t))$ should be in the vicinity of some constant over the same time interval (see Figure 6.1 for an illustration). Furthermore, as follows from (6.11), the value of θ_i is estimated by $h_{\theta,i}(\mathbf{x}(t),s(t))$. Hence information about the class and parameters of an input signal can be inferred from the values of $h_{f,i}(\mathbf{x}(t),s(t))$, $h_{\theta,i}(\mathbf{x}(t),s(t))$. If inequality (6.11) holds for multiple indices i, additional validation might be necessary. This, however, is beyond the scope of our study. In our current work we wish to answer the question of whether the desired behavior specified by (6.11) can in principle be realized in RNNs.

In general, Problem 6.1 has no solution for all possible functions $\xi(t)\in\mathcal{C}^1$, $f_i(\cdot,\cdot)\in\mathcal{C}^0$ and every $\theta_i\in\Omega_\theta$. Consider, for instance, the case in which $f_i(\xi(t),\theta_i)=\sin(\xi(t)\theta_i)$ and

$$\xi(t)=\begin{cases}\sin^2(\ln(t-t_0+1)), & \sin(\ln(t-t_0+1))\geq 0,\\ 0, & \sin(\ln(t-t_0+1))<0,\end{cases}\quad\forall t\geq t_0.$$

The time intervals within which $\xi(t) = 0$ are growing unboundedly in length with time. Hence for any fixed $\mathcal{T}$ and T^* there will always exist a time instant t_0' such that for all $t \geq t_0'$ the lengths of intervals when $\xi(t) = 0$ exceed $\mathcal{T} + T^*$. For all such intervals $\mathcal{T}_j$, $j = 1, \ldots, \infty$ and every $\theta_i \in \Omega_\theta$ it holds that $f_i(\xi(t), \theta_i) = 0$. This implies that solutions $s_i(t, s_{i,0}, \theta_i, \eta_i(t))$ do not depend on θ_i for all $t \in \mathcal{T}_j$.[5] Hence recovery of the actual values of θ_i from signal $s(t)$ cannot be achieved within a fixed time interval $[t_0, t_0 + \mathcal{T} + T^*]$ for all $t_0 \geq t_0'$. In order to enable a solution of the classification/recognition problem above, we must introduce an additional constraint on the functions $f_i(\xi(t), \theta_i)$. This constraint, which is a weak form of the nonlinear persistency-of-excitation condition (5.170), should ensure that variation in the parameter θ_i can be detected from the values of $f_i(\xi(t), \theta_i)$ within a finite time interval. We therefore require that the following property holds.

Assumption 6.2 *For the set of functions $f_i(\xi(t), \theta_i)$ specified by (6.1) and all $t \geq t_0$, θ_i, θ_i' there exist a constant $T \in \mathbb{R}_{>0}$ and a strictly increasing function $\rho : \mathbb{R}_{\geq 0} \to \mathbb{R}_{\geq 0}$, $\rho \in \mathcal{K}_\infty$ such that the following condition holds:*

$$\forall\, t \geq t_0\ \exists\, t' \in [t, t+T] : \ |f_i(\xi(t'), \theta_i) - f_i(\xi(t'), \theta_i')| \geq \rho\left(\|\theta_i\|_{E_i(\theta_i')}\right). \quad (6.12)$$

If the equivalence classes $E_i(\theta_i')$ consist of single elements, e.g. when there is a unique value of $\theta_i' = \theta_i$ satisfying (6.2), condition (6.12) will have a more transparent form:

$$\forall\, t \geq t_0\ \exists\, t' \in [t, t+T] : \ |f_i(\xi(t'), \theta_i) - f_i(\xi(t'), \theta_i')| \geq \rho(|\theta_i - \theta_i'|). \quad (6.13)$$

These conditions simply state that within a fixed time interval the values of $\|\theta_i\|_{E_i(\theta_i')}$ or $|\theta_i - \theta_i'|$ can be inferred from the differences $f_i(\xi(t), \theta_i) - f_i(\xi(t), \theta_i')$ for all $t \in \mathbb{R}_{\geq 0}$.

In the next section we show that a solution to Problem 6.1 can be obtained for the class $\mathcal{F}$ of functions $f_i(\xi(t), \theta_i)$ that are Lipschitz in θ_i. These results are presented in the form of sufficient conditions formulated in Theorem 6.1. In addition to showing the existence of an RNN and corresponding functions (6.10) satisfying the requirements of Problem 6.1, we demonstrate that the functions (6.10) can be made continuous and differentiable. Hence, in principle, they can be implemented as an extra feedforward component in the existing RNN structure (6.8).

[5] This is, of course, not true when the initial conditions $s_{i,0}$ are dependent on θ_i and such a dependence is known a priori. In our work, due to the presence of noise $\eta_i(t)$ and potential uncontrolled changes of θ_i for $t < t_0$, we do not consider this special case.

6.4 The existence result

As was suggested in Prokhorov *et al.* (2002a) and Younger *et al.* (1999) the reason why RNNs with fixed parameters, i.e. weights, demonstrate adaptive behavior could be found in their dynamics; supposedly, it is already sufficiently rich to have an adequate adaptation mechanism embedded within it. Finding a system that satisfies requirements (1) and (2) in Problem 6.1 and is, at the same time, realizable by an RNN, therefore, automatically constitutes an existence proof. As we will see later, this intuition is correct. The result is provided in Theorem 6.1 below.

Theorem 6.1 *Let the functions $\xi(t)$ and $f_i(\xi(t),\theta_i)$ be given and defined as in (6.1), and let Assumptions 6.1 and 6.2 hold. Furthermore, suppose that $f_i(\xi(t),\theta_i)$ are (locally) Lipschitz:*[6]

$$\exists\, D_\theta \in \mathbb{R}_{>0} : \ |f_i(\xi(t),\theta_i) - f_i(\xi(t),\theta_i')| \leq D_\theta \, \|\theta_i\|_{E_i(\theta_i')}, \quad \forall\, t > 0,\ \theta_i, \theta_i', \tag{6.14}$$

$$\exists\, D_\xi \in \mathbb{R}_{>0} : \ |f_i(\xi,\theta_i) - f_i(\xi',\theta_i)| \leq D_\xi |\xi - \xi'|, \quad \forall\, \theta_i, \xi, \xi', \tag{6.15}$$

and the time-derivative of $\xi(t)$ is bounded:

$$\left|\frac{d}{dt}\xi(t)\right| \leq \partial\xi_\infty \ \forall\, t \geq 0, \tag{6.16}$$

then for any $T^* \in \mathbb{R}_{>0}$, $\varepsilon \in \mathbb{R}_{>0}$ there is a recurrent neural network (6.8) satisfying the requirements of Problem 6.1, provided that the upper bound Δ_η for the $L_\infty[0,\infty]$-norms of the disturbance terms, $\eta_i(t)$, is sufficiently small.

Proof of Theorem 6.1. We prove the theorem in four steps. First, we present a dynamical system, which will be referred to as the *convergence prototype*. We have chosen this system to belong to the following class of differential-algebraic equations:

$$\dot{\hat{s}}_i = -\varphi_i(\hat{s}_i) + f_i(\xi(t),\hat{\theta}_i), \tag{6.17}$$

$$\hat{\theta}_i = a + \frac{b-a}{2}(x_i + 1), \tag{6.18}$$

[6] Property (6.14) can be understood as a generalized Lipschitz condition. When the equivalence sets $E_i(\theta_i')$ consist of single elements the property transforms into $|f_i(\xi(t),\theta_i) - f_i(\xi(t),\theta_i')| \leq D_\theta|\theta_i - \theta_i'|$.

$$\dot{x}_i = \gamma \|\hat{s}_i - s\|_\varepsilon \left(x_i - y_i - x_i(x_i^2 + y_i^2)\right),$$
$$\dot{y}_i = \gamma \|\hat{s}_i - s\|_\varepsilon \left(x_i + y_i - y_i(x_i^2 + y_i^2)\right), \tag{6.19}$$

where

$$\gamma \in \mathbb{R}_{>0},\ a, b \in \mathbb{R},\ a < \theta_{\min},\ b > \theta_{\max},\ \theta_0 \in [a, b],$$
$$i = 1, \ldots, N_f,\ \varepsilon \in \mathbb{R}_{>0}. \tag{6.20}$$

Notice that it is always possible to choose parameter values of the system (6.17)–(6.19) in accordance with (6.20). Indeed, the fact that $\theta_0 \in \mathbb{R}$ implies the existence of an interval $[a, b] \subset \mathbb{R}$ such that $\theta_0 \in [a, b]$ and $[\theta_{\min}, \theta_{\max}] \subset [a, b]$.

The system (6.17)–(6.19) has a locally Lipschitz right-hand side and its solutions are bounded for all initial conditions $\hat{s}_i(t_0)$, $x_i(t_0)$, $y_i(t_0) \in \mathbb{R}$. Equation (6.17) models the dynamics of the input signal $s(t)$; this requirement has been shown to be generally necessary for successful classification and adaptation (Conant and Ashby 1970; Sontag 2003). The intuition behind equations (6.18) and (6.19) is as follows. For every $\theta_i \in \Omega_\theta$ and in the absence of $\eta_i(t)$ there will always exist a point $x_i(t_0) = p_x$, $y_i(t_0) = p_y$, $p_x^2 + p_y^2 = 1$ such that $\hat{\theta}_i(x_i(t, p_x, p_y)) \in E_i(\theta_i)$ for all $t \geq t_0$, provided that $s(t) = s_i(t, s_{i,0}, \theta_i, 0)$ (see in Figure 6.2(a)). When signals $\eta_i(t)$ are present we will show that $\hat{\theta}_i(x_i(t, p_x, p_y))$ will remain in a neighborhood of $E_i(\theta_i)$, of which the size is determined by Δ_η. When $|s(t) - \hat{s}_i(t)| > \varepsilon$, solutions of (6.18) and (6.19) starting from initial conditions $x_i(t_0) = x_0'$, $y_i(t_0) = y_0'$, $x_0'^2 + y_0'^2 = 1$ evolve along a closed orbit towards p_x, p_y. We show that there exist $\gamma > 0$ and $\varepsilon > 0$ and their respective domains, and a point $\hat{s}_i(t_0) = s_0'$, $x_i(t_0) = x_0'$, $y_i(t_0) = y_0'$, such that the trajectories passing through this point converge to the following target set:

$$\|\hat{s}_i - s_i\|_\varepsilon = 0, \qquad \left\|\hat{\theta}_i\right\|_{E_i(\theta_i)} \leq \varepsilon_\theta(\varepsilon). \tag{6.21}$$

Second, we prove that there is a point $x_i(t_0) = x_0'$, $y_i(t_0) = y_0'$ such that convergence is locally uniform with respect to the values of the uncertain parameters θ_i and $s_{i,0}$. In other words, for all $t_0 \geq 0$, $s_{i,0} \in \Omega_s$, and $\theta_i \in \Omega_\theta$ there exists $\tau > 0$ such that solutions of (6.17)–(6.19) with initial conditions $x_i(t_0) = x_0'$, $y_i(t_0) = y_0'$ will be in an arbitrarily small neighborhood of (6.21) for all $t \geq t_0 + \tau$ (see also Figure 6.3).

The system (6.17)–(6.19), however, is not structurally stable. That is, small perturbations of its right-hand side might change the asymptotic properties of the system drastically. Hence, due to the inevitable approximation errors, it is unlikely that an RNN realization of (6.17)–(6.19) would solve Problem 6.1. To continue our

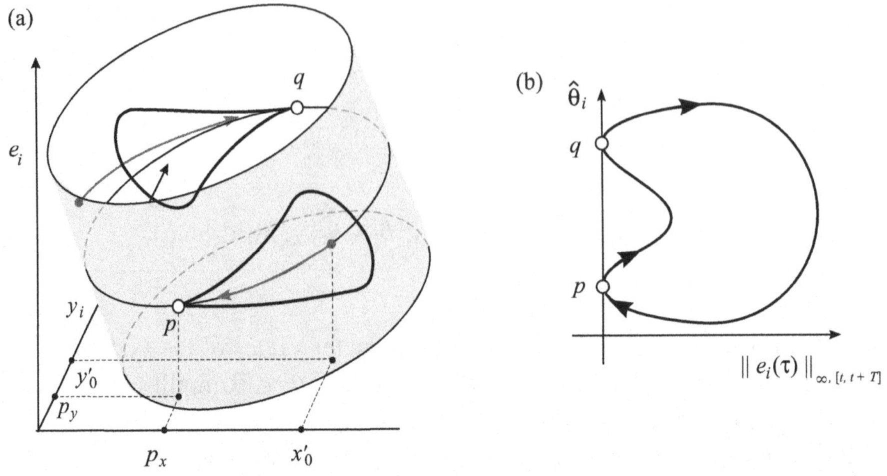

Figure 6.2 Diagrams illustrating the invariant and limit sets of the system (6.17)–(6.19) (a), $e_i = s_i - \hat{s}_i$, and a schematic representation of the system's dynamics in terms of $\hat{\theta}_i(x_i(t))$ versus $\|e_i(\tau)\|_{\infty,[t,t+T]}$ (b). In (a) are depicted the invariant and limit sets of (6.17)–(6.19) shown as circles p, q. The gray curves show trajectories that converge to these sets asymptotically. The domains of attraction of these sets, restricted to the cylinder $x_i^2 + y_i^2 = 1$, are shown by thick lines on the surface of the cylinder. Notice that these domains are not neighborhoods of p, q. In (b) the dynamics of (6.17)–(6.19) is depicted schematically: traveling along an attracting closed orbit until the domain of attraction of either p or q is reached.

argument, we need to modify (6.17)–(6.19) such that the resulting system becomes structurally stable.

For this reason, the third stage involves considering the perturbed version of the system (6.17)–(6.19):

$$\dot{\hat{s}}_i = -\varphi_i(\hat{s}_i) + f_i(\xi(t), \hat{\theta}_i),$$
$$\hat{\theta}_i = a + \frac{b-a}{2}(x_i + 1), \tag{6.22}$$

$$\dot{x}_i = \gamma(\|\hat{s}_i - s\|_\varepsilon + \delta)\left(x_i - y_i - x_i(x_i^2 + y_i^2)\right),$$
$$\dot{y}_i = \gamma(\|\hat{s}_i - s\|_\varepsilon + \delta)\left(x_i + y_i - y_i(x_i^2 + y_i^2)\right), \quad \delta \in \mathbb{R}_{>0}. \tag{6.23}$$

Our aim is to achieve structural stability of an otherwise structurally unstable system. We show that trajectories of the system (6.22) and (6.23) periodically visit a small vicinity of (6.21) and stay there for an arbitrary long time, depending on the value of δ.

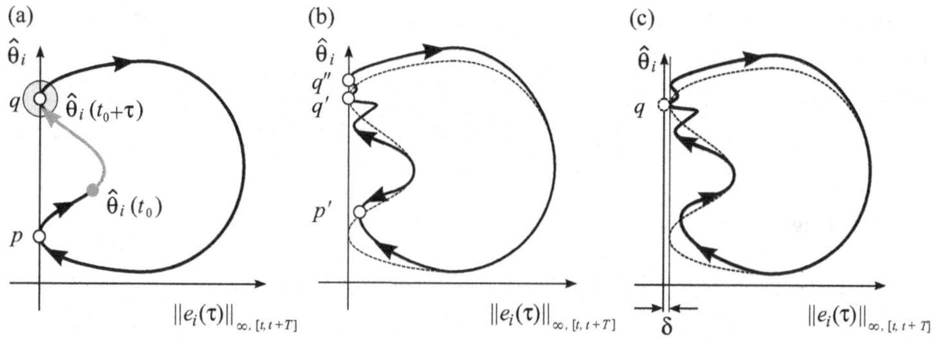

Figure 6.3 Diagrams illustrating the main steps of the proof. (a) The case, without perturbations and approximation errors, in which the estimate $\hat{\theta}_i(t)$, shown as a gray curve, asymptotically approaches points from a neighborhood of the equivalence class $E_i(\theta_i)$ for a given set of initial conditions and all θ_i. Demonstrating the existence of such systems (e.g. (6.17)–(6.19)) constitutes the *first step* of our proof. In the *second step* we show that the time required to reach this neighborhood for a given initial condition and all $\theta_i \in \Omega_\theta$ is bounded from above by some positive number τ. (b) The case in which the convergence prototype is approximated by an RNN. Because the system (6.17)–(6.19) is not structurally stable, approximation errors might lead to the emergence of new attracting invariant sets (points p', q', and q'') that do not belong to the neighborhood of $E_i(\theta_i)$. The system's behavior in this case is shown by a thick, solid line. (c) The dynamics of the perturbed yet structurally stable system (6.22) and (6.23). In this case convergence to invariant sets p, q is replaced with arbitrarily slow relaxation in small neighborhoods of these points (ghost attractors). Even in the presence of disturbances due to approximation errors, the system's state still visits these sets (shown by the dashed circle) and stays there as long as needed. The *third step* of the proof demonstrates this.

Fourth, given that the system (6.22) and (6.23) is structurally stable, we apply the results from Cybenko (1989) to demonstrate that solutions of (6.22) and (6.23) can be approximated in forward time over the semi-infinite interval $[0, \infty]$ by the state of an RNN specified by (6.8).

Figure 6.3 provides diagrams linking the main steps of the proof.

1. Convergence prototype. According to Assumption 6.1 there exist $i \in \{1, \ldots, N_f\}$, $s_{i,0}$, θ_i such that $s(t) = s_i(t, s_{i,0}, \theta_i, \eta_i(t))$ for all $t \geq 0$. Consider the ith subsystem of (6.17)–(6.19) and analyze the dynamics of the difference $s_i(t) - \hat{s}_i(t)$:

$$\frac{d}{dt}(s_i(t) - \hat{s}_i(t)) = -(\varphi_i(s_i) - \varphi_i(\hat{s}_i)) + f_i(\xi(t), \theta_i) \\ - f_i(\xi(t), \hat{\theta}_i(x_i(t))) + \eta_i(t).$$

According to our assumptions, the functions $\varphi_i(\cdot)$ are differentiable. Hence, invoking Hadamard's lemma,

$$\varphi_i(s_i) - \varphi_i(\hat{s}_i) = \left(\int_0^1 \frac{\partial \varphi(s_i r + (1-r)\hat{s}_i)}{\partial (s_i r + (1-r)\hat{s}_i)} dr \right) (s_i - \hat{s}_i), \tag{6.24}$$

and denoting

$$\begin{aligned} e_i(t) &= s_i(t) - \hat{s}_i(t), \\ \alpha_i(t) &= \int_0^1 \frac{\partial \varphi(s_i(t) r + (1-r)\hat{s}_i(t))}{\partial (s_i(t) r + (1-r)\hat{s}_i(t))} dr, \\ \Delta f_i(t) &= f_i(\xi(t), \theta_i) - f_i(\xi(t), \hat{\theta}_i(x_i(t))), \end{aligned} \tag{6.25}$$

we can obtain the following equivalent representation of (6.24):

$$\dot{e}_i = -\alpha(t) e_i + \Delta f_i(t) + \eta_i(t). \tag{6.26}$$

Solutions of equation (6.26) satisfy the expression

$$e_i(t) = e^{-\int_0^t \alpha(\tau) d\tau} e_i(0) + e^{-\int_0^t \alpha(\tau) d\tau} \int_0^t e^{\int_0^\tau \alpha(\tau_1) d\tau_1} (\Delta f_i(\tau) + \eta_i(\tau)) d\tau.$$

According to the mean-value theorem the following equivalence holds: $\alpha(t) = \partial \varphi(s_i(t) r' + (1-r')\hat{s}_i(t)) / \partial (s_i(t) r' + (1-r')\hat{s}_i(t))$ for some $r' \in [0,1]$. Hence, taking condition (6.7) into account, we can obtain that $\varphi_{\min} \leq \alpha(t) \leq \varphi_{\max}$. Therefore, using the inequality

$$e^{-\int_0^t \alpha(\tau) d\tau} \leq e^{-\varphi_{\min} t},$$

we can derive the following estimate:

$$\begin{aligned} |e_i(t)| &\leq e^{-\int_0^t \alpha_i(\tau) d\tau} |e_i(0)| + \left| e^{-\int_0^t \alpha(\tau) d\tau} \int_0^t e^{\int_0^\tau \alpha(\tau_1) d\tau_1} (\Delta f_i(\tau) + \eta_i(\tau)) d\tau \right| \\ &\leq e^{-\varphi_{\min} t} |e_i(0)| + e^{-\int_0^t \alpha(\tau) d\tau} \int_0^t e^{\int_0^\tau \alpha(\tau_1) d\tau_1} |\Delta f_i(\tau) + \eta_i(\tau)| d\tau \\ &= e^{-\varphi_{\min} t} |e_i(0)| + \int_0^t e^{-\int_\tau^t \alpha(\tau_1) d\tau_1} |\Delta f_i(\tau) + \eta_i(\tau)| d\tau \\ &\leq e^{-\varphi_{\min} t} |e_i(0)| + \int_0^t e^{-\varphi_{\min}(t-\tau)} |\Delta f_i(\tau) + \eta_i(\tau)| d\tau \\ &\leq e^{-\varphi_{\min} t} |e_i(0)| + \frac{1}{\varphi_{\min}} (1 - e^{-\varphi_{\min} t}) (\|\Delta f_i(\tau)\|_{\infty,[0,t]} + \|\eta_i(\tau)\|_{\infty,[0,\infty]}). \end{aligned} \tag{6.27}$$

Given that $\|\eta_i(\tau)\|_{\infty,[0,\infty]} \leq \Delta_\eta$ for all $t \geq 0$, inequality (6.27) implies that

$$
\begin{aligned}
|e_i(t)| &\leq e^{-\varphi_{\min} t}|e_i(0)| + \frac{1}{\varphi_{\min}}\left(1 - e^{-\varphi_{\min} t}\right)(\|\Delta f_i(\tau)\|_{\infty,[0,t]} + \Delta_\eta) \\
&= e^{-\varphi_{\min} t}\left(|e_i(0)| - \frac{\Delta_\eta}{\varphi_{\min}}\right) + \frac{1}{\varphi_{\min}}(\|\Delta f_i(\tau)\|_{\infty,[0,t]} + \Delta_\eta). \qquad (6.28)
\end{aligned}
$$

Regrouping terms in (6.28) yields

$$
\left(|e_i(t)| - \frac{\Delta_\eta}{\varphi_{\min}}\right) \leq e^{-\varphi_{\min} t}\left(|e_i(0)| - \frac{\Delta_\eta}{\varphi_{\min}}\right) + \frac{1}{\varphi_{\min}}\|\Delta f_i(\tau)\|_{\infty,[0,t]}.
$$

Let us denote $\varepsilon = \Delta_\eta/\varphi_{\min}$ and consider the values of $\|e_i(t)\|_\varepsilon$. When $|e_i(t)| - \Delta_\eta/\varphi_{\min} > 0$ we have

$$
\|e_i(t)\|_\varepsilon = \left(|e_i(t)| - \frac{\Delta_\eta}{\varphi_{\min}}\right) \leq e^{-\varphi_{\min} t}\|e_i(0)\|_\varepsilon + \frac{1}{\varphi_{\min}}\|\Delta f_i(\tau)\|_{\infty,[0,t]}.
$$

When $|e_i(t)| - \Delta_\eta/\varphi_{\min} \leq 0$ then

$$
\|e_i(t)\|_\varepsilon = 0 \leq e^{-\varphi_{\min} t}\|e_i(0)\|_\varepsilon + \frac{1}{\varphi_{\min}}\|\Delta f_i(\tau)\|_{\infty,[0,t]}.
$$

Hence we can conclude that the following estimate holds along the trajectories of (6.17):

$$
\|e_i(t)\|_\varepsilon \leq e^{-\varphi_{\min} t}\|e_i(0)\|_\varepsilon + \frac{1}{\varphi_{\min}}\|\Delta f_i(\tau)\|_{\infty,[0,t]}, \quad \varepsilon = \frac{\Delta_\eta}{\varphi_{\min}}. \qquad (6.29)
$$

On taking into account (6.14) and (6.29) plus the fact that $\left\|\hat{\theta}_i\right\|_{E_i(\theta_i)} = \inf_{\bar{\theta}_i \in E_i(\theta_i)} |\hat{\theta}_i - \bar{\theta}_i|$ we can conclude that the following inequality holds:

$$
\begin{aligned}
&\|e_i(t)\|_\varepsilon \leq e^{-\varphi_{\min} t}\|e_i(0)\|_\varepsilon + \frac{D_\theta}{\varphi_{\min}}\|\bar{\theta}_i - \hat{\theta}_i(\tau)\|_{\infty,[0,t]}, \\
&\bar{\theta}_i \in E_i(\theta_i) \cap [a,b]. \qquad (6.30)
\end{aligned}
$$

Let us now consider (6.18) and (6.19). We select a point x', y' that satisfies the following condition:

$$
x'^2 + y'^2 = 1. \qquad (6.31)
$$

Solutions of (6.19) passing through this point can be defined as follows:

$$\begin{aligned} x_i(t,x',y') &= \cos\left(\int_0^t \gamma\|\hat{s}_i(\tau)-s(\tau)\|_\varepsilon\, d\tau + \nu_x\right), \\ x' &= \cos(\nu_x),\ \nu_x \in [0,2\pi], \\ y_i(t,x',y') &= \sin\left(\int_0^t \gamma\|\hat{s}_i(\tau)-s(\tau)\|_\varepsilon\, d\tau + \nu_y\right), \\ y' &= \sin(\nu_y),\ \nu_y \in [0,2\pi]. \end{aligned} \tag{6.32}$$

This observation can easily be verified when writing (6.19) in polar coordinates: $x_i = r\cos(\nu)$, $y_i = r\sin(\nu)$ (Guckenheimer and Holmes 2002):

$$\begin{aligned} \dot{r} &= \gamma\|\hat{s}_i - s\|_\varepsilon \cdot r(1-r), \\ \dot{\nu} &= \gamma\|\hat{s}_i - s\|_\varepsilon \cdot 1. \end{aligned} \tag{6.33}$$

Given that $\bar{\theta}_i$ belongs to the interval $[a,b]$, there is a number $\bar{h}(\bar{\theta}_i) \in [0,\pi]$ such that for all $k \in \mathbb{Z}$ the following equivalence holds:

$$\bar{\theta}_i = a + \frac{b-a}{2}\left(\cos(\bar{h}(\bar{\theta}_i) + 2\pi k) + 1\right). \tag{6.34}$$

Hence, according to (6.18) and (6.32), the norm $\|\bar{\theta}_i - \hat{\theta}_i(\tau)\|_{\infty,[0,t]}$ can be estimated from the above as follows:

$$\|\bar{\theta}_i - \hat{\theta}_i(\tau)\|_{\infty,[0,t]} \leq \frac{b-a}{2}\left\|\bar{h}(\bar{\theta}_i) - \nu_x + 2\pi k - \int_0^t \gamma\|\hat{s}_i(\tau)-s(\tau)\|_\varepsilon\, d\tau\right\|_{\infty,[0,t]}. \tag{6.35}$$

Denoting

$$c = \frac{D_\theta}{\varphi_{\min}}\frac{b-a}{2}, \qquad h(t,\bar{\theta}_i,k) = \bar{h}(\bar{\theta}_i) - \nu_x + 2\pi k - \int_0^t \gamma\|\hat{s}_i(\tau)-s(\tau)\|_\varepsilon\, d\tau$$

and taking into account (6.30) and (6.35), we can conclude that the following holds along the solutions of (6.17)–(6.19):

$$\begin{aligned} \|e_i(t)\|_\varepsilon &\leq e^{-\varphi_{\min} t}\|e_i(0)\|_\varepsilon + c\|h(\tau,\bar{\theta}_i,k)\|_{\infty,[0,t]}, \\ h(0,\bar{\theta}_i,k) - h(t,\bar{\theta}_i,k) &= \int_0^t \gamma\|e_i(\tau)\|_\varepsilon\, d\tau. \end{aligned} \tag{6.36}$$

According to Theorem 4.7 (see also Corollaries 4.3–4.5) there exist $\gamma^* \in \mathbb{R}_{>0}$ and h^* such that, for a given bounded $e_i(0)$, all $\gamma \in \mathbb{R}_{>0}$, $\gamma < \gamma^*$, and $h(0,\bar{\theta}_i,k) \geq h^*$, the norm $\|e_i(\tau)\|_{\infty,[0,\infty]}$ is bounded and

$$\lim_{t\to\infty} h(t,\bar{\theta}_i,k) \in [0, h(0,\bar{\theta}_i,k)]. \tag{6.37}$$

The value of γ^* can be determined, according to Corollary 4.5, from the following inequality:

$$0 < \gamma^* < \frac{\varphi_{\min}}{c}\left(\ln\left(\frac{\kappa}{d}\right)\frac{\kappa}{\kappa-1}\left(2+\frac{\kappa}{1-d}\right)\right)^{-1},$$
$$\kappa \in \mathbb{R}_{>1},\ d \in (0,1) \subset \mathbb{R}. \tag{6.38}$$

The value of h^* can be estimated from

$$\|e_i(t_0)\|_{\varepsilon} \leq \left(\frac{\varphi_{\min}}{\gamma^*}\left(\ln\left(\frac{\kappa}{d}\right)\right)^{-1}\frac{\kappa-1}{\kappa} - c\left(2+\frac{\kappa}{1-d}\right)\right)h^*. \tag{6.39}$$

Given that $\|e_i(t_0)\|_{\varepsilon}$ in (6.39) is bounded from above for all $t_0 \geq 0$, $\|e_i(t_0)\|_{\varepsilon} \leq s_{\max} - s_{\min} + D_\theta/\varphi_{\min}(b-a)$, the condition

$$h^* \geq \left(s_{\max} - s_{\min} + \frac{D_\theta(b-a)}{\varphi_{\min}}\right)\left(\frac{\varphi_{\min}}{\gamma^*}\left(\ln\left(\frac{\kappa}{d}\right)\right)^{-1}\frac{\kappa-1}{\kappa}\right.$$
$$\left. - c\left(2+\frac{\kappa}{1-d}\right)\right)^{-1}, \tag{6.40}$$

together with (6.38), implies that for all $\hat{s}_i(t_0) \in \Omega_s$ and $h(0,\bar{\theta}_i,k) \geq h^*$ the norm $\|e_i(\tau)\|_{\infty,[0,\infty]}$ is bounded and property (6.37) holds.

Notice that in the definition of $h(0,\bar{\theta}_i,k)$,

$$h(0,\bar{\theta}_i,k) = \bar{h}(\bar{\theta}_i) - \nu_x + 2\pi k, \tag{6.41}$$

the value of k can be chosen arbitrarily large. Moreover, $\bar{h}(\bar{\theta}_i) \in [0,\pi]$ for all $\bar{\theta}_i \in [a,b]$. This implies that there exists a finite k' such that the condition $h(0,\bar{\theta}_i,k') \geq h^*$ will be satisfied for any fixed h^* (i.e. for all γ^* satisfying (6.38)) and all $\bar{\theta}_i \in [a,b]$. In addition, the following will hold:

$$\lim_{t\to\infty} h(t,\bar{\theta}_i,k') \in [0, h(0,\bar{\theta}_i,k')] \subset [0, \pi - \nu_x + 2\pi k']\ \forall \in \bar{\theta}_i \in [a,b]. \tag{6.42}$$

Taking (6.32) into account, we can conclude that solutions $x_i(t,x',y')$ converge to a point within the interval $[-1,1]$ as $t \to \infty$, and, furthermore, the vector $(x_i(t,x',y'), y_i(t,x',y'))$ makes no more than k' full rotations around the origin for all $\theta_i \in [\theta_{\min}, \theta_{\max}]$. Hence, for a given initial condition $x_i(0) = x'$, $y_i(0) = y'$,

$\hat{s}_{i,0} \in \Omega_s$, and $\theta_i \in [\theta_{\min}, \theta_{\max}]$, the estimate $\hat{\theta}_i(t) = a + (b-a)/2 \cdot (x_i(t, x', y') + 1)$ converges to a point in $[a, b]$ as $t \to \infty$. We denote this point by $\hat{\theta}_i^*$.

Given that $\hat{\theta}_i(t)$ converges to a limit, there exists a time instant t^* such that for all $t \geq t^*$ the following condition holds: $|\hat{\theta}_i(t) - \hat{\theta}_i^*| < \mu_\infty$, where $\mu_\infty \in \mathbb{R}_{>0}$ is an arbitrarily small constant. Therefore, taking condition (6.14) into account, we can conclude that for all $t \geq t^*$ the derivative $\dot{e}_i$ satisfies the following equation:

$$\dot{e}_i = -\alpha(t)e_i + f_i(\xi(t), \theta_i) - f_i(\xi(t), \hat{\theta}_i^*) + \mu_i(t) + \eta_i(t), \tag{6.43}$$

where $|\mu_i(t)| \leq D_\theta \, \mu_\infty$ is a continuous function.

Now we will show that the norm $\|\theta_i\|_{E_i(\hat{\theta}_i^*)}$ can be bounded from above by a $\mathcal{K}_\infty$-function of Δ_η. Consider the term $f_i(\xi(t), \theta_i) - f_i(\xi(t), \hat{\theta}_i^*)$. According to (6.12) there exists a sequence of monotonically increasing time instants t_j, $j = 1, 2, \ldots$ such that $t_{j+1} - t_j \leq 2T$ and $|f_i(\xi(t_j), \theta_i) - f_i(\xi(t_j), \hat{\theta}_i^*)| \geq \rho(\|\theta_i\|_{E_i(\hat{\theta}_i^*)})$. Furthermore, according to (6.15) and (6.16), the time-derivative of $f_i(\xi(t), \theta_i) - f_i(\xi(t), \hat{\theta}_i^*)$ is bounded:

$$\left| \frac{d}{dt} f_i(\xi(t), \theta_i) - f_i(\xi(t), \hat{\theta}_i^*) \right| \leq 2D_\xi \cdot \partial\xi_\infty = D_f.$$

Hence the following estimate holds:

$$\int_t^{t+L} |f_i(\xi(\tau), \theta_i) - f_i(\xi(\tau), \hat{\theta}_i^*)| \geq \frac{\rho(\|\theta_i\|_{E_i(\hat{\theta}_i^*)})^2}{2D_f},$$

$$L = \max\left\{2T, \frac{\rho(b-a)}{D_f}\right\}. \tag{6.44}$$

In order to proceed further we will need the following lemma.

Lemma 6.1 *Consider the following differential equation:*

$$\dot{z} = -\varphi(t, z) + u(t) + \eta(t), \ z_0 = z(0) \in [z_{\min}, z_{\max}] \subset \mathbb{R}. \tag{6.45}$$

Let us suppose that

(1) $\varphi(z)z \geq 0$, $\varphi_{\min} \leq \partial\varphi(t, z)/\partial z \leq \varphi_{\max}$;
(2) $u(t) \in L_\infty[0, \infty] \cap \mathcal{C}^1$, $\|u(t)\|_{\infty,[0,\infty]} \leq u_\infty$, $\|\dot{u}(t)\|_{\infty,[0,\infty]} \leq \partial u_\infty$;
(3) $\eta(t) \in L_\infty[0, \infty]$, $\|\eta(t)\|_{\infty,[0,\infty]} \leq \Delta$;
(4) there exist constants L and δ *such that for all* $t \geq 0$

$$\int_t^{t+L} |u(\tau)| d\tau \geq \delta; \tag{6.46}$$

(5) the following inequality holds:

$$\left(\frac{\delta}{L}\right)^2 - \Delta u_\infty > 0. \tag{6.47}$$

Then, for any $p \in \mathbb{R}_{>0}$, there exist constants $L^* > 0$ and $\delta^* \geq ((\delta/L)^2 - \Delta u_\infty)/p$, such that

$$\int_t^{t+L^*} |z(\tau)| d\tau \geq \delta^* \geq \frac{1}{p}\left(\frac{\delta^2}{L} - \Delta u_\infty L\right) \ \forall\, t \geq 0. \tag{6.48}$$

Proof of Lemma 6.1. We prove the lemma along the lines of an argument provided in Loria *et al.* (2003) (Property 1). Consider the time-derivative of zu:

$$\frac{d}{dt}(zu) = (-\varphi(t,z) + u + \eta)u + z\dot{u} \geq u^2 - |z|(\varphi_{\max} + \partial u_\infty) - |u|\Delta. \tag{6.49}$$

According to (6.49), for all $t, t_0 \in \mathbb{R}_{\geq 0}$, $t \geq t_0$ the following inequality holds:

$$\begin{aligned} z(t)u(t) - z(t_0)u(t_0) \geq & \int_{t_0}^t u^2(\tau)d\tau - (\varphi_{\max} + \partial u_\infty)\int_{t_0}^t |z(\tau)|d\tau \\ & - \Delta \int_{t_0}^t |u(\tau)| d\tau. \end{aligned} \tag{6.50}$$

Rearranging terms in (6.50) yields

$$\begin{aligned} (\varphi_{\max} + \partial u_\infty)\int_{t_0}^t |z(\tau)|d\tau \geq & \ z(t_0)u(t_0) - z(t)u(t) \\ & + \int_{t_0}^t u^2(\tau)d\tau - \Delta\int_{t_0}^t |u(\tau)|d\tau. \end{aligned}$$

Notice that $z(t_0)u(t_0) - z(t)u(t)$ is bounded from below for all $t \geq 0$. We denote this bound by the symbol M. Furthermore, according to the Hölder inequality and property (6.46), the following estimate holds for all $t \geq 0$:

$$\frac{\delta^2}{L} \leq \frac{1}{L}\left(\int_t^{t+L} |u(\tau)|d\tau\right)^2 \leq \int_t^{t+L} u^2(\tau)d\tau.$$

Hence for all time instances t: $(n+1)L \geq t - t_0 \geq nL$, where n is a positive integer, we have

$$\begin{aligned} (\varphi_{\max} + \partial u_\infty)\int_{t_0}^t |z(\tau)|d\tau & \geq M + n\frac{\delta^2}{L} - \Delta\int_{t_0}^t |u(\tau)|d\tau \\ & \geq M + n\frac{\delta^2}{L} - (n+1)\Delta u_\infty = (M - \Delta u_\infty L) + n\left(\frac{\delta^2}{L} - \Delta u_\infty L\right). \end{aligned} \tag{6.51}$$

According to the requirements of the lemma, inequality (6.47), the difference $\delta^2/L - \Delta u_\infty L > 0$ is a positive constant. Therefore, there exists $n = n'$ such that the right-hand side of (6.51) exceeds some $\delta' = (\delta^2/L - \Delta u_\infty L)/p' \in \mathbb{R}_{>0}$, $p' \in \mathbb{R}_{>0}$. On choosing $t' = \min_t\{t - t_0\} \geq n'L$ we can conclude that

$$(\varphi_{\max} + \partial u_\infty) \int_{t_0}^{t'} |z(\tau)| d\tau \geq \delta'. \tag{6.52}$$

Given that we could choose the value of t_0 arbitrarily in the domain $\mathbb{R}_{\geq 0}$, inequality (6.52) is equivalent to

$$\int_t^{t+L^*} |z(\tau)| d\tau \geq \delta^*,$$

where $L^* = t' - t_0$, $\delta^* = \delta' / (\varphi_{\max} + \partial u_\infty) = (\delta^2/L - \Delta u_\infty L)/p$, and $p = p'(\varphi_{\max} + \partial u_\infty)$. □

On denoting $f_i(\xi(t), \theta_i) - f_i(\xi(t), \hat{\theta}_i^*) = u(t)$ and $\eta_i(t) + \mu_i(t) = \eta(t)$ we can observe that (6.43) is of the same class as (6.45) in the formulation of Lemma 6.1. Furthermore, the following inequalities hold:

$$\Delta \leq \Delta_\eta + D_\theta \mu_\infty, \qquad \|u(t)\|_{\infty,[0,\infty]} \leq D_\theta \, \|\theta_i\|_{E_i(\hat{\theta}_i^*)} \leq D_\theta(b - a). \tag{6.53}$$

Notice that the value of μ_∞ in (6.53) can be made arbitrarily small because $\hat{\theta}_i(t)$ converges to a limit, and $\hat{\theta}_i^*$ can be chosen from its arbitrarily small vicinity. Let us therefore choose $\hat{\theta}_i^*$ such that $D_\theta \mu_\infty \leq \Delta_\eta$. Hence, in accordance with Lemma 6.1, the condition

$$\left(\frac{\rho^2(\|\theta_i\|_{E_i(\hat{\theta}_i^*)})}{2D_f L}\right)^2 > 2\Delta_\eta D_\theta(b - a) \tag{6.54}$$

implies the existence of constants L^*, $p \in \mathbb{R}_{>0}$ such that

$$\int_t^{t+L^*} |e_i(\tau)| d\tau \geq \frac{1}{p}\left(\left(\frac{\rho^2(\|\theta_i\|_{E_i(\hat{\theta}_i^*)})}{2D_f}\right)^2 \frac{1}{L} - \Delta u_\infty L\right)$$
$$= \delta^* > 0 \;\; \forall \; t \geq t^*. \tag{6.55}$$

We will now show that the norm $\|\theta_i\|_{E_i(\hat{\theta}_i^*)}$ is bounded from above by a function $\varepsilon_\theta(\Delta_\eta) \in \mathcal{K}_\infty$ for all sufficiently small Δ_η. Let us parametrize Δ_η as follows:

$$\Delta_\eta = \left(\frac{\rho^2(\varepsilon^*)}{2D_f L}\right)^2 \frac{1}{2D_\theta(b - a)}, \; \varepsilon^* \in \mathbb{R}_{>0}. \tag{6.56}$$

The parametrization (6.56) is always possible because $\rho(\cdot) \in \mathcal{K}_\infty$. Moreover, for all $\|\theta_i\|_{E_i(\hat{\theta}_i^*)} > \varepsilon^*$ condition (6.54) is satisfied. Hence, according to Lemma 6.1 there exist constants L^* and p such that inequality (6.55) holds. Given that δ^*, L^*, $\varphi_{\min} \in \mathbb{R}_{>0}$ there will always exist a number $\Delta_\eta^* \in \mathbb{R}_{>0}$ such that $\Delta_\eta^* < (L^*)^{-1}\delta^*\varphi_{\min}/2$. This implies that for all $\Delta_\eta \leq \Delta_\eta^*$ the following inequality holds:

$$\int_t^{t+L^*} \|e_i(\tau)\|_\varepsilon \, d\tau \geq \frac{\delta^*}{2}, \; \varepsilon = \frac{\Delta_\eta}{\varphi_{\min}}. \tag{6.57}$$

Let us suppose that the norm $\|\theta_i\|_{E_i(\hat{\theta}_i^*)}$ is greater than ε^*. In this case (6.54) and (6.57) hold and the integral

$$\int_{t^*}^{t} \|e_i(\tau)\|_\varepsilon \, d\tau \tag{6.58}$$

grows unboundedly with t. On the other hand, according to (6.36) and (6.37) the integral (6.58) is bounded. Hence we have reached a contradiction. This implies that $\|\theta_i\|_{E_i(\hat{\theta}_i^*)} \leq \varepsilon^*$. Given that $\rho(\cdot) \in \mathcal{K}_\infty$, the inverse $\rho^{-1}(\cdot)$ is well defined and is a $\mathcal{K}_\infty$-function. Therefore, taking (6.56) into account, we can conclude that the latter inequality is equivalent to

$$\|\theta_i\|_{E_i(\hat{\theta}_i^*)} \leq \rho^{-1}\left(\left(8\Delta_\eta D_\theta (b-a) D_f^2 L^2\right)^{1/4}\right). \tag{6.59}$$

Thus we have just shown that there exists a point x', y' in the state space of system (6.17)–(6.19) and parameters γ and ε, such that for all $s_{i,0} \in \Omega_s$ and every $\theta_i \in [\theta_{\min}, \theta_{\max}]$ the estimate $\hat{\theta}_i(x_i(t, x', y'))$ converges into a small neighborhood of $E_i(\theta_i)$ in finite time and stays there for an arbitrarily long time. The size of this neighborhood can be characterized by a $\mathcal{K}_\infty$-function of Δ_η, when Δ_η is sufficiently small. Let us now show that this convergence is uniform with respect to θ_i.

2. *Uniformity.* Consider (6.42). According to (6.36) and (6.42) trajectories passing through a point (x', y') satisfying (6.31) at $t = 0$ also satisfy the following constraint:

$$\begin{aligned} \exists k' \in \mathbb{Z}: \;\; h(0) - h(\infty) &= \gamma \int_0^\infty \|e_i(\tau, e_i(0), \theta_i, \eta_i(\tau))\|_\varepsilon \, d\tau \\ &\leq \pi - \nu_x + 2\pi k' < \infty \end{aligned} \tag{6.60}$$

for all $\theta_i \in [\theta_{\min}, \theta_{\max}]$ and $e_i(0)$. We will use this property to demonstrate that there is a point (x', y'), $\sqrt{x'^2 + y'^2} = 1$, $\|\hat{\theta}_i(x')\|_{E_i(\theta_i)} \geq \Delta_0$, $\Delta_0 \in \mathbb{R}_{>0}$, such that for any $\theta_i \in [\theta_{\min}, \theta_{\max}]$ the estimate $\hat{\theta}_i(x_i(t, x', y'))$ converges into a set

$$\|\theta_i\|_{E_i(\hat{\theta}_i)} \leq \rho^{-1}\left(\left(8\Delta_\eta D_\theta (b-a) D_f^2 L^2\right)^{1/4}\right) \tag{6.61}$$

in finite time $T'(\theta_i)$ for all t_0, $\hat{s}_{i,0} \in \Omega_s$, and stays there for all $t \geq t_0 + T'(\theta_i)$. Furthermore, the value of $T'(\theta_i)$ is bounded from above for all $\theta_i \in [\theta_{\min}, \theta_{\max}]$. In other words, there exists $T'_{\max} \in \mathbb{R}_{>0}$:

$$T'(\theta_i) \leq T'_{\max} \ \forall\, \theta_i \in [\theta_{\min}, \theta_{\max}]. \tag{6.62}$$

The fact that the estimate $\hat{\theta}_i$ converges into a set specified by (6.61) in finite time $T'(\theta_i)$ and stays there for $t \geq t_0 + T'(\theta_i)$ for all x', y' : $\sqrt{x'^2 + y'^2} = 1$ follows immediately from (6.59). We must show, however, that (6.62) holds.

According to (6.4) and (6.20) there is a point $\theta_0 \in [a, b]$ such that $\|\theta_0\|_{E_i(\theta_i)} \geq \Delta_\theta$ for every $\theta_i \in \Omega_\theta$. Hence, there exists a point $\theta_{i,1} \in [a, b]$ such that

$$\inf_{\bar{\theta}_i \in E_i(\theta_i) \cap [a,b]} \|\bar{\theta}_i - \theta_{i,1}\| = \Delta_\theta.$$

Without loss of generality, suppose that the set

$$\Omega_1 = \{\bar{\theta}_i \in E_i(\theta_i) \cap [a, b] | \theta_{i,1} > \bar{\theta}_i\}$$

is not empty.[7] By $\theta_{i,\max}$ we denote $\theta_{i,\max} = \sup\{\Omega_1\}$. Let us pick a point $\theta_{i,2} \in [a, b]$ according to the constraints

$$|\theta_{i,2} - \theta_{i,1}| = |\theta_{i,2} - \theta_{i,\max}| = \Delta_\theta/2, \quad \theta_{i,1} > \theta_{i,2} > \theta_{i,\max}, \tag{6.63}$$

and choose the value of ν_x in (6.32) such that

$$\theta_{i,2} = a + \frac{b-a}{2}(\cos(\nu_x) + 1), \ \nu_x \in [0, \pi].$$

According to (6.34) there exist $\bar{h}(\theta_{i,\max})$, k such that

$$\theta_{i,\max} = a + \frac{b-a}{2}(\cos(\bar{h}(\theta_{i,\max}) + 2\pi k) + 1), \ \bar{h}(\theta_{i,\max}) \in [0, \pi], \ k \in \mathbb{N}.$$

Given that $\theta_{i,2} > \theta_{i,\max}$, we set the value of $k = 0$ and choose $\bar{h}(\theta_{i,\max})$ in accordance with the following inequality:

$$\nu_x < \bar{h}(\theta_{i,\max}). \tag{6.64}$$

Because $|\hat{\theta}_i(\cos(\nu_x)) - \hat{\theta}_i(\cos(\nu'_x))| \leq ((b-a)/2)|\nu_x - \nu'_x|$ for all $\nu_x, \nu'_x \in \mathbb{R}$, conditions (6.63) and (6.64) ensure the existence of a constant $\nu'_x \leq \bar{h}(\theta_{i,\max})$, $\nu'_x = \nu_x + \Delta_\theta/(2(b-a))$ such that

$$|\hat{\theta}_i(\cos(\nu_x)) - \hat{\theta}_i(\cos(\nu''_x))| \leq \Delta_\theta/4 \ \forall\, \nu''_x \in [\nu_x, \nu'_x]. \tag{6.65}$$

[7] If Ω_1 is empty then $\Omega_2 = \{\bar{\theta}_i \in E_i(\theta_i) \cap [a, b] | \theta_{i,1} < \bar{\theta}_i\}$ is not empty. We can proceed with the same argument, replacing the interval $[0, \pi]$ with $[\pi, 2\pi]$ and sup with inf when appropriate.

Hence,

$$\|\hat{\theta}_i(\cos(\nu_x''))\|_{E_i(\theta_i)} \geq \frac{\Delta_\theta}{4} \quad \forall\, \nu_x'' \in [\nu_x, \nu_x'].$$

The inequality above implies that the values of $\hat{\theta}_i(\cos(\nu_x''))$ are outside of the $\Delta_\theta/4$-neighborhood of $E_i(\theta_i)$ for all $\nu_x'' \in [\nu_x, \nu_x']$. Furthermore, because $\hat{\theta}_i(\cos(\cdot))$ is monotone (non-increasing) over $[\nu_x, \bar{h}(\theta_{i,\max}))$, and $\theta_{i,2} > \theta_{i,\max}$, there are no values of $\nu_x'' \in [\nu_x, \bar{h}(\theta_{i,\max}))$ such that $\|\hat{\theta}_i(\cos(\nu_x''))\|_{E_i(\theta_i)} = 0$.

Let us consider those solutions of the system (6.17)–(6.19) that pass through the point $x_i(0) = \cos(\nu_x)$, $y_i(0) = \sin(\nu_x)$, $\hat{s}_i(0) \in \Omega_s$. Suppose that $0 < \gamma < \gamma^*$, and γ^* satisfies (6.40) with $h^* = \Delta_\theta/(2(b-a))$. Then the sum $\nu_x + \gamma \int_0^t \|e_i(\tau)\|_\varepsilon \, d\tau$ converges to a point in $[\nu_x, \bar{h}(\theta_{i,\max})]$. Taking the monotonicity and continuity of the function $\hat{\theta}_i(\cos(\nu_x''))$ for $\nu_x'' \in [\nu_x, \bar{h}(\theta_{i,\max})]$ into account, we can conclude that the trajectory $\hat{\theta}_i(x_i(t, x'(\theta_i)))$ enters the ε^*-neighborhood of $\theta_{i,\max}$ only once for all $t \in [0, \infty]$.

Let us show that the amount of time required for the system to enter this neighborhood is bounded from above for all $\theta_i \in \Omega_\theta$. Given that the trajectory $\hat{\theta}_i(x_i(t, x', y'))$ enters the ε^*-neighborhood of $\theta_{i,\max}$ only once, we shall show that the amount of time the system spends outside of this neighborhood is bounded from above for all $\theta_i \in \Omega_\theta$. We prove this by contradiction. Suppose that for any fixed $T_0' \in \mathbb{R}_{>0}$ there is a $\theta_i \in [\theta_{\min}, \theta_{\max}]$ such that $T'(\theta_i) \geq T_0'$. Consider the dynamics of (6.17)–(6.19) when $s(t) = s_i(t, s_{i,0}, \theta_i, \eta_i(t))$. Let us pick a sequence of time instances $\{t_j\}_{j=1}^\infty$, such that $t_{j+1} - t_j = D_T$, and $D_T \geq L^*$. For each interval $[t_j, t_{j+1}]$ we consider two alternative possibilities:

(1) the norm $\|\hat{\theta}_i(t_j) - \hat{\theta}_i(\tau)\|_{\infty,[t_j,t_{j+1}]} \leq \epsilon$, $\epsilon \in \mathbb{R}_{>0}$, $\epsilon \leq D_\theta^{-1}\Delta_\eta$;
(2) the norm $\|\hat{\theta}_i(t_j) - \hat{\theta}_i(\tau)\|_{\infty,[t_j,t_{j+1}]} > \epsilon$.

When the first alternative applies, according to (6.57) the following estimate holds: $\int_{t_j}^{t_{j+1}} \|e_i(\tau)\|_\varepsilon \, d\tau \geq \delta^*$. Hence $h(t_j) - h(t_{j+1}) > \gamma\delta^*$. When the second alternative holds, e.g. $\|\hat{\theta}_i(t_i) - \hat{\theta}_i(\tau)\|_{\infty,[t_j,t_{j+1}]} > \epsilon$, we can conclude, using inequality (6.35), that

$$\left\| \gamma \int_{t_j}^{\tau} \|e_i(\tau_1)\|_\varepsilon \, d\tau_1 \right\|_{\infty,[t_j,t_{j+1}]} > \epsilon \frac{2}{b-a}.$$

Given that $h(t)$ is monotone with respect to t, we obtain that $h(t_j) - h(t_{j+1}) > \epsilon 2/(b-a)$. Thus we have shown that

$$h(t_j) - h(t_{j+1}) > \min\{\gamma\delta^*, \epsilon 2/(b-a)\} = \Delta_h$$

for all j such that $\left\|\hat{\theta}_i(\tau)\right\|_{E_i(\theta_i)} \geq \varepsilon^*$ for all $\tau \in [t_j, t_{j+1}]$. Given that $h(t)$ is non-increasing and T' is arbitrarily large, there would be a time instant $t_m \leq T'$ when

$\sum_j^m h(t_j) - h(t_{j+1}) \geq m\Delta_h > \pi - v_x + 2\pi k'$. This, however, would contradict (6.60). Hence property (6.62) is proven.

3. Structurally stable prototype. So far we have shown that for the given system (6.17)–(6.19) there exists a non-empty set of parameters γ, ε, and x', y' : $\sqrt{x'^2 + y'^2} = 1$ such that trajectories $x_i(t, x', y')$, $y_i(t, x', y')$ converge to a point on the unit circle in $\mathbb{R}^2$, and the variable $\hat{\theta}_i(x_i(t, x', y'))$ reaches a given small neighborhood of $E_i(\theta_i)$ (see (6.61)) within finite time $T'_{\max}$ for all $\theta_i \in [\theta_{\min}, \theta_{\max}]$.

Let us now consider the perturbed system (6.22) and (6.23), where $\delta \in \mathbb{R}_{>0}$ and the initial conditions are selected in a neighborhood of x', y':

$$
\begin{aligned}
&(x_i(0), y_i(0)) \in \Omega(x', y'),\\
&\Omega(x', y') = \left\{(x, y) \in \mathbb{R}^2 | \sqrt{(x - x')^2 + (y - y')^2} \leq \delta_r\right\}, \ \delta_r \in \mathbb{R}_{>0}.
\end{aligned} \tag{6.66}
$$

In order to distinguish the solutions of (6.22) and (6.23) from the solutions of the unperturbed system (6.17)–(6.19), we denote the latter by $x_i^*(t, x_i(0), y_i(0))$, $y_i^*(t, x_i(0), y_i(0))$, and $\hat{s}_i^*(t, \theta_i, s_{i,0}, \eta_i(t))$. For the sake of notational compactness we also denote the state vector of the ith subsystem of (6.17)–(6.19) as $\mathbf{q}_i^* = (\hat{s}_i^*, x_i^*, y_i^*)$ and the state vector of the ith subsystem of (6.22) and (6.23) as $\mathbf{q}_i$.

Solutions of (6.22) and (6.23) are bounded:

$$
\begin{aligned}
&\|\hat{s}_i(t, \hat{s}_{i,0}, \eta_i(t))\|_{\infty,[0,\infty]} \leq |\hat{s}_{i,0}| + (\max\{|a|, |b|\}D_\theta + \Delta_\eta)/\varphi_{\min},\\
&\|x_i(t, x_i(0), y_i(0))\|_{\infty,[0,\infty]} \leq \max\left\{1, \sqrt{x_i(0)^2 + y_i(0)^2}\right\},\\
&\|y_i(t, x_i(0), y_i(0))\|_{\infty,[0,\infty]} \leq \max\left\{1, \sqrt{x_i(0)^2 + y_i(0)^2}\right\}.
\end{aligned} \tag{6.67}
$$

Hence for all $\hat{s}_i(0), x_i(0), y_i(0) \in \Omega_s \times \Omega(x', y')$ there exists a constant D_0 such that $\|\mathbf{q}_i(t)\|_{\infty,[0,\infty]} \leq D_0$ for all θ_i. Let us rewrite (6.22) and (6.23) as follows:

$$
\begin{aligned}
\dot{\hat{s}}_i &= -\varphi_i(\hat{s}_i) + f_i(\xi(t), \hat{\theta}_i(x_i)),\\
\dot{x}_i &= \gamma \|\hat{s}_i - s\|_\varepsilon \left(x_i - y_i - x_i(x_i^2 + y_i^2)\right) + \gamma\delta \cdot \varepsilon_x(x_i, y_i),\\
\dot{y}_i &= \gamma \|\hat{s}_i - s\|_\varepsilon \left(x_i + y_i - y_i(x_i^2 + y_i^2)\right) + \gamma\delta \cdot \varepsilon_y(x_i, y_i),
\end{aligned} \tag{6.68}
$$

where

$$
\begin{aligned}
\varepsilon_x(x_i(t), y_i(t)) &= x_i(t) - y_i(t) - x_i(t)(x_i^2(t) + y_i^2(t)),\\
\varepsilon_y(x_i(t), y_i(t)) &= x_i(t) + y_i(t) - y_i(t)(x_i^2(t) + y_i^2(t)).
\end{aligned}
$$

The right-hand side of (6.17)–(6.19) is locally Lipschitz in $\hat{s}_i, x_i$, and y_i (and so is the right-hand side of (6.22) and (6.23)). We denote its corresponding Lipschitz constant in the domain specified by (6.67) by symbol $L_i(D_0)$. Furthermore, provided that (6.67) holds, $\varepsilon_x(x_i(t), y_i(t))$ and $\varepsilon_y(x_i(t), y_i(t))$ are globally bounded with respect to t. Let us denote this bound by B:

$$\max\left\{\|\varepsilon_x(x_i(t), y_i(t))\|_{\infty,[0,\infty]}, \|\varepsilon_y(x_i(t), y_i(t))\|_{\infty,[0,\infty]}\right\} = B.$$

For the sake of notational compactness let us rewrite (6.68) as follows:

$$\dot{\mathbf{q}}_i = \mathbf{f}(\mathbf{q}_i, s(t), \xi(t)) + \gamma\delta \cdot \mathbf{g}(\mathbf{q}_i), \tag{6.69}$$

where $\mathbf{f}(\mathbf{q}_i, s(t), \xi(t))$ and $\mathbf{g}(\mathbf{q}_i)$ are defined to copy the right-hand side of (6.68). Notice that $\|\mathbf{f}(\mathbf{q}_i, s(t), \xi(t))\| \leq L_i(D_0)\|\mathbf{q}_i\|$ and $\|\mathbf{g}(\mathbf{q}_i)\| \leq B\sqrt{2}$.

According to the theorem on the continuous dependence of the solutions of an ordinary differential equation on the parameters and initial conditions (see, for instance, Khalil (2002), Theorem 3.4, page 96) the following holds:

$$\begin{aligned}\|\mathbf{q}_i(t) - \mathbf{q}_i^*(t)\| \leq \|\mathbf{q}_i(t_0) - \mathbf{q}_i^*(t_0))\|e^{L_i(D_0)(t-t_0)}\\ + \frac{\delta\gamma B\sqrt{2}}{L_i(D_0)}\left(e^{L_i(D_0)(t-t_0)} - 1\right).\end{aligned} \tag{6.70}$$

When the values of $\hat{s}_{i,0}$ and $\hat{s}_{i,0}^*$ coincide, estimate (6.70) implies that

$$\|\mathbf{q}_i(t) - \mathbf{q}_i^*(t)\| \leq \delta_r e^{L_i(D_0)(t-t_0)} + \frac{\delta\gamma B\sqrt{2}}{L_i(D_0)}\left(e^{L_i(D_0)(t-t_0)} - 1\right). \tag{6.71}$$

This assures the existence of $\delta_r \in \mathbb{R}_{>0}$ and $\delta \in \mathbb{R}_{>0}$ such that for a fixed, yet arbitrarily large, time $T''(\delta_r, \delta) > T'_{\max}$ solutions of the system (6.22) and (6.23) passing through a point from $\Omega(x', y')$ at $t = t_0$ will remain within a fixed, yet arbitrarily small, neighborhood of a solution of the system (6.17)–(6.19) with initial conditions $x_i(t_0) = x'$ and $y_i(t_0) = y'$. The value of $T'_{\max}$ does not depend on δ_r and δ.

Taking (6.33) into account, we can conclude that the set $x_i^2 + y_i^2 = 1$ is globally attracting in the state space of the system (6.22) and (6.23) for almost all initial conditions (except when $x_i(t_0) = 0$ and $y_i(t_0) = 0$). This implies that solutions starting in $\Omega(x', y')$ will remain there. In addition, according to (6.32), for any $t_0 \geq 0$ a δ_r-vicinity of (x', y') will be visited within at least time $t' \leq t_0 + 2\pi/(\gamma\cdot\delta)$. Hence we have just shown that for all $t_0 \geq 0$ solutions starting at $\Omega_s \times \Omega(x', y')$ approach the target set within a fixed time $T'_{\max}$ and stay in its vicinity for an arbitrarily long time $T''(\delta_r, \delta)$. The latter time is a function of δ_r and δ: the smaller the values of δ_r and δ, the larger the value of $T''(\delta_r, \delta)$.

4. Realizability. Let us finally show that the system (6.22) and (6.23) can be realized by an RNN. More precisely, we wish to prove that there exists a system (6.8) such that $\mathbf{x} = \boldsymbol{\zeta}_1 \oplus \boldsymbol{\zeta}_2 \oplus \cdots \oplus \boldsymbol{\zeta}_{N_f}$, $\boldsymbol{\zeta}_i \in \mathbb{R}^3$, $\boldsymbol{\zeta}_i = \zeta_{i,1} \oplus \zeta_{i,2} \oplus \zeta_{i,3}$, $i = \{1, \ldots, N_f\}$ and solutions $\boldsymbol{\zeta}_i(t, \mathbf{q}_{i,0})$ are sufficiently close to $\mathbf{q}_i(t, \mathbf{q}_{i,0})$, where $\mathbf{q}_{i,0} \in \Omega_s \times \Omega(x', y') \subset \mathbb{R}^3$.

It is clear that the right-hand side of (6.22) and (6.23) is a continuous and locally Lipschitz function. According to Cybenko (1989), for any arbitrarily small $\varepsilon_N \in \mathbb{R}_{>0}$, any given bounded intervals $\Omega_x \subset \mathbb{R}$ and $\Omega_y \subset \mathbb{R}$, and any

$$s(t), \xi(t): \ \max\{\|s(t)\|_{\infty,[0,\infty]}, \|\xi(t)\|_{\infty,[0,\infty]}\} < M, \ M \in \mathbb{R}_{>0},$$

there exist $N \in \mathbb{N}$, $\boldsymbol{\omega}_{j,m} \in \mathbb{R}^5$, $\alpha_{j,m} \in \mathbb{R}$, $\beta_{j,m} \in \mathbb{R}$, $j = 1, 2, \ldots, N$ such that

$$\begin{aligned} &\left| \sum_{m=1}^{N} \alpha_{j,m} \sigma(\boldsymbol{\omega}_{j,m}^{\mathrm{T}} \cdot \mathbf{u}(s(t), \xi(t), \boldsymbol{\zeta}_i) + \beta_{j,m}) \right. \\ &\qquad \left. - \mathbf{f}_j(\boldsymbol{\zeta}_i, s(t), \xi(t)) - \gamma\delta \cdot \mathbf{g}_j(\boldsymbol{\zeta}_i) \right| < \frac{\varepsilon_N}{3}, \\ &\mathbf{u}(s(t), \xi(t), \boldsymbol{\zeta}_i) = s(t) \oplus \xi(t) \oplus \boldsymbol{\zeta}_i, \end{aligned} \tag{6.72}$$

where $\boldsymbol{\zeta}_i \in \Omega_s \times \Omega_x \times \Omega_y$, and $\mathbf{f}_j(\boldsymbol{\zeta}_i, s(t), \xi(t))$ and $\mathbf{g}_j(\boldsymbol{\zeta}_i)$, $j = 1, 2, 3$ denote the jth components of the vector fields $\mathbf{f}(\boldsymbol{\zeta}_i, s(t), \xi(t))$ and $\mathbf{g}(\boldsymbol{\zeta}_i)$, respectively. It follows from (6.72) that there exist N, $\boldsymbol{\omega}_{j,m}$, $\alpha_{j,m}$, and $\beta_{j,m}$ such that

$$\begin{aligned} &\sum_{m=1}^{N} \alpha_{j,m} \sigma(\boldsymbol{\omega}_{j,m}^{\mathrm{T}} \cdot \mathbf{u}(s(t), \xi(t), \boldsymbol{\zeta}_i) + \beta_{j,m}) \\ &\quad = \mathbf{f}_j(\boldsymbol{\zeta}_i, s(t), \xi(t)) + \gamma\delta \cdot \mathbf{g}_j(\boldsymbol{\zeta}_i) + \Delta_j(\boldsymbol{\zeta}_i, s(t), \xi(t)), \\ &\quad \mathbf{u}(s(t), \xi(t), \boldsymbol{\zeta}_i) = s(t) \oplus \xi(t) \oplus \boldsymbol{\zeta}_i, \end{aligned} \tag{6.73}$$

where $\Delta_j(\boldsymbol{\zeta}_i, s(t), \xi(t))$ are continuous and

$$|\Delta_j(\boldsymbol{\zeta}_i, s(t), \xi(t))| < \frac{\varepsilon_N}{3} \ \ \forall\, j = 1, 2, 3.$$

Let us choose $\Omega_x = [-v, v]$ and $\Omega_y = [-v, v]$, where $v \in \mathbb{R}_{>0}$, $v > 1$ and consider the dynamics of

$$\begin{aligned} \dot{\boldsymbol{\zeta}}_i &= \mathbf{f}(\boldsymbol{\zeta}_i, s(t), \xi(t)) + \gamma\delta \cdot \mathbf{g}(\boldsymbol{\zeta}_i) + \Delta(\boldsymbol{\zeta}_i, s(t), \xi(t)), \\ \Delta(\boldsymbol{\zeta}_i, s(t), \xi(t)) &= \Delta_1(\boldsymbol{\zeta}_i, s(t), \xi(t)) \oplus \Delta_2(\boldsymbol{\zeta}_i, s(t), \xi(t)) \oplus \Delta_3(\boldsymbol{\zeta}_i, s(t), \xi(t)), \\ \|\Delta(\boldsymbol{\zeta}_i, s(t), \xi(t))\| &\le \varepsilon_N. \end{aligned} \tag{6.74}$$

System (6.74) has a globally attracting invariant set (for almost all initial conditions), which can be characterized as follows:

$$\{\boldsymbol{\zeta}_i \in \mathbb{R}^3 | 1 - \rho(\varepsilon_N) \leq \zeta_{i,2}^2 + \zeta_{i,3}^2 \leq 1 + \rho(\varepsilon_N)\}, \ \rho \in \mathcal{K}_\infty.$$

This follows immediately from the fact that (6.69) is structurally stable and has a globally attracting invariant set (for almost all initial conditions). Furthermore, for any given ε_N and a bounded set of initial conditions $\Omega_\zeta(r) = \{\boldsymbol{\zeta}_i \in \mathbb{R}^3 | \, \|\boldsymbol{\zeta}_i\| \leq r, \ r \in \mathbb{R}_{>0}\}$ there exists a constant B_1 such that $\|\boldsymbol{\zeta}_i(t)\|_{\infty,[0,\infty]} < B_1$. Hence solutions of the system

$$\begin{aligned} \dot{\zeta}_{i,j} &= \sum_{m=1}^{N} \alpha_{j,m}\sigma(\boldsymbol{\omega}_{j,m}^{\mathrm{T}} \cdot \mathbf{u}(s(t), \xi(t), \boldsymbol{\zeta}_i) + \beta_{j,m}), \\ \mathbf{u}(s(t), \xi(t), \boldsymbol{\zeta}_i) &= s(t) \oplus \xi(t) \oplus \boldsymbol{\zeta}_i, \ j = 1, 2, 3 \end{aligned} \tag{6.75}$$

are bounded for all initial conditions from $\Omega_\zeta(r)$, provided that inequality (6.72) holds over sufficiently large intervals Ω_x and Ω_y (for sufficiently large v). Furthermore, given that ε_N is sufficiently small, solutions of (6.75) enter the domain $\Omega_s \times \Omega(x', y')$ specified by (6.66) in finite time. Finally, according to equality (6.73) and Theorem 3.4 in Khalil (2002), solutions of (6.75) starting in $\Omega(x', y')$ satisfy the following inequality:

$$\begin{aligned} \|\mathbf{q}_i(t, \mathbf{q}_{i,0}) - \boldsymbol{\zeta}_i(t, \mathbf{q}_{i,0})\| &\leq \frac{\varepsilon_N}{L_i(D_0)} \left(e^{L_i(D_0)(t-t_0)} - 1 \right), \\ & \mathbf{q}_{i,0} \in \Omega_s \times \Omega(x', y'). \end{aligned} \tag{6.76}$$

Hence, for any $t \geq 0$, solutions of (6.75) starting from $\Omega_\zeta(r)$ approach the target set within a fixed time (dependent on δ) and stay in its vicinity arbitrarily long, provided that the values δ in (6.68) and ε_N in (6.72)–(6.74) are sufficiently small. The possibility of the latter follows from Cybenko (1989), i.e. the value of ε_N can be made arbitrarily small by appropriate choice of the parameters N, $\boldsymbol{\omega}_{j,m}$, $\alpha_{j,m}$, and $\beta_{j,m}$, and the value of δ can be made arbitrarily small because it is our design parameter.

Taking (6.76), (6.71), (6.25), and (6.22) into account, we conclude the proof by choosing $h_{f,i}(\mathbf{x}, s)$ and $h_{\theta,i}(\mathbf{x}, s)$ as follows:

$$\begin{aligned} h_{f,i}(\mathbf{x}, s) &= h_{f,i}(\boldsymbol{\zeta}_1 \oplus \cdots \oplus \boldsymbol{\zeta}_{N_f}, s) = s - \zeta_{i,1}, \\ h_{\theta,i}(\mathbf{x}, s) &= h_{\theta,i}(\boldsymbol{\zeta}_1 \oplus \cdots \oplus \boldsymbol{\zeta}_{N_f}, s) = a + \frac{b-a}{2}(\zeta_{i,2} + 1). \end{aligned} \tag{6.77}$$

□

Before concluding this section we would like to provide several remarks regarding Theorem 6.1.

Remark 6.1 As follows from the theorem, the class to which a given signal belongs can be determined from the values of $h_{f,j}(\mathbf{x}(t), s(t))$, $j = \{1, \ldots, N_f\}$ (specified, for example, by (6.77)) within a finite interval of time. When $s(t) = s_i(t, s_{i,0}, \theta_i, \eta_i(t))$ the values of $h_{f,i}(\mathbf{x}(t), s(t))$ should approach a small neighborhood of zero and stay there for a sufficiently long time. The estimate of θ_i up to its equivalence class is available from the values of $h_{\theta,i}(\mathbf{x}(t), s(t))$ over the same interval. As follows from our proof, the more accurately the RNN approximates system (6.68) the larger the interval of time during which $h_{f,i}(\mathbf{x}(t), s(t))$ is in the vicinity of zero. Indeed, if the approximation error is small, e.g. the value of ε_N in (6.72) is small, then for a given $\delta^* \in \mathbb{R}_{>0}$ the right-hand side of (6.76) should not exceed δ^* for all $t \in [t_0 + T]$:

$$0 \leq T = \frac{1}{L_i(D_0)} \ln \left(\frac{\delta^* L_i(D_0)}{\varepsilon_N} + 1 \right).$$

The smaller ε_N the larger the value of T and hence the longer the interval of time during which the trajectories of the RNN stay within the δ^*-neighborhood of the solutions of (6.68). The latter, as follows from (6.71), can be made arbitrarily close to that of the converging prototype (6.17)–(6.19) by choosing the value of δ sufficiently small. Thus the function $h_{f,i}(\mathbf{x}(t), s(t))$, as defined by (6.77), will asymptotically approach zero and stay in its close proximity for a sufficiently long time, subject to the choice of δ and ε_N.

From a practical viewpoint it might sometimes be preferable to readout from the RNN outputs directly, rather than having to satisfy ourselves with the existence of two sets of readout functions, for the state and input, respectively, of the RNN. Even though this option is not stated explicitly in Theorem 6.1, it can easily be shown that the preferred option can, indeed, be realized. Adding to the recurrent subsystem (6.8) a *feedforward* part realizing continuous "output" functions (6.77) enables direct readout from the RNN outputs.

Remark 6.2 Theorem 6.1 does not imply that recognition of the class of the input signal $s(t)$ involves convergence of the RNN state to an attractor. Yet its formulation does not exclude this option either. In fact, when $f_i(\xi(t), \theta_i)$ satisfies some additional restrictions (e.g. linear or monotone parametrization with respect to θ_i), it is possible to replace (6.18) and (6.19) with another prototype system: one that converges to a point attractor exponentially (see Theorem 5.9). This implies that it depends substantially on the properties of $f_i(\xi(t), \theta_i)$ whether the network state will behave intermittently or asymptotically converge to an attractor. It is important, however, that in both cases an RNN will successfully solve the recognition problem.

Remark 6.3 Even though the theorem applies to the case in which θ_i is a scalar, it can be trivially extended to the case in which the uncertain parameters are vectors from a bounded domain $\Omega_{\theta,d} \subset \mathbb{R}^d$. To do so one needs to find a Lipschitz mapping $\boldsymbol{\lambda} : \mathbb{R} \to \mathbb{R}^d$ such that for a given small $\varepsilon_\lambda \in \mathbb{R}_{>0}$ the following property holds:

$$\forall\, \boldsymbol{\theta}_i \in \Omega_{\theta,d}\ \exists\, \theta_i \in \Omega_\theta : \ \|\boldsymbol{\theta}_i - \boldsymbol{\lambda}(\theta_i)\| < \varepsilon_\lambda.$$

Hence the problem will reduce to the scalar case to which Theorem 6.1 applies.

Remark 6.4 Our proof of Theorem 6.1 not only demonstrates the possibility that an RNN can classify uncertain functions of time adaptively but also allows us to estimate the number of dynamic nodes, e.g. the value of N_x in (6.8), which is sufficient for successful classification. As follows immediately from (6.68), (6.74), and (6.75) the number of the dynamic states, N_x, can be as small as

$$N_x = 3N_f, \tag{6.78}$$

where N_f is the number of functions $f_i(\cdot,\cdot)$ to be classified. Further, suppose that N is the number of sigmoidal terms ensuring sufficiently accurate approximation of each function on the right-hand side of (6.68). Then the total number of sigmoidal functions in the network, N_{total}, can, in principle, be estimated as follows:

$$N_{\text{total}} = 3N_f N. \tag{6.79}$$

These estimates, despite inheriting a linear dependence on N_f, are still somewhat conservative. Simple numerical examples show that there is room for further improvements. Consider, for instance, the following set of signals:

$$\dot{s}_i = -s_i + f_i(\xi(t), \theta_i), \ i = \{1, 2\}, \ \theta_i \in [0, 4\pi], \tag{6.80}$$

where

$$\begin{aligned} f_1(\xi(t), \theta_1) &= \sin(\xi(t)\theta_1) + \cos(\xi(t)\theta_1), \\ f_2(\xi(t), \theta_i) &= \sin(\xi(t)\theta_2) + \cos^3(\xi(t)\theta_2), \ \xi(t) = \sin(t). \end{aligned} \tag{6.81}$$

According to (6.78) and (6.79) the number of the dynamic states in an RNN that adaptively classifies signals (6.80) and (6.81) could be as small as 6. Assuming that only two sigmoid functions are used to approximate the right-hand side of each equation of (6.68), the total number of sigmoidal nodes in the RNN is 12. In our numerical simulations we have found, however, that an RNN with as few as 10 recurrent nodes is able to solve the adaptive classification problem of signals $s_i(t)$ defined by (6.80) and (6.81). Results of this experiment are provided in Figure 6.4.

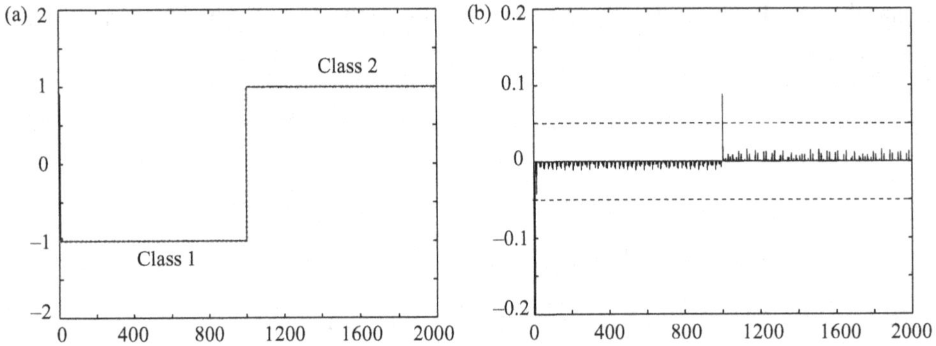

Figure 6.4 Adaptive classification of signals (6.78) and (6.79) by an RNN with 10 dynamic states. The network had two inputs, one for the signals $s_i(t)$ and the other for the signal $\xi(t)$, and one output. When $s_1(t)$ is present on the first input, the output should converge to -1; when signal $s_2(t)$ is present, the output should converge to 1. (a) The trajectories of the actual output (black solid line) and desired performance (gray dashed line) of the network as functions of model time. We started with signal $s_1(t)$, in which the value of θ_1 was set to $\theta_1 = 2\pi$. In the middle of the simulation we replaced $s_1(t)$ with $s_2(t)$, in which the value of θ_2 was set to $\theta_2 = 3\pi$. Even though the parameters and signals change, the network clearly solves the classification problem correctly. The same happens for other values of $\theta_i \in [0, 4\pi]$ for which the network was trained. (b) The difference between the actual and desired responses of the network. Always, after a short period of transient behavior, the error settles well within the standard 5% zone marked by two dashed lines.

In this example we used a network with 10 recurrent nodes. This number obviously exceeds the value provided by the estimate (6.78). On the other hand, adaptive classification is achieved by an RNN with a smaller total number of sigmoidal nodes than that predicted by (6.79). These results motivate further attention to this topic.

6.5 Summary

The result of this chapter is that we have shown how synthesis and analysis methods from the domain of adaptive control and regulation can be used to derive a formal proof that an RNN with fixed weights can serve as a universal adaptive classifier of both static and dynamic inputs. The number of dynamical states in an RNN recognizing N_f different classes of signals $s_i(t)$ can, according to our analysis, be as small as $3N_f$, i.e. it grows linearly with the size of the set of signals to be classified.

We stated the classification and recognition problems in a behavioral context in which, over time, the desired input–output relationship is achieved. Finding a solution corresponds to a network dynamics in which the state reaches a given neighborhood of the a-priori-specified set and stays there for a sufficiently long time, provided that the input to the network belongs to a given class (Problem 6.1). With these ramifications, RNNs solve the problem of adaptively classifying time-dependent signals. We did not set out to guarantee, however, that the state of the RNN will asymptotically converge to an equilibrium or its small vicinity as a result of recognition. On the other hand, the amount of time a network would spend in the vicinity of a target set can be made sufficiently large for this approach to qualify as a practical solution to the classification problem. For classification, after all, asymptotic convergence is not actually needed.

7

Adaptive template matching in systems for processing of visual information

Consider spatiotemporal pattern representation in the framework of template matching, the oldest and most common method for detecting an object in an image. According to this method the image is searched for items that match a template. A template consists of one or more local arrays of values representing the object, e.g. intensity, color, or texture. A similarity value between these templates and certain domains of the image is calculated,[1] and a domain is associated with the template once their similarity exceeds a given threshold.

Despite the simple and straightforward character of this method, its implementation requires us to consider two fundamental problems. The first relates to *what* features should be compared between the image $S_0(x, y)$ and the template $S_i(x, y)$, $i \in \mathcal{I}$. The second problem is *how* this comparison should be done.

The normative answer to the question of *what* features should be compared invokes solving the issue of optimal image representation, ensuring the most effective utilization of available resources and, at the same time, minimal vulnerability to uncertainties. Solutions in principle to this problem are well known from the literature and can be characterized as *spatial sampling*. For example, when the resource is the frequency bandwidth of a single measurement mechanism, the optimality of spatially sampled representations is proven in Gabor's seminal work (Gabor 1946).[2] In classification problems, the advantage of spatially sampled image representations is demonstrated in Ullman *et al.* (2002). In general, these representations are obtained naturally on balancing the system's resources and uncertainties in the measured signal.

[1] Traditionally a correlation measure is commonly used for this purpose (Jain *et al.* 2000).

[2] Consider, for instance, a system that measures an image $S_i(x, y)$ using a set of sensors $\{m_1, \ldots, m_n\}$. Each sensor m_i is capable of measuring signals within the given frequency band Δ_i at the location x_i in the corresponding spatial dimension x. Then, according to Gabor (1946), sensor m_i can measure both the frequency content of a signal and its spatial location with minimal uncertainty only if the signal has a Gaussian envelope in x: $S_i(x, y) \sim e^{\sigma_i^{-2}(x-x_i)^2}$. In other words, the signal should be practically spatially bounded. This implies that the image must be spatially sampled.

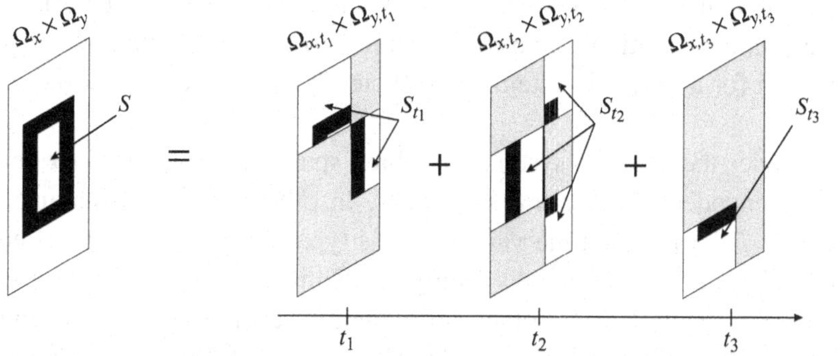

Figure 7.1 Spatial sampling of an image $S(x, y) : \Omega_x \times \Omega_y \to \mathbb{R}_{\geq 0}$ according to the factorization of $\Omega_x \times \Omega_y$ into subsets $\Omega_{x,t_1} \times \Omega_{y,t_1}$, $\Omega_{x,t_2} \times \Omega_{y,t_2}$, and $\Omega_{x,t_3} \times \Omega_{y,t_3}$.

A variety of sophisticated spatial sampling methods exists (Gabor 1946; Blake *et al.* 1994; Bueso *et al.* 1999; Lee and Yuille 2006). Here we limit ourselves to spatial sampling in its elementary form, which is achieved by factorizing both the domain $\Omega_x \times \Omega_y$ of the image S_0 and the templates S_i, $i \in \mathcal{I}$ into subsets:

$$\Omega_x \times \Omega_y = \bigcup_t \Omega_{x,t} \times \Omega_{y,t}, \; t \in \Omega_t, \; \Omega_{x,t} \subseteq \Omega_x, \; \Omega_{y,t} \subseteq \Omega_y. \tag{7.1}$$

Factorization (7.1) induces sequences $\{S_{i,t}\}$, where $S_{i,t}$ are the restrictions of mappings S_i to the domains $\Omega_{x,t} \times \Omega_{y,t}$. These sequences constitute sampled representations of S_i, $i \in \mathcal{I}^+$ (see Figure 7.1). Notice that the sampled image and template representations $\{S_{i,t}\}$ are, strictly speaking, sequences of functions. In order to compare them, scalar values $f(S_{i,t})$ are normally assigned to each $S_{i,t}$. Examples include various functional norms, correlation functions, spectral characterizations (average frequency or phase), or simply weighted sums of the values of $S_{i,t}$ over the entire domain $\Omega_{x,t} \times \Omega_{y,t}$. Formally, f could be defined as a functional, which maps restrictions $S_{i,t}$ into the field of real numbers:

$$f : L_\infty(\Omega_{x,t} \times \Omega_{y,t}) \to \mathbb{R}. \tag{7.2}$$

This formulation allows a simple representation of images and templates as sequences of scalar values $\{f(S_{i,t})\}$, $i \in \mathcal{I}^+$, $t \in \Omega_t$. We will therefore adopt this method here.

The answer to the second question, *how* the comparison is done, involves finding the best and simplest way possible to utilize the information that a given image representation provides, while at the same time ensuring invariance with respect to basic distortions. Despite the fact that considerable attention has been given to this problem, a unified solution in principle is not yet available. The primary goal of our current contribution is to present a unified framework within which to solve this problem for a class of systems of sufficiently broad theoretical and practical relevance.

We consider the class of systems in which spatially sampled image representations are encoded as temporal sequences. In other words, the parameter t in the notation $f(S_{i,t})$ is the time variable. This type of representation is frequently encountered in neuronal networks (Gutig and Sompolinsky 2006) (see also references therein). Examples of similar representation schemes are widely reported in the neuroscience literature. For example, Alonso *et al.* (1996) show that patches of visual stimuli that are perceived as spatially close by the processing system (e.g. when the receptive fields of individual cells overlap) are encoded by similar firing spike patterns and vice versa. In our model spatially non-overlapping patches are represented by different sequences $\{f(S_{i,t})\}$, and identical images have identical temporal representations. Hence, such systems have a claim to biological plausibility. In addition, they enable a simple solution to a well-known dilemma. This is about whether comparison between templates and image domains should be made on a large, i.e. global, or on a small, i.e. local, scale. The solution to this dilemma consists in temporal integration. Let, for instance, $\Omega_t = [0, T]$, $T \in \mathbb{R}_{>0}$. Then an example of a temporally integral, yet spatially sampled, representation is

$$f(S_{i,t}) \mapsto \phi_i(t) = \int_0^t f(S_{i,\tau})d\tau, \ t \in [0, T], \ i \in \mathcal{I}^+. \tag{7.3}$$

The temporal integral $\phi_i(t)$ contains both spatially local and global image characterizations. Whereas its time-derivative at t equals to $f(S_{i,t})$ and corresponds to the spatially sampled, local representation $S_{i,t}$, the global representation $\phi_i(T)$ is equal to the integral, cumulative characterization of S_i. An example illustrating these properties is provided in Figure 7.2. A further advantage of spatiotemporal representations $\phi_i(t)$ is that they offer powerful mechanisms for comparison, processing, and matching of $\phi_i(t)$, $i \in \mathcal{I}$. These mechanisms can generally be characterized in terms of dynamic oscillator networks that synchronize when their inputs are converging to the same function.

Despite advantages such as optimality, simplicity, and biological plausibility, there are theoretical issues that have prevented wide application of spatiotemporal representations to template matching. The most important issues, from our point of view, are, first, how to achieve effective recognition in the presence of modeled disturbances, of which the most common ones are blur, luminance, and rotational and translational distortions; and second, how to take into account inevitable unmodeled perturbations.

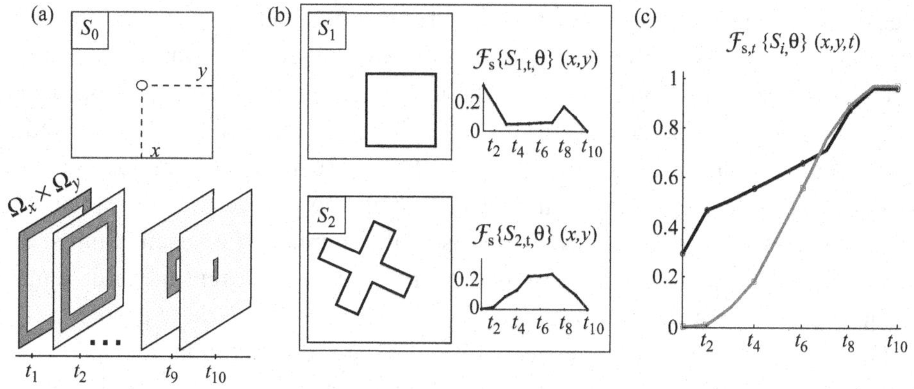

Figure 7.2 Spatiotemporal image representation via spatial sampling and temporal integration. Image (a) contains the original object, S_0; (x, y) marks a point on the image with respect to which the correlation is calculated; factorization of the domain $\Omega_x \times \Omega_y$ into 10 arbitrary (here chosen to be non-intersecting) subsets $\Omega_x \times \Omega_y = \cup_{j=1}^{10} \Omega_{x,t_j} \times \Omega_{y,t_j}$. In (b) templates S_1 and S_2 and plots of $f_{t_j}(S_{1,t_j})(x, y)$ and $f_{t_j}(S_{2,t_j})(x, y)$ – the values of the normalized correlation between $S_{i,t_j} = S_i(\Omega_{x,t_j} \times \Omega_{y,t_j})$ and $S_0(\Omega_{x,t_j} \times \Omega_{y,t_j})$ – are shown. (c) Plots of the values of (7.3) as a function of the parameter t for templates S_1 (gray line) and S_2 (black line).

The first class of problems amounts to finding a possible transformation of the template that can model the disturbance. Similarly to the framework of deformable templates (Amit 2002; Amit *et al.* 1991), we assume a disturbance model to be a mapping that maps the template, S_i, into the image, S_0. Unlike in traditional deformable-template approaches (Miller and Younes 2001), we do not wish to assume, however, that this mapping is invertible, or forms a group action. This is because we would like to enable multiple solutions to the matching problem, as is often the case in biological vision. Furthermore, even when the transformation is invertible, the inverse operation could be highly susceptible to small image noise, which, for instance, is the case for integration/differentiation operations. Finally, for the sake of computational effectiveness we would like to refrain from posing the matching problem as an optimization problem in the space of functions (transformations).

For these reasons we will consider modeled disturbances as known, yet *nonlinearly parametrized mappings*. The parameters of these mappings, however, are allowed to be uncertain. This enables us to consider standard group actions like rotation, translation, or scaling as a special case of perturbations. In addition, it allows us to consider non-invertible and generally nonlinear transformations. Last, but not least, in the context of our current approach finding a suitable transformation of the template amounts to designing a *dynamic* identification/adaptation algorithm with proven efficiency in reconstructing parameters of generally nonlinear perturbations. The latter is an optimization problem in finite (low)-dimensional space compared with optimization in the infinite-dimensional space of functions. The currently available approaches to designing such algorithms are restricted to linear parametrization of disturbances, involve overparametrization, or use domination feedback. Yet, linear parametrization is too restricted to be plausible, overparametrization is expensive in terms of the number of adjustable units, and domination lacks adequate sensitivity. For these reasons current methods remain unsatisfactory. We will, therefore, propose a novel solution to these problems.

The second class of problems, recognition in the presence of unmodeled perturbations, calls for procedures for recognizing an image from its perturbed temporal representation $\phi_i(t)$. At this level the system is facing contradictory requirements of ensuring robust performance while being highly sensitive to minor changes in the stimulation. Here, too, we will advocate a solution.

The proposed solution to both types of problem diverges from current approaches, which invoke the concept of Lyapunov-stable attractors. We concur that, by allowing the system to converge on an attractor, these methods are able to eliminate modeled and unmodeled distortions and thus, for instance, complete an incomplete pattern in the input (Amit *et al.* 1985; Fuchs and Haken 1988; Herz *et al.* 1989; Hopfield 1982; Ritter and Kohonen 1989). The strength of these systems resides in the robustness inherent to uniform asymptotic Lyapunov stability. There is, however, a corresponding weakness: such systems are generally lacking in flexibility. Each stable attractor in the system represents one pattern, but often an image contains more than one pattern. When the system is steered to one template, the other is lost from the representation. It would, therefore, be preferable to have a system that allows flexible switching between alternative patterns. Yet, the very notion of stable convergence to an attractor provides an obstacle to switching and exploration across patterns. Furthermore, as we will show, for a class of images with multiple representations and various symmetries globally stable solutions to the problem of invariant template matching might not even exist.

We propose a unifying framework capable of combining robustness and flexibility. In contrast to common intuitions, which aim at achieving the desired robustness by means of stable attractors, we advocate instability as an advantageous substitute. More precisely, we consider a specific type of instability inherent to solutions converging to proper weakly attracting sets and Milnor attractors (Milnor 1985). The utility of the notion of weakly attracting sets has already been acknowledged in the general context of modeling brain activity and decision-making. For example, in networks of nonlinear oscillators and coupled maps the emergence of Milnor attractors is considered as a precursor for chaotic itinerancy – the system's dynamic state corresponding to sporadic chaotic switching of trajectories from one quasi-attractor

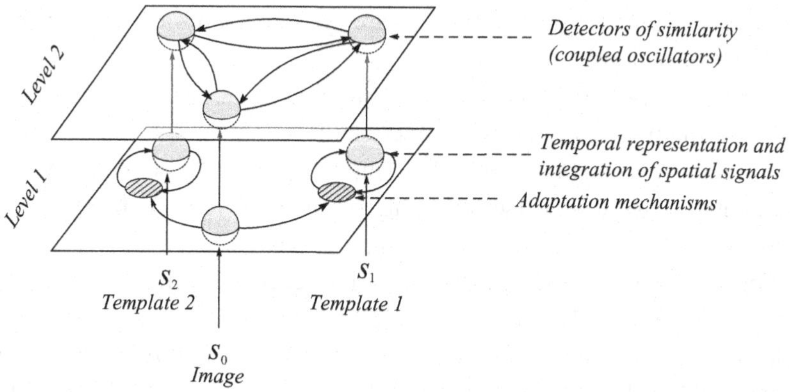

Figure 7.3 The general scheme of a system for adaptive template matching using temporal codes. Level 1 contains the adaptive compartments. Its functional role is to ensure invariance with respect to modeled uncertainties. Level 2 corresponds to the comparison compartments and consists of coupled nonlinear oscillators. The solid arrows represent the information flow in the system.

(ghost attractor) to another (Tsuda and Fujii 2007).[3] Here we demonstrate that the concept of weakly attracting sets provides a unifying framework for solving pattern-recognition problems such as that of template matching. We show that trading the habitual requirement of Lyapunov stability for a more relaxed property of convergence to weakly attracting sets provides both the necessary invariance and the flexibility needed. Doing so, furthermore, allows us to overcome challenging technical issues related to nonlinearity and non-convexity of modeled uncertainties with respect to the tuned parameters. Finally, the relevance of Milnor attractors has been argued extensively for models of information processing in the human brain (see van Leeuwen (2008) for a review). We will briefly illustrate with examples how Milnor attractors could instill functionality in the brain.

To illustrate these principles we designed a recognition system consisting of two major components (see Figure 7.3). The first is an adaptive component in which information is processed by a class of spatiotemporal filters. These filters represent internal models of distortions. The models of most common distortions, including rotation, translation, and blur, are often nonlinearly parametrized. Until recently adequate compensatory mechanisms for nonlinear parametrized uncertainties have been unexplored territory. In Tyukin and van Leeuwen (2005) it has been shown that the problem of non-dominating adaptation could, in principle, be solved within the concept of Milnor or weak, unstable attractors. Here we provide a solution to this problem that will enable systems to deal with specific nonlinearly

[3] See also the related concepts of heteroclinic channels (Rabinovic *et al.* 2008) and relaxation times (Gorban 2004).

parametrized models of distortions that are typical for a variety of optical and geometric perturbations.

The second major component of our system consists of a network of coupled nonlinear oscillators. These operate as coincidence detectors. Each oscillator in our system represents a Hindmarsh–Rose-model neuron. These model neurons are generally believed to provide a good qualitative approximation to biological neuron behavior. At the same time they are computationally cost-effective (Izhikevich 2004). For networks of these oscillators we prove, first of all, the boundedness of the state of the perturbed solutions. In addition, we specify the parameter values which lead to the emergence of globally stable invariant manifolds in the system's state space. Although we do not provide explicit criteria for the meta-stability that quasi-attractors provide in this class of networks, the conditions presented allow us to narrow substantially the domain of relevant parameter values in which this behavior is to be found.

There is an interesting consequence to the unstable character of the compensation for modeled perturbations. When the system negotiates multiple classes of uncertainties simultaneously (e.g. focal/contrast and intensity/luminance), different types of compensatory adjustments are made at different time scales. Adaptation at different time scales is a well-known phenomenon in biological visual systems (Baccus and Meister 2002; Demontis and Cervetto 2002; Sharpe and Stockman 1999; Smirnakis *et al.* 1997), in particular when light/dark adaptation is combined with optical/neuronal blur (Hofer and Williams 2002; Mather 2006; Mon-Williams *et al.* 1998; Rodieck 1998). Our analysis below suggests that this difference in time scales emerges naturally as a sufficient condition for the proper operation of our system.

7.1 Preliminaries and problem formulation

We assume that the values $S_0(x, y)$ of the original image are not available explicitly to the system; the system is able to measure only perturbed values of $S_0(x, y)$. Perturbation is defined as a mapping $\mathcal{F}$:

$$\mathcal{F}[S_0, \boldsymbol{\theta}] : L_\infty(\Omega_x \times \Omega_y) \times \mathbb{R}^d \to L_\infty(\Omega_x \times \Omega_y),$$

where $\boldsymbol{\theta}$ is the vector of parameters of the perturbation. The values of $\boldsymbol{\theta}$ are assumed to be unknown a priori, whereas the mapping $\mathcal{F}$ is known.

In systems for processing spatial information, mappings $\mathcal{F}$ often belong to a specific class that can be defined as follows:

$$\begin{aligned} \mathcal{F}[S_0, \boldsymbol{\theta}] &= \theta_1 \cdot \bar{\mathcal{F}}[S_0, \theta_2],\ \theta_1 \in \mathbb{R},\ \theta_2 \in \mathbb{R}, \\ \bar{\mathcal{F}}[S_0, \theta_2] &: L_\infty(\Omega_x \times \Omega_y) \times \mathbb{R} \to L_\infty(\Omega_x \times \Omega_y), \\ \boldsymbol{\theta} &= (\theta_1, \theta_2). \end{aligned} \tag{7.4}$$

Table 7.1. *Examples of typical nonlinear perturbations of S_0. The parameter Δ_θ in the right-hand column is a positive constant*

Physical meaning	Mathematical model of $\bar{\mathcal{F}}[S_0, \theta_2]$	Domain of relevance
Translation (in x), θ_2 – shift	$\bar{\mathcal{F}}[S_0, \theta_2] = S_0(x + \theta_2, y)$	$\theta_2 \in [-\Delta_\theta, \Delta_\theta]$
Scaling (in x), θ_2 – scaling factor	$\bar{\mathcal{F}}[S_0, \theta_2] = S_0(\theta_2 \cdot x, y)$	$0 < \theta_2 \leq \Delta_\theta$
Rotation around the origin, θ_2 – rotation angle	$\bar{\mathcal{F}}[S_0, \theta_2] = S_0(x_r(x, y, \theta_2), y_r(x, y, \theta_2))$, $x_r(x, y, \theta_2) = \cos(\theta_2)x - \sin(\theta_2)y$, $y_r(x, y, \theta_2) = \sin(\theta_2)x + \cos(\theta_2)y$	$0 \leq \theta_2 \leq 2\pi$
Image blur (not normalized), θ_2 – blur parameter	$\bar{\mathcal{F}}[S_0, \theta_2] = \int_{\Omega_x \times \Omega_y} h \cdot S_2(\xi, \gamma) d\xi \, d\gamma$ 1) Gaussian: $h = e^{-\frac{1}{\theta_2^2}((x-\xi)^2 + (y-\gamma)^2)}$ 2) Out-of-focus: $h = \begin{cases} 1/(\pi\theta_2^2), & r(x, y) \leq \theta_2, \\ 0, & \text{else}, \end{cases}$ $r(x, y) = \sqrt{(x - \xi)^2 + (y - \gamma)^2)}$	$0 < \theta_2 \leq \Delta_\theta$

The parameter $\theta_1 \in [\theta_{1,\min}, \theta_{1,\max}] \subset \mathbb{R}$ in (7.4) models *linear* perturbations, for instance variations of overall brightness or intensity of the original image S_0. The parameter can also be interpreted as an a-priori-unknown gain in the measurement channel of a sensor. The mapping $\bar{\mathcal{F}}(S_0, \theta_2)$ in (7.4), parametrized by $\theta_2 \in [\theta_{2,\min}, \theta_{2,\max}] \subset \mathbb{R}$, corresponds to typical *nonlinear* perturbations of image S_0 such as image blur (Banham and Katsaggelos 1997). Table 7.1 provides examples of these perturbations, their mathematical models, and the physical meaning of the parameter θ_2. Throughout this discussion we assume that the mappings $\bar{\mathcal{F}}[S_0, \theta_2]$ are Lipschitz in θ_2:

$$\exists\, D \in \mathbb{R}_{>0}: \ \left|\bar{\mathcal{F}}[S_0, \theta_2'](x, y) - \bar{\mathcal{F}}[S_0, \theta_2''](x, y)\right| \leq D|\theta_2' - \theta_2''|,$$
$$\forall\, (x, y) \in \Omega_x \times \Omega_y, \theta_2', \theta_2'' \in \mathbb{R}. \tag{7.5}$$

Notice that, strictly speaking, several typical transformations, such as translation, scaling, and rotation, are not always Lipschitz. This is because the image S_0 can, for instance, have sharp edges corresponding to discontinuities in x and y. In practice, however, prior application of a blurring linear filter will render sharp edges in an image smooth, thus assuring that condition (7.5) applies.[4]

The image $\mathcal{F}[S_0, \boldsymbol{\theta}]$ is assumed to be spatially sampled according to factorization (7.1):

$$\mathcal{F}_t[S_0, \boldsymbol{\theta}](x, y) = \begin{cases} \mathcal{F}[S_0, \boldsymbol{\theta}](x, y), & (x, y) \in \Omega_{x,t} \times \Omega_{y,t}, \\ 0, & \text{else}, \end{cases} \quad t \in \Omega_t. \tag{7.6}$$

Because the index t in (7.6) is assumed to be a time variable we let $\Omega_t = [0, \infty)$. To each $\mathcal{F}_t[S_0, \boldsymbol{\theta}]$ a value $f(\mathcal{F}_t[S_0, \boldsymbol{\theta}]) \in \mathbb{R}$ is assigned. Formally this procedure can be defined by a functional that maps mappings $\mathcal{F}_t[S_0, \boldsymbol{\theta}]$ into the real values:

$$f : L_\infty(\Omega_x \times \Omega_y) \to \mathbb{R}. \tag{7.7}$$

In the singular case, when $\Omega_{x,t} \times \Omega_{y,t}$ is a point (x_t, y_t), the functional f and mapping $\mathcal{F}_t[S_0, \boldsymbol{\theta}](x, y)$ will be defined as $f = \mathcal{F}_t[S_0, \boldsymbol{\theta}](x_t, y_t) = \mathcal{F}[S_0, \boldsymbol{\theta}](x_t, y_t)$.

We concentrated our efforts on obtaining a solution in principle to the problem of invariant template matching in systems with spatiotemporal processing of information. For this reason we prefer not to provide a specific description of the functionals f. We do, however, restrict our consideration to *linear* and Lipschitz functionals, e.g. the functionals satisfying the following constraints:

$$f(\kappa \mathcal{F}) = \kappa f(\mathcal{F}), \ \forall \kappa \in \mathbb{R}, \ \left| f(\mathcal{F}) - f(\mathcal{F}') \right| \le D_2 \|\mathcal{F} - \mathcal{F}'\|_\infty, \\ D_2 \in \mathbb{R}_{>0}. \tag{7.8}$$

Examples of functionals f satisfying conditions (7.8) and their physical interpretations are provided in Table 7.2.

Taking into account (7.4), (7.6), and the fact that f is linear, the following equality holds:

$$f(\mathcal{F}_t[S_0, \boldsymbol{\theta}]) = \theta_1 f(\bar{\mathcal{F}}_t[S_0, \theta_2]), \\ \bar{\mathcal{F}}_t[S_0, \theta_2] = \begin{cases} \bar{\mathcal{F}}[S_0, \theta_2](x, y), & (x, y) \in \Omega_{x,t} \times \Omega_{y,t}, \\ 0, & \text{else}. \end{cases} \tag{7.9}$$

For the sake of compactness, in what follows we replace $f(\bar{\mathcal{F}}_t[S_0, \theta_2])$ in the definition of $f(\mathcal{F}_t[S_0, \boldsymbol{\theta}])$ in (7.9) with the following notation:

$$f(\bar{\mathcal{F}}_t[S_0, \theta_2]) = f_0(t, \theta_2), \ f_0 : \Omega_t \times \mathbb{R} \to \mathbb{R}. \tag{7.10}$$

[4] In biological vision discontinuity of S_0 in x and y corresponds to images with abrupt local changes in brightness in the spatial dimensions x and y. Although this is a rather common situation in nature, in visual systems actual images S_0 rarely reach a sensor in their spatially discontinuous form. In fact, prior to reaching the sensory receptors, they are subject to optical linear filtering. Therefore the images that reach the sensor are always smooth. Hence condition (7.5) will generally be satisfied.

Table 7.2. *Examples of spatially-sampled representations of* S_0

Physical meaning	Mathematical model of f
Spectral power within the given frequency bands: $\omega_x \in [\omega_a, \omega_b]$, $\omega_y \in [\omega_c, \omega_d]$	$f = \int_{\omega_a}^{\omega_b} \int_{\omega_c}^{\omega_d} \left\| \int_{\Omega_x \times \Omega_y} \mathcal{F}_t[S_0, \boldsymbol{\theta}](x, y) \times e^{-j(\omega_x x + \omega_y y)} dx\, dy \right\| d\omega_x\, d\omega_y$
Weighted sum (for instance, convolution with exponential kernel)	$f = \int_{\Omega_x \times \Omega_y} \mathcal{F}_t[S_0, \boldsymbol{\theta}](x, y) e^{-\lvert x - x_0 \rvert - \lvert y - y_0 \rvert} dx\, dy$ (x_0, y_0) is the reference, "attention" point
Scanning the image along a given trajectory $(x(t), y(t)) = (\xi(t), \gamma(t))$	$\Omega_{x,t} \times \Omega_{y,t} = (\xi(t), \gamma(t))$ $f = \mathcal{F}[S_0, \boldsymbol{\theta}](\xi(t), \gamma(t))$

The notation $f_0(t, \theta_2)$ in (7.10) allows us to emphasize the dependence of f on the unknown θ_2, time variable t, and original image S_0. The subscript "0" in (7.10) indicates that $f_0(t, \theta_2)$ corresponds to the sampled and perturbed S_0 ((7.4), (7.7), and (7.8)), and the argument θ_2 is the nonlinear parameter of the perturbation applied to the image. Adhering to this logic, we introduce the notation

$$f(\mathcal{F}_t[S_i, \boldsymbol{\theta}]) = \theta_1 f(\bar{\mathcal{F}}_t[S_i, \theta_2]) = \theta_1 f_i(t, \theta_2),$$

where the subscript "i" indicates that $f_i(t, \theta_2)$ corresponds to the perturbed and sampled template S_i, and θ_2 is the nonlinear parameter of the perturbation applied to the template.

Let us now specify the class of schemes realizing temporal integration of spatially sampled image representations. Explicit realization of the temporal integration (7.3) is not feasible because it may lead to unbounded outputs for a wide class of relevant signals, for instance signals that are constant or periodic with a non-zero average. The behavior of a temporal integrator (7.3), however, can be fairly well approximated by a first-order linear filter. For the sampled image and template representations $\theta_1 f_i(t, \theta_2)$, these filters can be defined as follows:

$$\begin{aligned} \dot{\phi}_0 &= -\frac{1}{\tau}\phi_0 + k \cdot \theta_1 f_0(t, \theta_2), \\ \dot{\phi}_i &= -\frac{1}{\tau}\phi_i + k \cdot \theta_1 f_i(t, \theta_2), \ k, \tau \in \mathbb{R}_{>0}, \ i \in \mathcal{I}. \end{aligned} \tag{7.11}$$

In contrast to (7.3), for filters (7.11) it is ensured that their state remains bounded for bounded inputs. In addition, to a first approximation, the equations in (7.11) present a simple model of neural sensors collecting and encoding spatially distributed information in the form of a function of time.[5] With respect to the physical realizability of (7.11), in addition to requirements (7.5) and (7.8) we shall assume only that spatially sampled representations $\theta_1 f_i(t, \theta_2)$, $i \in \mathcal{I}^+$ of S_i ensure the existence of solutions for the system (7.11).

Consider the dynamics of variables $\phi_0(t)$ and $\phi_i(t)$, $i \in \mathcal{I}$ defined by (7.11). We say that the ith template matches the image iff for some given $\varepsilon \in \mathbb{R}_{\geq 0}$ the following condition holds:

$$\limsup_{t \to \infty} |\phi_0(t) - \phi_i(t)| \leq \varepsilon. \tag{7.12}$$

The problem, however, is that the parameters θ_1 and θ_2 in (7.11) are unknown a priori. While perturbations affect the image directly, they do not necessarily influence the templates. Rather to the contrary, for consistent recognition the templates are better kept isolated from external perturbations – at least within the time frame of pattern recognition, although they may, of course, be affected by adaptive learning on a larger time scale. Having fixed, unmodified templates in comparison with perturbed image representations implies that, even when objects corresponding to the templates are present in the image, the temporal image representation $\phi_0(t)$ will likely be different from any of the templates, $\phi_i(t)$. This will render the chance that condition (7.12) is satisfied very small, so a template would hardly ever be detected in an image.

We propose that the proper way for a system to meet requirement (7.12) is to mimic the effect of disturbances in the template. In order to achieve this, a template-matching system should be able to track the unknown values of the parameters θ_1 and θ_2. Hence the original equations for temporal integration (7.11) will be replaced with the following:

$$\begin{aligned} \dot{\phi}_0 &= -\frac{1}{\tau}\phi_0 + k \cdot \theta_1 f_0(t, \theta_2), \\ \dot{\phi}_i &= -\frac{1}{\tau}\phi_i + k \cdot \hat{\theta}_{i,1} f_i(t, \hat{\theta}_{i,2}), \ k, \tau \in \mathbb{R}_{>0}, \ i \in \mathcal{I}, \end{aligned} \tag{7.13}$$

where $\hat{\theta}_{i,1}$ and $\hat{\theta}_{i,2}$ are the estimates of θ_1 and θ_2. The estimates $\hat{\theta}_{i,1}$ and $\hat{\theta}_{i,2}$ must track instantaneous changes of θ_1 and θ_2. The information required for such an estimation should be kept to a minimum. An acceptable solution would be a simple mechanism capable of tracking the perturbations from the measurements of the image alone. The formal statement of this problem is provided below.

[5] In principle, (7.11) can be replaced with a more plausible model of temporal integration such as integrate-and-fire, Fitzhugh–Nagumo, or Hodgkin–Huxley-model neurons. These extensions, however, are not immediately relevant for the purpose of our current study. We decided to keep the mathematical description of the system as simple as possible, keeping in mind the possibility of extension to a wider class of temporal integrators (7.11).

Problem 7.1 For a given image S_0, template S_i, and their spatiotemporal representations satisfying (7.5), (7.8), and (7.13), find estimates

$$\hat{\theta}_{i,1} = \hat{\theta}_{i,1}(t, \tau, \kappa, \phi_0, \phi_i), \qquad \hat{\theta}_{i,2} = \hat{\theta}_{i,2}(t, \tau, \kappa, \phi_0, \phi_i) \tag{7.14}$$

as functions of time t, variables ϕ_0 and ϕ_i and parameters τ and κ such that for all possible values of parameters $\theta_1 \in [\theta_{1,\min}, \theta_{1,\max}]$, $\theta_2 \in [\theta_{2,\min}, \theta_{2,\max}]$

(1) solutions of system (7.13) are bounded;
(2) in the case $f_0 = f_i$ property (7.12) is ensured;
(3) the following holds for some $\theta_1' \in [\theta_{1,\min}, \theta_{1,\max}]$, $\theta_2' \in [\theta_{2,\min}, \theta_{2,\max}]$:

$$\begin{aligned} \limsup_{t\to\infty} |\hat{\theta}_{i,1}(t, \tau, \kappa, \phi_0(t), \phi_i(t)) - \theta_{i,1}'| &\leq \varepsilon_{\theta,1}, \ \varepsilon_{\theta,1} \in \mathbb{R}_{\geq 0}, \\ \limsup_{t\to\infty} |\hat{\theta}_{i,2}(t, \tau, \kappa, \phi_0(t), \phi_i(t)) - \theta_{i,2}'| &\leq \varepsilon_{\theta,2}, \ \varepsilon_{\theta,2} \in \mathbb{R}_{\geq 0}. \end{aligned} \tag{7.15}$$

Once the solution to Problem 7.1 has been found, the next step is to ensure that similarities (7.12) are registered in the system. In line with Figure 7.3, we propose that the detection of similarities is realized by a system of coupled oscillators. In particular, we require that states of oscillators i and 0 converge as soon as the signals $\phi_0(t)$ and $\phi_i(t)$ become sufficiently close.

In the present discussion we restrict ourselves to the class of systems composed of linearly coupled Hindmarsh–Rose-model neurons (Hindmarsh and Rose 1984). This choice is motivated by the fact that these oscillators can reproduce a broad class of behaviors observed in real neurons while being computationally efficient (Izhikevich 2004). A network of these neural oscillators can be mathematically described as follows:

$$S_{D_i} : \left\{ \begin{aligned} \dot{x}_i &= -a x_i^3 + b x_i^2 + y_i - z_i + I + u_i + \phi_i(t), \\ \dot{y}_i &= c - d x_i^2 - y_i, \\ \dot{z}_i &= \varepsilon(s(x_i + x_0) - z_i), \end{aligned} \right. \qquad i \in \mathcal{I}^+. \tag{7.16}$$

The variables x_i, y_i, and z_i correspond to the membrane potential, and aggregated fast and slow adaptation currents, respectively. The coupling u_i in (7.16) is assumed to be linear and symmetric:

$$\mathbf{u} = \begin{pmatrix} u_0 \\ u_1 \\ \vdots \\ u_n \end{pmatrix} = \Gamma \begin{pmatrix} x_0 \\ x_1 \\ \vdots \\ x_n \end{pmatrix}, \ \Gamma = \gamma \begin{pmatrix} -n & 1 & \cdots & 1 \\ 1 & -n & \cdots & 1 \\ \cdots & \cdots & \cdots & \cdots \\ 1 & 1 & \cdots & -n \end{pmatrix}, \tag{7.17}$$

and parameter $\gamma \in \mathbb{R}_{\geq 0}$. Our choice of the coupling function in (7.17) is motivated by the following considerations. First, we wish to preserve the intrinsic dynamics of the neural oscillators when they synchronize, e.g. when $x_i = x_j$, $y_i = y_j$, and $z_i = z_j, i, j \in \{0, \ldots, n\}$. For this reason it is desirable that the coupling vanishes when the synchronous state is reached. Second, we seek a system in which synchronization between two arbitrary nodes, say the ith and jth nodes, is determined exclusively by the degree of (mis)matches in $\phi_i(t)$ and $\phi_j(t)$, independently of the activity of other units in the system. Third, the coupling should "pull" the system's trajectories towards the synchronous state. The coupling function (7.17) satisfies all these requirements.

We set the parameters of (7.16) to the following values:

$$\begin{aligned} a &= 1, \quad b = 3, \quad c = 1, \quad d = 5, \\ s &= 4, \quad x_0 = 1.6, \quad \varepsilon = 0.001, \end{aligned} \tag{7.18}$$

which correspond to the regime of chaotic bursting in each uncoupled element in (7.16) (Hansel and Sompolinsky 1992).

The problem of detection of similarities in $\phi_0(t)$ and $\phi_i(t)$ can now be stated as follows.

Problem 7.2 Let the system (7.16) and (7.17) be given and let there exist $i \in \mathcal{I}$ such that condition (7.12) is satisfied. Determine the coupling parameter γ as a function of the parameters of system (7.16) such that

(1) solutions of the system are bounded for all bounded ϕ_i, $i \in \mathcal{I}$;
(2) states $(x_0(t), y_0(t), z_0(t))$ and $(x_i(t), y_i(t), z_i(t))$ asymptotically converge to a vicinity of the synchronization manifold $x_0 = x_i$, $y_0 = y_i$, $z_0 = z_i$. In particular,

$$\limsup_{t\to\infty} |x_0(t) - x_i(t)| \leq \delta(\varepsilon),$$

$$\limsup_{t\to\infty} |y_0(t) - y_i(t)| \leq \delta(\varepsilon),$$

$$\limsup_{t\to\infty} |z_0(t) - z_i(t)| \leq \delta(\varepsilon),$$

where $\delta(\cdot)$ is a non-decreasing function vanishing at zero.

In the next section we present solutions to the problems of invariance and detection. We start from considerations of what an adequate concept of analysis would be. Our considerations will lead us to the conclusion that, for solving the problem of invariance, using the concept of Milnor attractors is advantageous over traditional concepts resting on the notion of Lyapunov stability. This implies that the sets to which the estimates $\hat{\theta}_{i,1}$ and $\hat{\theta}_{i,2}$ converge should be weakly attracting rather than Lyapunov-stable. We present a simple mechanism realizing this requirement for a wide class of models of disturbances. With respect to the second problem, the problem of detection, we provide sufficient conditions for asymptotic synchrony in system (7.16).

7.2 A simple adaptive system for invariant template matching

Consider a system of temporal integrators, (7.13), in which the template subsystem (the second equation in (7.13)) is designed to mimic the temporal code of an image using adjustment mechanisms (7.14). Ideally, the template subsystem should have a single adjustment mechanism, which is structurally simple and yet capable of handling a broad class of perturbations. In addition, it should require the least possible amount of a priori information about images and templates.

In our search for a possible adaptation mechanism let us first explore the available theoretical concepts which can be used in its derivation. The problem of invariance, as stated in Problem 7.1, can generally be understood as a specific optimization task. Particular solutions to such tasks, as well as the choice of appropriate mathematical tools, depend significantly on the following characteristics: uniqueness of the solutions, convexity with respect to parameters, and sensitivity to the input data (images and templates). Let us consider whether the invariant template-matching problem meets these requirements.

Uniqueness. Solutions to the problem of invariant template matching are generally *not unique*. The image may contain multiple instances of the template. Even if there is only a single unique object, the template may fit it in multiple ways, for instance because it has rotational symmetry. Both cases are illustrated in Figure 7.4. A similar argument applies to translational invariance in the images with multiple instances of the template (the right-hand picture in Figure 7.4).

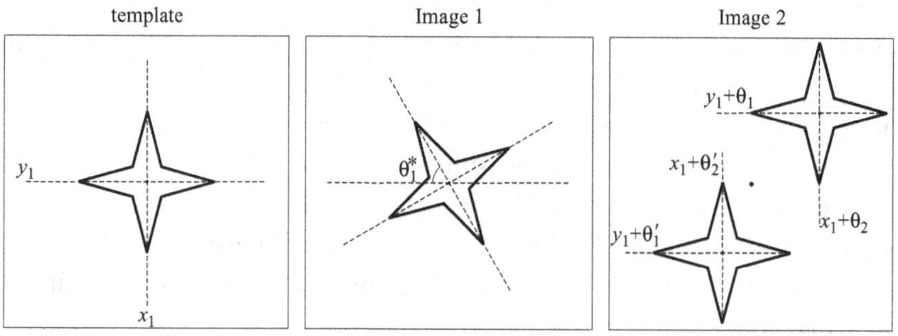

Figure 7.4 An example of a template and images that lead to non-unique solutions in the problem of invariant template matching. Image 1 is a rotated version of the template. Because the template has rotational symmetry, the angles $\theta_2 = \theta_2^* \pm (\pi/2)n$ and $n = 0, 1, \ldots$ at which the template and the image match each other are not unique. Image 2 contains two instances of the template, which also leads to non-uniqueness.

Nonlinearity and non-convexity. The problem of invariant template matching is generally *nonlinear and non-convex in* θ_1 *and* θ_2. Nonlinearity is already evident from Table 7.1. For illustration of non-convexity consider the following function,

$$\theta_1 f_i(t,\theta_2) = \theta_1 \int_{\Omega_{x,t}\times\Omega_{y,t}} e^{-|x-x_0|-|y-y_0|} \times \left(\int_{\Omega_x \times \Omega_y} e^{-\frac{1}{\theta_2}((x-\xi)^2+(y-\gamma)^2)} S_i(\xi,\gamma) d\xi\, d\gamma \right) dx\, dy, \quad (7.19)$$

which is a composition of Gaussian blur (the fourth row in Table 7.1) with spatial sampling and subsequent exponential weighting (the second row in Table 7.2). In the literature on adaptive systems two versions of the convexity requirement are available.[6] The first version applies to the case in which the difference $\theta_1 f_i(t,\theta_2) - \hat{\theta}_{1,i} f_i(t,\hat{\theta}_{i,2})$ is not accessible for explicit measurement, and the variables $\phi_0(t)$ and $\phi_i(t)$ should be used instead. In this case the convexity condition will have the following form (Fradkov 1979):

$$e_i(\phi_0,\phi_i)\left[(\theta_1 - \hat{\theta}_{i,1})\frac{\partial}{\partial\hat{\theta}_{i,1}}\hat{\theta}_{i,1} f_i(t,\hat{\theta}_{i,2}) + (\theta_2 - \hat{\theta}_{i,2})\frac{\partial}{\partial\hat{\theta}_{i,2}}\hat{\theta}_{i,1} f_i(t,\hat{\theta}_{i,2})\right] \geq e_i(\phi_0,\phi_i)\left[\theta_1 f_i(t,\theta_2) - \hat{\theta}_{i,1} f_i(t,\hat{\theta}_{i,2})\right]. \quad (7.20)$$

The term $e_i(\phi_0,\phi_i)$ in (7.20) is usually the difference $e_i(\phi_0,\phi_i) = \phi_0 - \phi_i$ and has the meaning of error. For the same pairs of points θ_1, θ_2 and $\hat{\theta}_{i,1}$, $\hat{\theta}_{i,2}$ condition (7.20) may hold or fail depending on the sign of $e_i(\phi_0(t),\phi_i(t))$ at the particular time instant t. Hence it is not always satisfied, not even for convex $\theta_{i,1} f_i(t,\theta_{i,2})$.

The second version of the convexity requirement applies when the difference $\theta_1 f_i(t,\theta_2) - \hat{\theta}_{i,1} f_i(t,\hat{\theta}_{i,2})$ can be measured explicitly. In this case the condition is formulated as definiteness of the Hessian of $\theta_1 f_i(t,\theta_2)$. It can easily be verified, however, that in (7.19) satisfaction of this requirement depends on the values of $S_i(\xi,\gamma)$. Hence both versions of the convexity conditions generally fail in invariant template matching.

Critical dependence on stimulation. An important feature of the invariant template-matching problem is that its solutions *critically depend* on particular images and templates. The presence of rotational symmetries in the templates affects

[6] See e.g. the velocity-gradient algorithm (Fradkov 1979) described in Chapter 3, or Assumptions 5.3 in Chapter 5.

the number of solutions. Hence objects with different numbers of symmetries will be characterized by sets of solutions with different cardinalities.

We conclude that the problem of invariant template matching generally assumes multiple alternative solutions, involving nonlinearity and non-convexity with respect to parameters, and the structure of its solutions depends critically on a-priori-unknown stimulation. What would be a suitable way to approach this class of problems in principle?

Traditionally, processes of matching and recognition are associated with convergence of the system's state to an attracting set. In our case the system's state is defined by the vector $\mathbf{x}$:

$$\mathbf{x} = (\phi_0, \phi_1, \ldots, \phi_i, \ldots, \hat{\theta}_{1,1}, \hat{\theta}_{2,1}, \ldots \hat{\theta}_{i,1}, \hat{\theta}_{i,2}, \ldots).$$

The attracting set, $\mathcal{A}$, is normally understood as a set satisfying attractivity in the sense of Definition 2.1.3. Traditional techniques for proving attractivity employ the concept of Lyapunov asymptotic stability. Although the notion of set attractivity is wider, the method of Lyapunov functions is constructive and, in addition, Lyapunov asymptotic stability implies the desired attractiviy. For these reasons it is highly practical, and the tandem of set attractivity in Definition 2.1.3 and Lyapunov stability has been used extensively in recognition systems, including Hopfield networks and RNNs.

The problem of *invariant* template matching, however, challenges the universality of these concepts. First, because of the inherent non-uniqueness of the solutions, there are multiple invariant sets in the system's state space. Hence, global Lyapunov asymptotic stability cannot be ensured. Second, when each solution is treated as a locally stable invariant set, it is essentially important to know the domain of its attractivity. This domain, however, depends on the properties of the function $\theta_1 f_0(t, \theta_2)$ in (7.13), which vary with stimulation. Third, no method exists for solving Problem 7.1 for general nonlinearly parametrized $\theta_1 f_0(t, \theta_2)$ that assures Lyapunov stability of the system.

In order to solve the problem of invariant template matching we therefore propose to replace the standard notion of an attracting set with a less restrictive concept. In particular we advocate the concept of weak or Milnor attracting sets (see Definition 2.1.4). The advantage of using unstable attractors in the present framework is illustrated in Figure 7.5.

In the next section we present technical details of how Problem 7.1 could be solved within the framework of Milnor attractors.

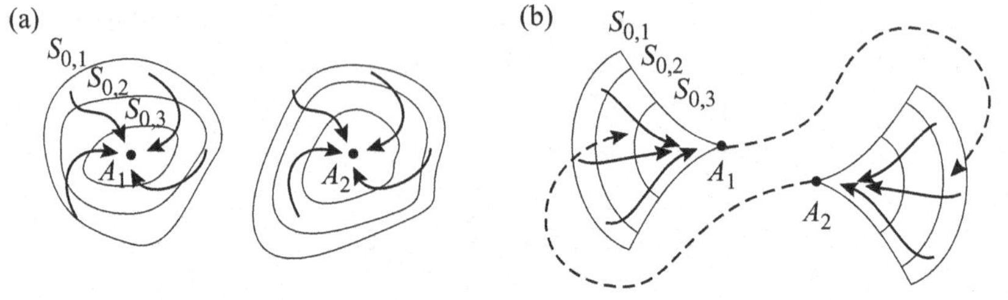

Figure 7.5 Standard stable attractors, (a), vs. weak attractors, (b). Domains of stable attractors are neighborhoods containing $\mathcal{A}_1$ and $\mathcal{A}_2$. Estimates of the sizes of these domains depend on particular images $S_{0,1}$, $S_{0,2}$, and $S_{0,3}$. These estimates are depicted as closed curves around $\mathcal{A}_1$ and $\mathcal{A}_2$. Once the state has converged to either of the attractors it stays there except, probably, when the image changes. In contrast to this, the domains of attraction for Milnor attracting sets are not neighborhoods. Hence, even the slightest perturbation in the image induces a finite probability of escape from the attractor. Hence multiple alternative representations of the image could eventually be recovered.

7.2.1 Invariant template matching by Milnor attractors

Consider system (7.13):

$$\dot{\phi}_0 = -\frac{1}{\tau}\phi_0 + k \cdot \theta_1 f_0(t, \theta_2),$$
$$\dot{\phi}_i = -\frac{1}{\tau}\phi_i + k \cdot \hat{\theta}_{i,1} f_i(t, \hat{\theta}_{i,2}), \; k, \tau \in \mathbb{R}_{>0}, \; i \in \mathcal{I}$$

and assume that the ith template is present in the image. This implies that both the image and the template will have, at least locally in space, sufficiently similar spatiotemporal representations. Formally this can be stated as follows:

$$\exists\, \Delta \in \mathbb{R}_{>0}: \; |\theta_1 f_0(t, \theta_2) - \theta_1 f_i(t, \theta_2)| \leq \Delta, \; \forall\, \theta_1, \theta_2, \; t \geq 0. \tag{7.21}$$

Hence, without loss of generality, we can replace (7.13) with the following:

$$\begin{aligned} \dot{\phi}_0 &= -\frac{1}{\tau}\phi_0 + k \cdot \theta_1 f_i(t, \theta_2) + \epsilon(t), \\ \dot{\phi}_i &= -\frac{1}{\tau}\phi_i + k \cdot \hat{\theta}_{i,1} f_i(t, \hat{\theta}_{i,2}), \; k, \tau \in \mathbb{R}_{>0}, \; i \in \mathcal{I}, \end{aligned} \tag{7.22}$$

where $\epsilon(t) \in L_\infty[0, \infty]$, $\|\epsilon(t)\|_\infty \leq \Delta$ is a bounded disturbance. Solving Problem 7.1, therefore, amounts to finding adjustment mechanisms (7.14) such that trajectories $\phi_0(t)$ and $\phi_i(t)$ in (7.22) converge and the limiting relations (7.15) hold.

The main idea of our proposed solution to this problem is very similar to what has been presented in Chapter 6, and can informally be summarized as follows. First, we introduce an auxiliary system

$$\dot{\boldsymbol{\lambda}} = \mathbf{g}(\boldsymbol{\lambda}, \phi_0, \phi_i, t),\ \boldsymbol{\lambda} \in \mathbb{R}^{\lambda},\ \mathbf{g} : \mathbb{R}^{\lambda} \times \mathbb{R} \times \mathbb{R} \times \mathbb{R}_{\geq 0} \to \mathbb{R}^{\lambda} \tag{7.23}$$

and define $\hat{\theta}_{i,1}$ and $\hat{\theta}_{i,2}$ as functions of $\boldsymbol{\lambda}$, ϕ_0, and ϕ_i:

$$\hat{\theta}_{i,1} = \hat{\theta}_{i,1}(\boldsymbol{\lambda}, \tau, \kappa, \phi_0, \phi_i), \qquad \hat{\theta}_{i,2} = \hat{\theta}_{i,2}(\boldsymbol{\lambda}, \tau, \kappa, \phi_0, \phi_i). \tag{7.24}$$

Second, we show that for some $\varepsilon \in \mathbb{R}_{>0}$ and $\Omega_\lambda \subset \mathbb{R}^\lambda$ the set

$$\Omega^* = \{\phi_0, \phi_i \in \mathbb{R}, \boldsymbol{\lambda} \in \mathbb{R}^\lambda \mid |\phi_0(t) - \phi_i(t)| \leq \varepsilon,\ \boldsymbol{\lambda} \in \Omega_\lambda \subset \mathbb{R}^\lambda\}$$

is forward-invariant in the extended system (7.22), (7.23), and (7.24). Third, we restrict our attention to systems that have a subset Ω in their state space such that trajectories starting in Ω converge to Ω^*. Finally, we guarantee that the state will eventually visit the domain Ω, thus ensuring that (7.12) holds.

Choosing extension (7.23) in the class of simple third-order bilinear systems

$$\begin{aligned} \dot{\lambda}_1 &= \frac{\gamma_1}{\tau} \cdot (\phi_0 - \phi_i), \\ \dot{\lambda}_2 &= \gamma_2 \cdot \lambda_3 \cdot \|\phi_0 - \phi_i\|_\varepsilon,\ \gamma_1, \gamma_2 \in \mathbb{R}_{>0}, \\ \dot{\lambda}_3 &= -\gamma_2 \cdot \lambda_2 \cdot \|\phi_0 - \phi_i\|_\varepsilon,\ \sqrt{\lambda_2^2(t_0) + \lambda_3^2(t_0)} = 1, \end{aligned} \tag{7.25}$$

ensures the solution of Problem 7.1. Specific technical details and conditions are provided in Theorem 7.1

Theorem 7.1 *Let the system (7.22) and (7.25) be given, and let the function $f_i(t, \theta_2)$ be separated from zero and bounded. In other words, there exist constants $D_3, D_4 \in \mathbb{R}_{>0}$ such that for all $t \geq 0$, $\theta_2 \in [\theta_{2,\min}, \theta_{2,\max}]$ the following condition holds:*

$$D_3 \leq f_i(t, \theta_2) \leq D_4. \tag{7.26}$$

Then there exist positive γ_1, γ_2, and ε (see Table 7.3, shown later for the particular values)

$$\gamma_2 \ll \gamma_1,\ \varepsilon > \tau\Delta\left(1 + \frac{D_4}{D_3}\right) \tag{7.27}$$

such that adaptation mechanisms

$$\begin{aligned} \hat{\theta}_{i,1} &= e_i\gamma_1 + \lambda_1, \\ \hat{\theta}_{i,2}(t) &= \theta_{2,\min} + (\lambda_2(t) + 1)\,\frac{\theta_{2,\max} - \theta_{2,\min}}{2} \end{aligned} \tag{7.28}$$

deliver a solution to Problem 7.1. In particular, for all $\theta_1 \in [\theta_{1,\min}, \theta_{1,\max}]$, $\theta_2 \in [\theta_{2,\min}, \theta_{2,\max}]$ *the following properties are guaranteed:*

$$\limsup_{t\to\infty} |\phi_0(t) - \phi_i(t)| \le \varepsilon; \ \ \exists\, \theta_2' \in [\theta_{2,\min}, \theta_{2,\max}] : \lim_{t\to\infty} \hat{\theta}_{i,2}(t) = \theta_2',$$

where the value of ε, *depending on the choice of parameters* γ_2 *and* γ_1, *can be made arbitrarily close to* $\tau\Delta\,(1 + D_4/D_3)$.

Proof of Theorem 7.1. We prove the theorem in three steps. First, we show that the solution of the extended system (7.22), (7.25), and (7.28) is bounded. Second, we prove that there are constants ρ, b, and ε and a time instant $t' > 0$ such that the following holds for system solutions:

$$\|\phi_0(t) - \phi_i(t)\|_\varepsilon \le e^{-\rho(t-t_0)} \|\phi_0(t_0) - \phi_i(t_0)\|_\varepsilon + b\|\theta_2 - \hat{\theta}_{i,2}(\tau)\|_{\infty,[t_0,t]}$$
$$\forall\, t \ge t_0 > t'. \tag{7.29}$$

Third, using this representation, we invoke results and demonstrate that the conclusions of the theorem follow.

1. Boundedness. To prove the boundedness of solutions of the extended system in forward time let us first consider the difference $e_i(t) = \phi_0(t) - \phi_i(t)$. According to (7.22), the dynamics of $e_i(t)$ will be defined as

$$\dot{e}_i = -\frac{1}{\tau} e_i + k\left(\theta_1 f_i(t, \theta_2) - \hat{\theta}_{i,1} f_i(t, \hat{\theta}_{i,2})\right) + \epsilon(t). \tag{7.30}$$

On noticing that

$$\theta_1 f_i(t, \theta_2) - \hat{\theta}_{i,1} f_i(t, \hat{\theta}_{i,2}) = [\theta_1 f_i(t, \theta_2) - \theta_1 f_i(t, \hat{\theta}_{i,2})]$$
$$+ [\theta_1 f_i(t, \hat{\theta}_{i,2}) - \hat{\theta}_{i,1} f_i(t, \hat{\theta}_{i,2})]$$

and denoting $\delta_1 = \hat{\theta}_{i,1} - \theta_1$ and $\delta_2(t, \theta_1, \theta_2, \hat{\theta}_{i,2}) = \theta_1 f_i(t, \theta_2) - \theta_1 f_i(t, \hat{\theta}_{i,2})$ we can rewrite (7.30) as follows:

$$\dot{e}_i = -\frac{1}{\tau} e_i - \delta_1 [k f_i(t, \hat{\theta}_{i,2})] + \delta_2(t, \theta_1, \theta_2, \hat{\theta}_{i,2}) k + \epsilon(t). \tag{7.31}$$

Let us now write the equations for $\hat{\theta}_{i,1} - \theta_1$ in (7.24) in differential form. To do so we differentiate the variable $\hat{\theta}_{i,1} - \theta_1 = \delta_1$ with respect to time, taking into account (7.30) and (7.31):

$$\dot{\delta}_1 = -\gamma_1 \left(\delta_1 [k f_i(t, \hat{\theta}_{i,2})] - \delta_2(t, \theta_1, \theta_2, \hat{\theta}_{i,2}) k - \epsilon(t)\right). \tag{7.32}$$

The variable $\epsilon(t)$ in (7.32) is bounded according to (7.21). Let us show that $\delta_2(t,\theta_1,\theta_2,\hat{\theta}_{i,2})$ is also bounded. First of all, notice that the positive definite function

$$V_\lambda = 0.5\left(\lambda_2^2+\lambda_3^2\right)$$

is not growing with time:

$$\dot{V} = \lambda_2\gamma_2\lambda_3\|\phi_0(t)-\phi_i(t)\|_\varepsilon - \lambda_3\gamma_2\lambda_2\|\phi_0(t)-\phi_i(t)\|_\varepsilon = 0.$$

Furthermore,

$$\begin{aligned}\lambda_2(t) &= r\cdot\sin\left(\gamma_2\int_{t_0}^{t}\|\phi_0(\tau)-\phi_i(\tau)\|_\varepsilon\,d\tau+\varphi_0\right)\\ \lambda_3(t) &= r\cdot\cos\left(\gamma_2\int_{t_0}^{t}\|\phi_0(\tau)-\phi_i(\tau)\|_\varepsilon\,d\tau+\varphi_0\right),\quad r,\varphi_0\in\mathbb{R}.\end{aligned}\tag{7.33}$$

Choosing initial conditions $\lambda_2^2(t_0)+\lambda_3^2(t_0)=1$ ensures that $r=1$. Hence, according to (7.28), the variable $\hat{\theta}_{i,2}$ belongs to the interval $[\theta_{2,\min},\theta_{2,\max}]$.

Consider the variable $\delta_2(t,\theta_1,\theta_2,\hat{\theta}_{i,2})$:

$$\begin{aligned}\delta_2(t,\theta_1,\theta_2,\hat{\theta}_{i,2}) &= \theta_1 f_i(t,\theta_2)-\theta_1 f_i(t,\hat{\theta}_{i,2}(t))\\ &= \theta_1\left(f_i(t,\theta_2)-f_i(t,\hat{\theta}_{i,2}(t))\right).\end{aligned}\tag{7.34}$$

Taking into account the notational agreement (7.9), and properties (7.5) and (7.8), we conclude that the following estimate holds:

$$\begin{aligned}|\delta_2(t,\theta_1,\theta_2,\hat{\theta}_{i,2})| &\le \theta_1|f_i(t,\theta_2)-f_i(t,\hat{\theta}_{i,2}(t))|\\ &\le \theta_{1,\max}DD_2|\theta_2-\hat{\theta}_{i,2}(t)|.\end{aligned}\tag{7.35}$$

Given that $\hat{\theta}_{i,2}(t)\in[\theta_{2,\min},\theta_{2,\max}]$ and using (7.35), we can provide the following estimate for $\delta_2(t,\theta_1,\theta_2,\hat{\theta}_{i,2})$:

$$|\delta_2(t,\theta_1,\theta_2,\hat{\theta}_{i,2})| \le \theta_{1,\max}DD_2|\theta_{2,\max}-\theta_{2,\min}|.\tag{7.36}$$

Let us consider equality (7.32). According to condition (1) of the theorem, the term

$$\alpha(t) = kf_i(t,\hat{\theta}_{i,2}(t))$$

is non-negative and bounded from below:

$$\alpha(t) = kf_i(t,\hat{\theta}_{i,2}(t)) \ge kD_3,\ \forall\, t\ge 0.\tag{7.37}$$

Taking into account (7.32) and (7.37), we can estimate $|\delta_1(t)|$ as follows:

$$|\delta_1(t)| \leq e^{-\gamma_1 \int_{t_0}^{t} \alpha(\tau)d\tau} |\delta_1(t_0)| + \gamma_1 e^{-\gamma_1 \int_{t_0}^{t} \alpha(\tau)d\tau} \int_{t_0}^{t} e^{\gamma_1 \int_{t_0}^{\tau} \alpha(\tau_1)d\tau_1} |\epsilon(\tau) + \delta_2(\tau)k| d\tau. \quad (7.38)$$

According to (7.21) and (7.36) we have that for all $t \geq t_0 \geq 0$

$$|\epsilon(t) + \delta_2(t)k| \leq \|\epsilon(\tau) + k\delta_2(\tau)\|_{\infty,[t_0,t]} \leq \Delta + k\theta_{1,\max} D D_2 |\theta_{2,\max} - \theta_{2,\min}| = M_1. \quad (7.39)$$

Furthermore,

$$\int_{t_0}^{t} e^{\gamma_1 \int_{t_0}^{\tau} \alpha(\tau_1)d\tau_1} d\tau = \frac{1}{\gamma_1} \left(\frac{1}{\alpha(t)} e^{\gamma_1 \int_{t_0}^{t} \alpha(\tau)d\tau} - \frac{1}{\alpha(t_0)} \right) \leq \frac{1}{\gamma_1 D_3 k} e^{\gamma_1 \int_{t_0}^{t} \alpha(\tau)d\tau}. \quad (7.40)$$

Taking into account (7.38), (7.39), and (7.40), we can obtain the following estimate:

$$|\delta_1(t)| \leq e^{-\gamma_1 k D_4 (t-t_0)} |\delta_1(t_0)| + \frac{M_1}{D_3 k}. \quad (7.41)$$

Inequality (7.41) proves that $\delta_1(t)$ is bounded.

In order to complete this step of the proof it is sufficient to show that $e_i(t)$ is bounded. This would automatically imply boundedness of $\phi_i(t)$, thus confirming the boundedness of the state of the extended system. To show the boundedness of $e_i(t)$ let us write the closed-form solution of (7.30):

$$e_i(t) = e^{-\frac{t-t_0}{\tau}} e_i(t_0) + e^{-\frac{t}{\tau}} \int_{t_0}^{t} e^{\frac{\tau_1}{\tau}} (\delta_1(\tau_1)\alpha(\tau_1) + k\delta_2(\tau_1) + \epsilon(\tau_1)) d\tau_1. \quad (7.42)$$

Using (7.39) and (7.41) we can derive that

$$|e_i(t)| \leq e^{-\frac{t-t_0}{\tau}} |e_i(t_0)| + M_1 \tau \left(1 + \frac{D_4}{D_3}\right) + \epsilon_1(t), \quad (7.43)$$

where $\epsilon_1(t)$ is an exponentially decaying term:

$$|\epsilon_1(t)| \leq e^{-\gamma_1 k D_3 (t-t_0)} \left(\frac{1 - e^{-\left(\frac{1}{\tau} - \gamma_1 k D_3\right)(t-t_0)}}{1/\tau - \gamma_1 k D_3} \right) |\theta_1 - \hat{\theta}_{i,1}(t_0)|. \quad (7.44)$$

As follows from (7.33), (7.41), (7.43), and (7.44), the variables $e_i(t)$, $\hat{\theta}_{i,1}(t)$, and $\hat{\theta}_{i,2}(t)$ are bounded. Hence, the state of the extended system is bounded in forward time.

2. *Transformation.* Let us now show that there exists a time instant t' and constants $\rho, c \in \mathbb{R}_{>0}$ such that the dynamics of $e_i(t) = \phi_0(t) - \phi_i(t)$ satisfies inequality (7.29). In order to do so we first show that the term

$$\delta_1(t)kf_i(t, \hat{\theta}_{i,2}(t))$$

in (7.31) can be estimated as

$$|\delta_1(t)kf_i(t, \hat{\theta}_{i,2}(t))| \leq M_2|\theta_2 - \hat{\theta}_{i,2}(t)| + \Delta_2 + \epsilon_2(t), \tag{7.45}$$

where M_2 and Δ_2 are positive constants and $\epsilon_2(t)$ is a function of time that converges to zero asymptotically with time.

According to (7.38) the following holds:

$$\begin{aligned}|\delta_1(t)| &\leq e^{-\gamma_1 \int_{t_0}^t \alpha(\tau)d\tau}|\delta_1(t_0)| \\ &\quad + \gamma_1 e^{-\gamma_1 \int_{t_0}^t \alpha(\tau)d\tau} \int_{t_0}^t e^{\gamma_1 \int_{t_0}^\tau \alpha(\tau_1)d\tau_1}|\epsilon(\tau) + \delta_2(\tau)k|d\tau.\end{aligned}$$

Taking into account (7.21) and (7.37), we can conclude that

$$\begin{aligned}|\delta_1(t)| &\leq e^{-\gamma_1 k D_3(t-t_0)}|\delta_1(t_0)| + \frac{\Delta}{kD_3} \\ &\quad + \gamma_1 e^{-\gamma_1 \int_{t_0}^t \alpha(\tau)d\tau} \int_{t_0}^t e^{\gamma_1 \int_{t_0}^\tau \alpha(\tau_1)d\tau_1}|\delta_2(\tau)k|d\tau.\end{aligned} \tag{7.46}$$

Substituting (7.35) into (7.46) results in

$$\begin{aligned}|\delta_1(t)| &\leq e^{-\gamma_1 k D_3(t-t_0)}|\delta_1(t_0)| + \frac{\Delta}{kD_3} \\ &\quad + \gamma_1 e^{-\gamma_1 \int_{t_0}^t \alpha(\tau)d\tau} \int_{t_0}^t e^{\gamma_1 \int_{t_0}^\tau \alpha(\tau_1)d\tau_1}|\theta_2 - \hat{\theta}_{i,2}(\tau)|d\tau \cdot \left(k\theta_{1,\max} D D_2\right).\end{aligned} \tag{7.47}$$

Consider the following term in (7.47):

$$\int_{t_0}^t e^{\gamma_1 \int_{t_0}^\tau \alpha(\tau_1)d\tau_1}|\theta_2 - \hat{\theta}_{i,2}(\tau)|d\tau. \tag{7.48}$$

Integration of (7.48) by parts yields

$$\begin{aligned}
&\int_{t_0}^{t} e^{\gamma_1 \int_{t_0}^{\tau} \alpha(\tau_1)d\tau_1} |\theta_2 - \hat{\theta}_{i,2}(\tau)| d\tau \\
&= \frac{1}{\gamma_1}\left(\frac{1}{\alpha(t)} e^{\gamma_1 \int_{t_0}^{t} \alpha(\tau)d\tau} |\theta_2 - \hat{\theta}_{i,2}(t)| - \frac{|\theta_2 - \hat{\theta}_{i,2}(t_0)|}{\alpha(t_0)}\right) \\
&\quad - \frac{1}{\gamma_1}\int_{t_0}^{t} \frac{1}{\alpha(\tau)} e^{\gamma_1 \int_{t_0}^{\tau} \alpha(\tau_1)d\tau_1} \left(\frac{d|\theta_2 - \hat{\theta}_{i,2}(\tau)|}{d\tau}\right) d\tau \\
&\leq \frac{1}{\gamma_1 k D_3} e^{\gamma_1 \int_{t_0}^{t} \alpha(\tau)d\tau} |\theta_2 - \hat{\theta}_{i,2}(t)| \\
&\quad + \frac{1}{\gamma_1 k D_3} \int_{t_0}^{t} e^{\gamma_1 \int_{t_0}^{\tau} \alpha(\tau_1)d\tau_1} \left|\frac{d|\theta_2 - \hat{\theta}_{i,2}(\tau)|}{d\tau}\right| d\tau. \qquad (7.49)
\end{aligned}$$

Given that

$$\hat{\theta}_{2,i} = \theta_{2,\min} + \frac{\theta_{2,\max} - \theta_{2,\min}}{2}(\lambda_2(t) + 1),$$

we can estimate the derivative $d|\theta_2 - \hat{\theta}_{2,i}(t)|/dt$ as follows:

$$\frac{d|\theta_2 - \hat{\theta}_{i,2}(t)|}{dt} \leq \frac{\theta_{2,\max} - \theta_{2,\min}}{2} \cdot \gamma_2 \cdot |\phi_0(t) - \phi_i(t)|. \qquad (7.50)$$

Notice that the value of $|\phi_0(t) - \phi_i(t)| = e_i(t)$ in (7.50) can be estimated according to (7.43) as

$$|\phi_0(t) - \phi_i(t)| \leq M_1 \tau \left(1 + \frac{D_4}{D_3}\right) + \mu_1(t),$$

where $\mu_1(t) \sim \epsilon_1(t) + e_i(t) e^{-\frac{t-t_0}{\tau}}$ is an asymptotically decaying term.

Hence, taking into account (7.40), (7.43), (7.47), (7.49), and (7.50), we may conclude that the following inequality holds:

$$\begin{aligned}
|\delta_1(t)| &\leq \frac{\theta_{1,\max} D D_2}{D_3} |\theta_2 - \hat{\theta}_{i,2}(t)| + \frac{\Delta}{D_3 k} \\
&\quad + \frac{\gamma_2}{\gamma_1} \frac{\theta_{1,\max} D D_2}{D_3^2 k} \frac{\theta_{2,\max} - \theta_{2,\min}}{2} M_1 \tau \left(1 + \frac{D_4}{D_3}\right) + \mu(t),
\end{aligned}$$

where $\mu(t)$ is an asymptotically vanishing term. Therefore (7.45) holds with the following values of M_2 and Δ_2:

$$\begin{aligned}
M_2 &= \frac{k \theta_{1,\max} D D_2 D_4}{D_3}, \\
\Delta_2 &= \frac{\gamma_2}{\gamma_1}\left[\frac{\theta_{1,\max} D D_2 D_4}{(D_3)^2} M_1 \tau \left(1 + \frac{D_4}{D_3}\right) \frac{\theta_{2,\max} - \theta_{2,\min}}{2}\right] + \frac{\Delta D_4}{D_3}.
\end{aligned} \qquad (7.51)$$

To finalize this step of the proof, consider the variable $e_i(t)$ for $t \in [t_1, \infty]$, $t_1 \geq t_0$. According to (7.35) and (7.42) we have that

$$\begin{aligned}|e_i(t)| \leq e^{-\frac{t-t_1}{\tau}} |e_i(t_1)| + \tau M_2 \|\theta_2 - \hat{\theta}_{i,2}(t)\|_{\infty,[t_1,t]} \\ + \tau \Delta_2 \left(1 - e^{-\frac{t-t_1}{\tau}}\right) + \tau \|\epsilon_2(t)\|_{\infty,[t_1,\infty]} \left(1 - e^{-\frac{t-t_1}{\tau}}\right) \\ + \tau k \theta_{1,\max} D D_2 \|\theta_2 - \hat{\theta}_{i,2}(t)\|_{\infty,[t_1,t]} + \tau \Delta \left(1 - e^{-\frac{t-t_1}{\tau}}\right).\end{aligned} \tag{7.52}$$

Regrouping terms in (7.52) yields

$$\begin{aligned}|e_i(t)| - \tau \left(\Delta_2 + \Delta + \|\epsilon_2(t)\|_{\infty,[t_1,\infty]}\right) \leq e^{-\frac{t-t_1}{\tau}} \left(|e_i(t_1)|\right. \\ \left. - \tau \left(\Delta_2 + \Delta + \|\epsilon_2(t)\|_{\infty,[t_1,\infty]}\right)\right) \\ + \tau (M_2 + k \theta_{1,\max} D D_2) \|\theta_2 \\ - \hat{\theta}_{i,2}(t)\|_{\infty,[t_1,t]}.\end{aligned}$$

On denoting

$$\Delta' = \tau \left(\Delta_2 + \Delta + \|\epsilon_2(t)\|_{\infty,[t_1,\infty]}\right) \tag{7.53}$$

we can obtain

$$\begin{aligned}|e_i(t)| - \Delta' &\leq e^{-\frac{t-t_1}{\tau}} \left(|e_i(t_1)| - \Delta'\right) \\ &\quad + \tau (M_2 + k \theta_{1,\max} D D_2) \|\theta_2 - \hat{\theta}_{i,2}(t)\|_{\infty,[t_1,t]} \\ &\leq e^{-\frac{t-t_1}{\tau}} \|e_i(t_1)\|_{\Delta'} \\ &\quad + \tau (M_2 + k \theta_{1,\max} D D_2) \|\theta_2 - \hat{\theta}_{i,2}(t)\|_{\infty,[t_1,t]}.\end{aligned} \tag{7.54}$$

Given that

$$\|e_i(t)\|_{\Delta'} = \begin{cases} |e_i(t)| - \Delta', & |e_i(t)| > \Delta', \\ 0, & |e_i(t)| \leq \Delta', \end{cases}$$

and taking into account inequality (7.54), we can conclude that

$$\begin{aligned}\|e_i(t)\|_{\Delta'} \leq e^{-\frac{t-t_1}{\tau}} \|e_i(t_1)\|_{\Delta'} \\ + \tau (M_2 + k \theta_{1,\max} D D_2) \|\theta_2 - \hat{\theta}_{i,2}(t)\|_{\infty,[t_1,t]}.\end{aligned} \tag{7.55}$$

Because equations (7.52)–(7.55) hold for any $t_1 \in (t_0, \infty]$ and

$$\limsup_{t_1 \to \infty} \|\epsilon_2(t)\|_{\infty,[t_1,\infty]} = 0$$

for every

$$\varepsilon > \tau(\Delta + \Delta_2),$$

there exists a time instant $t' \geq t_0$ such that the inequality

$$\|e_i(t)\|_\varepsilon \leq e^{-\frac{t-t_1}{\tau}} \|e_i(t_1)\|_\varepsilon + \tau(M_2 + k\theta_{1,\max} D D_2)\|\theta_2 - \hat{\theta}_{i,2}(t)\|_{\infty,[t_1,t]} \quad (7.56)$$

is satisfied for all $t \geq t_1 \geq t'$. This proves (7.29) for $\rho = 1/\tau$ and $b = \tau(M_2 + k\theta_{1,\max} D D_2)$. Hence, the second step of the proof has been completed.

3. Convergence. In order to prove convergence we employ Corollary 4.5 from Chapter 4. In order to apply Corollary 4.5 we need to further transform (7.33) and (7.56) and

$$\hat{\theta}_{i,2}(t) = \theta_{2,\min} + \frac{\theta_{2,\max} - \theta_{2,\min}}{2}(\lambda_2(t) + 1) \quad (7.57)$$

into the form of (4.76). Notice that for every $\theta_2 \in [\theta_{2,\min}, \theta_{2,\max}]$ there always exists a real number $\lambda^* \in [-1, 1]$ such that

$$\theta_2 = \theta_{2,\min} + \frac{\theta_{2,\max} - \theta_{2,\min}}{2}(\lambda_2^* + 1).$$

Hence, denoting

$$c = \tau(M_2 + k\theta_{1,\max} D D_2)\frac{\theta_{2,\max} - \theta_{2,\min}}{2}$$

and using (7.56), we ascertain that the following holds for solutions of the system (7.22), (7.25), and (7.28):

$$\|e_i(t)\|_\varepsilon \leq e^{-\frac{t-t_1}{\tau}} \|e_i(t_1)\|_\varepsilon + c\|\lambda_2^* - \lambda_2(t)\|_{\infty,[t_1,t]} \quad (7.58)$$

for $\varepsilon > \tau(\Delta + \Delta_2)$, and $t \geq t_1 \geq t'$.

Consider the difference $\lambda_2^* - \lambda_2(t)$. According to (7.33) we have

$$|\lambda_2^* - \lambda_2(t)| \leq |\sigma^* - \int_{t_1}^{t} \gamma_2 \|e_i(\tau)\|_\varepsilon - \varphi_0|, \ \lambda_2^* = \sin(\sigma^*). \quad (7.59)$$

On denoting

$$h(t) = \sigma^* - \int_{t_1}^{t} \gamma_2 \|e_i(\tau)\|_\varepsilon - \varphi_0 \quad (7.60)$$

and taking into account (7.58), we therefore obtain the following equations:

$$\begin{aligned} \|e_i(t)\|_\varepsilon &\leq e^{-\frac{t-t_1}{\tau}} \|e_i(t_1)\|_\varepsilon + c\|h(t)\|_{\infty,[t_1,t]}, \\ h(t_1) - h(t) &= \int_{t_1}^{t} \gamma_2 \|e_i(\tau)\|_\varepsilon \, d\tau. \end{aligned} \quad (7.61)$$

Equations (7.61) are a particular case of (4.76) to which Corollary 4.5 applies. In system (7.61), however, the function $\beta_t(t)$ is defined as $\beta_t(t) = e^{-\frac{t}{\tau}}$. Hence

$$\beta_t^{-1}(t) = -\tau \ln(t).$$

Therefore, according to Corollary 4.5, satisfying the inequality

$$\gamma_2 \cdot c \cdot \tau \ln\left(\frac{\kappa}{d}\right)\frac{k}{k-1}\left(\left(1+\frac{\kappa}{1-d}\right)+1\right) < 1 \tag{7.62}$$

for some $\kappa \in (1,\infty)$, $d \in (0,1)$ ensures the existence of initial conditions $e_i(t_1)$ and $h(t_1)$ such that $h(t)$ is bounded. Given that

$$\min_{\kappa\in(1,\infty),d\in(0,1)} \ln\left(\frac{\kappa}{d}\right)\frac{k}{k-1}\left(\left(1+\frac{\kappa}{1-d}\right)+1\right) \approx 15.6886 < 16$$

we can rewrite condition (7.62) in a more conservative, yet simpler form:

$$\gamma_2 \cdot c \cdot \tau < \frac{1}{16}.$$

Taking into account notations (7.51) and (7.60) we can rewrite this inequality as follows:

$$\gamma_2 < \left(\frac{1}{4\tau}\right)^2 \left[k\theta_{1,\max} D D_2 \left(1+\frac{D_4}{D_3}\right)\left(\frac{\theta_{2,\max}-\theta_{2,\min}}{2}\right)\right]^{-1}.$$

Notice that, because the function $\sin(\cdot)$ is periodic, the value of σ^* in (7.59) and consequently the value of $h(t_1)$ can be chosen arbitrarily large. Hence, for any finite $e_i(t_1)$ and φ_0 there will always exist σ^* and $h(t_1)$ such that the variable $h(t)$ is bounded.

Taking into account that $h(t)$ is monotone and bounded, we can conclude that according to the Bolzano–Weierstrass theorem the function $h(t)$ has a limit in $[0, h(t_1)]$:

$$\exists\, h^* \in [0, h(t_1)] : \lim_{t\to\infty} h(t) = h^*.$$

This in turn implies that

$$\lim_{t\to\infty} \int_{t_1}^{t} \gamma_2 \|e_i(\tau)\|_\varepsilon \, d\tau = \sigma^* - \varphi_0 - h^* < \infty.$$

Therefore,

$$\exists\, \theta_2' \in [\theta_{2,\min}, \theta_{2,\max}] : \lim_{t\to\infty} \hat{\theta}_{i,2}(t)$$
$$= \theta_{2,\min} + \frac{\theta_{2,\max}-\theta_{2,\min}}{2}(\sin(\sigma^* - \varphi_0 - h^*) + 1) = \theta_2'.$$

Moreover, because $\|e_i(t)\|_\varepsilon$ is uniformly continuous in t, convergence of $\|e_i(t)\|_\varepsilon$ to zero as $t \to \infty$ follows immediately from Barbalat's lemma. □

Let us comment on the conclusions and conditions of Theorem 7.1. First of all, the theorem shows that each ith subsystem ensuring invariance of spatiotemporal image

Table 7.3. *Parameters of the compensatory mechanisms (7.28)*

Parameter	Values
γ_1	$\frac{\gamma_1}{\gamma_2} = q,\ q \in \mathbb{R}_{>0}$
ε	$\varepsilon > \tau\left(\Delta\left(1+\frac{D_4}{D_3}\right) + \frac{\gamma_2}{\gamma_1}\left[\frac{\theta_{1,\max} D D_2 D_4}{(D_3)^2} M_1 \tau \left(1+\frac{D_4}{D_3}\right)\frac{\theta_{2,\max}-\theta_{2,\min}}{2}\right]\right),$ $M_1 = \Delta + k\theta_{1,\max} D D_2 \lvert\theta_{2,\max}-\theta_{2,\min}\rvert$
γ_2	$\gamma_2 < \left(\frac{1}{4\tau}\right)^2 \left[k\theta_{1,\max} D D_2 \left(1+\frac{D_4}{D_3}\right)\left(\frac{\theta_{2,\max}-\theta_{2,\min}}{2}\right)\right]^{-1}$

representation with respect to the given modeled perturbations can be composed of no more than four differential equations:

$$\text{temporal integration}: \dot{\phi}_i = -\frac{1}{\tau}\phi_i + k\cdot\hat{\theta}_{i,1} f_i(t,\hat{\theta}_{i,2}), \tag{7.63a}$$

$$\text{fast adaptation dynamics}: \dot{\lambda}_1 = \frac{\gamma_1}{\tau}\cdot(\phi_0 - \phi_i), \tag{7.63b}$$

$$\text{slow adaptation dynamics}: \begin{array}{l} \dot{\lambda}_2 = \gamma_2\cdot\lambda_3\cdot\|\phi_0-\phi_i\|_\varepsilon, \\ \dot{\lambda}_3 = -\gamma_2\cdot\lambda_2\cdot\|\phi_0-\phi_i\|_\varepsilon. \end{array} \tag{7.63c}$$

Notice that the time scales of temporal integration, (7.63a), adaptation to linearly parametrized uncertainties, (7.63b), and adaptation to nonlinearly parametrized uncertainties, (7.63c), are all different. Because of these differences in time scales, subsystem (7.63b) is referred to as *slow adaptation dynamics* and subsystem (7.63c) as *fast adaptation dynamics*. The difference between the time scales determines the degree of invariance and precision in the resulting system. For instance, as follows from Table 7.3, the ratio γ_2/γ_1 affects the value of ε. This value defines the acceptable level of mismatches between an image and a template. In other words, it regulates the sensitivity of the system. The smaller the ratio γ_2/γ_1, the higher the sensitivity. The ratio $\gamma_2/(1/\tau)$ (see the proof for details) affects the conditions for convergence.

The slow adaptation dynamics, (7.63c), can be interpreted as a *searching, or wandering* dynamics in the interval $[\theta_{2,\min}, \theta_{2,\max}]$. Its function is to explore the interval $[\theta_{2,\min}, \theta_{2,\max}]$ for possible values of $\hat{\theta}_{i,2}$ when models of perturbation are inherently nonlinear and no other choice except that of explorative search is available. The solutions of the searching dynamics in (7.63c) are harmonic signals with time-varying frequency $\gamma_2\|\phi_0(t)-\phi_i(t)\|_\varepsilon$. The larger the error, the higher the frequency of oscillation. When $\gamma_2\|\phi_0(t)-\phi_i(t)\|_\varepsilon$ is constant, for instance equal to unity, the equations in (7.63c) reduce to

$$\begin{aligned}\dot{\lambda}_2 &= \lambda_3,\\ \dot{\lambda}_3 &= -\lambda_2.\end{aligned} \tag{7.64}$$

In general, every subsystem

$$\begin{aligned}\dot{\lambda}_2 &= g_2(\lambda_2,\lambda_3,t),\\ \dot{\lambda}_3 &= g_3(\lambda_2,\lambda_3,t), \quad g_2, g_3 \in C^0\end{aligned} \tag{7.65}$$

generating dense trajectories $\lambda_2(t)$ in $[\theta_{2,\min},\theta_{2,\max}]$ for some initial conditions $\lambda_2(t_0)$ and $\lambda_3(t_0)$ and at the same time ensuring boundedness of $\lambda_2(t)$ and $\lambda_3(t)$ for all $t\in\mathbb{R}_{\geq 0}$ could be a replacement for (7.64) in (7.63c). The conclusions of the theorem in this case will remain the same except, probably, with respect to the choice of the particular values of γ_1, γ_2, and ε in Table 7.3. Our present choice of subsystem (7.64) in (7.63c) as a prototype for the searching trajectory was motivated primarily by its simplicity of realization and linearity in state.

The fast adaptation dynamics, (7.63b), corresponds to exponentially stable mechanisms. This can easily be verified by differentiating the difference $\hat{\theta}_{i,1}(t)-\theta_1$ with respect to time. The function of the fast adaptation subsystem is to track instantaneous changes in θ_1 exponentially fast in such a way that the difference $\hat{\theta}_{i,1}(t)-\theta_1$ is determined mostly by mismatches $\hat{\theta}_{i,2}(t)-\theta_2$.

The problem of template matching is solved through the interplay of the searching dynamics $\hat{\theta}_{i,2}(t)-\theta_2$ and the stable, contracting dynamics expressed by $\phi_0(t)-\phi_i(t)$. We use the results from Chapter 4 to prove the emergence of weakly (Milnor) attracting sets in the system's state space.

In principle, linearity of the uncertainty models in θ_1 is not necessary to guarantee exponential stability of $\hat{\theta}_{i,1}(t)-\theta_1$. As has been shown in Chapter 5, exponential stability of $\hat{\theta}_{i,1}(t)-\theta_1$ can be ensured by the same function $\hat{\theta}_{i,1}(t)$ as in (7.24) if we replace $\theta_1 f_i(t,\theta_2)$ with $\tilde{f}_i(t,\theta_1,\theta_2):\mathbb{R}_{\geq 0}\times\mathbb{R}\times\mathbb{R}\to\mathbb{R}$. Nonlinearities $\tilde{f}_i(t,\theta_1,\theta_2)$, however, should be monotone in θ_1. In this case condition (7.26) is to be replaced with the following:

$$D_3 \leq \frac{\tilde{f}_i(t,\hat{\theta}_{i,1},\theta_2)-\tilde{f}_i(t,\theta_1,\theta_2)}{\hat{\theta}_{i,1}-\theta_1} \leq D_4,\ \forall\,\theta_2\in[\theta_{2,\min},\theta_{2,\max}]. \tag{7.66}$$

The general line of the proof remains unaffected by this extension.

The proposed compensatory mechanisms (7.22), (7.25), and (7.28) are nearly optimal in terms of the dimension of the state of the whole system. Indeed, in order to track uncertain and independent θ_1 and θ_2 two extra variables need to be introduced. This implies that the minimal dimension of the state of a system which solves Problem 7.1 is three. Our four-dimensional system is therefore close to the optimal configuration. Furthermore, as follows from the proof of the theorem, the dimension of the slow subsystem could be reduced to one. Thus, in principle, a minimal realization could be achieved. In this case, however, boundedness of the state for every initial condition is no longer guaranteed.

Theorem 7.1 establishes conditions for convergence of the trajectories of our prototype system (7.22), (7.25), and (7.28) to an invariant set in the system's state space. In particular, when the matching condition (7.21) is met, the theorem assures that the temporal representation $\phi_i(t)$ of the template tracks the temporal representation $\phi_0(t)$ of the image. In the next subsection we discuss how the similarity between these temporal representations can be detected by a system of coupled spiking oscillators. In particular, we will consider coincidence detectors (7.16), (7.17), and (7.18) modeled by a system of coupled Hindmarsh–Rose oscillators.

7.2.2 Conditions for synchronization of coincidence detectors

The goal of this section is to provide a constructive solution to Problem 7.2. First, we seek conditions ensuring global exponential stability of the synchronization manifold of $\phi_0(t) = \phi_i(t)$ when $\phi_i(t)$ are identical for each i. We do this by showing that solutions of the system are globally bounded, and that for each pair of indices $i, j \in \{0, \dots, n\}$ there exists a differentiable positive definite function $V(x_i, y_i, z_i, x_j, y_j, z_j)$, $\partial V/\partial x_i = -\partial V/x_j$ such that V grows towards infinity with increasing distance from the synchronization manifold and for all bounded continuous $\phi_i(t) = \phi_j(t)$ the following holds:

$$\dot{V} \leq -\alpha V, \ \alpha \in \mathbb{R}_{>0}. \tag{7.67}$$

When $\phi_i(t) \neq \phi_j(t)$, (7.67) implies that

$$\dot{V} \leq -\alpha V + \frac{\partial V}{x_i}(\phi_i(t) - \phi_j(t)). \tag{7.68}$$

Then, using (7.68) and the comparison lemma from Khalil (2002), we show that convergence of $\phi_i(t)$ to $\phi_j(t)$ at $t \to \infty$ implies convergence of the variables $x_i(t)$, $y_i(t)$, $z_i(t)$, $x_j(t)$, $y_j(t)$, and $z_j(t)$ to the synchronization manifold. The formal statement of this result is provided in Theorem 7.2.

Theorem 7.2 *Let system (7.16) be given, let the function* **u** *be defined as in (7.17), and let the functions $\phi_i(t)$, $i \in \{0, \dots, n\}$ be bounded. Then*

(1) solutions of the system are bounded for all $\gamma \in \mathbb{R}_{\geq 0}$;

(2) if, in addition, the condition

$$\gamma > \frac{1}{(n+1)\cdot a}\left(\frac{d^2}{2} + b^2\right) \tag{7.69}$$

is satisfied, then for all $i, j \in \{0, \ldots, n\}$ the condition

$$\limsup_{t\to\infty} |\phi_i(t) - \phi_j(t)| \leq \varepsilon$$

implies that

$$\begin{aligned}
&\limsup_{t\to\infty} |x_i(t) - x_j(t)| \leq \delta(\varepsilon), \\
&\limsup_{t\to\infty} |y_i(t) - y_j(t)| \leq \delta(\varepsilon), \\
&\limsup_{t\to\infty} |z_i(t) - z_j(t)| \leq \delta(\varepsilon).
\end{aligned} \tag{7.70}$$

where $\delta : \mathbb{R}_{\geq 0} \to \mathbb{R}_{\geq 0}$ is a monotone function vanishing at zero.

Proof of Theorem 7.2. The proof is similar to Oud and Tyukin (2004) and consists of three major steps. First, we show that a single Hindmarsh–Rose oscillator is a semi-passive system with a radially unbounded storage function (Pogromsky 1998). In other words, the system

$$\begin{aligned}
\dot{x} &= -ax^3 + bx^2 + y - z + I + u, \\
\dot{y} &= c - dx^2 - y, \\
\dot{z} &= \varepsilon(s(x + x_0) - z), \quad a, b, c, d, \varepsilon, s > 0
\end{aligned} \tag{7.71}$$

obeys the following inequality:

$$\begin{aligned}
&V(x(t), y(t), z(t)) - V(x(0), y(0), z(0)) \\
&\quad \leq \int_0^t x(\tau)u(\tau) - H(x(\tau), y(\tau), z(\tau))d\tau,
\end{aligned} \tag{7.72}$$

where the function $H(\cdot)$ is non-negative outside a ball in $\mathbb{R}^3$, and the function V is positive definite and radially unbounded. Second, similarly to Pogromsky (1998), we show that semi-passivity of (7.71) implies that solutions of coupled system (7.16) are bounded. Third, for an arbitrary pair (i, j) of oscillators we present a non-negative function such that properties (7.67) and (7.68) hold for sufficiently large values of γ. Then we use the comparison lemma from Khalil (2002) to complete the proof.

(1) Semi-passivity of the Hindmarsh–Rose oscillator. Let us consider the following class of functions V:

$$V(x, y, z) = \frac{1}{2}\left(c_1 x^2 + c_2 y^2 + c_3 z^2\right).$$

Then showing the existence of a function V from the above class that, in addition satisfies the inequality

$$\dot{V} \leq xu - H(x, y, z), \tag{7.73}$$

where H is non-negative outside some ball in $\mathbb{R}^3$, would imply semi-passivity of (7.71).

Consider the time-derivative of V:

$$\begin{aligned}\dot{V}(x, y, z) = &-c_1 a x^4 - c_2 d x^2 y - c_2 y^2 + c_1 xy - c_3 \varepsilon z^2 + (c_3 \varepsilon s - c_1) xz \\ &+ c_1 b x^3 + c_1 I x + c_2 c y + c_3 \varepsilon s x_0 z + c_1 x u.\end{aligned} \tag{7.74}$$

Let us rewrite (7.74) such that the cross terms xy, xz, and $x^2 y$ are expressed in terms of the powers of x, y, and z and their sums. In order to do this we employ the following three equalities:

$$\begin{aligned}-c_2 y^2 + c_1 xy = &-c_2 \lambda_2 y^2 - c_2(1 - \lambda_2)\left(y - \frac{c_1}{2c_2(1 - \lambda_2)} x\right)^2 \\ &+ \frac{c_1{}^2}{4c_2(1 - \lambda_2)} x^2,\end{aligned} \tag{7.75}$$

$$\begin{aligned}-c_3 \varepsilon z^2 + (c_3 \varepsilon s - c_1) xz = &-c_3 \varepsilon \lambda_3 z^2 - c_3 \varepsilon (1 - \lambda_3)\left(z - \frac{c_3 \varepsilon s - c_1}{2c_3 \varepsilon (1 - \lambda_3)} x\right)^2 \\ &+ \frac{(c_3 \varepsilon s - c_1)^2}{4c_3 \varepsilon (1 - \lambda_3)} x^2,\end{aligned} \tag{7.76}$$

$$\begin{aligned}-c_1 a x^4 - c_2 d x^2 y = &-c_1 a \lambda_1 x^4 - c_1 a (1 - \lambda_1)\left(x^2 + \frac{c_2 d}{2c_1 a (1 - \lambda_1)} y\right)^2 \\ &+ \frac{(c_2 d)^2}{4c_1 a (1 - \lambda_1)} y^2.\end{aligned} \tag{7.77}$$

In what follows we will assume that the constants λ_1, λ_2, and λ_3 in (7.75)–(7.77) are chosen arbitrarily in the interval $(0, 1)$: $0 < \lambda_i < 1,\ i = 1, 2, 3$.

Taking the equalities (7.75)–(7.77) into account, we can rewrite the time-derivative of V, (7.74), in the following form:

$$\begin{aligned}\dot{V}(x,y,z) = &-c_1a(1-\lambda_1)\left(x^2+\frac{c_2d}{2c_1a(1-\lambda_1)}y\right)^2\\ &-c_2(1-\lambda_2)\left(y-\frac{c_1}{2c_2(1-\lambda_2)}x\right)^2\\ &-c_3\varepsilon(1-\lambda_3)\left(z-\frac{c_3\varepsilon s-c_1}{2c_3\varepsilon(1-\lambda_3)}x\right)^2\\ &-c_2\left(\lambda_2-\frac{c_2d^2}{4c_1a(1-\lambda_1)}\right)y^2+c_2cy\\ &-c_3\varepsilon\lambda_3z^2+c_3\varepsilon sx_0z-c_1a\lambda_1x^4+c_1bx^3\\ &+\left(\frac{c_1{}^2}{4c_2(1-\lambda_2)}+\frac{(c_3\varepsilon s-c_1)^2}{4c_3\varepsilon(1-\lambda_3)}\right)x^2+c_1Ix+c_1xu. \end{aligned} \tag{7.78}$$

Our goal is to express the right-hand side of (7.78) in the following form:

$$\dot{V} \le c_1xu+(M-H_0(x,y,z)), \tag{7.79}$$

where $H_0(x,y,z)$ is a radially unbounded non-negative function outside a ball in $\mathbb{R}^3$, and M is a constant. For this reason we select the constants λ_2 and c_2 in (7.73) as follows:

$$\lambda_2-\frac{c_2d^2}{4c_1a(1-\lambda_1)}>0,\ \text{or}\ \frac{c_2}{c_1}<\frac{4a\lambda_2(1-\lambda_1)}{d^2}. \tag{7.80}$$

Noticing that

$$\begin{aligned}-c_2\left(\lambda_2-\frac{c_2d^2}{4c_1a(1-\lambda_1)}\right)y^2+c_2cy = &-c_2\left(\lambda_2-\frac{c_2d^2}{4c_1a(1-\lambda_1)}\right)\\ &\times\left(y-\frac{2cc_1a(1-\lambda_1)}{4\lambda_2c_1a(1-\lambda_1)-c_2d^2}\right)^2\\ &+\frac{c_1c_2c^2a(1-\lambda_1)}{4\lambda_2c_1a(1-\lambda_1)-c_2d^2}\end{aligned} \tag{7.81}$$

and

$$-c_3\varepsilon\lambda_3z^2+c_3\varepsilon sx_0z=-c_3\varepsilon\lambda_3\left(z-\frac{sx_0}{2\lambda_3}\right)^2+\frac{c_3\varepsilon s^2x_0{}^2}{4\lambda_3} \tag{7.82}$$

proves representation (7.79) for any fixed x = constant. In order to show that (7.79) holds with respect to the complete set of variables, e.g. (x,y,z), we use the following sequence of equalities:

$$\begin{aligned}
-c_1 a\lambda_1 x^4 + c_1 b x^3 + \left(\frac{c_1^2}{4c_2(1-\lambda_2)} + \frac{(c_3\varepsilon s - c_1)^2}{4c_3\varepsilon(1-\lambda_3)}\right)x^2 + c_1 I x& \\
= -a_0 x^4 + a_1 x^3 + a_2 x^2 + a_3 x + a_4& \\
= -b_0 x^4 - (x-b_1)^4 + b_2 x^2 + b_3 x + b_4& \\
= -b_0 x^4 - (x-b_1)^4 + (b_2+d_0)x^2 - d_0(x-d_1)^2 + d_2& \\
= -b_0(x^2-e_0)^2 - (x-b_1)^4 - d_0(x-d_1)^2 + e_1& \qquad (7.83)
\end{aligned}$$

with

$$\begin{aligned}
&a_0 = c_1 a\lambda_1, \quad a_1 = c_1 b, \quad a_2 = \frac{c_1^2}{4c_2(1-\lambda_2)} + \frac{(c_3\varepsilon s - c_1)^2}{4c_3\varepsilon(1-\lambda_3)}, \\
&a_3 = c_1 I, \quad a_4 = 0, \quad b_0 = a_0 - 1, \quad b_1 = \tfrac{1}{4}a_1, \quad b_2 = a_2 + \tfrac{3}{8}a_1^2, \\
&b_3 = a_3 - \tfrac{1}{16}a_1^3, \quad b_4 = a_4 + \tfrac{1}{256}a_1^4, \\
&d_0 = 1, \quad d_1 = \frac{b_3}{2d_0}, \quad d_2 = b_4 + d_1^2 d_0, \quad e_0 = \frac{b_2 + d_0}{2b_0}, \quad e_1 = d_2 + b_0 e_0^2.
\end{aligned} \qquad (7.84)$$

Notice that we want the value of b_0 in (7.83) and (7.84) to be positive. Hence the value of

$$a_0 = c_1 a\lambda_1$$

should be greater than 1. This can be ensured by choosing the value of c_1 in (7.73) to be sufficiently large. As a result of this choice, taking restrictions (7.80) into account, we conclude that the value of c_2 in (7.73) must be sufficiently small, i.e. it must satisfy the following inequality:

$$c_2 < c_1 \frac{4a\lambda_2(1-\lambda_1)}{d^2}.$$

The value for d_0 can be chosen arbitrarily; here $d_0 = 1$.

The time-derivative $\dot{V}$ can now be written as follows:

$$\begin{aligned}
\dot{V}(x,y,z) = -c_1 a(1-\lambda_1)\left(x^2 + \frac{c_2 d}{2c_1 a(1-\lambda_1)}y\right)^2& \\
- c_3\varepsilon(1-\lambda_3)\left(z - \frac{c_3\varepsilon s - c_1}{2c_3\varepsilon(1-\lambda_3)}x\right)^2&
\end{aligned}$$

$$
\begin{aligned}
&- c_2(1-\lambda_2)\left(y - \frac{c_1}{2c_2(1-\lambda_2)}x\right)^2 \\
&- c_3\varepsilon\lambda_3\left(z - \frac{sx_0}{2\lambda_3}\right)^2 + \frac{c_3\varepsilon s^2 x_0{}^2}{4\lambda_3} \\
&- c_2\left(\lambda_2 - \frac{c_2 d^2}{4c_1 a(1-\lambda_1)}\right)\left(y - \frac{2cc_1 a(1-\lambda_1)}{4\lambda_2 c_1 a(1-\lambda_1) - c_2 d^2}\right)^2 \\
&+ \frac{c_1 c_2 c^2 a(1-\lambda_1)}{4\lambda_2 c_1 a(1-\lambda_1) - c_2 d^2} \\
&- b_0\left(x^2 - e_0\right)^2 - \left(x - b_1\right)^4 - d_0\left(x - d_1\right)^2 + e_1 + c_1 x u. \qquad (7.85)
\end{aligned}
$$

It is straightforward to see that expression (7.85) is of the form (7.79), where

$$
\begin{aligned}
H_0(x,y,z) = {}& c_1 a(1-\lambda_1)\left(x^2 + \frac{c_2 d}{2c_1 a(1-\lambda_1)}y\right)^2 \\
&+ c_3\varepsilon(1-\lambda_3)\left(z - \frac{c_3\varepsilon s - c_1}{2c_3\varepsilon(1-\lambda_3)}x\right)^2 \\
&+ c_2(1-\lambda_2)\left(y - \frac{c_1}{2c_2(1-\lambda_2)}x\right)^2 + c_3\varepsilon\lambda_3\left(z - \frac{sx_0}{2\lambda_3}\right)^2 \\
&+ c_2\left(\lambda_2 - \frac{c_2 d^2}{4c_1 a(1-\lambda_1)}\right)\left(y - \frac{2cc_1 a(1-\lambda_1)}{4\lambda_2 c_1 a(1-\lambda_1) - c_2 d^2}\right)^2 \\
&+ b_0\left(x^2 - e_0\right)^2 + \left(x - b_1\right)^4 + d_0\left(x - d_1\right)^2
\end{aligned}
$$

and

$$
M = \frac{c_3\varepsilon s^2 x_0{}^2}{4\lambda_3} + \frac{c_1 c_2 c^2 a(1-\lambda_1)}{4\lambda_2 c_1 a(1-\lambda_1) - c_2 d^2} + e_1.
$$

Let us denote

$$
H_1(x,y,z) = H_0 - M
$$

and rewrite (7.79) as

$$
\dot{V} \leq c_1 x u - H_1(x,y,z).
$$

The function $H_1(x,y,z)$ is radially unbounded. Furthermore, it is non-negative outside a ball in $\mathbb{R}^3$. Hence, on choosing

$$
V^*(x,y,z) = \frac{1}{c_1}V(x,y,z)
$$

we assure the existence of a (radially unbounded) positive definite $V^*(x, y, z)$ such that

$$\dot{V}^* \leq xu - \frac{H_1(x, y, z)}{c_1}, \tag{7.86}$$

where $H_1(x, y, z)/c_1$ is radially unbounded and non-negative outside a ball in $\mathbb{R}^3$. Thus, according to (7.73), semi-passivity of the Hindmarsh–Rose system follows.

(2) Boundedness of the solutions. We aim to prove that boundedness of $\phi_i(t)$, $i \in \{0, \ldots, n\}$ implies boundedness of the state of the coupled system. Without loss of generality we assume that

$$\|\phi_i(\tau)\|_{\infty,[0,\infty]} \leq D_\phi.$$

Let us denote

$$V_i = V^*(x_i, y_i, z_i), \qquad H_{1,i} = \frac{1}{c_1} H_1(x_i, y_i, z_i).$$

Consider the following function:

$$V_\Sigma(\mathbf{x}, \mathbf{y}, \mathbf{z}) = \rho\left(\sum_{i=0}^{n} V_i(x_i, y_i, z_i), C\right), \tag{7.87}$$

where $\mathbf{x} = \mathrm{col}(x_0, \ldots, x_n)$, $\mathbf{y} = \mathrm{col}(y_0, \ldots, y_n)$, $\mathbf{z} = \mathrm{col}(z_0, \ldots, z_n)$, and

$$\rho(s, C) = \begin{cases} s - C, & s \geq C, \\ 0, & s < C. \end{cases}$$

The function V_Σ is non-negative for any $C \in \mathbb{R}$ and, furthermore, is radially unbounded. Hence, its boundedness for some $C \in \mathbb{R}$ implies boundedness of x_i, y_i, and z_i, $i \in \{0, \ldots, n\}$.

Let us pick $C \in \mathbb{R}$ such that the interior of the domain

$$\Omega_C = \left\{\mathbf{x}, \mathbf{y}, \mathbf{z} \in \mathbb{R} \middle| \sum_{i=0}^{n} V_i(x_i, y_i, z_i) \leq C\right\}$$

contains the domain

$$\sum_{i=0}^{n} H_{1,i}(x_i, y_i, z_i) - \kappa x_i^2 < M_i, \ M_i \in \mathbb{R}_{>0}, \ \kappa \in \mathbb{R}_{>0},$$

where M_i is an arbitrarily large and κ is an arbitrarily small positive constant. In other words, the following implication holds:

$$\sum_{i=0}^{n} V_i(x_i, y_i, z_i) \geq C \Rightarrow \sum_{i=0}^{n} H_{1,i}(x_i, y_i, z_i) - \kappa x_i^2 \geq M_i. \tag{7.88}$$

Such a C always exists because $H_{1,i}(x_i, y_i, z_i) - \kappa x_i^2$ can be expressed as a sum of a non-negative quadratic form in x_i, y_i, and z_i and non-negative functions of the higher order plus a constant, and $V_i(x_i, y_i, z_i)$ is a positive definite quadratic form.

Consider the time-derivative of the function $V_\Sigma(\mathbf{x}, \mathbf{y}, \mathbf{z})$. According to (7.87) and (7.86) it is zero for all $\mathbf{x}, \mathbf{y}, \mathbf{z} \in \Omega_C$, and satisfies the following inequality otherwise:

$$\dot{V}_\Sigma \leq \sum x_i u_i - \sum_{i=0}^{n} H_{1,i}(x_i, y_i, z_i) = \gamma \mathbf{x}^{\mathrm{T}} \Gamma \mathbf{x} + \sum_{i=0}^{n} x_i \phi_i(t) - \sum_{i=0}^{n} H_{1,i}(x_i, y_i, z_i).$$

Using Gershgorin's circle theorem, we can conclude that

$$\dot{V}_\Sigma \leq \sum_{i=0}^{n} x_i \phi_i(t) - \sum_{i=0}^{n} H_{1,i}(x_i, y_i, z_i).$$

Rewriting

$$x_i \phi_i(t) = -\kappa \left(x_i - \frac{\phi_i(t)}{2\kappa} \right)^2 + \kappa x_i^2 + \frac{1}{4\kappa} \phi_i^2(t), \ \kappa > 0$$

leads to the following inequality:

$$\begin{aligned} \dot{V}_\Sigma &\leq \kappa \sum_{i=0}^{n} x_i^2 - \sum_{i=0}^{n} \left(H_{1,i}(x_i, y_i, z_i) - \frac{D_\phi^2}{4\kappa} \right) \\ &= - \sum_{i=0}^{n} \left(H_{1,i}(x_i, y_i, z_i) - \frac{D_\phi^2}{4\kappa} - \kappa x_i^2 \right). \end{aligned}$$

Hence, by choosing the value of C such that $M_i \geq D_\phi^2/(4\kappa)$ in (7.88) we can ensure that

$$\dot{V}_\Sigma \leq 0.$$

This implies that $V_\Sigma(\mathbf{x}(t), \mathbf{y}(t), \mathbf{z}(t))$ is not growing with time. Hence the trajectories $x_i(t)$, $y_i(t)$, and $z_i(t)$ in the coupled system are bounded.

(3) Convergence to a vicinity of the synchronization manifold. Consider the ith and jth oscillators in (7.16), $i, j \in \{0, \ldots, n\}$, $i \neq j$. Let us introduce the following function:

$$V = 0.5 \left(C_x (x_i - x_j)^2 + C_y (y_i - y_j)^2 + C_z (z_i - z_j)^2 \right), \tag{7.89}$$

where C_x, $C_y > 0$ are to be defined and $C_z = C_x/(s\varepsilon)$.

Its time-derivative can be expressed as follows:

$$\dot{V} = -C_x(x_i - x_j)^2 \left(\frac{ax_i^2}{2} + \frac{ax_j^2}{2} + \frac{a(x_i + x_j)^2}{2} - b(x_i + x_j) + \gamma(n+1) \right)$$
$$+ C_x(y_i - y_j)(x_i - x_j) - C_y d(x_i - x_j)(x_i + x_j)(y_i - y_j)$$
$$- C_y(y_i - y_j)^2 - C_z\varepsilon(z_i - z_j)^2 + C_x(x_i - x_j)(\phi_i - \phi_j). \tag{7.90}$$

Consider the following term in (7.90):

$$C_x(y_i - y_j)(x_i - x_j) - C_y d(x_i - x_j)(x_i + x_j)(y_i - y_j) - C_y(y_i - y_j)^2.$$

It can be written as follows:

$$\frac{C_x^2}{4C_y\Delta_1}(x_i - x_j)^2 - \left(\left(\frac{C_x^2}{4C_y\Delta_1} \right)^{0.5} (x_i - x_j) - (\Delta_1 C_y)^{0.5} (y_i - y_j) \right)^2$$
$$+ \frac{C_y d^2}{4\Delta_2}(x_i - x_j)^2(x_i + x_j)^2 - C_y \left(\left(\frac{d^2}{4\Delta_2} \right)^{0.5} (x_i^2 - x_j^2) + \Delta_2^{0.5}(y_i - y_j) \right)^2$$
$$- (1 - \Delta_1 - \Delta_2)(y_i - y_j)^2, \tag{7.91}$$

where $\Delta_1, \Delta_2 \in \mathbb{R}_{>0}$ and $\Delta_1 + \Delta_2 \in (0, 1)$. Taking (7.91) into account, we rewrite (7.90) as

$$\dot{V} \leq -C_x(x_i - x_j)^2 \left(\frac{ax_i^2}{2} + \frac{ax_j^2}{2} + \frac{a(x_i + x_j)^2}{2} - \frac{C_y d^2}{C_x 4\Delta_2}(x_i + x_j)^2 \right.$$
$$\left. - b(x_i + x_j) + \gamma(n+1) - \frac{C_x}{4C_y\Delta_1} \right)$$
$$- C_z\varepsilon(z_i - z_{i+1})^2 - C_y(1 - \Delta_1 - \Delta_2)(y_i - y_j)^2$$
$$+ C_x(x_i - x_j)(\phi_i - \phi_j). \tag{7.92}$$

Let

$$\frac{C_y}{C_x} = \frac{2a\Delta_2}{d^2}.$$

Then

$$\dot{V} \leq -C_x(x_i - x_j)^2 \left(\frac{a}{2}\left(x_i - \frac{b}{a}\right)^2 + \frac{a}{2}\left(x_j - \frac{b}{a}\right)^2 + \gamma(n+1) - \frac{d^2}{8a\Delta_1\Delta_2} - \frac{b^2}{a} \right) - (1 - \Delta_1 - \Delta_2)C_y(y_i - y_j)^2 - C_z\varepsilon(z_i - z_{i+1})^2 + C_x(x_i - x_j)(\phi_i - \phi_j). \tag{7.93}$$

Hence, by choosing

$$\gamma > \frac{1}{(n+1)a}\left(\frac{d^2}{8\Delta_1\Delta_2} + b^2\right)$$

we can ensure that the first term in (7.93) is non-positive. The minimal value of γ ensuring this property can be calculated by minimizing the value

$$\frac{1}{8\Delta_1\Delta_2}$$

for all $\Delta_1, \Delta_2 \in \mathbb{R}_{>0} : \Delta_1 + \Delta_2 < 1$. This can be done by letting $\Delta_2 = r - \Delta_1$, $r \in (0, 1)$ and differentiating the term $1/(8\Delta_1(r - \Delta_1))$ with respect to Δ_1. This leads to the following solution: $\Delta_1 = \Delta_2 = r/2$. Taking this into account, we rewrite (7.93) as follows:

$$\dot{V} \leq -C_x(x_i - x_j)^2 \left(\frac{a}{2}\left(x_i - \frac{b}{a}\right)^2 + \frac{a}{2}\left(x_j - \frac{b}{a}\right)^2 + \gamma(n+1) - \frac{d^2}{2ar} - \frac{b^2}{a} \right) - (1 - r)C_y(y_i - y_j)^2 - C_z\varepsilon(z_i - z_{i+1})^2 + C_x(x_i - x_j)(\phi_i - \phi_j). \tag{7.94}$$

Let

$$\gamma = \frac{1}{(n+1)a}\left(\frac{d^2}{2} + b^2\right) + \varepsilon_1, \ \varepsilon_1 \in \mathbb{R}_{>0}.$$

Alternatively, we can rewrite this as

$$\gamma = \frac{1}{(n+1)a}\left(\frac{d^2}{2r} + b^2\right) + \varepsilon_2, \ r \in (0, 1), \ \varepsilon_2 \in \mathbb{R}_{>0}.$$

Hence, according to (7.94), the following inequality holds:

$$\dot{V} \leq -C_x\varepsilon_2(x_i - x_j)^2 - (1 - r)C_y(y_i - y_j)^2 - C_z\varepsilon(z_i - z_{i+1})^2 + C_x(x_i - x_j)(\phi_i - \phi_j).$$

Then, denoting $\alpha = 2\min\{\varepsilon_2, \varepsilon, (1 - r)\}$, we obtain

$$\dot{V} \leq -\alpha V + C_x(x_i - x_j)(\phi_i - \phi_j). \tag{7.95}$$

Consider the following differential equation:

$$\dot{\upsilon} = -\alpha\upsilon + C_x(x_i - x_j)(\phi_i - \phi_j). \tag{7.96}$$

Its solution can be estimated as follows:

$$|\upsilon(t)| \leq e^{-\alpha(t-t_0)}|\upsilon(t_0)| + e^{-\alpha t}\int_{t_0}^{t} e^{\alpha\tau} C_x(x_i(\tau) - x_j(\tau))(\phi_i(\tau) - \phi_j(\tau))d\tau$$

for all $t \geq t_0$. Given that $x_i(t)$ and $x_j(t)$ are bounded there exists a constant B such that

$$|\upsilon(t)| \leq e^{-\alpha(t-t_0)}|\upsilon(t_0)| + \frac{C_x B}{\alpha}\|\phi_i(\tau) - \phi_j(\tau)\|_{\infty,[t_0,t]}.$$

Then, applying the comparison lemma (see, for example, Khalil (2002), page 102), we can conclude that

$$V(t) \leq e^{-\alpha(t-t_0)}V(t_0) + \frac{C_x B}{\alpha}\|\phi_i(\tau) - \phi_j(\tau)\|_{\infty,[t_0,t]}.$$

Hence, conclusion (2) of the theorem follows. □

Theorem 7.2 specifies the boundaries for stable synchrony in the system of coupled neural oscillators (7.16) as a function of the coupling strength, γ, and the parameters a, b, and d of a single oscillator. The last three parameters represent properties of the membrane and, combined with x_0, ε, s, and I, completely characterize the dynamics of a single model neuron (Hindmarsh and Rose 1984), ranging from single spiking to periodic or chaotic bursts.

The distinctive feature of Theorem 7.2 is that it is suitable for analysis of systems with external time-dependent perturbations $\phi_i(t)$. This property is essential for the comparison task, wherein the oscillators are fed with time-varying inputs and the degree of their mutual synchrony is the measure of similarity between the inputs.

While the theorem provides us with conditions for stable synchrony, it allows us to estimate the domain of values of the coupling parameter γ corresponding to potential intermittent, itinerant (Kaneko and Tsuda 2000, 2003), or meta-stable regimes. In particular, as follows from Theorem 7.2, a necessary condition for unstable synchronization in system (7.16) is

$$\gamma < \frac{1}{(n+1)\cdot a}\left(\frac{d^2}{2} + b^2\right). \tag{7.97}$$

Notice that conditions (7.97) and (7.69) do not depend on the "bifurcation" parameter I which usually determines the type of bursting in a single oscillator. They also do not depend on the differences in the time scales defined by the parameter ε between the fast, x and y, and slow, z, variables. Hence these conditions apply for a wide range of system behavior on the synchronization manifold. This advantage also has a downside, because conditions (7.69) and (7.97) are too conservative. However, this may be a reasonable price to pay in return for invariance of the criteria (7.69) and (7.97) with respect to the full range of dynamical behavior of a generally nonlinear system.

7.2.3 Further extensions

We have provided a solution in principle to the issue of invariant recognition in the problem of template matching. Recognition occurs when mismatches in the temporal representations of image and templates converge to a small neighborhood of zero. This in turn leads to synchronized trajectories in a network of nonlinear oscillators serving as detectors of coincidences. Although our overall implementation of this idea might not be normative, we have tried to keep the number of relevant parameters to a minimum. In particular, the dimension of the state of a single adaptation compartment is three, which is minimal for the generation of spikes ranging from periodic to chaotic bursts. Moreover, conditions (7.69) and (7.97) allow us to choose the coupling strength γ as a single control parameter for regulating the stability/instability of the synchronous activity in the network.

In this section we provide further extensions of the basic results of Theorems 7.1 and 7.2, and discuss possible links between the normative part of our theory and some known results in vision.

Extension to frequency-encoding schemes

For the sake of notational simplicity we restricted our attention to temporal representations (7.6) and (7.9) of spatially sampled images. These encoding schemes can be interpreted as scanning of an image over time. Yet, the results of Section 7.2 apply to a broader class of encoding schemes. One example is the frequency-coding used in many neural systems. Let us consider the factorization (7.6), where in the notation $\mathcal{F}_t[S_0, \boldsymbol{\theta}](x, y)$ the symbol t is replaced with ν. In order to extend the initial encoding scheme to the domain of frequency/spike-rate encoding we introduce an additional linear functional f_ω as follows:

$$f_\omega(t, \mathcal{F}_\nu[S_0, \boldsymbol{\theta}]) = \sum_\nu h(\omega_\nu \cdot t) \cdot f(\mathcal{F}_\nu[S_0, \boldsymbol{\theta}]), \tag{7.98}$$

where $h : \mathbb{R} \to \mathbb{R}$ is a bounded periodic function, and ω_ν are distinct real numbers indexed by ν. The function $h(\omega_\nu \cdot t)$ in (7.98) serves as a basis for carrier-function-generating periodic impulses of various frequencies ω_ν. Thus, each νth spatial sample of the image is assigned a particular frequency, and the amplitude

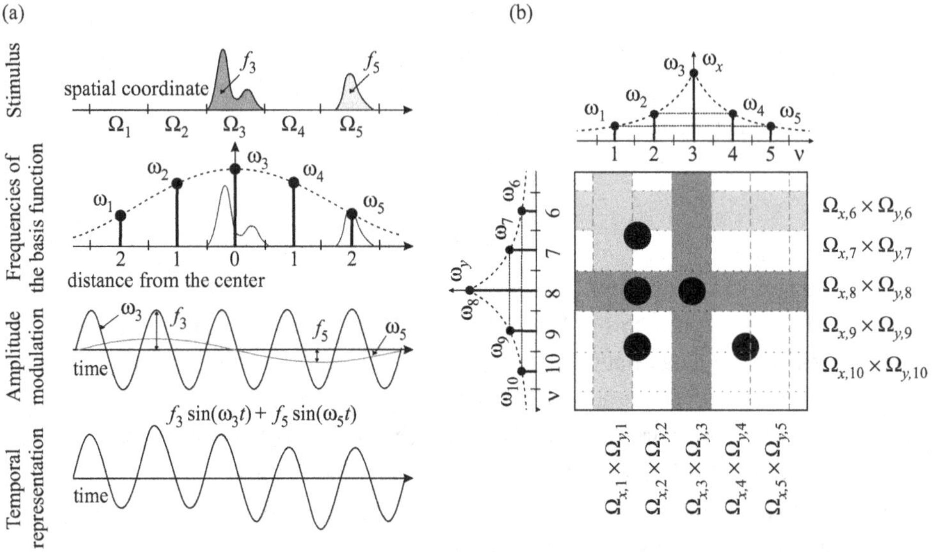

Figure 7.6 (a) Temporal representation of a spatially distributed stimulus using frequency encoding. A stimulus (upper row) $S(x)$ is spatially sampled by partitioning its domain into the union of intervals Ω_i. For each Ω_i an integral $f_i = f(\mathcal{F}_i) = \int_{\Omega_i} S(x)dx$ is calculated and a frequency ω_i is assigned. The resulting temporal representation (lower row) is expressed as the sum of two amplitude-modulated harmonic signals of frequencies, ω_3 and ω_5. (b) Temporal representation of a two-dimensional pattern. The pattern consists of black filled circles. The image domain is partitioned into a collection of horizontal and vertical strips. Dark domains correspond to higher frequencies.

of the oscillation is specified by $f(\mathcal{F}_\nu[S_0, \boldsymbol{\theta}])$. Temporal representation of a one-dimensional stimulus according to encoding scheme (7.98) is illustrated in Figure 7.6(a).

This encoding scheme is plausible to biological vision, when frequencies ω_ν are ordered according to the positions of domains $\Omega_{x,\nu}$ and $\Omega_{y,\nu}$ relative to the center of the image. This corresponds, in particular, to the receptive fields in cat retinal ganglion cells (Enroth-Cugell *et al.* 1983). Because the functional f_ω is linear in $f(\mathcal{F}_\nu[S_0, \boldsymbol{\theta}])$ and the function $h(\omega_\nu \cdot t)$ is bounded for all t, condition (7.8) will be satisfied for f_ω. Hence, the conclusions of Theorem 7.1 apply to these representations.

Multiple representations of uncertainties

Another property of the system (7.22), (7.25), and (7.28), in addition to its ability to accommodate relevant encoding schemes such as frequency/rate and sequential/random scanning, is that each single value of $\theta_2 \in (\theta_{2,\min}, \theta_{2,\max})$ induces at least two distinct attracting sets in the extended space. Indeed,

$$\lambda_2^2(t) + \lambda_3^2(t) = \text{constant} = 1$$

along the trajectories of (7.22), (7.25), and (7.28) (see also the proof of Theorem 7.1). Hence, for almost every value of λ_2 (i.e. except when $\lambda_2 = \pm 1$) in the definition of $\hat{\theta}_2(t)$ in (7.28) there will always be two distinct values of λ_3:

$$\lambda_{3,1} = \sqrt{1 - \lambda_2^2}, \qquad \lambda_{3,2} = -\sqrt{1 - \lambda_2^2}.$$

These two values give rise to distinct invariant sets in the system's state space for a single value of θ_2. The presence of two complementary encodings for the same figure is a plausible assumption that has been used in the perceptual-organization literature to explain a range of phenomena, including perceptual ambiguity, and modal and amodal completion (Hatfield and Epstein 1985; Leeuwenberg and Buffart 1983, p. 29; Shepard 1981). A consequence of the presence of multiple attractors corresponding to the single value of perturbation is that the time for convergence (the decision time) may change abruptly with small variations of initial conditions. The latter property is well documented in human subjects (Gilden 2001). Furthermore, the presence of two attractors with different basins for a single value of perturbation will lead to asymmetric distributions of decision times, which is typically observed in human and animal reaction-time data (Smith and Ratcliff 2004).

Multiple time scales for different modalities in vision

An important property of the proposed solution to the problem of invariance is that the time scales of adaptation to linearly and nonlinearly parametrized uncertainties are substantially different. This difference in time scales emerged naturally in the course of our mathematical argument as a consequence of splitting the system dynamics into a slow searching subsystem and a fast asymptotically stable one. This allowed us to prove the emergence of unstable yet attracting invariant sets, thus ensuring the existence of a solution to the problem of invariant template matching. In particular, Theorem 7.1 requires that the time constants of adaptation to linearly parametrized uncertainties (for instance, unknown intensity of the image) are substantially smaller than the time constants of adaptation to nonlinearly parametrized uncertainties (image blur, rotation, scaling etc.). Indeed, as follows from Table 7.3, the larger the difference in the time scales the higher the possible precision and the smaller the errors. This is not to say that successful adaptation is impossible if the time scales of adaptation are of the same order. Our results are sufficient and only suggest that having different adaptation time scales may be beneficial for convergence. On the other hand, a simple geometric argument can be used to demonstrate that the larger the value of γ_1/τ the larger the trapping regions of stimuli-induced attracting sets. In Section 7.3.1 we illustrate with an example how the time scales of adaptation to different modalities might affect the performance of a simple recognition system.

Even though the difference in time scales was motivated purely by theoretical considerations, there is strong evidence that biological systems adapt at different time scales to uncertainties from different modalities. For example, the time scale of adaptation to light is within tens of milliseconds (Wolfson and Graham 2000) whereas adaptation to "higher-order" modalities like rotation and image blur extends from hundreds of milliseconds to minutes (Webster *et al.* 2002). Evidence for the presence of slow and fast adaptation in motor learning is reported in Smith *et al.* (2006). These findings, therefore, motivate our belief that the system (7.22), (7.25) and (7.28) could serve as a simple, yet qualitatively realistic, model for adaptation mechanisms in vision, motor behavior, and decision making.

7.3 Examples

In this section we provide simple illustrations of how particular systems for invariant template matching can be constructed using the results presented above.

7.3.1 Rotation-invariant matching and mental rotation experiments

Let us illustrate how the results of Theorems 7.1 and 7.2 can be applied to template matching when an object is rotated over an unknown angle and its brightness is uncertain a priori. Below we consider three examples illustrating the performance of our system in different experimental settings. The first example corresponds to the case when only one object is present in the image. The task is to detect an object and infer its rotation angle and brightness. In the second example we consider an image that contains two different objects of which the rotation angle and brightness are uncertain. The system should be able to report the presence of templates and estimate the values of rotation angles and brightness. In the third experiment we illustrate the importance of separating the adaptation time scales.

Rotation-invariant matching in images with single objects

Without loss of generality and for the sake of notational convenience, suppose that the domain $\Omega_x \times \Omega_y$ is a square of the following dimensions: $\Omega_x \times \Omega_y = [0, y_{\max}] \times [0, x_{\max}]$. In order to obtain a temporal representation of the image we use the frequency-encoding scheme (7.98) which was illustrated in Figure 7.6(b).

In particular we use the following transformation

$$\theta_1 f_i(t,\theta_2) = \theta_1 \sum_\nu h(\omega_\nu \cdot t) \cdot f(\bar{\mathcal{F}}_\nu[S_i,\theta_2]), \tag{7.99}$$

where θ_2 is the rotation angle of image $S_i(x, y)$ around its central point, θ_1 is the image brightness, the function $h(\omega_\nu \cdot t) = \sin^2(\omega_\nu \cdot t)$, and

$$f(\bar{\mathcal{F}}_\nu[S_i,\theta_2]) = \int_{\Omega_{x,\nu}\times\Omega_{y,\nu}} \bar{\mathcal{F}}_\nu[S_i,\theta_2](\xi,\gamma)d\xi\, d\gamma \tag{7.100}$$

is simply an integral of the image S_i rotated by an angle θ_2 over the strip $\Omega_{x,\nu} \times \Omega_{y,\nu}$.

For instance, let us have m horizontal strips aligned along the x-coordinates of the templates, and n vertically aligned strips along the y-coordinates. Then

$$\Omega_{x,i} \times \Omega_{y,i} = \begin{cases} [a(i), b(i)] \times [0, y_{\max}], & a(i) < b(i),\ i = 1, \ldots, m, \\ [0, x_{\max}] \times [a(i), b(i)], & a(i) < b(i),\ i = m+1, \ldots, m+n, \end{cases}$$

and the integrals (7.100) transform into

$$f(\bar{\mathcal{F}}_\nu[S_i,\theta_2]) = \begin{cases} \displaystyle\int_{a(\nu)}^{b(\nu)} \int_0^{y_{\max}} \bar{\mathcal{F}}_\nu[S_i,\theta_2](\xi,\gamma)d\xi\, d\gamma, \\ \qquad \nu = 1, \ldots, m, \\ \displaystyle\int_0^{x_{\max}} \int_{a(\nu)}^{b(\nu)} \bar{\mathcal{F}}_\nu[S_i,\theta_2](\xi,\gamma)d\xi\, d\gamma, \\ \qquad \nu = m+1, \ldots, m+n. \end{cases} \tag{7.101}$$

Hence our temporal representation of the image and the templates is simply a weighted sum of periodic functions of time $\sin^2(\omega_\nu \cdot t)$, scaled by θ_1, with weights determined by (7.101).

According to (7.16), (7.22), (7.25), and (7.28) the recognition system (see Figures 7.3 and 7.8 for its general structure) can be described by the system of differential equations provided in Table 7.4. Implementation details, initial conditions, and the source files of a working MATLAB Simulink model can be found in Tyukin *et al.* (2007a).

We tested the system's performance for a variety of input images, in particular the class of Garner patterns (Garner 1962) (see Figure 7.7, first row; for ease of computation we used relatively low-resolution images of size 101×101, i.e. with $x_{\max}, y_{\max} = 101$). These patterns serve as an interesting benchmark because in a long line of behavioral experiments, most recently Lachmann and van Leeuwen (2005), the human pattern-recognition process of these patterns has been studied

Table 7.4. *Equations of the system for rotation and brightness-invariant template matching; N is the total number of templates*

Function	Image	Template
Temporal integration	$\dot{\phi}_0 = -\frac{1}{\tau}\phi_0 + k \cdot \theta_1 f(t, \theta_2)$	$\dot{\phi}_i = -\frac{1}{\tau}\phi_i + k \cdot \hat{\theta}_{1,i} f(t, \hat{\theta}_{2,i})$
Adaptation to brightness	No	$\hat{\theta}_{1,i} = (\phi_0 - \phi_i)\gamma_1 + \lambda_{i,1}$ $\dot{\lambda}_{i,1} = \frac{\gamma_1}{\tau}(\phi_0 - \phi_i)$
Adaptation to rotation	No	$\hat{\theta}_{2,i} = (\lambda_{2,i}(t) + 1)\pi$ $\dot{\lambda}_{i,2} = \gamma_2 \lambda_{i,3} \lVert \phi_0 - \phi_i \rVert_\varepsilon$ $\dot{\lambda}_{i,3} = -\gamma_2 \lambda_{i,2} \lVert \phi_0 - \phi_i \rVert_\varepsilon$
Detectors of similarity	$\dot{x}_0 = -ax_0^3 + bx_0^2 + y_0 - z_0 + \gamma u_0 + \phi_0(t) + I$ $\dot{y}_0 = c - dx_0^2 - y_0$ $\dot{z}_0 = \varepsilon(s(x_0 + \bar{x}_0) - z_0)$	$\dot{x}_i = -ax_i^3 + bx_i^2 + y_i - z_i + \gamma u_i + \phi_i(t) + I$ $\dot{y}_i = c - dx_i^2 - y_i$ $\dot{z}_i = \varepsilon(s(x_i + \bar{x}_0) - z_i)$
Coupling function	$u_0 = -(N+1)x_0 + \sum_{j \neq 0} x_j$	$u_i = -(N+1)x_i + \sum_{j \neq i} x_j$
Parameters	Temporal-integration subsystem: $\tau = 1$, $\theta_1 = 2.3$, $\theta_2 = \pi/4$, $k = 1/20$, $x_{\max} = 101$, $y_{\max} = 101$, $m = 33$, $n = 17$, $a(i) = 1, 4, 7, \ldots, 101 - 3$, $\omega_i = 1/80$, $i \leq m$; $a(i + m) = 1, 7, 13, \ldots$, $\omega_{i+m} = 1/30$, $i \leq n$, $b(i) = a(i) + 3$ Detectors of similarity: $a = 1$, $b = 3$, $c = 1$, $d = 5$, $\varepsilon = 0.001$, $s = 4$, $\gamma = 0.5$, $I = -6.2$, $\bar{x}_0 = 1.6$	Intensity adaptation subsystem: $\gamma_1 = 0.5$ Rotation adaptation subsystem: $\gamma_2 = 0.01$, $\varepsilon = 0.05$

in great detail. The overall intensity of these patterns does not vary from one pattern to another. At the same time, the numbers of symmetries increases from the first, through the second to the third pattern in Figure 7.7. Because of this, their complexity decreases proportionally (Garner 1962).

Using as templates the first row of Figure 7.7, we simulated our recognition system of which the description is given in Table 7.4. The second row of Figure 7.7

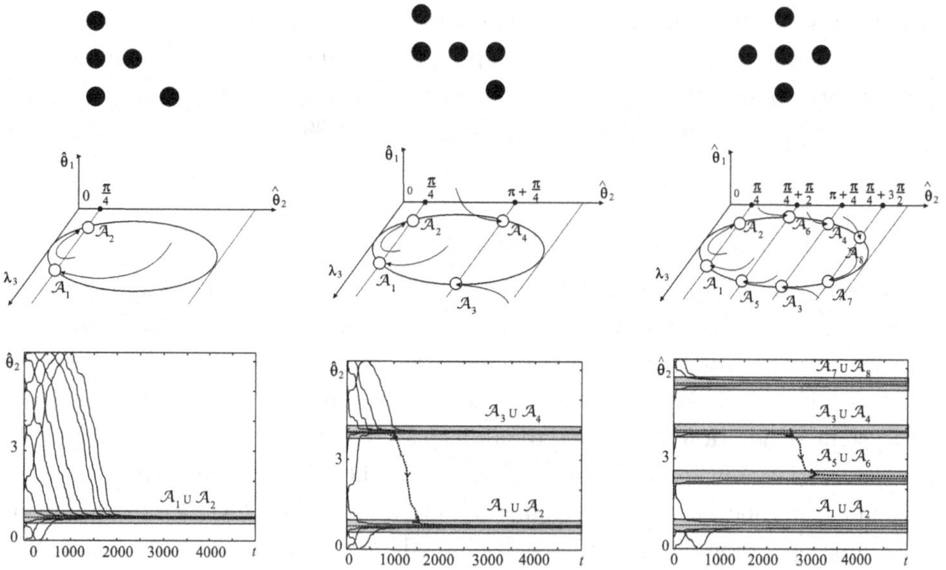

Figure 7.7 Template matching for Garner patterns with three levels of complexity. The patterns (upper row) were rotated by $\pi/4$ and had various intensities. Depending on the number of their rotational symmetries, they induced different numbers of invariant sets in the system's state space: two, four, and eight, respectively. The diagrams of corresponding phase plots are provided in the middle row. Estimates of the rotation angle as functions of time for different initial conditions are depicted in the third row.

illustrates the system dynamics involved in template matching. The diagrams represent phase plots of the successful node j (for the template subsystem in which the matching occurs). The third row contains trajectories of the estimates of the rotation angle $\hat{\theta}_{j,2}(t)$. Each object induces various numbers of invariant sets in the template subsystems. The number of these invariant sets is inversely proportional to the complexity of the stimulus. Hence, the higher the complexity the more time the system requires to reach one. Thus the time needed for recognition increases monotonically with the complexity of the stimulus. This is consistent with empirical results reported in experimental studies, for instance Lachmann and van Leeuwen (2005).

An additional property of our system is that it is capable of reporting multiple representations of the same object. This is indicated by the dashed trajectories in Figure 7.7. Even though the system parameters are chosen such that trajectories converge to an attractor, we can still observe meta-stable behavior. This is because the attractors in our system are of Milnor type, which implies that trajectories starting in the vicinity of one attractor may actually belong to the basin of another attractor. Furthermore, it is even possible to tune the system in such a way that it will always switch from one representation to another. The latter property suggests that our simple system in Table 7.4 can provide a simple model for visual perception, where spontaneous switching and perceptual multi-stability are commonly observed (Attneave 1971; Leopold and Logothetis 1999).

Rotation-invariant matching in images with multiple objects

Let us now consider the case in which two patterns are simultaneously present in an image. As an image we chose a concatenation of two rotated Garner patterns. The values of the rotation angles and intensities of the patterns are supposed to be unknown. In order to be able to detect and recognize multiple patterns in the image, the system, in addition to ensuring rotation and intensity adaptation, should be able to scan the image in space. Therefore, we extend the system for invariant template matching as proposed in the previous subsection (see Table 7.4) by introducing an additional operation, i.e. a (moving) frame, which projects part of the image into a spatiotemporal code, similarly to (7.99). In particular, instead of (7.99)–(7.101) we will deal with the following spatiotemporal representation:

$$\theta_1 f_i(t, \theta_2, p) = \theta_1 \sum_{\nu} h(\omega_\nu \cdot t) \cdot f(\bar{\mathcal{F}}_\nu[S_i, \theta_2, p]), \tag{7.102}$$

$$f(\bar{\mathcal{F}}_\nu[S_i, \theta_2]) = \begin{cases} \displaystyle\int_{a(\nu)+p}^{b(\nu)+p} \int_0^{y_{\max}} \bar{\mathcal{F}}_\nu[S_i, \theta_2](\xi, \gamma) d\xi \, d\gamma, \\ \quad \nu = 1, \ldots, m, \\ \displaystyle\int_0^{x_{\max}} \int_{a(\nu)}^{b(\nu)} \bar{\mathcal{F}}_\nu[S_i, \theta_2](\xi, \gamma) d\xi \, d\gamma, \\ \quad \nu = m+1, \ldots, m+n, \end{cases} \tag{7.103}$$

where p is the position of the frame in an image. The spatial configuration of the frame is chosen to be identical to that of the templates.

The values of p corresponding to the true positions of the objects in the image are unknown. Therefore, an extra adaptation mechanism is needed. An economical way to include such an adaptation mechanism into the system is shown in Table 7.5. Instead of having an additional compartment in each template subsystem to realize the scanning of an image, we propose that the frame of the recognition system moves along the x-coordinate of the image (see also Figure 7.8). The equations

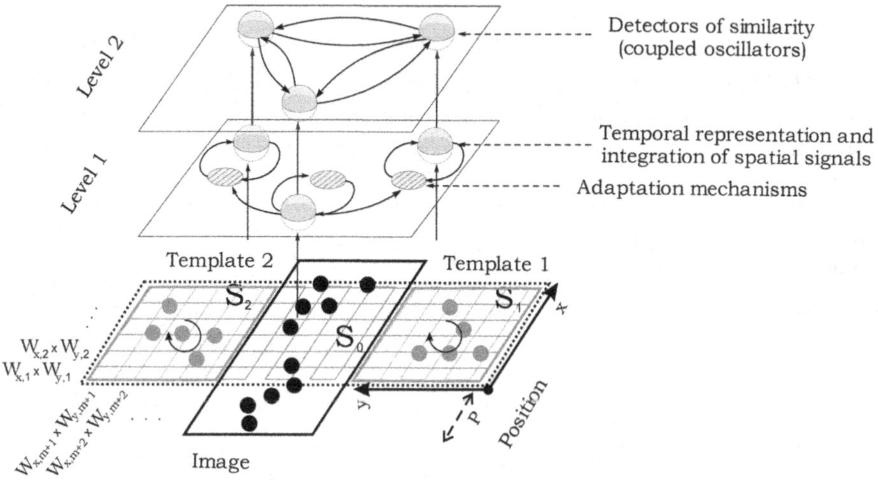

Figure 7.8 A diagram of the template-matching system detecting multiple objects in an image. The image is viewed through a frame (depicted as a dotted rectangle) generating a temporal representation of the corresponding portion of the image in accordance with (7.102) and (7.103). This temporal representation is provided to the template subsystems, which search for the values of parameters θ_1 and θ_2 (level 1). The results are passed on to the subsystem detecting coincidence between temporal representations of the templates and the portion of the image accessible through the frame.

that govern this motion can be defined as

$$\begin{aligned} p &= p_{\max} \frac{\lambda_{0,1} + 1}{2}, \\ \dot{\lambda}_{0,1} &= \gamma_0 \lambda_{0,2} \min_{i \in \{1,2\}} \{\|\phi_0 - \phi_i\|_\varepsilon\}, \\ \dot{\lambda}_{0,2} &= -\gamma_0 \lambda_{0,1} \min_{i \in \{1,2\}} \{\|\phi_0 - \phi_i\|_\varepsilon\}, \end{aligned} \tag{7.104}$$

where $p_{\max}$ determines the range of scanning. The value of γ_0 needs to be chosen small enough and is set to be rationally independent of the value of γ_2 to assure that the conditions of Theorem 7.1 apply. This ensures that every combination of rotation angles and positions of the templates in the image will be visited during the process of searching.

If we were to detect just one template in the image then adding these new equations (7.102)–(7.104) to the previous system would suffice; it would then behave similarly to Figure 7.7. However, if all patterns need to be recovered, a slight modification of the adaptation algorithms is required.

In systems with weakly attracting sets connected by homoclinic trajectories (see e.g. Figure 7.7, the second row) intermittent switching between the attractors can

Table 7.5. *Equations of the system for rotation, position, and brightness-invariant template matching in images with multiple objects; N is the total number of templates*

Function	Image	Template
Temporal integration	$\dot{\phi}_0 = -\frac{1}{\tau}\phi_0 + k\cdot\theta_1 f(t, \theta_2, p)$	$\dot{\phi}_i = -\frac{1}{\tau}\phi_i + k\cdot\hat{\theta}_{1,i} f(t, \hat{\theta}_{2,i}, 0)$
Adaptation to brightness	No	$\hat{\theta}_{1,i} = (\phi_0 - \phi_i)\gamma_1 + \lambda_{i,1}$ $\dot{\lambda}_{i,1} = \frac{\gamma_1}{\tau}(\phi_0 - \phi_i)$
Adaptation to rotation	No	$\hat{\theta}_{2,i} = (\lambda_{2,i}(t) + 1)\pi$ $\dot{\lambda}_{i,2} = \gamma_2\lambda_{i,3}\|\phi_0 - \phi_i\|_\varepsilon$ $\dot{\lambda}_{i,3} = -\gamma_2\lambda_{i,2}\|\phi_0 - \phi_i\|_\varepsilon$
Adaptation position	$p = p_{\max}\frac{\lambda_{0,1}+1}{2}$ $\dot{\lambda}_{0,1} = \gamma_0\lambda_{0,2}e$ $\dot{\lambda}_{0,2} = -\gamma_0\lambda_{0,1}e$ $e = \min_{i\in\{1,2\}}\{\|\phi_0 - \phi_i\|_\varepsilon\} + \delta$	No
Detectors of similarity	$\dot{x}_0 = -ax_0^3 + bx_0^2 + y_0 - z_0 + \gamma u_0 + I$ $\dot{y}_0 = c - dx_0^2 - y_0$ $\dot{z}_0 = \varepsilon(s(x_0 + \bar{x}_0) - z_0)$	$\dot{x}_i = -ax_i^3 + bx_i^2 + y_i - z_i + \gamma u_i + \phi_i(t) - \phi_0(t) + I$ $\dot{y}_i = c - dx_i^2 - y_i$ $\dot{z}_i = \varepsilon(s(x_i + \bar{x}_0) - z_i)$
Coupling function	$u_0 = -(N+1)x_0 + \sum_{j\neq 0} x_j$	$u_i = -(N+1)x_i + \sum_{j\neq i} x_j$
Parameters (additional or having different values from those provided in the previous examples)	Detectors of similarity: $\gamma = 0.4$, $I = 2.2$ Position adaptation subsystem: $\gamma_0 = 0.005/\pi$, $\delta = 0.05$, $p_{\max} = 101$	Rotation adaptation subsystem: $\varepsilon = 0.2$

be achieved by arbitrarily small perturbations applied to the solutions. In our case we implement these perturbations by adding a small positive constant δ to the error variables $\|\phi_i(t) - \phi_0(t)\|_\varepsilon$. This is reflected in the adaptation equations of Table 7.5.

We simulated this simple template-matching system, taking as an input image a combination of two rotated and shifted Garner patterns (Figure 7.9, on the right).

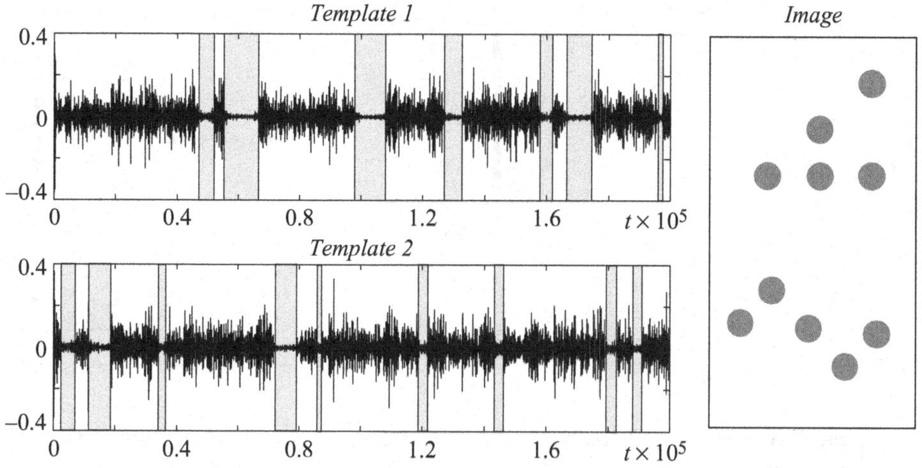

Figure 7.9 On the left: trajectories $x_0(t) - x_1(t)$ (top) and $x_0(t) - x_2(t)$ (bottom) as functions of time. The intervals of time corresponding to synchronized parts of the trajectories $x_0(t)$, $x_1(t)$, and $x_2(t)$ are marked by gray rectangles. Synchronization between $x_0(t)$ and $x_1(t)$ corresponds to detection of Template 1 (Garner pattern with no symmetry); and synchronization between $x_0(t)$ and $x_2(t)$ indicates detection of Template 2 (Garner pattern with symmetry). The system switches from one synchronized state to another, revealing the complete set of patterns present in the image. On the right: the input image presented to the system.

The system persistently reports the presence of two patterns (Figure 7.9, on the left), and successfully estimates their positions and rotation angles. Figure 7.10 shows the accuracy of estimation and demonstrates how much time the system spends in the states corresponding to successful recognition. The latter amounts to about 50 percent of the total time spent; the rest is due to transients. These values can be controlled by parameter $\delta > 0$: the smaller the value of δ the more time the system will spend in the state of successful recognition.

An additional observation can be made about this model: in both template subsystems there are domains (e.g. gray areas on the right in the left-hand picture, and the areas on the left in the right-hand picture) that are visited relatively often despite the absence of templates in these regions. These are phantom states induced by the concurrent presence of several template subsystems. These phantom states occur because the value of p is shared between the template subsystems. If Template 1 is detected then the value of $\min_i \|\phi_i(t) - \phi_0(t)\|$ is small, and position p is not changing much. Because the value of p is shared, the second template subsystem is forced to search for Template 2 in the same position. This property is reflected in the picture on the right.

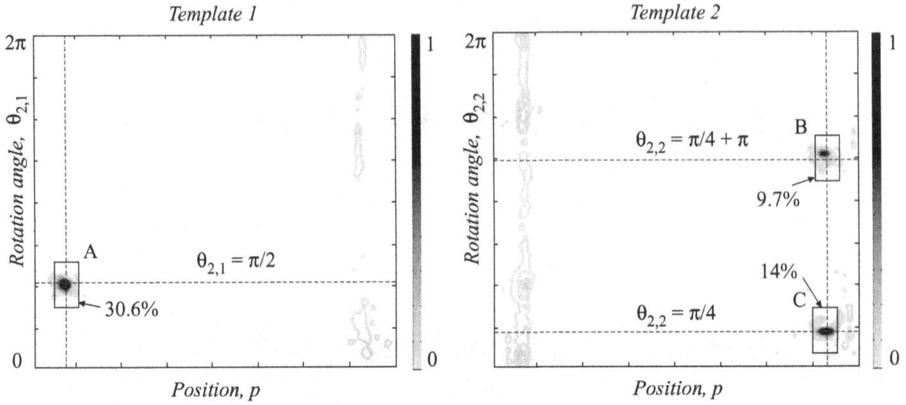

Figure 7.10 Normalized contour plots of the frequencies with which trajectories $(\hat{\theta}_{2,i}(t), p(t))$ explore the parameter space of the modeled perturbations (rotation and translation). The left panel corresponds to the Garner pattern with no symmetries (Template 1), and the right panel corresponds to the pattern with one symmetry (Template 2). As these pictures demonstrate, the system spends most of the time in close proximity to the true values of θ_2 and p. In the case of Template 2 the most visited set consists of two separated domains corresponding to two values of θ_2 with identical temporal representation due to the symmetry.

The effect of differences in time scales

As we mentioned earlier, the difference in time scales between adaptation to linearly and nonlinearly parametrized uncertainties is likely to affect recognition performance. In particular, our sufficient conditions suggest that the accuracy is lower when adaptation to linearly parametrized uncertainties is faster than adaptation to nonlinearly parametrized ones. Here we check computationally whether this prediction holds for the template-matching system considered in Section 7.3.1.

In order to avoid the potential influence of image complexity, the Garner pattern without symmetries was used. We computed system solutions starting from identical initial conditions but for different values of $\gamma_1 \in [0.005, 0.1]$. In total 200 equally spaced points in this interval were tested. The value of ε, regulating the accuracy of inferring the true value of the rotation angle, was set to $\varepsilon = 0.01$. All other parameters were kept identical to those of the setup described in Section 7.3.1.

For every γ_1 we calculated the amount of time $T(\gamma_1)$ needed for the estimate of the rotation angle, $\hat{\theta}_{2,1}$, to converge into a 5-percent neighborhood of its true value, $\pi/4$. This value was chosen as an estimate of the convergence time. The results of these simulations are summarized in Figure 7.11. As we observe, very small values of γ_1, chosen in the interval [0.005, 0.01], result in large convergence times. When the value of γ_1 exceeds a certain threshold, $\gamma_c = 0.02$, the convergence times $T(\gamma_1)$ reduce substantially and remain relatively constant, with slight fluctuations, for all

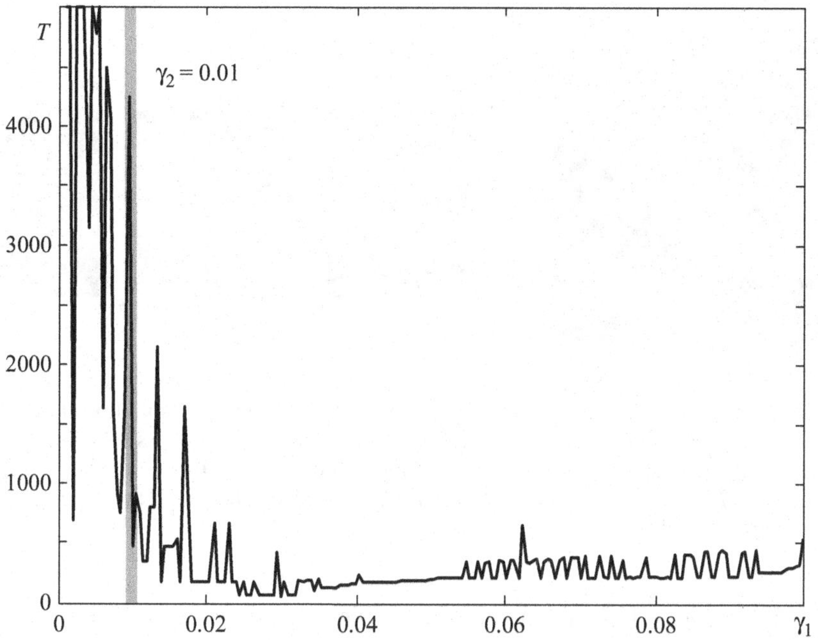

Figure 7.11 The convergence time T as a function of γ_1. The vertical gray line delimits the domain of γ_1 into two regions, $\gamma_1 \geq \gamma_2$ and $\gamma_1 < \gamma_2$.

$\gamma_1 \in [\gamma_c, 0.1]$. Notice that γ_c is twice the value of γ_2 corresponding to the time scale of adaptation to rotation. This supports our initial hypothesis that separate time scales of adaptation may be advantageous for the system's performance.

7.3.2 Tracking disturbances in scanning microscopes

We next consider the application of a template-matching system with weakly attracting sets to a problem of realistic complexity. We applied our approach to the problem of tracking morphological changes in dendritic spines on the basis of measurements received from a multiphoton scanning microscope *in vitro*. A distinctive property of laser microscopy is that in order to "see" an object one needs, first, to inject it with a photo-sensitive dye (fluorophore). The particles of the fluorophore emit photons of light under external stimulation, thus illuminating an object from inside the tissue. Typical data from a two-photon microscope are provided in Figure 7.12.[7]

We addressed the problem of how to register fast dynamical changes in spine geometry after application of chemical stimulation. The measurements were

[7] The images were provided by S. Grebenyuk, of the group working on neuronal circuit mechanisms, RIKEN BSI.

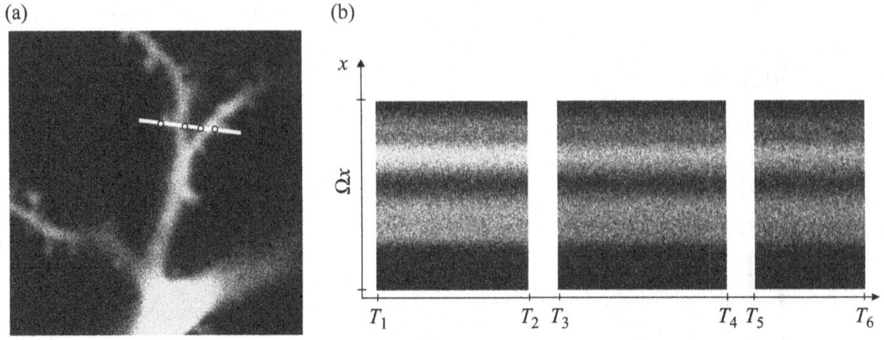

Figure 7.12 Typical images from the two-photon microscope. (a) A dendrite; the domain of scanning (white line) is in the vicinity of two spines (small protrusions on the dendrite). The size of the domain is 5.95 μ and the speed of scanning, v_s, is one pixel every 2 μs. (b) Results of scanning as a function of time at the beginning (interval $[T_1, T_2]$), in the middle (domain $[T_3, T_4]$), and at the end of the experiment (domain $[T_5, T_6]$).

performed on slices. Here the need for unstable convergence is motivated by *non-linearly and non-convexly parametrized models* of uncertainty. Measurements of this kind suffer from effects of photobleaching and diffusion of the dye (see Figure 7.12), and dependence of the scattering of the emitted light on the a-priori-unknown position of the object in the slice. On-line estimation and tracking of the effects of photobleaching (intensity) and changes of the object position (blur) in the slice are therefore necessary.

The measured signal is already a temporal sequence, which fits nicely with our approach. An inherent feature of scanning microscopy is that the object is measured using a sequence of scans along one-dimensional domains (see Figure 7.12(a)). Hence, the objects in this case are one-dimensional mappings, and the domain Ω_x is an interval $\Omega_x = [x_{\min}, x_{\max}]$. For the particular images we set $x_{\min} = 1$ and $x_{\max} = 176$, which corresponds to a scanning line of 176 pixels and 5.95 μm. In order to reduce measurement noise we consider the averaged data in the scanning line over n successive subsequent trials.

The measured image, S_0, was chosen to be the averaged data along the scanning line over n successive subsequent trials. The template, S_1, substituted the averaged measurements of the object at the initial time T_1.

Samples of data used to generate S_1 are provided in Figure 7.13(a). These correspond to the intensity of the radiation emitted from the object in the red part of the spectrum for the data shown in the first fragment of Figure 7.12(b). The measured objects, S_0, are the averaged samples of data at the time instants $T_i \neq T_1$ (proportional to T_s). Focal distortions were simulated using conventional filters from Photoshop applied to S_1. These fragments are shown in Figures 7.13(b) and (c).

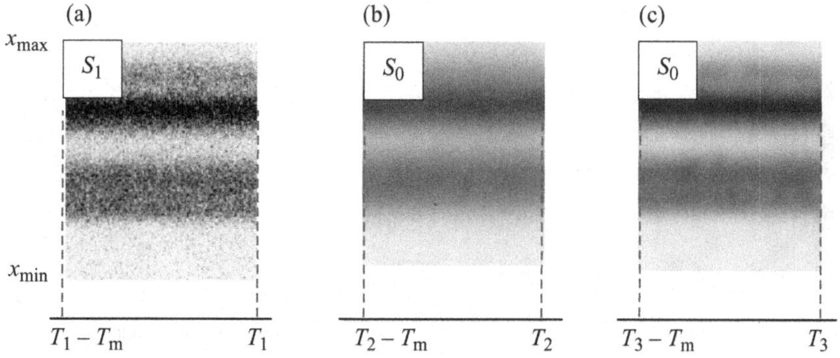

Figure 7.13 The data used to generate the template, S_1 (a), and perturbed measurements S_0 at time instants T_2 (b) and T_3 (c).

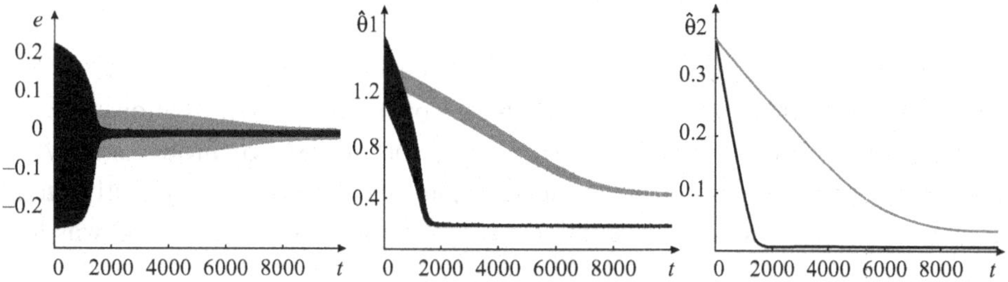

Figure 7.14 The trajectories $e(t)$, $\hat{\theta}_1(t)$, and $\hat{\theta}_2(t)$ as functions of time. Black lines and regions correspond to measurements in Figure 7.13(b). Gray lines and regions correspond to the data in Figure 7.13(c).

The sources of perturbation are the effects of photobleaching (affecting brightness) and deviations in the object position in the slice (affecting scattering and leading to blurred images). Therefore the following model of uncertainty was used:

$$\theta_1 f_1(x, \theta_2, t) = \theta_1 \int_{\Omega_x} e^{-\theta_2(\xi - x(t))^2} S_1(\xi) d\xi, \tag{7.105}$$

where $x(t)$, the scanning trajectory in (7.105), is defined as

$$x(t) = \begin{cases} x_{\min} + k_s \cdot t, & t \leq x_{\max} - x_{\min}, \\ x(t - (x_{\max} - x_{\min})), & t > x_{\max} - x_{\min}, \end{cases} \quad k_s = 1.$$

Figures 7.14 and 7.15 show the performance of the system (7.16), (7.22), (7.25), and (7.28) in tracking focal/brightness perturbations for two measurements S_0. Figure 7.14 shows the tracking of unknown modeled perturbations in the images. Figure 7.15 shows the synchronization errors of the detection subsystem. The symbol t_{syn} denotes the "synchronization time" spent in the vicinity of the invariant

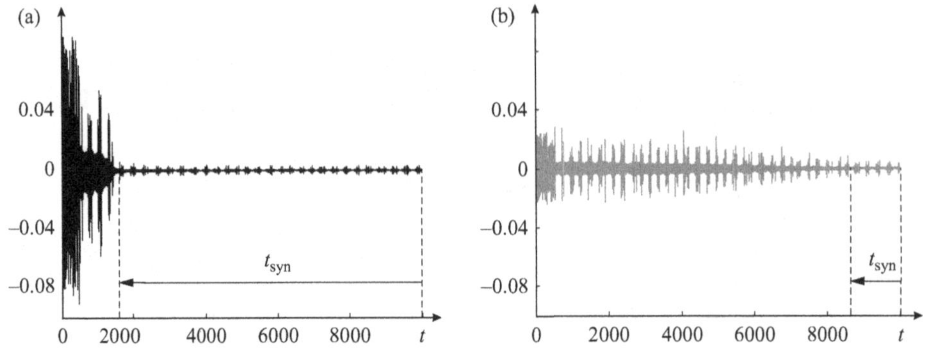

Figure 7.15 Plots of the synchronization error $x_0(t) - x_1(t)$ as a function of time: (a) corresponds to the data depicted in Figure 7.13(b) and (b) corresponds to the measurements shown in Figure 7.13(c).

synchronization manifold. As can be seen from both these figures, the system successfully tracks/reconstructs the estimates of unknown perturbations applied to the object (Figure 7.14). Coincidence detectors report synchrony only when the error between the profiles of the template and object is sufficiently small (Figure 7.15). The actual time required for recognition on a standard PC was less than 5 s.

7.4 Summary

In this chapter we illustrated how the theory and methods from the second part of the book can be used for deriving a solution in principle to the problem of invariant template matching on the basis of temporal coding of spatial information. Our analysis of the problem shows that the traditional requirement of stable dynamics in the sense of Lyapunov need not necessarily be compatible with the desired behavior of the system. As a substitute we proposed the concept of Milnor attracting sets, and at the level of implementation we provided systems in which such attractors emerge as a result of external stimulation. The results provided are normative in the sense that we require a minimal number of additional variables and consider as simple structures as possible.

Even though the proposed systems stem from theoretical considerations, they capture qualitatively a wide range of phenomena observed in biological visual perception. These phenomena include multiple time scales for different modalities during adaptation (Wolfson and Graham 2000; Webster *et al.* 2002; Smith *et al.* 2006), switching and perceptual multi-stability (Attneave 1971; Leopold and

Logothesis 1999), perceptual ambiguity (van Leeuwen 1990; Hatfield and Epstein 1985), empirical observations in mental rotation (Lachmann and van Leeuwen 2005) and decision-time distributions (Gilden 2001). This motivates our belief that the present results may contribute to the further understanding of visual perception in biological systems, including humans.

8

State and parameter estimation of neural oscillators

In this chapter we show how our approach for developing adaptation algorithms in dynamical systems can be applied to the problem of quantitative modeling of individual neural cells. The available measurements are restricted to the values of input currents, as a function of time, and evoked potentials measured at a point on the surface of the cell's membrane.

Most of the available models of individual biological neurons consist of systems of ordinary differential equations describing the cell's response to stimulation; their parameters characterize variables such as time constants, conductances, and response thresholds, which are important for relating the model responses to the behavior of biological cells. Typically two general classes of models co-exist: phenomenological and mathematical ones. Models of the first class, such as e.g. the Hodgkin–Huxley equations, claim biological plausibility, whereas models of the second class are more abstract mathematical reductions without explicit relation of all of their variables to physical quantities such as conductances and ionic currents (see Table 8.1).

Despite these differences, these models admit a common general description, which will be referred to as *canonic*. In particular, the dynamics of a typical neuron are governed by the following set of equations:

$$\begin{aligned} \dot{v} &= \sum\nolimits_j \varphi_j(v,t)p_j(r)\theta_j + I(t), \\ \dot{r}_i &= -a_i(v,\theta,t)r_i + b_i(v,\theta,t), \\ \theta &= (\theta_1,\theta_2,\ldots), \end{aligned} \tag{8.1}$$

in which the variable v is the membrane potential, and r_i are the gating variables of which the values are not available for direct observation. The functions $\varphi_j(\cdot,\cdot)$, $p_j(\cdot) \in \mathcal{C}^1$ are assumed to be known; they model components of specific ionic conductances. The functions $a_i(\cdot)$, $b_i(\cdot) \in \mathcal{C}^1$ are also known, yet they depend

Table 8.1. *Examples of typical mathematical models of spiking single neurons, namely biologically plausible equations of membrane-potential generation in a giant axon of a squid (Hodgkin–Huxley model) and a reduction of these equations to an oscillator with polynomial right-hand side (Hindmarsh–Rose model)*

Hodgkin–Huxley model	Hindmarsh–Rose model
$\dot{v} = I(t) - (\theta_1 m^3 h(v+\theta_2) + \theta_3 n^4 (v+\theta_4) + \theta_5 v + \theta_6)$ $\dot{m} = (1-m)\varphi\left(\frac{v+\theta_7}{\theta_8}\right) - m\theta_9 \exp\left(\frac{v}{\theta_{10}}\right)$ $\dot{n} = (1-n)\theta_{11}\varphi\left(\frac{v+\theta_{12}}{\theta_{13}}\right) - n\theta_{14} \exp\left(\frac{v}{\theta_{15}}\right)$ $\dot{h} = (1-h)\theta_{16} \exp\left(\frac{v}{\theta_{17}}\right) - h\Big/\left(1+\exp\left(\frac{v+\theta_{18}}{\theta_{19}}\right)\right)$ $\varphi(x) = x/(\exp x - 1)$, $\theta_1, \ldots, \theta_{19}$ – parameters, $I : \mathbb{R} \to \mathbb{R}$ – input current	$\dot{v} = \theta_1(\theta_2 r - f(v)) + I(t)$ $\dot{r} = \theta_3(g(v) - \theta_4 r)$ $f(v)$ and $g(v)$ are polynomials: $f(v) = \theta_5 + \theta_6 v + \theta_7 v^2 + \theta_8 v^3$ $g(v) = \theta_9 + \theta_{10} v + \theta_{11} v^2$ $I : \mathbb{R} \to \mathbb{R}$ is the external input, $\theta_1, \ldots, \theta_{11}$ – parameters

on the unknown parameter vector θ. System (8.1) is a typical conductance-based description of the evoked potential generation in neural membranes (Izhikevich 2007). It is also an obvious generalization of many purely mathematical models of spike generation such as the FitzHugh–Nagumo (FitzHugh 1961) and Hindmarsh–Rose (Hindmarsh and Rose 1984) equations. In this sense system (8.1) represents typical building blocks in the modeling literature.

In order to be able to model the behavior of large numbers of individual cells of which the input–output responses are described by (8.1), computational tools for automated fitting of models of neurons to data are needed. These tools are the algorithms for state and parameter reconstruction of (8.1) from the available measurements of $v(t)$ and $I(t)$ over time.

Fitting parameters of nonlinear ordinary differential equations to data is recognized as a hard computational problem (Brewer *et al.* 2008) that "has not been yet treated in full generality" (Ljung 2008). Within the field of neuroscience, conventional methods for fitting parameters of model neurons to measured data are often restricted to hand-tuning or exhaustive trial-and-error searching in the space of model parameters (Prinz *et al.* 2003). Even though these strategies allow careful and detailed exploration in the space of parameters, they suffer from the same problem – the curse of dimensionality.

In available alternatives, recognizing the obvious nonlinearity of the original problem, it is proposed that one reformulate the original estimation problem as that of searching for the parameters of a system of difference equations approximating solutions of (8.1) (Abarbanel *et al.* 2009); or predominantly search-based optimization heuristics (see van Geit *et al.* (2008) for a detailed review) are advocated as the main tool for automated fitting of neural models. Straightforward exhaustive-search approaches, however, are limited to varying only a few model parameters over sparse grids, e.g. as in Prinz *et al.* (2003), where eight parameters were split into six bands. The coarseness of this parametrization leads to non-uniqueness of signal representation, leaving room for uncertainty and inability to distinguish between subtle changes in the cell. Finer-grained search algorithms are currently not feasible, technically speaking. Other heuristics, such as evolutionary algorithms, are examined in Achard and de Schutter (2006). According to Achard and de Schutter (2006), replacing exhaustive search with evolutionary algorithms allows one to increase the number of varying parameters to 24. Yet, the computational complexity of the problem still limits the search to sparse grids (six bands per single parameter) and requires days of simulation by a cluster of ten Apple 2.3-GHz nodes. Furthermore, because all these strategies are heuristic, the accuracy of the final results is not guaranteed.

Here we take advantage of results presented in Chapters 4 and 5, and provide a feasible substitute for heuristic strategies for automatic reconstruction of the state and parameters of canonic neural models (8.1). Generally, we will follow the line of our earlier works (Fairhurst *et al.* 2010; Tyukin *et al.* 2008a, 2010). In particular, to develop computationally efficient procedures for state and parameter reconstruction of (8.1) we propose to exploit the wealth of existing system-identification and estimation approaches from the domain of control theory. These approaches are based on the system-theoretic concepts of observability and identifiability (Nijmeijer and van der Schaft 1990; Isidori 1989; Ljung 1999), and the notions of Lyapunov stability and weakly attracting sets (see Chapter 2). The advantage of using these approaches is that an abundance of algorithms (observers) has already been developed within the domain of control. These algorithms guarantee asymptotic and stable reconstruction of unmeasured quantities from the available observations, provided that the system equations are in an *adaptive-observer canonical form*. Moreover, this reconstruction can be made exponentially fast without any need for substantial computational recourses. We examine whether system (8.1) is at all observable with respect to the output v, that is, whether its state and parameters can be reconstructed from observations of v. We present and analyze typical algorithms (adaptive observers) that are available in the literature. We show that for a large class of mathematical models of neural oscillators at least a part of the model parameters can be reconstructed exponentially fast.

In order to deal with more general classes of models and also to recover the rest of the model parameters we exploit a novel observer scheme presented in Chapter 5. This scheme benefits from (1) the efficiency of uniformly converging estimation procedures (stable observers), (2) the success of explorative search strategies in global optimization, by allowing unstable convergence along dense trajectories, and (3) the power of qualitative analysis of dynamical systems. We present a general description of this observer and list its asymptotic properties. The performance of these algorithms is demonstrated with examples.

8.1 Observer-based approaches to the problem of state and parameter estimation

Let us consider the following class of dynamical systems:

$$\begin{aligned} &\dot{\mathbf{x}} = \mathbf{f}(\mathbf{x}, \boldsymbol{\theta}) + \mathbf{g}(\mathbf{x}, \boldsymbol{\theta})u(t), \quad \mathbf{x}(t_0) \in \Omega_x \subset \mathbb{R}^n, \\ &y = h(\mathbf{x}), \quad \mathbf{x} \in \mathbb{R}^n, \quad \boldsymbol{\theta} \in \mathbb{R}^d, \quad y \in \mathbb{R}, \end{aligned} \tag{8.2}$$

where $\mathbf{f}, \mathbf{g} : \mathbb{R}^n \times \mathbb{R}^m \to \mathbb{R}^n$ and $h : \mathbb{R}^n \to \mathbb{R}$ are smooth functions, and $u : \mathbb{R} \to \mathbb{R}$. The variable x stands for the state vector, $u \in \mathcal{U} \subset C^1[t_0, \infty)$ is the known input, $\boldsymbol{\theta} \in \mathbb{R}^m$ is the vector of unknown parameters, and y is the output of (8.2). System (8.2) includes equations (8.1) as a subclass and in this respect can be considered a plausible generalization. Obviously, conclusions about (8.2) should be valid for system (8.1) as well.

Given that the right-hand side of (8.2) is differentiable, for any $\mathbf{x}' \in \Omega_x$, $u \in \mathcal{C}^1[t_0, \infty)$ there exists a time interval $\mathcal{T} = [t_0, t_1]$, $t_1 > t_0$ such that a solution $\mathbf{x}(t, \mathbf{x}')$ of (8.2) passing through $\mathbf{x}'$ at t_0 exists for all $t \in \mathcal{T}$. Hence, $y(t) = h(\mathbf{x}(t))$ is defined for all $t \in \mathcal{T}$. For the sake of convenience we will assume that the interval $\mathcal{T}$ of the solutions is large enough or even coincides with $[t_0, \infty)$ when necessary.

We are interested in finding an answer to the following question: suppose that we are able to measure the values of $y(t)$ and $u(t)$ precisely; then, can the values of $\mathbf{x}'$ and the parameter vector $\boldsymbol{\theta}$ be recovered from the observations of $y(t)$ and $u(t)$ over a finite subinterval of $\mathcal{T}$, and, if so, how? A natural framework within which to answer to these questions is offered by the concept of *observability* (Nijmeijer and van der Schaft 1990).

Definition 8.1.1 Two states $\mathbf{x}_1, \mathbf{x}_2 \in \mathbb{R}^n$ are said to be indistinguishable (denoted by $\mathbf{x}_1 \mathcal{I} \mathbf{x}_2$) for (8.2) if for every admissible input function u the output function $t \mapsto y(t, 0, \mathbf{x}_1, u)$, $t \geq 0$ of the system for initial state $\mathbf{x}(0) = \mathbf{x}_1$, and the output function $t \mapsto y(t, 0, \mathbf{x}_2, u)$, $t \geq 0$ of the system for initial state $\mathbf{x}(0) = \mathbf{x}_2$, are identical on their common domain of definition. The system is called observable if $\mathbf{x}_1 \mathcal{I} \mathbf{x}_2$ implies $\mathbf{x}_1 = \mathbf{x}_2$.

According to Definition 8.1.1, observability of a dynamical system implies that the values of its state, $\mathbf{x}(t)$, $t \in [t_1, t_2]$ are completely determined by inputs and outputs $u(t)$ and $y(t)$ over $[t_1, t_2]$. Although this definition does not account for any unknown parameter vectors, one can easily see that the very same definition can be used for parametrized systems as well. Indeed, extending the original equations (8.2) by including the parameter vector $\boldsymbol{\theta}$ as a component of the extended state vector $\tilde{\mathbf{x}} = (\mathbf{x}, \boldsymbol{\theta})^{\mathrm{T}}$ results in

$$\begin{aligned} \dot{\mathbf{x}} &= f(\mathbf{x}, \boldsymbol{\theta}) + \mathbf{g}(\mathbf{x}, \boldsymbol{\theta}) u(t), \\ \dot{\boldsymbol{\theta}} &= 0, \\ y &= h(\mathbf{x}), \quad \mathbf{x}(t_0) \in \Omega_x \subset \mathbb{R}^n, \end{aligned} \tag{8.3}$$

or, similarly, in

$$\begin{aligned} \dot{\tilde{\mathbf{x}}} &= \tilde{\mathbf{f}}(\tilde{\mathbf{x}}) + \tilde{\mathbf{g}}(\tilde{\mathbf{x}}) u(t), \\ y &= \tilde{h}(\tilde{\mathbf{x}}), \quad \tilde{\mathbf{x}}(t_0) = (\mathbf{x}(t_0), \boldsymbol{\theta})^{\mathrm{T}} \in \Omega_{\tilde{x}} \subset \mathbb{R}^{n+d}, \end{aligned} \tag{8.4}$$

where $\tilde{\mathbf{f}}(\tilde{\mathbf{x}}) = (\mathbf{f}(\mathbf{x}, \boldsymbol{\theta}), 0)^{\mathrm{T}}$, $\tilde{\mathbf{g}}(\tilde{\mathbf{x}}) = (\mathbf{g}(\mathbf{x}, \boldsymbol{\theta}), 0)^{\mathrm{T}}$, and $\tilde{h}(\tilde{\mathbf{x}}) = (h(\mathbf{x}), 0)$. All of the uncertainties in (8.2) and (8.3), including the parameter vector $\boldsymbol{\theta}$, are now combined into the state vector of (8.4). Hence, the problem of state and parameter reconstruction of (8.2) can be viewed as that of recovering the values of state for (8.4).

Definition 8.1.1 characterizes observability as a global property of a dynamical system. Sometimes, however, global observability of a system in $\mathbb{R}^n$ is not necessarily needed. Instead of asking whether *every* point in the system's state space is distinguishable from any other point it may be sufficient to know whether the system's states are distinguishable in some neighborhood of a given point. This necessitates the notion of local observability (Nijmeijer and van der Schaft 1990).

Let V be an open subset of $\mathbb{R}^n$. Two states $\mathbf{x}_1, \mathbf{x}_2 \in V$ are said to be indistinguishable (denoted by $\mathbf{x}_1 \mathcal{I}^V \mathbf{x}_2$) on V for (8.2) if for every admissible input function $u : [0, T] \to \mathbb{R}$ with the property that the solutions $\mathbf{x}(t, 0, \mathbf{x}_1, u)$ and $\mathbf{x}(t, 0, \mathbf{x}_2, u)$ both remain in V for $t \leq T$ the output function $t \mapsto y(t, 0, \mathbf{x}_1, u)$, $t \geq 0$ of the system for initial state $\mathbf{x}(0) = \mathbf{x}_1$ and the output function $t \mapsto y(t, 0, \mathbf{x}_2, u)$, $t \geq 0$ of the system for initial state $\mathbf{x}(0) = \mathbf{x}_2$ are identical for $0 \leq t \leq T$ on their common domain of definition.

Definition 8.1.2 The system is called locally observable, at $\mathbf{x}_0$ if there exists a neighborhood $W \subset \mathbb{R}^n$ of $\mathbf{x}_0$ such that for every neighborhood $V \subset W$ of $\mathbf{x}_0$ the relation $\mathbf{x}_0 \mathcal{I}^{\mathcal{V}} \mathbf{x}_1$ implies $\mathbf{x}_0 = \mathbf{x}_1$. The system is locally observable if it is observable at each $\mathbf{x}_0$.

Observability tests are available that, given the functions h and $\mathbf{f}$ on the right-hand side of (8.4), indicate whether a given system is observable. Particular formulations of these tests may vary depending on whether e.g. the functions $\mathbf{f}$, and $\mathbf{g}$ are analytic or time-invariant (the inputs are constants).

Here we will restrict our attention to those systems (8.3) in which the inputs $u(t)$ are constants. In this case we can replace the function $u(t)$ with an unknown parameter, and system (8.3) can be viewed as a system of type (8.4) yet without inputs. One of the most common observability tests for this class of autonomous systems[1] is given below (see also Nijmeijer and van der Schaft (1990), Theorem 3.32).

Proposition 8.1 *System (8.4) is locally observable at a point* $\mathbf{x}^0 \in U \subset \mathbb{R}^{n+d}$ *if*

$$\operatorname{rank} \frac{\partial}{\partial \tilde{\mathbf{x}}} \left(\tilde{h}(\tilde{\mathbf{x}}) \quad L_{\tilde{\mathbf{f}}}\tilde{h}(\tilde{\mathbf{x}}) \quad L_{\tilde{\mathbf{f}}}^{2}\tilde{h}(\tilde{\mathbf{x}}) \quad \ldots \quad L_{\tilde{\mathbf{f}}}^{n+d-1}\tilde{h}(\tilde{\mathbf{x}}) \right)^{\mathrm{T}} = n + d,$$

$$\forall\, \tilde{\mathbf{x}} \in U. \tag{8.5}$$

In what follows we shall use the test above to determine whether the models of neural dynamics are at all observable.

8.1.1 Local observability of neural oscillators

We start our observability analysis by applying the local observability test (8.5) to the Hindmarsh–Rose model (in Table 8.1). In order to do so we shall extend the system's state space so that the unknown parameters are the components of the extended state vector. In the case of the Hindmarsh–Rose model this procedure leads to the following extended system of equations:

$$\begin{pmatrix} \dot{x}_1 \\ \dot{x}_2 \\ \dot{\theta}_{13} \\ \dot{\theta}_{12} \\ \dot{\theta}_{11} \\ \dot{\theta}_{10} \\ \dot{\theta}_{22} \\ \dot{\theta}_{21} \\ \dot{\lambda} \end{pmatrix} = \begin{pmatrix} \theta_{13}x_1^3 + \theta_{12}x_1^2 + \theta_{11}x_1 + \theta_{10} + x_2 \\ -\lambda x_2 + \theta_{22}x_1^2 + \theta_{21}x_1 \\ 0 \\ 0 \\ 0 \\ 0 \\ 0 \\ 0 \\ 0 \end{pmatrix} \tag{8.6}$$

[1] Given that the functions u are known and constant, for the sake of notational convenience, system (8.4) can be considered autonomous.

To test whether there are points of local observability of system (8.6) it suffices to find a point in the state space of (8.6) at which the rank condition (8.5) holds. Here we computed the determinant,

$$D(\tilde{\mathbf{x}}) = \frac{\partial}{\partial \tilde{\mathbf{x}}} \Big(h(\tilde{\mathbf{x}}) \quad L_{\mathbf{f}} h(\tilde{\mathbf{x}}) \quad L_{\mathbf{f}}^2 h(\tilde{\mathbf{x}}) \quad \ldots \quad L_{\mathbf{f}}^{n-1} h(\tilde{\mathbf{x}}) \Big)^{\mathrm{T}},$$
$$\tilde{\mathbf{x}} = (x_1, x_2, \theta_{13}, \theta_{12}, \dot{\theta}_{11}, \theta_{10}, \theta_{22}, \theta_{21}, \lambda)^{\mathrm{T}}, \tag{8.7}$$

on a sparse grid (of 101×101 pixels) and plotted those regions for which the determinant is less than a certain value, δ. The neuron parameters were set to $L = -2, \theta_{13} = -10, \theta_{12} = -4, \theta_{11} = 6, \theta_{10} = 1, \theta_{22} = -32$, and $\theta_{21} = -32$. Figure 8.1 shows results (obtained using Maple) for various values of δ. The shaded regions correspond to the domains where $D(\tilde{\mathbf{x}}) < \delta$. According to these results, when the value of δ is made sufficiently small, the condition $D(\tilde{\mathbf{x}}) > \delta$ holds for almost all points in the grid. This suggests that there are domains in which the Hindmarsh–Rose model is indeed at least locally observable.

Let us now consider a more realistic (with respect to biological plausibility) set of equations. One of the simplest models of this type is the Morris–Lecar system (Morris and Lecar 1981):

$$\begin{aligned}
\dot{v}(t) &= -g_{\mathrm{Ca}} \left(\frac{1}{2} + \frac{1}{2} \tanh\left(\frac{v(t)+1}{E_4} \right) (v(t) - E_1) \right) \\
&\quad - g_{\mathrm{K}} w(t)(v(t) - E_2) - g_{\mathrm{m}}(v(t) - E_3), \\
\dot{w}(t) &= \frac{1}{5} \left(\frac{1}{2} + \frac{1}{2} \tanh\left(\frac{v(t)+1}{E_5} \right) - w(t) \right) \cosh\left(\frac{v(t)}{E_6} \right), \\
& E_1 = 100, \quad E_2 = -70, \quad E_3 = -50, \quad E_4 = 15, \quad E_5 = 30, \\
& E_6 = 60, \quad g_{\mathrm{Ca}} = 1.1, \quad g_{\mathrm{K}} = 2.0, \quad g_{\mathrm{m}} = 0.5.
\end{aligned} \tag{8.8}$$

As in the previous example, we extend the system's state space by considering unknown parameters as components of the extended state vector. This extension procedure results in the following set of equations:

$$\begin{pmatrix} \dot{v}(t) \\ \\ \dot{w}(t) \\ \dot{g}_{\mathrm{Ca}}(t) \\ \dot{g}_{\mathrm{K}}(t) \\ \dot{g}_{\mathrm{m}}(t) \\ \dot{\lambda}(t) \end{pmatrix} = \begin{pmatrix} -g_{\mathrm{Ca}} \left[\frac{1}{2} + \frac{1}{2} \tanh((v+1)/15)(v - 100) \right] \\ - g_{\mathrm{k}} w(v + 70) - g_{\mathrm{m}}(v + 50) \\ \frac{1}{5} \left[\frac{1}{2} + \frac{1}{2} \tanh((v+1)/30)) - w \right] \cosh(v/60) \\ 0 \\ 0 \\ 0 \\ 0 \end{pmatrix}. \tag{8.9}$$

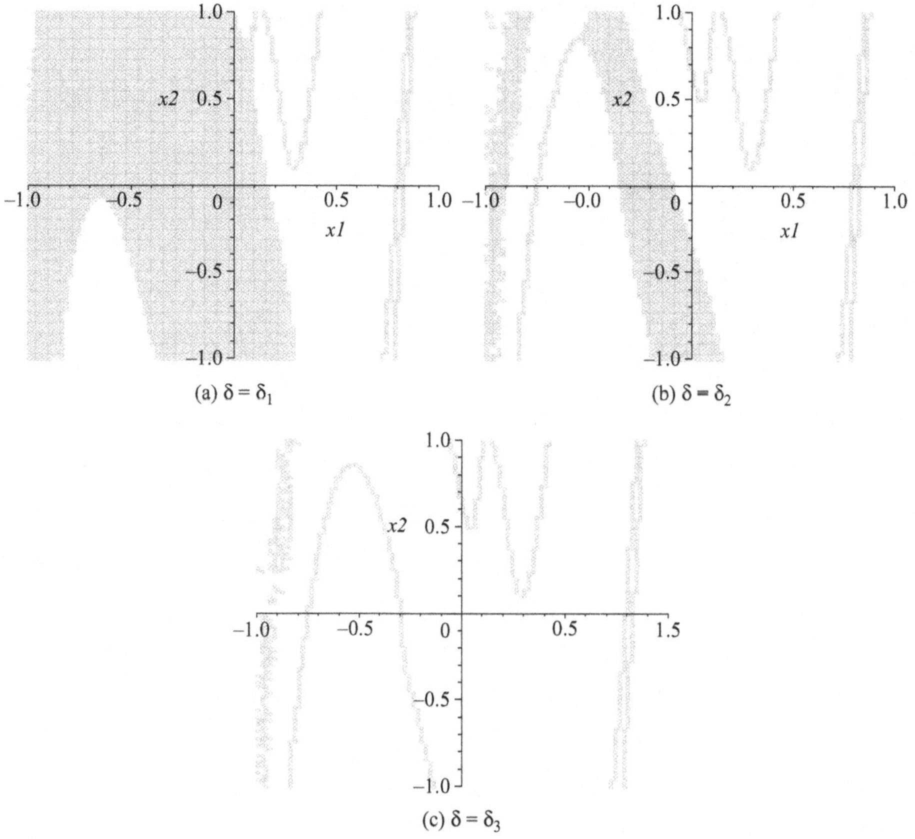

Figure 8.1 Observability tests for a Hindmarsh–Rose-model neuron (see Table 8.1).

For this extended set of equations we estimated the regions where the value of $D(\mathbf{x})$ exceeds some given $\delta > 0$. These regions are presented in Figure 8.2 for various values of δ. These results demonstrate that the Morris–Lecar system (8.8) is also locally observable.

As we have seen above, a fairly wide class of canonical mathematical and conductance-based models of evoked responses in neural membranes satisfy local observability conditions. We may thus expect to be able to solve the reconstruction problem for these models. In fact, as we show below, the reconstruction problem can indeed be resolved efficiently at least for a part of the unmeasured variables of the system. However, before we proceed with detailed description of these reconstruction algorithms, let us first review classes of systems for which solutions to the problem of exponentially fast reconstruction of *all components* of state and parameter vectors are already available in the literature.

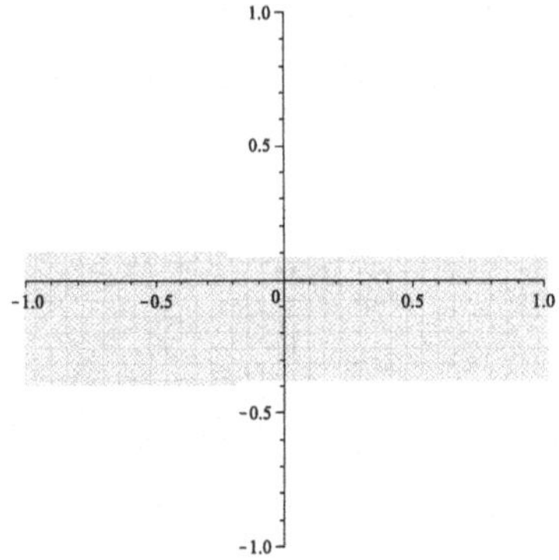

(a) $\delta = 10000$

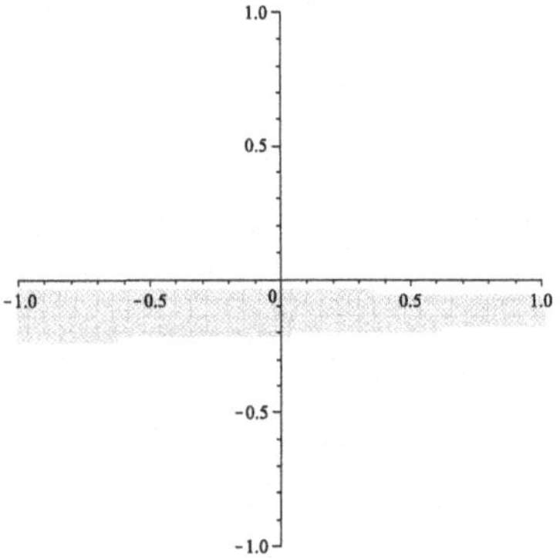

(b) $\delta = 1000$

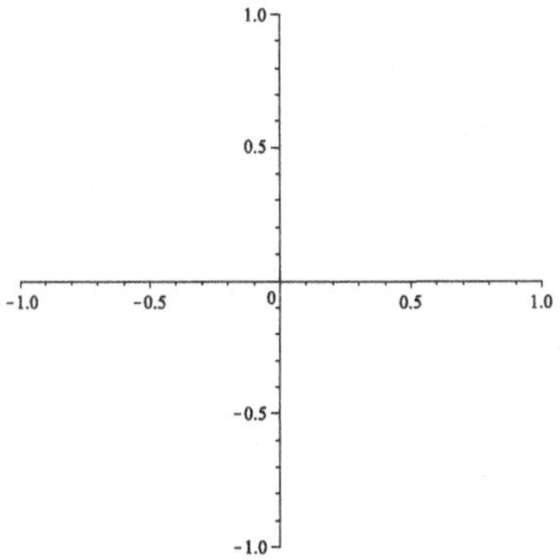

(c) $\delta = 100$

Figure 8.2 Observability tests for a Morris–Lecar-model neuron.

8.1.2 Bastin–Gevers canonical form

We start with a class of systems consisting of a linear time-invariant part of which the equations are known and an additive time-varying component with linear parametrization. The parameters of this time-varying component are assumed to be uncertain. This class of systems was presented in Bastin and Gevers (1988), and its general form is as follows:

$$\dot{\mathbf{x}} = R\mathbf{x} + \Omega(t)\boldsymbol{\theta} + \mathbf{g}(t),$$
$$R = \begin{pmatrix} 0 & \mathbf{k}^{\mathrm{T}} \\ 0 & F \end{pmatrix}, \qquad \Omega(t) = \begin{pmatrix} \Omega_1(t) \\ \overline{\Omega}(t) \end{pmatrix}, \tag{8.10}$$
$$y(t) = x_1(t).$$

In (8.10), $\mathbf{x} \in \mathbb{R}^n$ is the state vector with $y = x_1$ assigned to be the output. $\boldsymbol{\theta} \in \mathbb{R}^p = (\theta_1, \ldots, \theta_p)^{\mathrm{T}}$ is the vector of unknown parameters. R is a known matrix of constants where $\mathbf{k}^{\mathrm{T}} = (k_2, \ldots, k_n)$ and F has dimension $(n-1) \times (n-1)$ with eigenvalues in the open left half plane. $\Omega(t) \in \mathbb{R}^{n \times p}$ is an $n \times p$ matrix of known functions of t; the first row is designated Ω_1 and the remaining $n-1$ rows $\overline{\Omega}$. The vector function $\mathbf{g}(t) : \mathbb{R} \to \mathbb{R}^n$ is known.

Equations (8.10) are often referred to as an adaptive-observer canonical form. This is because, subject to some mild non-degeneracy conditions, it is always possible to reconstruct the vector of unknown parameters $\boldsymbol{\theta}$ and state $\mathbf{x}$ from observations of y over time. Moreover, the reconstruction can be done exponentially fast. Shown below is the adaptive observer presented in Bastin and Gevers (1988). The system to be observed, state estimator, parameter adaption, auxiliary filter, and regressor are given in (8.10), (8.11), (8.12), (8.13), and (8.14), respectively:

$$\dot{\hat{\mathbf{x}}}(\mathbf{t}) = R\hat{\mathbf{x}}(t) + \Omega(t)\hat{\boldsymbol{\theta}} + \mathbf{g}(t) + \begin{pmatrix} c_1\tilde{y} \\ V(t)\dot{\hat{\boldsymbol{\theta}}} \end{pmatrix}, \tag{8.11}$$

$$\dot{\hat{\boldsymbol{\theta}}}(t) = \Gamma\boldsymbol{\varphi}(t)\tilde{y}(t), \tag{8.12}$$

$$\dot{V}(t) = FV(t) + \overline{\Omega}(t), \quad V(0) = 0, \tag{8.13}$$

$$\boldsymbol{\varphi}(t) = V^{\mathrm{T}}(t)\mathbf{k} + \Omega_1^{\mathrm{T}}(t). \tag{8.14}$$

The output is $y = x_1$, its estimate is $\hat{y} = \hat{x}_1$ and its error is $\tilde{y} = y - \hat{y}$. This observer contains some parameters of its own, which are at the design's disposal. $\Gamma = \Gamma^{\mathrm{T}}$ is an arbitrary positive definite matrix, normally chosen as $\Gamma = \mathrm{diag}(\gamma_1, \gamma_2, \ldots, \gamma_p)$, $\gamma_i > 0$, $c_1 > 0$. The auxiliary filter $V(t)$ is an $(n-1) \times p$ matrix and $\boldsymbol{\varphi}(t)$ is a p vector.

Using the transformation

$$\tilde{\mathbf{x}}^* = \tilde{\mathbf{x}} - \begin{pmatrix} 0 \\ V\tilde{\boldsymbol{\theta}} \end{pmatrix}, \quad \tilde{\mathbf{x}} = \mathbf{x} - \hat{\mathbf{x}}, \tag{8.15}$$

the following error system is obtained:

$$\dot{\tilde{\mathbf{x}}}^* = \begin{pmatrix} -c_1 & \mathbf{k}^{\mathrm{T}} \\ 0 & F \end{pmatrix}\tilde{\mathbf{x}}^* + \begin{pmatrix} \boldsymbol{\varphi}^{\mathrm{T}}\tilde{\boldsymbol{\theta}} \\ 0 \end{pmatrix},$$

$$\dot{\tilde{\boldsymbol{\theta}}} = -\Gamma\boldsymbol{\varphi}\tilde{x}_1^*. \tag{8.16}$$

It is shown in Bastin and Gevers (1988), for constant unknown parameters, that the solution $\mathbf{x}(t) = \hat{\mathbf{x}}(t)$, $\boldsymbol{\theta} = \hat{\boldsymbol{\theta}}(t)$ of the extended system (8.11)–(8.14) is globally exponentially stable, provided that the regressor vector $\boldsymbol{\varphi}(t)$ is bounded for all $t \geq 0$, and is persistently exciting (see Definitions 2.5.1 and 5.6.1). Formally, the asymptotic properties of the observer (8.11) and (8.12) are specified in the theorem below (Bastin and Gevers 1988).[2]

Theorem 8.1 *Suppose that*

(1) $c_1 > 0$, and F is a Hurwitz matrix, that is its eigenvalues belong to the left half of the complex plane;
(2) the function $\boldsymbol{\varphi}(t)$ is globally bounded in t, and its time-derivative exists and is globally bounded for all $t \geq 0$;
(3) the function $\boldsymbol{\varphi}(t)$ is persistently exciting.

Then the origin of (8.16) is globally exponentially asymptotically stable.

Adaptive-observer canonical form (8.10) applies to systems in which the regressor $\Omega(t)\boldsymbol{\theta}$ does not depend explicitly on the unmeasured components of the state vector. The question, however, is that of when a rather general nonlinear system can be transformed into the proposed canonical form. This question was addressed in Marino and Tomei (1992), in which a modified adaptive-observer canonical form was proposed together with necessary and sufficient conditions describing when a given system can be transformed into such a form via a diffeomorphic coordinate transformation. This canonical form is described below.

8.1.3 Marino–Tomei canonical form

The canonical form presented in Marino and Tomei (1992) is now shown here. The system to be observed (8.17), state estimator (8.18), and parameter adaptation (8.19) are given below:

$$\begin{aligned}
\dot{\mathbf{x}}(t) &= A_1\mathbf{x}(t) + \boldsymbol{\phi}_0(y(t), u(t)) + \mathbf{b}\sum_{i=1}^{p} \beta_i(y(t), u(t))\theta_i, \\
y(t) &= C_1\mathbf{x}(t), \\
A_1 &= \begin{pmatrix} 0 & 1 & 0 & \dots & 0 \\ 0 & 0 & 1 & \dots & 0 \\ \cdot & \cdot & \cdot & \dots & 0 \\ 0 & 0 & 0 & \dots & 1 \\ 0 & 0 & 0 & \dots & 0 \end{pmatrix}, \\
C_1 &= \begin{pmatrix} 1 & 0 & 0 & \dots & 0 \end{pmatrix}, \\
&\mathbf{x}(t) \in \mathbb{R}^n, \qquad y(t) \in \mathbb{R}, \qquad \beta_i(\cdot,\cdot) : \mathbb{R} \times \mathbb{R} \to \mathbb{R}.
\end{aligned} \tag{8.17}$$

[2] Here we provide a slightly reduced formulation of the main statement of Bastin and Gevers (1988) corresponding to the case in which the values of $\boldsymbol{\theta}$ do not change over time.

In (8.17) $\mathbf{x}(t) \in \mathbb{R}^n$ is the state vector with $x_1(t)$ assigned to be the output y. Matrices A_1, are C_1 are in canonical observer form, $\boldsymbol{\theta} \in \mathbb{R}^p = (\theta_1, \cdots, \theta_p)^{\mathrm{T}}$ is the vector of unknown parameters. The functions β_i are known, bounded, and piecewise continuous functions of $y(t)$, and $u(t)$. The column vector $\mathbf{b} \in \mathbb{R}^n$ is assumed to be Hurwitz[3] with $b_1 \neq 0$.

The adaptive observer presented in Marino and Tomei (1992) is

$$\dot{\hat{\mathbf{x}}}(\mathbf{t}) = (A_1 - KC_1)\hat{\mathbf{x}}(t) + \boldsymbol{\phi}_0(y(t), u(t)) + \mathbf{b}\sum_{i=1}^{p} \beta_i(y(t), u(t))\hat{\theta}_i + Ky(t), \tag{8.18}$$

$$\dot{\hat{\boldsymbol{\theta}}} = \Gamma\beta(t)(y - C_1\hat{\mathbf{x}})\mathrm{sign}(b_1), \tag{8.19}$$

$$K = \frac{1}{b_n}(A_1\mathbf{b} + \lambda\mathbf{b}) = (k_1, \cdots, k_n)^{\mathrm{T}}, \tag{8.20}$$

with Γ an arbitrary symmetric positive definite matrix and λ an arbitrary positive real. The $n \times 1$ vector, $\mathbf{b}$, is Hurwitz with $b_1 \neq 0$.

For the more general case in which the vector $\mathbf{b}$ is an arbitrary vector, an observer is presented in Marino and Tomei (1995a).

Theorem 8.2 *Marino (1990). There exists a local change of coordinates, $z = \boldsymbol{\Phi}(x)$, transforming*

$$\dot{\mathbf{x}} = \mathbf{f}(\mathbf{x}) + \sum_{i=1}^{p} \theta_i(t)\mathbf{q}_i(\mathbf{x}), \quad y = x_1,$$
$$\mathbf{x} \in \mathbb{R}^n, \quad y \in \mathbb{R}, \quad \theta_i \in \mathbb{R}, \quad \mathbf{q}_i : \mathbb{R}^n \to \mathbb{R}^n, \quad n \geq 2, \tag{8.21}$$

with $h(\mathbf{x}^{\mathrm{o}}) = 0$ and $(\mathbf{f}, h)$ an observable pair, into the system

$$\dot{\mathbf{z}} = A_1\mathbf{z} + \boldsymbol{\psi}(y) + \sum_{i=1}^{p} \theta_i(t)\boldsymbol{\psi}_i(y), \quad y = C_1\mathbf{z},$$
$$\mathbf{z} \in \mathbb{R}^n, \quad \boldsymbol{\psi}_i : \mathbb{R} \to \mathbb{R}^n \tag{8.22}$$

with A_1 and C_1 in canonical observer form (8.17), if and only if

(i) $[ad_{\mathbf{f}}^i\mathbf{g}, ad_{\mathbf{f}}^j\mathbf{g}] = 0,\ 0 \leq i, j \leq n-1;$
(ii) $[\mathbf{q}_i, ad_{\mathbf{f}}^j\mathbf{g}] = 0,\ 0 \leq j \leq n-2,\ 1 \leq i \leq p;$

[3] We recall that a vector $\mathbf{b} = (b_1, \ldots, b_n)^{\mathrm{T}} \in \mathbb{R}^n$ is Hurwitz if all roots of the corresponding polynomial $b_1 p^{n-1} + \cdots + b_{n-1}p + b_n$ have negative real parts.

where the vector field, $\mathbf{g}(\mathbf{x})$, *is uniquely defined by*

$$\left\langle \frac{\partial}{\partial \mathbf{x}} \begin{pmatrix} h(\mathbf{x}) \\ L_{\mathbf{f}} h(\mathbf{x}) \\ \cdots \\ L_{\mathbf{f}}^{n-1} h(\mathbf{x}) \end{pmatrix}, \mathbf{g}(\mathbf{x}) \right\rangle = \begin{pmatrix} 0 \\ 0 \\ \cdots \\ 1 \end{pmatrix}. \tag{8.23}$$

The proof of this result is constructed along the following lines. Suppose we use the change of coordinates $\mathbf{z} = \mathbf{\Phi}(\mathbf{x})$, then we have

$$\dot{\mathbf{z}} = \frac{\partial \mathbf{\Phi}}{\partial \mathbf{x}} \dot{\mathbf{x}} \tag{8.24}$$

$$= \left(\frac{\partial \mathbf{\Phi}}{\partial \mathbf{x}} \mathbf{f}(\mathbf{x}) \right)_{\mathbf{x}=\mathbf{\Phi}^{-1}(\mathbf{z})} + \frac{\partial \mathbf{\Phi}}{\partial \mathbf{x}} \sum_{i=1}^{p} \theta_i \mathbf{q}_i(\mathbf{x}). \tag{8.25}$$

It is shown in Marino (1990) that, provided we meet the constraint

$$[ad_{\mathbf{f}}^{i}\mathbf{g}, ad_{\mathbf{f}}^{j}\mathbf{g}] = 0, \quad \forall \mathbf{x} \in U, \quad 0 \le i, \quad j \le n-1, \tag{8.26}$$

we can cast the system into the adaptive-observer canonical form

$$\dot{\mathbf{z}} = A_1 \mathbf{z} + \boldsymbol{\psi}(y) + \sum_{i=1}^{p} \frac{\partial \mathbf{\Phi}}{\partial \mathbf{x}} \theta_i \mathbf{q}_i(\mathbf{x}), \quad y = C_1 \mathbf{z} \tag{8.27}$$

with A_1 and C_1 in canonical observer form (8.17). Furthermore, if

$$[\mathbf{q}_i, ad_{\mathbf{f}}^{j}\mathbf{g}] = 0, \quad \forall \mathbf{x} \in U, \quad 0 \le j \le n-2 \tag{8.28}$$

then we can put the system into

$$\dot{\mathbf{z}} = A_1 \mathbf{z} + \boldsymbol{\psi}(y) + \sum_{i=1}^{p} \theta_i \boldsymbol{\psi}_i(y), \quad y = C_1 \mathbf{z}. \tag{8.29}$$

This representation is linear in the unknown variables, $z_1(t), z_2(t), \ldots, z_n(t)$, and $\theta_1(t), \theta_2(t), \ldots, \theta_p(t)$, while it is nonlinear only in the output, $y(t)$, which is available for measurement.

8.2 The feasibility of conventional adaptive-observer canonical forms

Here we consider technical difficulties preventing straightforward application of conventional adaptive observers for solving the state- and parameter-reconstruction problems for typical neural oscillators. We start with the most simple polynomial systems such as the Hindmarsh–Rose equations. We show that even for this relatively simple class of linearly parametrized models the problem of reconstructing all parameters of the system is a difficult theoretical challenge. Whether complete reconstruction is possible depends substantially on what part of the system's right-hand side is corrupted with uncertainties. Although in the most general case reconstruction of all components of the parameter vector by using standard techniques might not be possible, in some special yet relevant cases estimation of *a part* of the model parameters is still achievable in principle.

Let us consider, for example, the problem of fitting the parameters of the conventional Hindmarsh–Rose oscillator to measured data. In particular we wish to be able to model a single spike from the measured train of spikes evoked by a constant current injection. The classical two-dimensional Hindmarsh–Rose model is defined by the following system:

$$\begin{aligned} \dot{x} &= -ax^3 + bx^2 + y + I, \\ \dot{y} &= c - dx^2 - y, \ a = 1, \ b = 3, \ c = 1, \ d = 5, \end{aligned} \tag{8.30}$$

in which $I \in \mathbb{R}$ stands for the stimulation current. Trajectories $x(t)$ of this model are known to be able to reproduce a wide range of typical responses of actual neurons qualitatively. Quantitative modeling, however, requires the availability of a linear transformation of $(x(t), y(t))$ so that the amplitude and the frequency of oscillations $x(t)$ can be made consistent with data.

In what follows we will consider (8.30) subject to the class of transformations

$$\begin{pmatrix} x_1 \\ x_2 \end{pmatrix} = \begin{pmatrix} k_1 & 0 \\ k_2 & k_3 \end{pmatrix} \begin{pmatrix} x \\ y \end{pmatrix} + \begin{pmatrix} p_x \\ p_y \end{pmatrix}, \ k_2 < k_3, \tag{8.31}$$

where $k_i > 0$ and $p_x, p_y \in \mathbb{R}$ are unknown. Transformations (8.31) include stretching and translations as a special case. In addition to (8.31) we will also allow the time constants on the right-hand side of (8.30) to be slowly time-varying. This will allow us to adjust the scaling of the system trajectories with respect to time.

Taking these considerations into account, we obtain the following re-parametrized description of model (8.30):

$$\begin{aligned} \dot{x}_1 &= \sum_{i=0}^{3} x_1^i \theta_{1,i} + x_2, \\ \dot{x}_2 &= -\lambda x_2 + \sum_{i=1}^{3} x_1^i \theta_{2,i}. \end{aligned} \tag{8.32}$$

Alternatively, in vector-matrix notation we obtain

$$\begin{pmatrix} \dot{x}_1 \\ \dot{x}_2 \end{pmatrix} = A(\lambda) \begin{pmatrix} x_1 \\ x_2 \end{pmatrix} + \varphi(x_1)\boldsymbol{\theta}, \tag{8.33}$$

where

$$\begin{aligned} A(\lambda) &= \begin{pmatrix} 0 & 1 \\ 0 & -\lambda \end{pmatrix}, \\ \varphi(x_1) &= \begin{pmatrix} x_1^3 & x_1^2 & x_1 & 1 & 0 & 0 & 0 \\ 0 & 0 & 0 & 0 & x_1^3 & x_1^2 & x_1 \end{pmatrix}, \\ \boldsymbol{\theta} &= (\theta_{1,3}, \theta_{1,2}, \theta_{1,1}, \theta_{1,0}, \theta_{2,3}, \theta_{2,2}, \theta_{2,1}). \end{aligned} \tag{8.34}$$

One of the main obstacles is that the original equations of neural dynamics are not written in any of the canonical forms for which the reconstruction algorithms are available. The question, therefore, is whether there exists an invertible coordinate transformation such that the model equations can be rendered canonical. Below we demonstrate that this is generally not the case if the transformation is parameter-independent. This is formally stated in Section 3.1. However, if we allow our transformation to be both parameter- and time-dependent, a relevant class of models with polynomial right-hand sides can be transformed into one of the canonical forms.

8.2.1 Parameter-independent time-invariant transformations

Let us consider a class of systems that can be described by (8.33). Clearly this system is not in a canonical adaptive-observer form because $A(\lambda)$ depends on the unknown parameter λ explicitly. The question, however, is whether there exists a differentiable coordinate transformation

$$\mathbf{z} = \boldsymbol{\Phi}(\mathbf{x}), \; z_1 = x_1$$

such that in the new coordinates the equations of system (8.33) satisfy one of the canonical descriptions. The answer to this question is negative, and it follows from the slightly more general statement in the theorem below.

Theorem 8.3 *The system*

$$\begin{aligned} \dot{\mathbf{x}} &= \mathbf{f}(\mathbf{x}) + \sum_{i=1}^{p} \theta_i \mathbf{q}_i(\mathbf{x}), \quad y = h(\mathbf{x}), \\ \mathbf{x} &\in \mathbb{R}^n, \quad y \in \mathbb{R}, \quad \mathbf{q}_i : \mathbb{R}^n \to \mathbb{R}^n, \quad n \geq 2 \end{aligned} \tag{8.35}$$

with

$$\mathbf{f}(\mathbf{x}) = \begin{pmatrix} 0 & 1 & 1 & \dots & 1 \\ 0 & 0 & 0 & \dots & 0 \\ . & . & . & \dots & . \\ 0 & 0 & 0 & \dots & 0 \end{pmatrix} \mathbf{x}, \quad h(\mathbf{x}) = \begin{pmatrix} 1 & 0 & 0 & \dots & 0 \end{pmatrix} \mathbf{x} \tag{8.36}$$

cannot be transformed by a diffeomorphic change of coordinates, $\mathbf{z} = \boldsymbol{\Phi}(\mathbf{x})$, *into*

$$\dot{\mathbf{z}} = A_1 \mathbf{z} + \boldsymbol{\psi}_0(y) + \sum_{i=1}^{p} \theta_i \boldsymbol{\psi}_i(y), \quad y = C_1 \mathbf{z},$$

$$\mathbf{z} \in \mathbb{R}^n, \qquad y \in \mathbb{R}, \qquad \boldsymbol{\psi}_i : \mathbb{R} \to \mathbb{R}^n \tag{8.37}$$

with A_1 *and* C_1 *in canonical observer form (8.17), if either (i)* $n > 2$ *or (ii) there exists* $i \in \{1, \dots, p\}$, $j \in \{2, \dots, n\}$ *such that* $\partial q_i / \partial x_j \neq 0$.

Proof The proof is straightforward. Indeed, for the observability test we have

$$\begin{pmatrix} h(\mathbf{x}) \\ L_{\mathbf{f}} h(\mathbf{x}) \\ L_{\mathbf{f}}^2 h(\mathbf{x}) \\ . \\ . \\ . \\ L_{\mathbf{f}}^{n-1} h(\mathbf{x}) \end{pmatrix} = \begin{pmatrix} x_1 \\ x_2 + x_3 + \dots + x_n \\ 0 \\ . \\ 0 \end{pmatrix} \tag{8.38}$$

and

$$\frac{\partial}{\partial \mathbf{x}} \begin{pmatrix} h(\mathbf{x}) \\ L_{\mathbf{f}} h(\mathbf{x}) \\ L_{\mathbf{f}}^2 h(\mathbf{x}) \\ . \\ . \\ . \\ L_{\mathbf{f}}^{n-1} h(\mathbf{x}) \end{pmatrix} = \begin{pmatrix} 1 & 0 & 0 & 0 & 0 & \dots & 0 \\ 0 & 1 & 1 & 1 & 1 & \dots & 1 \\ 0 & 0 & 0 & 0 & 0 & \dots & 0 \\ . & . & . & . & . & \dots & . \\ 0 & 0 & 0 & 0 & 0 & \dots & 0 \end{pmatrix}. \tag{8.39}$$

It follows that observability is lost for $n > 2$. This proves condition (i). In order to demonstrate condition (ii) we use Theorem 8.2 as follows. From (8.23) we have

$$\left\langle \begin{pmatrix} 1 & 0 \\ 0 & 1 \end{pmatrix}, \begin{pmatrix} g_1 \\ g_2 \end{pmatrix} \right\rangle = \begin{pmatrix} 0 \\ 1 \end{pmatrix}, \tag{8.40}$$

giving

$$g = \begin{pmatrix} 0 \\ 1 \end{pmatrix}. \tag{8.41}$$

Condition (i) of Theorem 8.2 will be satisfied since the system is linear, and from condition (ii) of Theorem 8.2 we have

$$[\mathbf{q}_1, ad^0_{\mathbf{f}}\mathbf{g}] = [\mathbf{q}_1, \mathbf{g}] \tag{8.42}$$

$$= \frac{\partial \mathbf{g}}{\partial \mathbf{x}}\mathbf{q}_1 - \frac{\partial \mathbf{q}_1}{\partial \mathbf{x}}\mathbf{g} \tag{8.43}$$

$$= -\frac{\partial \mathbf{q}_1}{\partial \mathbf{x}}\begin{pmatrix} 0 \\ 1 \end{pmatrix} \tag{8.44}$$

$$= -\frac{\partial \mathbf{q}_1}{\partial x_2}. \tag{8.45}$$

Thus we satisfy condition (ii) of Theorem 8.2 if and only if $\partial \mathbf{q}_1/\partial x_2 = 0$ for all $\mathbf{x} \in U$. □

8.2.2 Parameter-dependent and time-varying transformations

Let us now consider the case in which the transformation $\mathbf{z} = \mathbf{\Phi}(\mathbf{x}, \lambda, \boldsymbol{\theta}, t)$ is allowed to depend on unknown parameters and time. As we show below, this class of transformations is much more flexible. In principle, it allows us to solve the problem of *partial* state and parameter reconstruction for an important class of oscillators with polynomial right-hand sides and time-invariant time constants.

We start by searching for a transformation $\mathbf{\Phi}$:

$$\mathbf{\Phi} : \mathbf{q} = T(\lambda)\mathbf{x}, \; |T(\lambda)| \neq 0$$

such that

$$T(\lambda)A(\lambda)T^{-1}(\lambda) = \begin{pmatrix} \star & 1 \\ \star & 0 \end{pmatrix}, \tag{8.46}$$

where the matrix $A(\lambda)$ is defined as in (8.34). It is easy to see that the transformation satisfying this constraint exists, and it is determined by

$$T(\lambda) = \begin{pmatrix} 1 & 0 \\ \lambda & 1 \end{pmatrix}. \tag{8.47}$$

According to (8.46), (8.47), and (8.34) the equations of (8.33) in the coordinates q can be written as

$$\dot{\mathbf{q}} = A_1\mathbf{q} + \psi(q_1)\eta(\boldsymbol{\theta}, \lambda), \tag{8.48}$$

where

$$A_1 = \begin{pmatrix} 0 & 1 \\ 0 & 0 \end{pmatrix},$$

$$\psi(q_1) = \begin{pmatrix} q_1^3 & q_1^2 & q_1 & 1 & 0 & 0 & 0 & 0 \\ 0 & 0 & 0 & 0 & q_1^3 & q_1^2 & q_1 & 1 \end{pmatrix}, \quad (8.49)$$

$$\boldsymbol{\eta}(\boldsymbol{\theta}, \lambda) = (\theta_{1,3}, \theta_{1,2}, \theta_{1,1} - \lambda, \theta_{1,0}, \lambda\theta_{1,3} + \theta_{2,3}, \lambda\theta_{1,2} + \theta_{2,2}, \lambda\theta_{1,1} + \theta_{2,1}, \lambda\theta_{1,0})^{\mathrm{T}}.$$

Remark 8.1 Notice that

- availability of the parameter vector $\boldsymbol{\eta}$ in (8.48) and (8.49), expressed as a function of $\boldsymbol{\theta}$ and λ, implies the availability of $\boldsymbol{\theta}$ and λ if $\theta_{1,0} \neq 0$ (indeed, in this case the value of $\lambda = \eta_8/\eta_4$ and the values of all $\theta_{i,j}$ are uniquely defined by η_i);
- the condition $\theta_{1,0} \neq 0$ is sufficient for reconstructing the values of x_2, provided that $\mathbf{q}$ and $\boldsymbol{\eta}$ are available (indeed, in this case $\mathbf{x} = T^{-1}(\lambda)\mathbf{q}$).

As follows from Remark 8.1, the problem of state and parameter reconstruction of (8.33) from measured data $x_1(t)$ amounts to solving the problem of state and parameter reconstruction of (8.48). In order to solve this problem we shall employ yet another coordinate transformation:

$$\begin{aligned} z_1 &= q_1, \\ z_2 &= q_2 + \boldsymbol{\zeta}^{\mathrm{T}}(t)\boldsymbol{\eta}, \end{aligned} \quad (8.50)$$

in which the functions $\boldsymbol{\zeta}(t)$ are some differentiable functions of time. Coordinate transformation (8.50) is clearly time-dependent. The role of this additional transformation is to transform the equations of system (8.48) into the form for which a solution already exists.

Definitions of these functions, specific estimation algorithms, and their convergence properties are discussed in detail in the next section.

8.2.3 Observers for transformed equations

Bastin–Gevers adaptive observer

Proceeding from (8.48) and (8.49) and applying a second change of coordinates given by

$$\begin{pmatrix} z_1 \\ z_2 \end{pmatrix} = \begin{pmatrix} 1 & 0 \\ f/k & 1/k \end{pmatrix} \begin{pmatrix} q_1 \\ q_2 \end{pmatrix},$$

where $f \in \mathbb{R}_{<0}$ and $k \in \mathbb{R}$ are some design parameters, we obtain the canonical form (8.10) presented in Bastin and Gevers (1988)

$$\begin{aligned}
\begin{pmatrix} \dot{z}_1 \\ \dot{z}_2 \end{pmatrix} &= R \begin{pmatrix} z_1 \\ z_2 \end{pmatrix} + \mathbf{g}(t) + \Omega(y)\boldsymbol{\eta}(\boldsymbol{\theta}, \lambda), \\
R &= \begin{pmatrix} 0 & k \\ 0 & f \end{pmatrix}, \\
\mathbf{g}(t) &= -y \begin{pmatrix} f \\ f^2/k \end{pmatrix}, \\
\Omega(y) &= \begin{pmatrix} y^3 & y^2 & y & 1 & 0 & 0 & 0 & 0 \\ f/ky^3 & f/ky^2 & f/ky & f/k & 1/ky^3 & 1/ky^2 & 1/ky & 1/k \end{pmatrix}, \\
\boldsymbol{\eta}(\boldsymbol{\theta}, \lambda) &= (\theta_{13}, \theta_{12}, \theta_{11} - \lambda, \theta_{10}, \theta_{23} + \lambda\theta_{13}, \theta_{22} + \lambda\theta_{12}, \theta_{21} + \lambda\theta_{11}, \lambda\theta_{10})^{\mathrm{T}}.
\end{aligned} \tag{8.51}$$

System (8.51) now is in the Bastin–Gevers adaptive-observer canonical form. Notice that the parameter vector $\boldsymbol{\eta}(\boldsymbol{\theta}, \lambda)$ remains unchanged and recall Remark 8.1. Let us proceed to the observer construction following the steps described in (8.11)–(8.16).

We start by introducing an auxiliary filter of which the general form is given by (8.13). According to (8.51) the auxiliary filter is defined as follows:

$$\begin{aligned}
\dot{v}_1 &= f v_1 + (f/k)y^3, \\
\dot{v}_2 &= f v_2 + (f/k)y^2, \\
\dot{v}_3 &= f v_3 + (f/k)y, \\
\dot{v}_4 &= f v_4 + (f/k), \\
\dot{v}_5 &= f v_5 + (1/k)y^3, \\
\dot{v}_6 &= f v_6 + (1/k)y^2, \\
\dot{v}_7 &= f v_7 + (1/k)y, \\
\dot{v}_8 &= f v_8 + 1/k.
\end{aligned} \tag{8.52}$$

Hence, in accordance with (8.14) the regressor vector $\boldsymbol{\varphi}(t)$ is written as

$$\begin{aligned}
\varphi_1 &= k v_1 + y^3, \\
\varphi_2 &= k v_2 + y^2, \\
\varphi_3 &= k v_3 + y, \\
\varphi_4 &= k v_4 + 1, \\
\varphi_5 &= k v_1, \\
\varphi_6 &= k v_2, \\
\varphi_7 &= k v_3, \\
\varphi_8 &= k v_4,
\end{aligned} \tag{8.53}$$

and the observer equations are as follows:

$$\begin{aligned}\dot{\hat{\mathbf{x}}} &= \begin{pmatrix} -c_1 & k \\ 0 & f \end{pmatrix}\hat{\mathbf{x}} + \begin{pmatrix} c_1 \\ 0 \end{pmatrix}\hat{x}_1 + \mathbf{g}(t) + \Omega(x_1)\hat{\eta} + \begin{pmatrix} 0 \\ V\dot{\hat{\eta}} \end{pmatrix}, \\ \dot{\hat{\eta}} &= \Gamma\boldsymbol{\varphi}(x_1 - \hat{x}_1), \\ V &= (v_1, v_2, v_3, v_4, v_5, v_6, v_7, v_8). \end{aligned} \tag{8.54}$$

On taking (8.52)–(8.54) and (8.16) into account, we obtain the following equations governing the dynamics of the estimation error, $(\tilde{\mathbf{x}}^*, \tilde{\eta})^{\mathrm{T}}$:

$$\begin{aligned}\dot{\tilde{\mathbf{x}}}^* &= \begin{pmatrix} -c_1 & k \\ 0 & f \end{pmatrix}\tilde{\mathbf{x}}^* + \begin{pmatrix} \boldsymbol{\varphi}^{\mathrm{T}}\tilde{\eta} \\ 0 \end{pmatrix}, \\ \dot{\tilde{\eta}} &= -\Gamma\boldsymbol{\varphi}\tilde{x}_1^*. \end{aligned} \tag{8.55}$$

The auxiliary filter (8.52) acts here as an inherent component of a time-varying coordinate transformation rendering the error dynamics into (8.16). This coordinate transformation is similar to that defined by (8.50), provided that $\mathbf{z}$ and η in (8.50) are replaced by estimation errors $\tilde{\mathbf{x}}^*$ and $\tilde{\eta}$.

Let us now explore asymptotic properties of the observer. First we notice that $v_4(t)$ and $v_8(t)$ both converge to constant values exponentially fast as $t \to \infty$. In fact,

$$\lim_{t\to\infty} v_4(t) = -\frac{1}{k}, \qquad \lim_{t\to\infty} v_8(t) = -\frac{1}{fk}.$$

Thus accordingly $\varphi_4(t)$ and $\varphi_8(t)$ both tend to constant values as $t \to \infty$:

$$\lim_{t\to\infty} \varphi_4(t) = 0, \qquad \lim_{t\to\infty} \varphi_8(t) = -\frac{1}{f}.$$

The latter fact implies that the persistency-of-excitation requirement is necessarily violated for regressor (8.53). Indeed, this condition does not hold if one of the components of $\boldsymbol{\varphi}(t)$ is exponentially converging to zero. The question, therefore, is whether this approach can be used at all to construct asymptotically converging estimators of the state and parameters of (8.51). The answer to this question is provided in the corollary below.

Corollary 8.1 *Consider the function*

$$\bar{\boldsymbol{\varphi}}(t) = (\varphi_1(t), \varphi_2(t), \varphi_3(t), \varphi_5(t), \varphi_6(t), \varphi_7(t), \varphi_8(t))^{\mathrm{T}}.$$

If it is globally bounded and persistently exciting then the following holds along the solutions of (8.52)–(8.54):

$$\lim_{t\to\infty} \hat{\mathbf{x}}(t) - \mathbf{x}(t) = 0, \qquad \lim_{t\to\infty} \hat{\eta}_i(t) - \eta_i, \; i \neq 4,$$

and the convergence is exponential.

Remark 8.2 Corollary 8.1 demonstrates that even though the original result of Bastin and Gevers (1988), i.e. Theorem 8.1, does not apply to system (8.51) directly one can still construct a reduced-order observer for this system. This reduced observer guarantees partial reconstruction of unmeasured parameters, and this reconstruction is exponentially fast. To recover the true values of unknown parameters one needs to solve the system

$$\begin{aligned}
\eta_1 &= \theta_{13}, \\
\eta_2 &= \theta_{12}, \\
\eta_3 &= \theta_{11} - \lambda, \\
\eta_5 &= \theta_{23} + \lambda\theta_{13}, \\
\eta_6 &= \theta_{22} + \lambda\theta_{12}, \\
\eta_7 &= \theta_{21} + \lambda\theta_{11}, \\
\eta_8 &= \lambda\theta_{10},
\end{aligned}$$

for θ_i and λ taking the values of $\hat{\eta}_i$ as the estimates of η_i. The solution to this system might not be unique, hence the reconstruction is generally possible only up to a certain scaling factor.

The Marino–Tomei observer

Let us define the vector-function $\zeta(t)$ in (8.50) as follows:

$$\dot{\zeta}_i = -k\zeta_i + k\psi_{1,i}(q_1) - \psi_{2,i}(q_1), \; k \in \mathbb{R}_{>0}.$$

In this case we have

$$\begin{aligned}
\dot{z}_2 &= \sum_{i=1}^{8} \psi_{2,i}(q_1)\eta_i + k\left(-\sum_{i=1}^{8} \zeta_i\eta_i + \psi_{1,i}(q_1)\eta_i\right) - \sum_{i=1}^{8} \psi_{2,i}(q_1)\eta_i \\
&= k\sum_{i=1}^{8}(-\zeta_i + \psi_{1,i}(q_1))\eta_i.
\end{aligned}$$

Hence, taking equality (8.50) into account and expressing q_2 as $q_2 = z_2 - \boldsymbol{\zeta}^{\mathrm{T}}(t)\boldsymbol{\eta}(\boldsymbol{\theta}, \lambda)$, we obtain

$$\begin{aligned}
\dot{z}_1 &= z_2 + \sum_{i=1}^{8}(-\zeta_i + \psi_{1,i}(q_1))\eta_i, \\
\dot{z}_2 &= k\sum_{i=1}^{8}(-\zeta_i + \psi_{1,i}(q_1))\eta_i.
\end{aligned} \tag{8.56}$$

Notice that $\psi_{1,4}(q_1) = \psi_{2,8}(q_1) = 1$, hence $-\zeta_4 + \psi_{1,4}(q_1)$ and $-\zeta_8 + \psi_{1,8}(q_1)$ converge to some constants in $\mathbb{R}$ exponentially fast as $t \to \infty$. Moreover, the sum $-\zeta_4 + \psi_{1,4}(q_1)$ converges to zero and the sum $-\zeta_8 + \psi_{1,8}(q_1)$ converges to $-1/k$ as $t \to \infty$. Taking these facts into account, we can conclude that system (8.56) can be rewritten in the following (reduced) form:

$$\begin{aligned}
\dot{\mathbf{z}} &= A_1\mathbf{z} + \mathbf{b}\boldsymbol{\phi}^{\mathrm{T}}(z_1,t)\boldsymbol{\upsilon}(\boldsymbol{\theta},\lambda) + \mathbf{b}\varepsilon(t),\\
A_1 &= \begin{pmatrix} 0 & 1 \\ 0 & 0 \end{pmatrix}, \qquad \mathbf{b} = \begin{pmatrix} 1 \\ k \end{pmatrix},\\
\boldsymbol{\phi}(z_1,t) &= \begin{pmatrix} -\zeta_1 + \psi_{1,1}(z_1) \\ -\zeta_2 + \psi_{1,2}(z_1) \\ -\zeta_3 + \psi_{1,3}(z_1) \\ -\zeta_8 + \psi_{1,8}(z_1) \\ -\zeta_5 + \psi_{1,5}(z_1) \\ -\zeta_6 + \psi_{1,6}(z_1) \\ -\zeta_7 + \psi_{1,7}(z_1) \end{pmatrix},\\
\boldsymbol{\upsilon}(\boldsymbol{\theta},\lambda) &= (\theta_{1,3}, \theta_{1,2}, \theta_{1,1}, \bar{\theta}_{1,0}, \lambda\theta_{1,3} + \theta_{2,3}, \lambda\theta_{1,2} + \theta_{2,2}, \lambda\theta_{1,1} + \theta_{2,1})^{\mathrm{T}},
\end{aligned} \tag{8.57}$$

where $\varepsilon(t)$ is an exponentially decaying term.

System (8.57) is clearly in the adaptive-observer canonical form. Hence, it admits the following adaptive observer:

$$\begin{aligned}
\dot{\hat{\mathbf{z}}} &= A_1\hat{\mathbf{z}} + L(\hat{\mathbf{z}}_1 - \mathbf{z}_1) + \mathbf{b}\boldsymbol{\phi}^{\mathrm{T}}(z_1,t)\hat{\boldsymbol{\upsilon}},\\
C_1 &= (1,0), \qquad \mathbf{z}_1 = C_1\mathbf{z},\\
L &= \begin{pmatrix} -l_1 \\ -l_2 \end{pmatrix}, \; l_1 = k+1, \quad l_2 = k,\\
\dot{\hat{\upsilon}}_i &= -\gamma(\hat{z}_1 - z_1)\phi_i(z_1,t), \; \gamma \in \mathbb{R}_{>0},
\end{aligned} \tag{8.58}$$

of which the asymptotic properties are specified in the following theorem.

Theorem 8.4 *Let us suppose that system (8.57) is given and its solutions are defined for all t. Then, for all initial conditions, solutions of the combined system (8.57) and (8.58) exist for all t and*

$$\lim_{t\to\infty} \hat{\mathbf{z}}(t) - \mathbf{z}(t) = 0.$$

Furthermore, if the function $\boldsymbol{\phi}(z_1,t)$ is persistently exciting and $\mathbf{z}(t)$ is bounded then

$$\lim_{t\to\infty} \hat{\boldsymbol{\upsilon}}(t) - \boldsymbol{\upsilon}(\boldsymbol{\theta},\lambda) = 0,$$

and the dynamics of $\hat{\mathbf{z}} - \mathbf{z}$ and $\hat{\boldsymbol{\upsilon}} - \boldsymbol{\upsilon}$ are exponentially stable in the sense of Lyapunov.

Proof The proof of the theorem is standard and can be constructed from many other more general results (see, for example, Marino (1990) and Narendra and Annaswamy (1989)). Here we present just a sketch of the argument for consistency. According to our assumptions, the matrix $A_1 + LC_1$ is Hurwitz. Moreover one can easily check that the transfer function $C_1(p - (A_1 + LC_1))^{-1}\mathbf{b}$ is strictly positive real. Hence, using the Meyer–Kalman–Yakubovich–Popov lemma, we can conclude that there exists a symmetric and positive definite matrix H such that

$$H(A_1 + LC_1) + (A_1 + LC_1)^{\mathrm{T}}H < -Q, \ H\mathbf{b} = (1, 0)^{\mathrm{T}}, \tag{8.59}$$

where Q is a positive definite matrix. Let us now consider the following function

$$V(\mathbf{z}, \hat{\boldsymbol{\upsilon}}, t) = \frac{1}{2}(\mathbf{z} - \hat{\mathbf{z}})^{\mathrm{T}}H(\mathbf{z} - \hat{\mathbf{z}}) + \frac{1}{2}\|\boldsymbol{\upsilon} - \hat{\boldsymbol{\upsilon}}\|^2\gamma^{-1} + D\int_t^{\infty} \varepsilon^2(\tau)d\tau,$$

where the value of D is to be specified later. Clearly, the function V is well defined for the term $\varepsilon(t)$ is continuous and exponentially decaying to zero as $t \to \infty$. Thus, the boundedness of V implies that $\|\hat{\mathbf{z}} - \mathbf{z}\|$ and $\|\hat{\boldsymbol{\upsilon}} - \boldsymbol{\upsilon}\|$ are bounded.

Consider the time-derivative of V:

$$\begin{aligned}\dot{V} &= (\mathbf{z} - \hat{\mathbf{z}})^{\mathrm{T}}(H(A_1 + LC_1) + (A_1 + LC_1)^{\mathrm{T}}H)(\mathbf{z} - \hat{\mathbf{z}})\tfrac{1}{2}\\ &\quad + (\mathbf{z} - \hat{\mathbf{z}})^{\mathrm{T}}H\mathbf{b}\boldsymbol{\phi}^{\mathrm{T}}(z_1, t)(\boldsymbol{\upsilon} - \hat{\boldsymbol{\upsilon}}) + (\mathbf{z} - \hat{\mathbf{z}})^{\mathrm{T}}H\mathbf{b}\varepsilon\\ &\quad - (\boldsymbol{\upsilon} - \hat{\boldsymbol{\upsilon}})^T(1, 0)(\mathbf{z} - \hat{\mathbf{z}})\boldsymbol{\phi}(z_1, t) - D\varepsilon^2.\end{aligned}$$

Taking (8.59) into account, we obtain

$$\begin{aligned}\dot{V} &\leq -(\mathbf{z} - \hat{\mathbf{z}})^{\mathrm{T}}Q(\mathbf{z} - \hat{\mathbf{z}})\tfrac{1}{2} + \|\mathbf{z} - \hat{\mathbf{z}}\||\varepsilon(t)|M - D\varepsilon^2\\ &\leq -\alpha\|\mathbf{z} - \hat{\mathbf{z}}\|^2 + \|\mathbf{z} - \hat{\mathbf{z}}\||\varepsilon(t)|M - D\varepsilon^2,\end{aligned} \tag{8.60}$$

where $\alpha > 0$ is the minimal eigenvalue of $Q\frac{1}{2}$, M is a fixed positive number, and D is a parameter of V that can be chosen arbitrarily and independently of M. By choosing D such that

$$\sqrt{\frac{\alpha D}{4}} = M,$$

we ensure that

$$\dot{V} \leq -\frac{\alpha}{2}\|z - \hat{z}\|^2.$$

Thus, the function V is bounded from above, and, given that the solution $\mathbf{z}(t)$ exists for all $t > t_0$, so does the solution of the combined system. The rest of the proof follows directly from Barbalat's lemma and the asymptotic-stability theorem for the class of skew-symmetric-time-varying systems presented in Morgan and Narendra (1992). □

Remark 8.3 Similarly to Corollary 8.1 for the Bastin–Gevers observer, Theorem 8.4 provides us with a computational scheme that, subject to the condition that $\boldsymbol{\phi}(z_1, t)$ is persistently exciting, can be used to estimate the values of the modified vector of uncertain parameters $\boldsymbol{\upsilon}(\boldsymbol{\theta}, \lambda)$. The question, however, is that of whether the values of $\boldsymbol{\theta}$ and λ can always be restored from $\boldsymbol{\upsilon}(\boldsymbol{\theta}, \lambda)$. In general, the answer to this question is negative. Indeed, according to (8.48) we have

$$\begin{aligned} \theta_{1,3} &= \upsilon_1, \\ \theta_{1,2} &= \upsilon_2, \\ \theta_{1,1} - \lambda &= \upsilon_3, \\ \bar{\theta}_{1,0} = \lambda\theta_{1,0} &= \upsilon_4, \\ \lambda\theta_{1,3} + \theta_{2,3} &= \upsilon_5, \\ \lambda\theta_{1,2} + \theta_{2,2} &= \upsilon_6, \\ \lambda\theta_{1,1} + \theta_{2,1} &= \upsilon_7. \end{aligned} \tag{8.61}$$

As follows from (8.61) one can easily recover the values of $\theta_{1,3}$, $\theta_{1,2}$, and $\theta_{1,1}$. However, recovering the values of the remaining parameters explicitly from the estimates of $\boldsymbol{\upsilon}(\boldsymbol{\theta}, \lambda)$ is possible only up to a certain scaling parameter. Indeed, the number of unknowns in (8.61) exceeds the number of equations by one.

Remark 8.4 Notice that in the relevant special cases, when the value of any one of either $\theta_{2,3}$, $\theta_{2,2}$, and $\theta_{2,1}$ is zero, such reconstruction is obviously possible. Let us suppose that $\theta_{2,3} = 0$. Hence, the value of λ can be expressed from (8.61) as

$$\lambda = \frac{\upsilon_5}{\upsilon_1}, \tag{8.62}$$

and thus the rest of the parameters can be reconstructed as well. Owing to the presence of division in (8.62), this scheme may be sensitive to persistent perturbations when $\upsilon_5 = \lambda\theta_{1,3}$ is small.

So far we have considered special cases of (8.1) in which the time constants of unmeasured variables were unknown yet constant and the parametrization of the right-hand side was linear. As we mentioned in Remark 8.4, even for this simpler class of systems solving the problem of parameter reconstruction might not be a straightforward operation. For example, if there are cubic, quadratic, and linear terms in the second equation of (8.33) then recovering all parameters of (8.33) by use of the observer (8.58) might not be possible. Nonlinear parametrization, time-varying time constants, and nonlinear coupling between equations on the right-hand side of (8.1) make the reconstruction problem even more complicated.

In the next section we show that for a large subclass of (8.1) there always exists an observer that solves the problem of state and parameter reconstruction from the measurements of V. Moreover the structure of this observer does not depend significantly on the specific equations describing the dynamics of the observed system.

8.3 Universal adaptive observers for conductance-based models

The ideas of universal adaptive observers for systems with nonlinearly parametrized uncertainty were introduced in Chapter 5. Here we discuss how they can be applied to the problem of state and parameter reconstruction of (8.1).

The following subclass of (8.1) is considered:

$$\begin{aligned}
\dot{x}_0 &= \theta_0^{\mathrm{T}} \phi_0(x_0, p_0, t) + \sum_{i=1}^{n} c_i(x_0, q_i, t) x_i + c_0(x_0, q_0, t) + \xi_0(t) + u(t), \\
\dot{x}_1 &= -\beta_1(x_0, \tau_1, t)\, x_1 + \theta_1^{\mathrm{T}} \phi_1(x_0, p_1, t) + \xi_1(t), \\
&\vdots \\
\dot{x}_i &= -\beta_i(x_0, \tau_i, t)\, x_i + \theta_i^{\mathrm{T}} \phi_i(x_0, p_i, t) + \xi_i(t), \\
&\vdots \\
\dot{x}_n &= -\beta_n(x_0, \tau_n, t)\, x_n + \theta_n^{\mathrm{T}} \phi_i(x_0, p_n, t) + \xi_n(t), \\
& y = (1, 0, \ldots, 0)\mathbf{x} = x_0, \qquad x_i(t_0) = x_{i,0} \in \mathbb{R}, \\
& \mathbf{x} = \mathrm{col}(x_0, x_1, \ldots, x_n), \qquad \theta_i = \mathrm{col}(\theta_{i,1}, \ldots, \theta_{i,d_i}),
\end{aligned} \tag{8.63}$$

where

$$\begin{aligned}
&\boldsymbol{\phi}_i : \mathbb{R} \times \mathbb{R}^{m_i} \times \mathbb{R}_{\geq 0} \to \mathbb{R}^{d_i},\ \boldsymbol{\phi}_i \in \mathcal{C}^0,\ d_i, m_i \in \mathbb{N},\ i = \{0, \ldots, n\}, \\
&\beta_i : \mathbb{R} \times \mathbb{R} \times \mathbb{R}_{\geq 0} \to \mathbb{R}_{>0},\ \beta_i \in \mathcal{C}^0,\ i = \{1, \ldots, n\}, \\
&c_i : \mathbb{R} \times \mathbb{R}^{r_i} \times \mathbb{R}_{\geq 0} \to \mathbb{R},\ c_i \in \mathcal{C}^0,\ r_i \in \mathbb{N},\ i = \{0, \ldots, n\},
\end{aligned}$$

are continuous and known functions, $u : \mathbb{R}_{\geq 0} \to \mathbb{R}$, $u \in \mathcal{C}^0$ is a known function of time modeling the control input, and $\xi_i : \mathbb{R}_{\geq 0} \to \mathbb{R}$, $\xi_i \in \mathcal{C}^0$ are functions that are unknown, yet bounded. The functions $\xi_i(t)$ represent *unmodeled dynamics*, external perturbations, residuals due to the coarse-graining procedures at the stage of reduction, etc.

The variable y in system (8.63) is the *output*, and the variables x_i, $i \geq 1$ are the components of state $\mathbf{x}$, which are not available for direct observation. The vectors $\theta_i \in \mathbb{R}^{d_i}$ consist of *linear* parameters of uncertainties on the right-hand side of the ith equation in (8.63). The parameters $\tau_i \in \mathbb{R}$, $i = \{1, \ldots, n\}$ are the unknown parameters of the time-varying relaxation rates, $\beta_i(x_0, \tau_i, t)$, of the state variables x_i, and the vectors $p_i \in \mathbb{R}^{m_i}$ and $q_i \in \mathbb{R}^{r_i}$, consist of the *nonlinear* parameters of the uncertainties. The functions $c_i(x_0, q_i, t)$ are supposed to be bounded.

Notice that system (8.63) is almost as general as (8.1). The only difference is that the variables x_i, $i \geq 1$ enter the first equation of (8.63) as

$$\sum_{i=1}^{n} c_i(x_0, q_i, t)x_i,$$

whereas the corresponding variables r_i in system (8.1) enter the first equation in a slightly more general way,

$$\sum\nolimits_j \varphi_j(v, t)p_j(r)\theta_j.$$

For notational convenience we denote:

$$\begin{aligned}
\boldsymbol{\theta} &= \mathrm{col}(\theta_0, \theta_1, \ldots, \theta_n), \\
\boldsymbol{\lambda} &= \mathrm{col}(p_0, q_0, \tau_1, p_1, q_1, \ldots, \tau_n, p_n, q_n), \\
s &= \dim(\boldsymbol{\lambda}) = n + \sum_{i=0}^{n}(m_i + r_i).
\end{aligned}$$

The symbols Ω_θ and Ω_λ denote the domains of admissible values for $\boldsymbol{\theta}$ and $\boldsymbol{\lambda}$, respectively.

The system's state $\mathbf{x} = \mathrm{col}(x_0, x_1, \ldots, x_n)$ is not measured; only the values of the input $u(t)$ and the output $y(t) = x_0(t)$, $t \geq t_0$ in (8.63) are accessible over any time interval $[t_0, t]$ that belongs to the history of the system. The actual values of the parameters $\boldsymbol{\theta}$ and $\boldsymbol{\lambda}$ are assumed to be unknown a priori. We assume however, that they belong to a set, e.g. a hypercube, with known bounds: $\theta_{i,j} \in [\theta_{i,\min}, \theta_{i,\max}]$, $\lambda_i \in [\lambda_{i,\min}, \lambda_{i,\max}]$.

Instead of imposing the traditional requirement of asymptotic estimation of the unknown parameters with arbitrarily small error we relax our demands to estimating the values of the state and parameters of (8.63) up to a certain tolerance. This is because we allow unmodeled dynamics, $\xi_i(t)$, on the right-hand side of (8.63). As a result of such a practically important addition there may exist a set of systems of which the solutions are relatively close to the measured data yet their parameters could be different. Instead of just one value of the unknown parameter vectors $\boldsymbol{\theta}$ and $\boldsymbol{\lambda}$ we therefore have to deal with a set of $\boldsymbol{\theta}$ and $\boldsymbol{\lambda}$ corresponding to the solutions of (8.63) which over time are sufficiently close. This set of model parameters is referred to as an *equivalence class* of (8.63).

Similarly to canonical observer schemes (Marino 1990; Bastin and Gevers 1988; Marino and Tomei 1995a), the method presented in Chapter 5, in Section 5.6.2, relies on the ability to evaluate the integrals

$$\mu_i(t, \tau_i, p_i) = \int_{t_0}^{t} e^{-\int_\tau^t \beta_i(x_0(\chi),\tau_i,\chi)d\chi} \boldsymbol{\phi}_i(x_0(\tau), p_i, \tau)d\tau \tag{8.64}$$

at a given time t and for the given values of τ_i and p_i within a given accuracy. In classical adaptive-observer schemes, the values of $\beta_i(x_0, \tau_i, t)$ are constant. This allows us to transform the original equations by a (possibly parameter-dependent) non-singular linear coordinate transformation, $\Phi : \mathbf{x} \mapsto \mathbf{z}, x_1 = z_1$, into an equivalent form in which the values of all time constants are known. In the new coordinates the variables $z_2, \ldots, z_n$ can be estimated by integrals (8.64) in which the values of $\beta_i(x_0, \tau_i, t)$ are constant and known. This is usually done by using auxiliary linear filters. In our case, the values of $\beta_i(x_0, \tau_i, t)$ are not constant and are unknown due to the presence of τ_i. Yet if the values of τ_i were known we could still estimate the values of integrals (8.64) as follows:

$$\begin{aligned}&\int_{t_0}^{t} e^{-\int_\tau^t \beta_i(x_0(\chi),\tau_i,\chi)d\chi} \boldsymbol{\phi}_i(x_0(\tau), p_i, \tau)d\tau \simeq \\ &\int_{t-T}^{t} e^{-\int_\tau^t \beta_i(x_0(\chi),\tau_i,\chi)d\chi} \boldsymbol{\phi}_i(x_0(\tau), p_i, \tau)d\tau = \bar{\mu}_i(t, \tau_i, p_i),\end{aligned} \tag{8.65}$$

where $T \in \mathbb{R}_{>0}$ is sufficiently large and $t \geq T + t_0$.

Alternatively, if $\boldsymbol{\phi}_i(x_0(t), p_i, t)$ and $\beta_i(x_0(t), \tau_i, t)$ are periodic with rationally dependent periods and satisfy the Dini condition in t, the integrals (8.64) can be estimated by invoking a Fourier expansion. Notice that for functions $\mu_i(t, \tau_i, p_i)$ that are continuous and Lipschitz in p_i the coefficients of their Fourier expansion remain continuous and Lipschitz with respect to p_i.

In the next sections we present the general structure of the observer for (8.63) and provide a list of its asymptotic properties.

8.3.1 Observer definition and assumptions

Consider the following function $\boldsymbol{\varphi}(x_0, \boldsymbol{\lambda}, t) : \mathbb{R} \times \mathbb{R}^s \times \mathbb{R}_{\geq 0} \to \mathbb{R}^d$, $d = \sum_{i=0}^n d_i$:

$$\begin{aligned}\boldsymbol{\varphi}(x_0, \boldsymbol{\lambda}, t) = \big(&\boldsymbol{\phi}_0(x_0, p_0, t), c_1(x_0, q_1, t)\mu_1(t, \tau_1, p_1), \ldots, \\ &c_n(x_0, q_n, t)\mu_n(t, \tau_n, p_n)\big)^{\mathrm{T}}.\end{aligned} \tag{8.66}$$

The function $\boldsymbol{\varphi}(x_0, \boldsymbol{\lambda}, t)$ is a concatenation of $\boldsymbol{\phi}_0(\cdot)$ and the integrals (8.64). We assume that the values of $\boldsymbol{\varphi}(x_0, \boldsymbol{\lambda}, t)$ can be efficiently estimated for all $x_0, \boldsymbol{\lambda}, t \geq 0$ up to a small mismatch. In other words, we suppose that there exists a function $\bar{\boldsymbol{\varphi}}(x_0, \boldsymbol{\lambda}, t)$ such that the following property holds:

$$\|\bar{\boldsymbol{\varphi}}(x_0, \boldsymbol{\lambda}, t) - \boldsymbol{\varphi}(x_0, \boldsymbol{\lambda}, t)\| \leq \Delta_\varphi, \ \Delta_\varphi \in \mathbb{R}_{>0}, \tag{8.67}$$

where the values of $\bar{\boldsymbol{\varphi}}(x_0, \boldsymbol{\lambda}, t)$ are efficiently computable for all x_0, $\boldsymbol{\lambda}$, and t (see e.g. (8.65) for an example of such approximations), and Δ_φ is sufficiently small.

If the parameters τ_i, p_i, and q_i on the right-hand side of (8.63) were known, and $c_i(x_0, q_i, t) = 1$ and $\beta_i(x_0, \tau_i, t) = \tau_i$, then the function $\boldsymbol{\varphi}(x_0, \boldsymbol{\lambda}, t)$ could be estimated by $(\boldsymbol{\phi}_0(x_0, t), \eta_1, \ldots, \eta_n)$, where η_i are the solutions of the auxiliary system (filter)

$$\dot{\eta}_i = -\tau_i \eta_i + \boldsymbol{\phi}_i(x_0, p_i, t) \tag{8.68}$$

with zero initial conditions. Systems like (8.68) are inherent components of standard adaptive observers (Kreisselmeier 1977; Bastin and Gevers 1988; Marino 1990). In our case we suppose that the values of τ_i, q_i, and p_i are not known a priori and that $c_i(x_0, q_i, t)$ and $\beta_i(x_0, \tau_i, t)$ are not constant. Therefore, we replace η_i with their approximations, e.g. as in (8.65):

$$\bar{\boldsymbol{\varphi}}(x_0, \boldsymbol{\lambda}, t) = \big(\boldsymbol{\phi}_0(x_0, p_0, t), c_1(x_0, q_1, t)\bar{\mu}_1(t, \tau_1, p_1), \ldots,\\ c_n(x_0, q_n, t)\bar{\mu}_n(t, \tau_n, p_n)\big)^{\mathrm{T}}.$$

For periodic $\boldsymbol{\phi}_i(x_0(t), p_i, t)$ and $\beta_i(x_0(t), \tau_i, t)$ a Fourier expansion can be employed to define $\bar{\boldsymbol{\varphi}}(x_0, \boldsymbol{\lambda}, t)$. The value of Δ_φ in (8.67) stands for the accuracy of approximation, and as a rule of thumb the more computational resources are devoted to approximate $\boldsymbol{\varphi}(x_0, \boldsymbol{\lambda}, t)$ the smaller is the value of Δ_φ.

With regard to the functions $\xi_i(t)$ in (8.63) we suppose that an upper bound, Δ_ξ, of the following sum is available:

$$\sum_{i=1}^{n} \frac{1}{\tau_i} \|\xi_i(\tau)\|_{\infty,[t_0,\infty]} + \|\xi_0(\tau)\|_{\infty,[t_0,\infty]} \leq \Delta_\xi, \ \Delta_\xi \in \mathbb{R}_{\geq 0}. \tag{8.69}$$

Denoting $c_0(x_0, q_0, t) = c_0(x_0, \boldsymbol{\lambda}, t)$, for notational convenience, we can now define the observer as

$$\begin{cases} \dot{\hat{x}}_0 = -\alpha(\hat{x}_0 - x_0) + \hat{\boldsymbol{\theta}}^{\mathrm{T}} \bar{\boldsymbol{\varphi}}(x_0, \hat{\boldsymbol{\lambda}}, t) + c_0(x_0, \hat{\boldsymbol{\lambda}}, t) + u(t), \\ \dot{\hat{\boldsymbol{\theta}}} = -\gamma_\theta(\hat{x}_0 - x_0)\bar{\boldsymbol{\varphi}}(x_0, \hat{\boldsymbol{\lambda}}, t), \ \gamma_\theta, \alpha \in \mathbb{R}_{>0}, \end{cases} \tag{8.70}$$

$$\dot{\hat{x}}_i = -\beta_i(x_0, \hat{\tau}_i, t)\hat{x}_i + \hat{\theta}_i^{\mathrm{T}} \boldsymbol{\phi}_i(x_0, \hat{p}_i, t), \ i = \{1, \ldots, n\}, \tag{8.71}$$

where

$$\hat{\boldsymbol{\theta}} = \mathrm{col}(\hat{\theta}_0, \hat{\theta}_1, \ldots, \hat{\theta}_n)$$

is the vector of estimates of $\boldsymbol{\theta}$. The components of the vector

$$\hat{\boldsymbol{\lambda}} = \mathrm{col}(\hat{p}_0, \hat{q}_0, \hat{\tau}_1, \hat{p}_1, \hat{q}_1, \ldots, \hat{\tau}_n, \hat{p}_n, \hat{q}_n) = \mathrm{col}(\hat{\lambda}_1, \ldots, \hat{\lambda}_s),$$

with $s = \dim(\boldsymbol{\lambda})$, evolve according to the following equations:

$$\begin{aligned} \dot{\hat{x}}_{1,j} &= \gamma \cdot \omega_j \cdot e \cdot \left(\hat{x}_{1,j} - \hat{x}_{2,j} - \hat{x}_{1,j}\left(\hat{x}_{1,j}^2 + \hat{x}_{2,j}^2\right)\right), \\ \dot{\hat{x}}_{2,j} &= \gamma \cdot \omega_j \cdot e \cdot \left(\hat{x}_{1,j} + \hat{x}_{2,j} - \hat{x}_{2,j}\left(\hat{x}_{1,j}^2 + \hat{x}_{2,j}^2\right)\right), \\ \hat{\lambda}_j(\hat{x}_{1,j}) &= \lambda_{j,\min} + [(\lambda_{j,\max} - \lambda_{j,\min})/2](\hat{x}_{1,j} + 1), \\ e &= \sigma(\|x_0 - \hat{x}_0\|_\varepsilon), \end{aligned} \tag{8.72}$$

with

$$j = \{1, \ldots, s\}, \qquad \hat{x}_{1,j}^2(t_0) + \hat{x}_{2,j}^2(t_0) = 1,$$

where $\sigma(\cdot) : \mathbb{R} \to \mathbb{R}_{\geq 0}$ is a bounded continuous function, i.e. $\sigma(\upsilon) \leq S \in \mathbb{R}_{>0}$, and $|\sigma(\upsilon)| \leq |\upsilon|$ for all $\upsilon \in \mathbb{R}$; $\omega_j \in \mathbb{R}_{>0}$ are set to be rationally independent (see (5.169)).

8.3.2 Asymptotic properties of the observer

The asymptotic properties of the observer are specified in Corollaries 8.2 and 8.3 from Theorem 5.10. Corollary 8.2 establishes conditions for state boundedness of the observer, and states its general asymptotic properties. Corollary 8.3 specifies a set of conditions for the possibility of asymptotic reconstruction of θ_i, τ_i, and p_i, up to their equivalence classes and small mismatch due to errors.

Proofs of these corollaries follow immediately from Theorem 5.10.

Corollary 8.2 *Let the system of (8.63) and (8.70)–(8.72) be given. Assume that the function $\bar{\boldsymbol{\varphi}}(x_0(t), \boldsymbol{\lambda}, t)$ is $\boldsymbol{\lambda}$-uniformly persistently exciting, and that the functions $\bar{\boldsymbol{\varphi}}(x_0(t), \boldsymbol{\lambda}, t)$ and $c_0(x_0(t), \boldsymbol{\lambda}, t)$ are Lipschitz in $\boldsymbol{\lambda}$:*

$$\begin{aligned} \|\bar{\boldsymbol{\varphi}}(x_0(t), \boldsymbol{\lambda}, t) - \bar{\boldsymbol{\varphi}}(x_0(t), \boldsymbol{\lambda}', t)\| &\leq D\|\boldsymbol{\lambda} - \boldsymbol{\lambda}'\|, \\ \|c_0(x_0(t), \boldsymbol{\lambda}, t) - c_0(x_0(t), \boldsymbol{\lambda}', t)\| &\leq D_c\|\boldsymbol{\lambda} - \boldsymbol{\lambda}'\|. \end{aligned} \tag{8.73}$$

Then there exist numbers $\varepsilon > 0$ and $\gamma^* > 0$ such that for all $\gamma \in (0, \gamma^*]$

(1) the trajectories of the closed-loop system (8.70)–(8.72) are bounded and

$$\lim_{t\to\infty} \|\hat{x}_0(t) - x_0(t)\|_\varepsilon = 0;$$

(2) there exists $\boldsymbol{\lambda}^* \in \Omega_\lambda$, $\kappa \in \mathbb{R}_{>0}$ such that

$$\lim_{t\to\infty} \hat{\boldsymbol{\lambda}}(t) = \boldsymbol{\lambda}^*. \tag{8.74}$$

Remark 8.5 Corollary 8.2 assures that the estimates $\hat{\boldsymbol{\theta}}(t)$ and $\hat{\boldsymbol{\lambda}}(t)$ asymptotically converge to a neighborhood of the actual values $\boldsymbol{\theta}$ and $\boldsymbol{\lambda}$. It does not specify, however, how close these estimates are to the true values of $\boldsymbol{\theta}$ and $\boldsymbol{\lambda}$.

The next result states that, if the values of Δ_φ and Δ_ξ,

$$\|\boldsymbol{\varphi}(x_0,\boldsymbol{\lambda},t) - \bar{\boldsymbol{\varphi}}(x_0,\boldsymbol{\lambda},t)\| \leq \Delta_\varphi,$$

$$\sum_{i=1}^{n} \frac{1}{\tau_i} \|\xi_i(\tau)\|_{\infty,[t_0,\infty]} + \|\xi_0(\tau)\|_{\infty,[t_0,\infty]} \leq \Delta_\xi,$$

in (8.67) and (8.69) are small, e.g. $\bar{\boldsymbol{\varphi}}(x_0(t),\boldsymbol{\lambda},t)$ approximates $\boldsymbol{\varphi}(x_0(t),\boldsymbol{\lambda},t)$ with sufficiently high accuracy and the unmodeled dynamics is negligible, then the estimates $\hat{\boldsymbol{\theta}}(t)$ and $\hat{\boldsymbol{\lambda}}(t)$ will converge to small neighborhoods of the equivalence classes of $\boldsymbol{\theta}$ and $\boldsymbol{\lambda}$. The sizes of these neighborhoods are shown to be bounded from above by monotone functions Δ:

$$\Delta = \|\boldsymbol{\theta}\|\Delta_\varphi + \Delta_\xi$$

vanishing at zero. Formally this result is stated below.

Corollary 8.3 *Let the assumptions of Corollary 8.2 hold, and assume that* $\bar{\varphi}_0(x_0,\boldsymbol{\lambda},t) \in \mathcal{C}^1$, *the derivative* $\partial\bar{\boldsymbol{\varphi}}_0(x_0(t),\boldsymbol{\lambda},t)/\partial t$ *is globally bounded, and* $\Delta = \|\boldsymbol{\theta}\|\Delta_\varphi + \Delta_\xi$ *is small. Then there exist numbers* $\varepsilon > 0$ *and* $\gamma^* > 0$ *such that for all* $\gamma_i \in (0,\gamma^*)$

(1) $\limsup_{t\to\infty} \|\hat{\boldsymbol{\theta}}(t) - \boldsymbol{\theta}\| = r_1(\Delta),\ r_1 \in \mathcal{K}$;
(2) if $\boldsymbol{\theta}^{\mathrm{T}}\bar{\boldsymbol{\varphi}}(x_0(t),\boldsymbol{\lambda},t) + c_0(x_0(t),\boldsymbol{\lambda},t)$ *is weakly nonlinearly persistently exciting with respect to* $\boldsymbol{\lambda}$*, then the estimates* $\hat{\boldsymbol{\lambda}}(t)$ *converge into a small vicinity of* $\mathcal{E}(\lambda)$*:*

$$\limsup_{t\to\infty} \left\|\hat{\boldsymbol{\lambda}}(t)\right\|_{\mathcal{E}(\boldsymbol{\lambda})} = r_2(\Delta),\ r_2 \in \mathcal{K}.$$

8.4 Examples

The canonical form (8.51) and the corresponding observer were implemented in MATLAB, and numerical results are presented in Figure 8.3. The measured trajectory $x_0(t)$ was generated by the Hindmarsh–Rose model with parameter values set as follows:

$$\lambda = 2.027,\quad \theta_{1,3} = -10.4,\quad \theta_{1,2} = -4.35,\quad \theta_{1,1} = 6.65,$$
$$\theta_{1,0} = 0.9125,\quad \theta_{2,2} = -32.45,\quad \theta_{1,1} = -32.15.$$

The parameters of the observer were chosen as

$$k = 1,\qquad c_1 = 1,\qquad \Gamma = \mathrm{diag}(1,1,1,1,1,1,1,1),\qquad F = -1.$$

In addition to simulating the Bastin–Gevers observer for the Hindmarsh–Rose model we also simulated the observer (8.58) derived within the framework of the

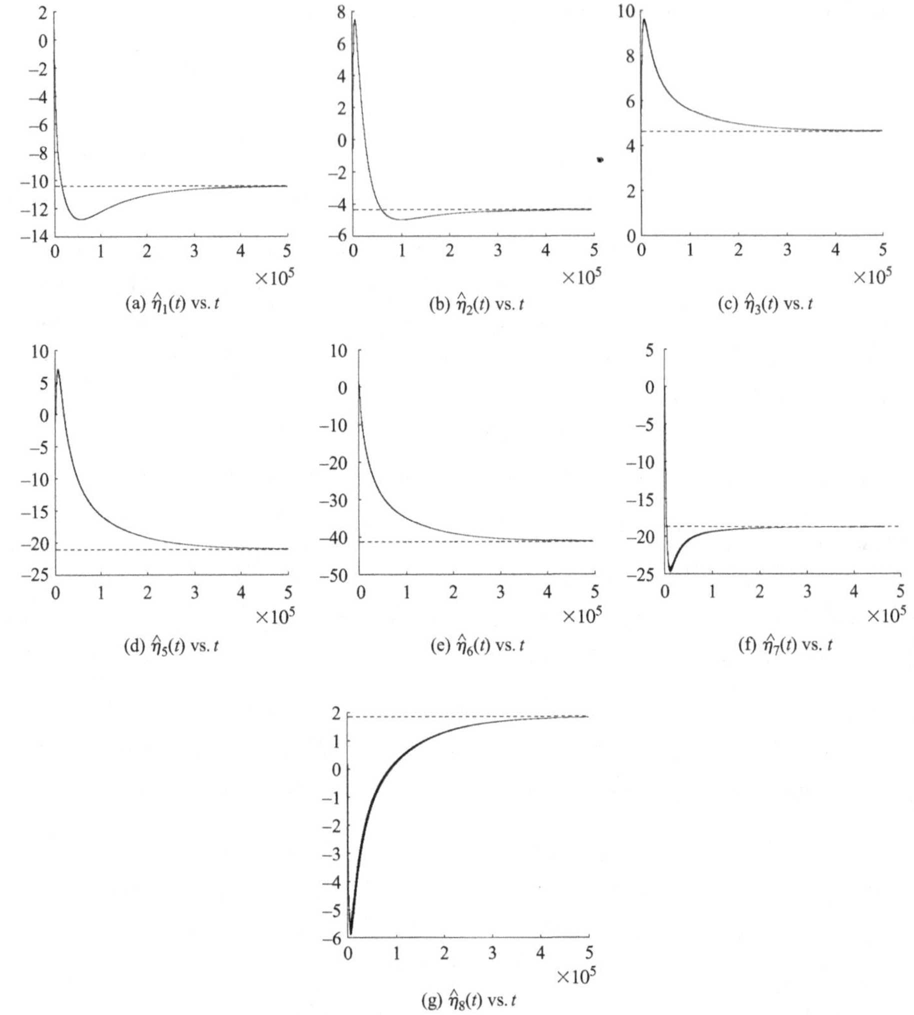

Figure 8.3 Simulation results for the Bastin–Gevers observer. Each $\hat{\eta}_i(t)$, $i \neq 4$, is shown with the true values indicated with a broken line. The periodic time of the spikes is about 10 s while the total time simulated is 5×10^5 s, thus the figures spans thousands of spike cycles. The visual effect of these extreme time scales is seen in the figures; hunting oscillations within the observer are seen as thickening of the graphs of certain estimated parameters.

approach presented in Marino (1990). In this case we set

$$\theta_{1,3} = -1, \qquad \theta_{1,2} = 3, \qquad \theta_{1,3} = 0, \qquad \theta_{1,0} = 1.5,$$
$$\theta_{2,3} = 0, \qquad \theta_{2,2} = -5, \qquad \theta_{2,1} = 0, \qquad \lambda = -1.$$

There is no particular reasoning behind such a choice of parameters in both this and the previous case apart from the fact that these parameters must induce persistent oscillatory dynamics of the solutions of (8.32). The parameters of the observer were chosen as follows:

$$l_1 = 2, \qquad l_2 = 1, \qquad \gamma = 1, \qquad k = 1.$$

Simulation results for this system are shown in Figure 8.4.

Let us now turn to a more realistic class of equations, i.e. conductance-based models. In particular, we consider the Morris–Lecar model (Morris and Lecar 1981):

$$\begin{aligned} \dot{V} &= \frac{1}{C}(-\bar{g}_{\mathrm{Ca}} m_\infty(V)(V - E_{\mathrm{Ca}}) - \bar{g}_{\mathrm{K}} w(V - E_{\mathrm{K}}) - \bar{g}_{\mathrm{L}}(V - V_0)) + I, \\ \dot{w} &= -\frac{1}{\tau(V)} w + \frac{w_\infty(V)}{\tau(V)}, \end{aligned} \tag{8.75}$$

where

$$\begin{aligned} m_\infty(V) &= 0.5\left(1 + \tanh\left(\frac{V - V_1}{V_2}\right)\right), \\ w_\infty(V) &= 0.5\left(1 + \tanh\left(\frac{V - V_3}{V_4}\right)\right), \\ \tau(V) &= T_0 \frac{1}{\cosh[(V - V_3)/(2V_4)]}. \end{aligned}$$

System (8.75) is a reduction of the standard four-dimensional Hodgkin–Huxley equations, and is one of the simplest models describing the dynamics of evoked membrane potential and, at the same time, claiming biological plausibility.

The parameters $\bar{g}_{\mathrm{Ca}}$, $\bar{g}_{\mathrm{K}}$, and $\bar{g}_{\mathrm{L}}$ stand for the maximal conductances of the calcium, potassium, and leakage currents, respectively; C is the membrane capacitance; V_1, V_2, V_3, and V_4 are the parameters of the gating variables; T_0 is the parameter regulating the time scale of ionic currents; E_{Ca} and E_{K} are the Nernst potentials of the calcium and potassium currents; and E_{L} is the rest potential. The variable I models an external stimulation current. In this example the value of I was set to $I = 10$.

The total number of parameters in system (8.75) is 12, excluding the stimulation current I. Some of these parameters, however, are already available or can be considered typical. For example the values of the Nernst potentials for calcium and potassium channels, E_{Ca} and E_{K}, are known and usually are set as $E_{\mathrm{Ca}} = 100$ and $E_{\mathrm{K}} = -70$. The value of the rest potential, V_0, can be estimated from the cell

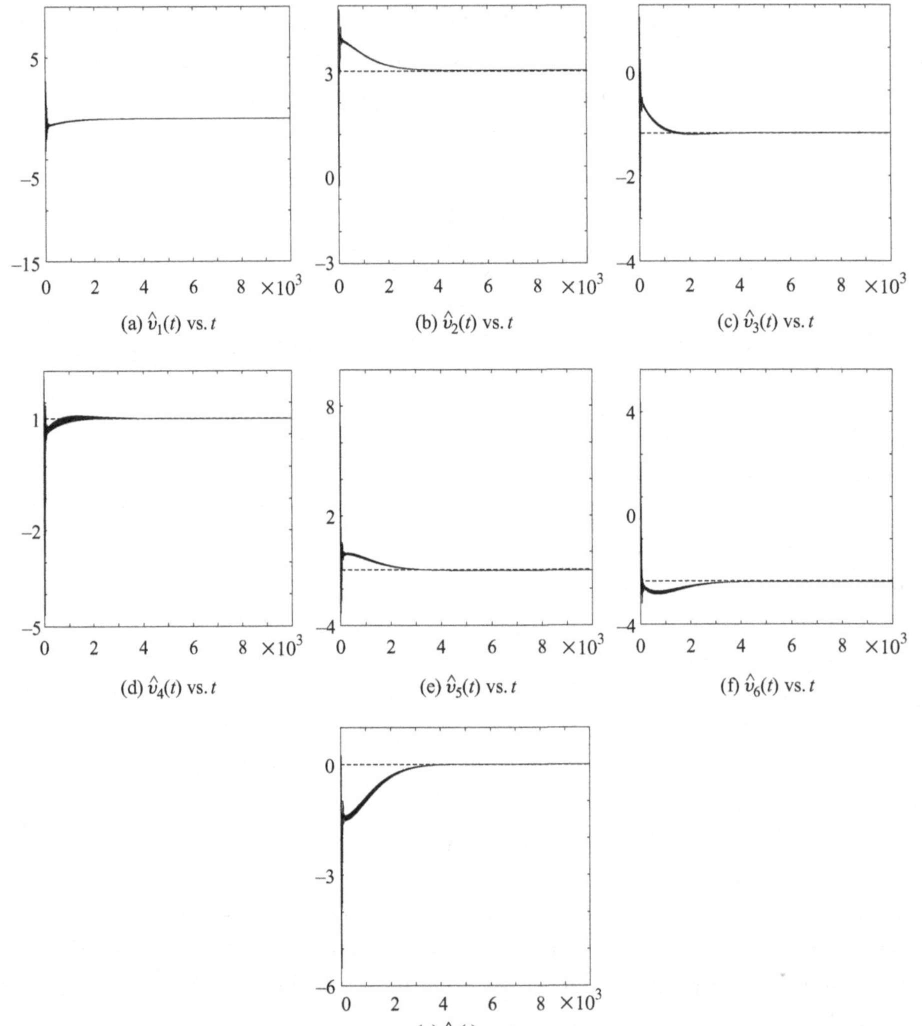

(a) $\hat{\upsilon}_1(t)$ vs. t (b) $\hat{\upsilon}_2(t)$ vs. t (c) $\hat{\upsilon}_3(t)$ vs. t

(d) $\hat{\upsilon}_4(t)$ vs. t (e) $\hat{\upsilon}_5(t)$ vs. t (f) $\hat{\upsilon}_6(t)$ vs. t

(g) $\hat{\upsilon}_7(t)$ vs. t

Figure 8.4 Simulation results for the Marino–Tomei observer. Each $\hat{\upsilon}_i(t)$ is shown with the true values indicated with a broken line. The periodic time of the spikes is about 10 s while the total time simulated is 10×10^3 s, thus the figures span thousands of spike cycles. The visual effect of these extreme time scales is seen in the figures; hunting oscillations within the observer are seen as thickening of the graphs of certain estimated parameters.

explicitly. Here we set $V_0 = -50$. Parameters V_1 and V_2 characterize the steady-state response curve of the activation gates corresponding to the calcium channels, and V_3 and V_4 are the parameters of the potassium channels. In the simulations we set these parameters to standard values as e.g. in Koch (2002): $V_1 = -1$, $V_2 = 15$, $V_3 = 10$, and $V_4 = 29$.

The values of the parameters $\bar{g}_{\mathrm{Ca}}, \bar{g}_{\mathrm{K}}, \bar{g}_{\mathrm{L}}$, and T_0, however, may vary substantially from one cell to another. For example, the values of $\bar{g}_{\mathrm{Ca}}$, $\bar{g}_{\mathrm{K}}$, $\bar{g}_{\mathrm{L}}$ depend on the density of ion channels in a patch of the membrane; and the value of T_0 is dependent on temperature. Hence, in order to model the dynamics of individual cells, we need to be able to recover these values from data.

As before, we suppose that the values of V over time are available for direct observation, and the values of w are not measured. System (8.75) has no linear time-invariant part, and the dynamics of w are governed by a nonlinear differential equation with the time-varying relaxation factor, $1/\tau(V)$. Therefore, the observers presented in Section 8.2 need not be applied explicitly to this system. This does not imply, however, that the parameters $\bar{g}_{\mathrm{Ca}}$, $\bar{g}_{\mathrm{K}}$, $\bar{g}_{\mathrm{L}}$, and T_0 cannot be recovered from the measurements of V. In fact, as we show below, one can successfully reconstruct these parameters by using observers defined in Section 8.3.

For the sake of notational consistency we denote $x_0 = V$, $x_1 = w$, and without loss of generality suppose that $C = 1$. Hence, system (8.75) can now be rewritten as follows:

$$
\begin{aligned}
\dot{x}_0 &= \theta_{0,1} m_\infty(x_0)(x_0 - E_{\mathrm{Ca}}) + \theta_{0,1} x_1 (x_0 - E_{\mathrm{K}}) + \theta_{0,3}(x_0 - V_0) + I, \\
\dot{x}_1 &= -\beta_1(x_0, \lambda) x_1 + \frac{1}{\lambda}\phi_1(x_0),
\end{aligned}
\tag{8.76}
$$

where

$$
\beta_1(x_0, \lambda) = \frac{1}{\lambda}\cosh\left(\frac{x_0 - V_3}{2V_4}\right), \quad \lambda = T_0,
$$

$$
\phi_1(x_0) = \cosh\left(\frac{x_0 - V_3}{2V_4}\right) w_\infty(x_0).
$$

Noticing that $\beta_1(x_0, \lambda)$ is separated away from zero for all bounded x_0 and positive λ, we substitute variable x_1 in (8.76) with its estimation

$$
\chi(\lambda, t) = \int_{t-T}^{t} \frac{1}{\lambda} e^{-\frac{1}{\lambda}\int_\tau^t \cosh\left(\frac{x_0(s) - V_3}{2V_4}\right) ds} \cosh\left(\frac{x_0(\tau) - V_3}{2V_4}\right) w_\infty(x_0(\tau)) d\tau.
$$

The larger the value of T the higher the accuracy of estimation for large t. After this substitution system (8.76) reduces to just one equation,

$$
\dot{x}_0 = \theta_{0,1}\phi_{0,1}(x_0) + \theta_{0,2}\phi_{0,2}(x_0, \lambda, t) + \theta_{0,3}\phi_{0,3}(x_0) + I + \xi_0(t), \tag{8.77}
$$

where

$$
\begin{aligned}
\phi_{0,1}(x_0) &= m_\infty(x_0)(x_0 - E_{\mathrm{Ca}}), \\
\phi_{0,2}(x_0, \lambda, t) &= (x_0 - E_{\mathrm{K}})\chi(\lambda, t), \\
\phi_{0,3}(x_0) &= (x_0 - V_0),
\end{aligned}
$$

and the term $\xi_0(t)$ is bounded.

Equation (8.77) is a special case of (8.63), and hence we can apply the results of Section 8.3 to construct an observer for asymptotic estimation of the values of $\theta_{0,1}$, $\theta_{0,2}$, $\theta_{0,3}$, and λ. In accordance with (8.70)–(8.72) we obtain the following observer equations:

$$\begin{aligned}
\dot{\hat{x}}_0 &= -\alpha(\hat{x}_0 - x_0) + \hat{\theta}_1\phi_{0,1}(x_0) + \hat{\theta}_2\phi_{0,2}(x_0, \hat{\lambda}, t) + \hat{\theta}_3\phi_{0,3}(x_0) + I, \\
\dot{\hat{\theta}}_1 &= -\gamma_\theta(\hat{x}_0 - x_0)\phi_{0,1}(x_0), \\
\dot{\hat{\theta}}_2 &= -\gamma_\theta(\hat{x}_0 - x_0)\phi_{0,2}(x_0, \hat{\lambda}, t), \\
\dot{\hat{\theta}}_3 &= -\gamma_\theta(\hat{x}_0 - x_0)\phi_{0,3}(x_0), \qquad (8.78)\\
\hat{\lambda} &= 3 + \hat{x}_{1,1},
\end{aligned}$$

$$\begin{aligned}
\dot{\hat{x}}_{1,1} &= \gamma e\left(\hat{x}_{1,1} - \hat{x}_{2,1} - \hat{x}_{1,1}\left(\hat{x}_{1,1}^2 + \hat{x}_{2,1}^2\right)\right), \\
\dot{\hat{x}}_{2,1} &= \gamma e\left(\hat{x}_{1,1} + \hat{x}_{2,1} - \hat{x}_{2,1}\left(\hat{x}_{1,1}^2 + \hat{x}_{2,1}^2\right)\right), \\
e &= \sigma(\|x_0 - \hat{x}_0\|_\varepsilon). \qquad (8.79)
\end{aligned}$$

The parameters of the observer were set as follows: $\alpha = 1$, $\varepsilon = 0.001$, $\gamma = 0.01$, and $\gamma_\theta = 0.05$.

According to Theorems 8.2 and 8.3, the observer (8.78) and (8.79) should ensure successful reconstruction of the model parameters, provided that the regressor is persistently exciting. This requirement is satisfied for model (8.76) generating periodic solutions. We simulated the system (8.76), (8.78), and (8.79) over a wide range of initial conditions. Figure 8.5 shows an example of the typical behavior of the observer over time. As we can see from this figure, all estimates converge to small neighborhoods of true values of the parameters.

8.5 Summary

In this chapter we have explored observer-based approaches to the problem of state and parameter reconstruction for classes of typical models of neural oscillators. The estimation procedure in this approach is defined as a system of ordinary differential equations of which the right-hand side does not depend explicitly on the unmeasured variables. The solution of this system (or functions of the solutions) should asymptotically converge to small neighborhoods of the actual values of the variables to be estimated. Until recently, due to nonlinear dependences of the vector fields of the models on unknown parameters and also due to uncertainties in the time scales of hidden variables, the observer-based approach to solving the problem

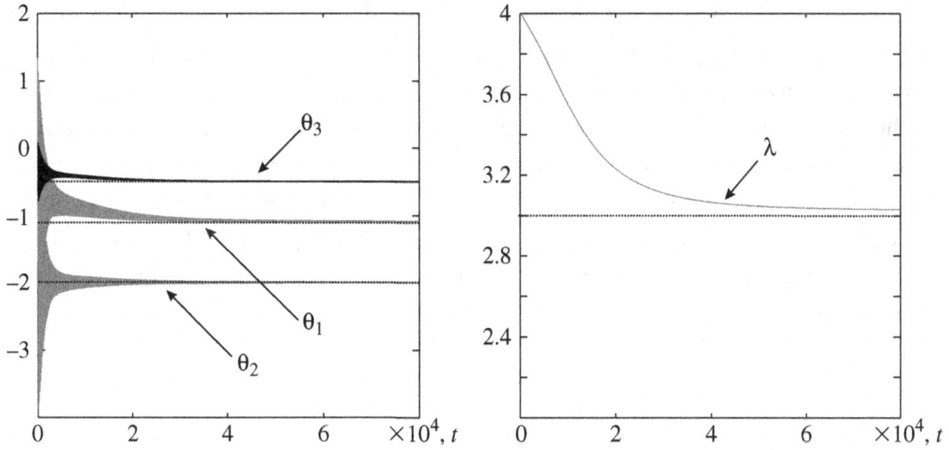

Figure 8.5 Trajectories of the estimates of parameters $\theta_1, \theta_2, \theta_3$, and λ as functions of time. The periodic time of the spikes is about 10 s while the total time simulated is 8×10^4 s, thus the figures span thousands of spike cycles. The visual effect of these extreme time scales is seen in the figures; hunting oscillations within the observer are seen as thickening of the graphs of certain estimated parameters.

of state and parameter estimation of neural oscillators was a relatively unexplored territory. Here we demonstrate that, despite these obvious difficulties, the approach can be successfully applied to a wide range of models.

Two different strategies for observer design have been investigated. The first strategy is based on the availability of canonical representations of the original system. The success of this strategy is obviously determined by whether one can find a suitable coordinate transformation such that the equations of the original model can be transformed into the canonical adaptive-observer form. Because a coordinate transformation is required, different classes of models are likely to lead to different observers. The second strategy is based on the ideas described in Chapters 4 and 5. The structure of observers obtained as a result of this design strategy does not change much from one model to another. The main difference between these design strategies is in the convergence rates: exponential for the first and asymptotical for the second. As long as merely the overall convergence time is accounted for there is no big difference caused by whether the convergence itself is exponential or not. Yet, the fact that it can be made exponential with known rates of convergence allows us to derive the a priori estimates of the amount of time needed to achieve a certain given accuracy of estimation.

We have shown that for linearly parametrized models such as e.g. the FitzHugh–Nagumo and Hindmarsh–Rose oscillators one can develop an observer for state and parameter estimation of which the convergence rate is exponential. For the nonlinearly parametrized and more realistic models such as the Morris–Lecar and Hodgkin–Huxley equations we presented an observer of which the convergence is asymptotic. In both cases the rate of convergence depends on the degree of excitation in the measured data. In the case of linearly parametrized systems this excitation can be measured in terms of the minimal eigenvalue of a certain matrix constructed explicitly from the data and the model. For the nonlinearly parametrized systems the degree of excitation is defined by a more complex expression. In principle, one can ensure arbitrarily fast convergence of the estimator, provided that the excitation is sufficiently high.

Appendix. The Meyer–Kalman–Yakubovich lemma

Definition 1 *A function $H : \mathbb{C} \to \mathbb{C}$ is said to be positive real if it is analytic in the open right-half of the complex plane, and*

$$\mathrm{Re}(H(s)) \geq 0 \; \forall \, s : \mathrm{Re}(s) \geq 0.$$

Definition 2 *A function $H : \mathbb{C} \to \mathbb{C}$ is said to be*

(i) positive real if it is analytic in the open right-half of the complex plane, zeros and poles of H on the imaginary axis are simple with the remaining residues ≥ 0, and $\mathrm{Re}(H(i\omega)) \geq 0$ for all $\omega \in (-\infty, \infty)$;
(ii) strictly positive real if the function $H(s)$ is not identically zero, and $H(s - \epsilon)$ is positive real for some $\epsilon > 0$.

It is easy to see that if the function $H(s)$ is positive real then so are the functions $1/H(s)$ and $H(1/s)$. Thus, if

$$H(s) = \frac{b_m s^m + b_{m-1} s^{m-1} + \cdots + b_0}{a_n s^n + a_{n-1} s^{n-1} + \cdots + a_0},$$

is positive real then so must be the function

$$G(s) = H(1/s) = s^{n-m} \frac{(b_m + b_{m-1} s + \cdots + b_0 s^m)}{a_n + a_{n-1} s + \cdots + a_0 s^n}.$$

Because zeros and poles of a positive real function on the imaginary axis must be simple, this implies that $|n - m| \leq 1$. This fact allows one to produce an important characterization of strictly positive real rational functions $H(i\omega)$ as $\omega \to \infty$. The result was first established by J. H. Taylor (1974), and it is summarized in the lemma below.

Lemma 1 *Let p and d be two polynomials with* $\deg(d) > \deg(p)$*, and let* $H(s) = p(s)/d(s)$ *be a strictly positive real rational function. Then* $\mathrm{Re}(H(i\omega))$ *can go to zero no more rapidly than* ω^{-2} *as* $\omega \to \infty$.

Proof of Lemma 1. Given that the function $H(s)$ is positive real, we can conclude that $\deg(d) = \deg(p) + 1$. Let $\deg(d) = n$, then

$$H(s) = \frac{b_{n-1}s^{n-1} + b_{n-2}s^{n-2} + \cdots + b_0}{s^n + a_{n-1}s^{n-1} + \cdots + a_0}, \ b_i > 0, \ a_i > 0 \, \forall \, 0 \le i \le n-1.$$

If the function $H(s)$ is strictly positive real then there must exist $\epsilon > 0$ such that the function $H(s - \epsilon)$,

$$H(s-\epsilon) = \frac{p(s-\epsilon)d(-s-\epsilon)}{d(s-\epsilon)d(-s-\epsilon)} = \frac{\omega^{2n-2}(a_{n-1}b_{n-1} - b_{n-2} - \epsilon b_{n-1}) + \cdots}{\omega^{2n} + \cdots},$$

is positive real. This, however implies the existence of ϵ such that $b_{n-1}a_{n-1} - b_{n-2} - \epsilon b_{n-1} > 0$. Thus $\lim_{\omega\to\infty} \mathrm{Re}(H(i\omega))\omega^2 > 0$, and statement of the lemma follows. □

Now we are ready to state and prove a version of the Lefschetz–Kalman–Yakubovich lemma.

Lemma 2 *Let $A \in \mathbb{R}^{n\times n}$ be a Hurwitz matrix, and $b, c \in \mathbb{R}^n$ be arbitrary vectors. Then there exist symmetric positive definite matrices P and Q:*

$$PA + A^{\mathrm{T}}P = -Q,$$
$$Pb = c$$

if and only if the function

$$H(s) = c^{\mathrm{T}}(sI - A)^{-1}b$$

is strictly positive real.

Proof of Lemma 2.

(i) Sufficiency. First of all notice that for any $A \in \mathbb{R}^{n\times n}$, $b \in \mathbb{R}^n$, and $P = P^{\mathrm{T}}$ the following identity holds:

$$(-i\omega - A)^{\mathrm{T}}P + P(i\omega - A) = -(PA + A^{\mathrm{T}}P).$$

Hence

$$\begin{aligned} &b^{\mathrm{T}}((-i\omega - A)^{-1})^{\mathrm{T}}[(-i\omega - A)^{\mathrm{T}}P + P(i\omega - A)](i\omega - A)^{-1}b \\ &= b^{\mathrm{T}}P(i\omega - A)^{-1}b + b^{\mathrm{T}}((-i\omega - A)^{-1})^{\mathrm{T}}Pb. \end{aligned} \qquad (1)$$

Case 1. Suppose that (A, b), $b \in \mathbb{R}^n$ is a controllable pair. Since the pair A, b is controllable there is a similarity transformation T: $A_0 = TAT^{-1}$, $b_0 = Tb$, $c_0 = (T^{\mathrm{T}})^{-1}c$, such that A_0 and b_0 are in the canonic controllable form

$$A_0 = \begin{pmatrix} a_1 & a_2 & a_3 & \cdots & a_n \\ 1 & 0 & 0 & \cdots & 0 \\ 0 & 1 & 0 & \cdots & 0 \\ \cdots & \cdots & \cdots & \cdots & \cdots \\ 0 & 0 & 0 & \cdots & 0 \\ 0 & 0 & 0 & \cdots & 0 \end{pmatrix}, \qquad b_0 = (1, 0, \ldots, 0)^{\mathrm{T}}. \tag{2}$$

Let $h(s) = (sI - A_0)^{-1}b_0$, and consider $g(s) = c_0^{\mathrm{T}}h(s) + h(s)^*c_0$. The function $H(s) = c_0^{\mathrm{T}}h(s)$ is clearly strictly positive real. Hence, according to Lemma 1, the function $g(i\omega) = 2\,\mathrm{Re}(c_0^{\mathrm{T}}h(i\omega))$ is real and can be represented as

$$\begin{aligned} g(i\omega) = \frac{r_{n-1}\omega^{2n-2} + r_{n-2}\omega^{2n-4} + \cdots}{\omega^{2n} + \cdots} &= \frac{\mathrm{Re}(p(i\omega)d_1(-i\omega))}{d_1(i\omega)d_1(-i\omega)} \\ &= \frac{\mathrm{Re}(p(i\omega)d_1(-i\omega))}{d(\omega)}, \end{aligned} \tag{3}$$

where $p(i\omega), d_1(i\omega)$, and $d(\omega)$ are polynomials with real coefficients, and $r_{n-1} > 0$. Let L be a symmetric positive definite matrix. Consider

$$w(i\omega) = g(i\omega) - \varepsilon h(i\omega)^*Lh(i\omega).$$

We claim that there exists $\varepsilon > 0$ such that $w(i\omega) > 0$ for all $\omega \in (-\infty, \infty)$. Since L is symmetric, we have that

$$w(i\omega) \geq g(i\omega) - \varepsilon\lambda_{\max}(L)h(i\omega)^*h(i\omega).$$

Notice that the function $h(i\omega)^*h(i\omega)$ is real, and that it can be expressed as

$$h(i\omega)^*h(i\omega) = \frac{p_1(\omega)}{d(\omega)},$$

where p_1 is a real polynomial of ω of degree $2n - 2$. According to (3) $r_{n-1} > 0$. Hence there is an $\omega_0 > 0$ such that $(r_{n-1}/2)\omega^{2n-2} > \epsilon_1\lambda_{\max}(L)p_1(\omega) + r_{n-2}\omega^{2n-4} + \cdots$ for all $|\omega| \geq |\omega_0|$. On the other hand, since the function $H(s)$ is strictly positive real, there must exist an $\epsilon_2 > 0$ such that $g(i\omega) > \epsilon_2\lambda_{\max}(L)p_1(\omega)$ for $|\omega| < |\omega_0|$. Taking $\varepsilon = \min\{\epsilon_1, \epsilon_2\}$ ensures that $w(i\omega) > 0$ for all $\omega \in (-\infty, \infty)$.

Let us fix $\varepsilon > 0$ such that $w(i\omega) > 0$ for all ω. The function $w(i\omega)$ is a rational function:

$$\frac{p_2(\omega)}{d_1(i\omega)d_1(-i\omega)},$$

where the polynomial p_2 has no real roots. Hence

$$p_2 = \prod_{i=1}^{m} r_n(\omega^2 + k_i) \prod_{i=m+1}^{n-1} (\omega^2 + \beta_i)(\omega^2 + \beta_i^*),\ k_i > 0,\ \beta_i \in \mathbb{C}.$$

Noticing that

$$\begin{aligned}
\omega^2 + k_i &= (i\omega + k_i^{1/2})(-i\omega + k_i^{1/2}),\\
(\omega^2 + \beta_i) &= (i\omega + \beta_i^{1/2})(-i\omega + \beta_i^{1/2}),\\
(\omega^2 + \beta_i^*) &= (i\omega + \beta_i^{*1/2})(-i\omega + \beta_i^{*1/2}),\ \beta_i^{*1/2} = \beta_i^{1/2^*}
\end{aligned}$$

and that

$$\begin{aligned}
&(i\omega + k_i^{1/2})(i\omega + \beta_i^{1/2})(i\omega + \beta_i^{*1/2})\\
&\quad = (i\omega + k_i^{1/2})((i\omega)^2 + i\omega(\beta_i^{1/2} + \beta_i^{*1/2}) + \beta_i^{1/2}\beta_i^{*1/2})
\end{aligned}$$

and

$$\begin{aligned}
&(-i\omega + k_i^{1/2})(-i\omega + \beta_i^{1/2})(-i\omega + \beta_i^{*1/2})\\
&\quad = (-i\omega + k_i^{1/2})((-i\omega)^2 - i\omega(\beta_i^{1/2} + \beta_i^{*1/2}) + \beta_i^{1/2}\beta_i^{*1/2})
\end{aligned}$$

are identical polynomials in $i\omega$ and $-i\omega$ with real coefficients, we can conclude that there exists a polynomial p_3 such that

$$\frac{p_2(\omega)}{d(\omega)} = \frac{p_3(i\omega)p_3(-i\omega)}{d_1(i\omega)d_1(-i\omega)}.$$

Notice that $d_1(s)$ must necessarily be the characteristic polynomial of A_0, and that the degree of p_3 is $n-1$. Finally, we notice that

$$(sI - A_0)^{-1}b_0 = \frac{1}{d_1(s)}(s^{n-1}, s^{n-2}, \ldots, 1)^{\mathrm{T}}, \tag{4}$$

so that $p_3(s)/d_1(s)$ can be represented as

$$\frac{p_3(s)}{d_1(s)} = q^{\mathrm{T}}h(s),\ q \in \mathbb{R}^n. \tag{5}$$

Thus

$$w(i\omega) = h(s)^* qq^{\mathrm{T}} h(s) = c_0^{\mathrm{T}} h(s) + h^*(s) c_0 - \varepsilon h^*(s) L h(s) > 0 \tag{6}$$

and

$$c_0^{\mathrm{T}} h(s) + h^*(s) c_0 = \varepsilon h^*(s) L h(s) + h(s)^* qq^{\mathrm{T}} h(s) > 0. \tag{7}$$

Let P_0 be the solution of

$$P_0 A_0 + A_0^{\mathrm{T}} P_0 = -qq^{\mathrm{T}} - \varepsilon L. \tag{8}$$

Given that A_0 is Hurwitz, such a matrix always exists and is uniquely defined by q and L. According to (1) and (7) we have that

$$c_0^{\mathrm{T}} h(s) + h^*(s) c_0 = b^{\mathrm{T}} P_0 h(s) + h(s)^* P_0 b_0.$$

This implies that $\mathrm{Re}((c_0^{\mathrm{T}} - b^{\mathrm{T}} P_0) h(s)) = 0$ for all s. The latter fact ensures that $c_0^{\mathrm{T}} - b^{\mathrm{T}} P_0$ must necessarily be zero, and hence $P_0 b_0 = c_0$.

Finally, on pre-multiplying (8) by T^{T}, post-multiplying it by T, and denoting $P = T^{\mathrm{T}} P_0 T$ and $Q = T^{\mathrm{T}}(qq^{\mathrm{T}} + \varepsilon L)T$, we get

$$PA + A^{\mathrm{T}} P = -Q, \qquad Pb = c,$$

which proves sufficiency for the case of (A, b) being controllable.

Case 2. The pair (A, b) is not controllable. Since the pair does not consist of controllable identities (2), (4), and (5) might not hold, and hence a modification of the sufficiency part is needed. The modification was proposed by K. R. Meyer (1966), and we reproduce it below with minor changes.

First of all notice that strict positive realness of $H(s)$ implies that b and c are non-zero vectors. For any $A \in \mathbb{R}^{n\times n}$, $b, c \in \mathbb{R}^n$, $b, c \neq 0$, there is a similarity transformation T such that

$$TAT^{-1} = \tilde{A} = \begin{pmatrix} A_1 & A_2 \\ 0 & A_3 \end{pmatrix}, \qquad Tb = \tilde{b} = \begin{pmatrix} b_1 \\ 0 \end{pmatrix},$$

$$(T^{\mathrm{T}})^{-1} c = \tilde{c} = \begin{pmatrix} c_1 \\ c_2 \end{pmatrix},$$

where $A_1 \in \mathbb{R}^{n_1\times n_1}$, $A_2 \in \mathbb{R}^{n_1\times n_2}$, $A_3 \in \mathbb{R}^{n_2\times n_2}$, $b_1, c_1 \in \mathbb{R}^{n_1}$, $c_2 \in \mathbb{R}^{n_2}$, and (A_1, b_1) is a controllable pair.

Thus if we find a symmetric positive definite matrix

$$\tilde{P} = \begin{pmatrix} P_1 & P_2 \\ P_2^{\mathrm{T}} & P_3 \end{pmatrix}, \quad P_1 \in \mathbb{R}^{n_1\times n_1}, \quad P_2 \in \mathbb{R}^{n_1\times n_2}, \quad P_3 \in \mathbb{R}^{n_2\times n_2}$$

such that

$$\begin{aligned}
&\tilde{P}\tilde{A} + \tilde{A}^{\mathrm{T}}\tilde{P} = -\tilde{Q},\\
&\tilde{Q} = \tilde{q}\tilde{q}^{\mathrm{T}} + \varepsilon\tilde{L},\ \tilde{q} = \begin{pmatrix} q_1 \\ q_2 \end{pmatrix},\ q_1 \in \mathbb{R}^{n_1},\ q_2 \in \mathbb{R}^{n_2},\ \tilde{L} = \begin{pmatrix} L_1 & 0 \\ 0 & L_2 \end{pmatrix}, \\
&\qquad L_1 > 0, L_2 > 0,\\
&\tilde{P}\tilde{b} = \tilde{c},
\end{aligned} \tag{9}$$

then $T^{\mathrm{T}}\tilde{P}T$ is the desired solution for the original matrix A and vectors b and c.

Equation (9), if expressed in terms of individual blocks, leads to the following set of conditions:

$$P_1A_1 + A_1^{\mathrm{T}}P_1 = -q_1q_1^{\mathrm{T}} - \varepsilon L_1, \qquad P_1b_1 = c_1, \tag{10}$$

$$P_1A_2 + P_2A_3 + A_1^{\mathrm{T}}P_2 = -q_1q_2^{\mathrm{T}}, \tag{11}$$

$$P_2^{\mathrm{T}}A_2 + P_3A_3 + A_2^{\mathrm{T}}P_2 + A_3^{\mathrm{T}}P_3 = -q_2q_2^{\mathrm{T}} - \varepsilon L_2, \tag{12}$$

$$P_2^{\mathrm{T}}b_1 = c_2. \tag{13}$$

Notice that

$$\begin{aligned}
H(s) &= c^{\mathrm{T}}(Is - A)^{-1}b = \tilde{c}^{\mathrm{T}}T(Is - A)^{-1}T^{-1}\tilde{b} = \tilde{c}(Is - TAT^{-1})^{-1}\tilde{b}\\
&= \tilde{c}(Is - \tilde{A})^{-1}\tilde{b},
\end{aligned}$$

where

$$(Is - \tilde{A})^{-1} = \begin{pmatrix} (Is - A_1)^{-1} & (Is - A_1)A_2(Is - A_3)^{-1} \\ 0 & (Is - A_3)^{-1} \end{pmatrix}.$$

This implies that

$$H(s) = \tilde{c}(Is - \tilde{A})^{-1}\tilde{b} = c_1^{\mathrm{T}}(Is - A_1)^{-1}b_1.$$

Hence the existence of $P_1 = P_1^{\mathrm{T}} > 0$ and q_1 satisfying (10) is already proven because the pair (A_1, b_1) is controllable and the function $H(s) = c_1^{\mathrm{T}}(Is - A_1)^{-1}b_1$ is strictly positive real (see Case 1). Condition (11) allows us to express an unknown matrix P_2 as a function of q_2:

$$\begin{aligned}
&P_2A_3 + A_1^{\mathrm{T}}P_2 = -q_1q_2^{\mathrm{T}} - P_1A_2 \Rightarrow\\
&\qquad P_2 = \int_0^{\infty} e^{A_1^T t}(q_1q_2^T + P_1A_2)e^{A_3 t}\,dt.
\end{aligned} \tag{14}$$

Given that A_1 and A_3 are Hurwitz, such a matrix is always defined. Substituting (14) into (13) results in

$$\left(\int_0^\infty e^{A_3^{\mathrm{T}}t}(q_2 q_1^{\mathrm{T}} + A_2^{\mathrm{T}} P_1)e^{A_1 t}\, dt\right) b_1 = c_2.$$

On noticing that $q_1^{\mathrm{T}} e^{A_1 t} b_1$ is a scalar we can rewrite the expression above as

$$\left(\int_0^\infty q_1^{\mathrm{T}} e^{A_1 t} b_1 e^{A_3^{\mathrm{T}} t}\, dt\right) q_2 = c_2 - \int_0^\infty e^{A_3^{\mathrm{T}} t} A_2^{\mathrm{T}} P_1 e^{A_1 t} b_1\, dt. \qquad (15)$$

It is clear that the vector q_2, and hence the matrix P_2, are uniquely defined, provided that the matrix

$$R = \left(\int_0^\infty q_1^{\mathrm{T}} e^{A_1 t} b_1 e^{A_3^{\mathrm{T}} t}\, dt\right)$$

is non-singular. Let M be a non-singular matrix, possibly complex, such that $MA_3^{\mathrm{T}}M^{-1}$ is in the Jordan canonical form. Such a matrix always exists, and $\det(MRM^{-1}) = \det R$. Thus, if we show that $\det(MRM^{-1}) \neq 0$, then this will ensure the existence of q_2 satisfying (15). Taking into account that the matrix MRM^{-1} is upper-triangular, non-singularity of R would follow if we could show that each element on the main diagonal of MRM^{-1} is non-zero. Notice that $Me^{A_3^{\mathrm{T}} t}M^{-1} = e^{MA_3^{\mathrm{T}}M^{-1}t}$. Hence the main diagonal of $e^{MA_3^{\mathrm{T}}M^{-1}t}$ is comprised of terms $e^{\lambda_i t}$, with λ_i being the eigenvalues of A_3. Thus the diagonal elements of MRM^{-1} can be expressed as

$$m_i = \int_0^\infty q_1^{\mathrm{T}} e^{A_1 t} b_1 e^{\lambda_i t}\, dt = q_1^{\mathrm{T}}(-I\lambda_i - A_1)^{-1} b_1.$$

According to (6), the vector q_1 can be chosen such that $q_1^{\mathrm{T}}(Is - A_1)^{-1}b_1$ is non-zero for all s with $\mathrm{Re}(s) \geq 0$. Given that $\mathrm{Re}(\lambda_i) < 0$, we can therefore conclude that $m_i \neq 0$. Hence there exist q_1, q_2, P_1 and P_2 satisfying conditions (10), (11), and (13). Matrices P_3 and L_2 satisfying (12) can now be immediately defined. This proves sufficiency.

(ii) Necessity follows from (1) and the fact that $Pb = c$. Indeed, let $\epsilon > 0$ be a small positive number, and consider

$$\begin{aligned} c^{\mathrm{T}}(i\omega - \epsilon - A)^{-1}b + b^{\mathrm{T}}(-i\omega - \epsilon - A)^{-1^{\mathrm{T}}}c &= 2\,\mathrm{Re}(H(i\omega - \epsilon)) \\ &= b^{\mathrm{T}}P(i\omega - \epsilon - A)^{-1}b \\ &\quad + b^{\mathrm{T}}(-i\omega - \epsilon - A)^{-1^{\mathrm{T}}}Pb. \end{aligned}$$

Now, using this fact, and that

$$(-i\omega - \epsilon - A)^{\mathrm{T}} P + P(i\omega - \epsilon - A) = -(PA + A^{\mathrm{T}} P) - 2\epsilon P,$$

we can conclude that

$$\begin{aligned}
2\,\mathrm{Re}(H(i\omega - \epsilon)) &= b^{\mathrm{T}} P(i\omega - \epsilon - A)^{-1} b + b^{\mathrm{T}} (-i\omega - \epsilon - A)^{-1^{\mathrm{T}}} P b \\
&= b^{\mathrm{T}} (-i\omega - \epsilon - A)^{-1^{\mathrm{T}}} \\
&\quad \left[(-i\omega - \epsilon - A)^{\mathrm{T}} P + P(i\omega - \epsilon - A) \right] \\
&\quad \times (i\omega - \epsilon - A)^{-1} b \Rightarrow
\end{aligned}$$

$$\begin{aligned}
2\,\mathrm{Re}(H(i\omega - \epsilon)) &= b^{\mathrm{T}} (-i\omega - \epsilon - A)^{-1^{\mathrm{T}}} (Q - 2\epsilon P)(i\omega - \epsilon - A)^{-1} b \\
&= b^{\mathrm{T}} (-i\omega - \epsilon - A)^{-1^{\mathrm{T}}} Q (i\omega - \epsilon - A)^{-1} b \\
&\quad - b^{\mathrm{T}} (-i\omega - \epsilon - A)^{-1^{\mathrm{T}}} 2\epsilon P (i\omega - \epsilon - A)^{-1} b \\
&\geq \|(i\omega - \epsilon - A)^{-1} b\|^2 \lambda_{\min}(Q) \\
&\quad - 2\epsilon \|(i\omega - \epsilon - A)^{-1} b\|^2 \lambda_{\max}(P).
\end{aligned}$$

Finally, since $\lambda_{\min}(Q) > 0$, there will always exist a $\epsilon > 0$ such that $\mathrm{Re}(H(i\omega - \epsilon)) \geq 0$ for all $\omega \in (-\infty, \infty)$. Thus the function $H(s)$ has to be strictly positive real. □

References

Abarbanel, H. D. I., Creveling, D. R., Farsian, R., and Kostuk, M. 2009. Dynamical state and parameter estimation. *SIAM J. Appl. Dynamical Systems*, **8**(4), 1341–1381.

Achard, P., and de Schutter, E. 2006. Complex parameter landscape for a complex neuron model. *PLOS Computational Biol.*, **2**(7), 794–804.

Alonso, J. M., Usrey, W. M., and Reid, R. C. 1996. Precisely correlated firing in cells of the lateral geniculate nucleus. *Nature*, **383**, 815—819.

Amit, D. J., Gutfreund, H., and Sompolinsky, H. 1985. Spin-glass models of neural networks. *Phys. Rev. A*, **32**, 1007–1018.

Amit, Y. 2002. *2D Object Detection and Recognition: Models, Algorithms and Networks*. Cambridge, MA: MIT Press.

Amit, Y., Grenader, U., and Piccioni, M. 1991. Structural image restoration through deformable templates. *J. American Statistical Association*, **86**(414), 376–387.

Angeli, D., Ingalls, B., Sontag, E., and Wang, Y. 2004. Separation principles for input–output and integral-input-to-state-stability. *SIAM J. Control and Optimization*, **43**, 256–276.

Annaswamy, A. M., Skantze, F. P., and Loh, A.-P. 1998. Adaptive control of continuous time systems with convex/concave parametrization. *Automatica*, **34**(1), 33–49.

Arcak, M., Angeli, D., and Sontag, E. 2002. A unifying integral ISS framework for stability of nonlinear cascades. *SIAM J. Control and Optimization*, **40**, 1888–1904.

Armstrong-Helouvry, B. 1991. *Control of Machines with Friction*. Dordrecht: Kluwer.
1993. Stick slip and control in low-speed motion. *IEEE Trans. Automatic Control*, **38**(10), 1483–1496.

Arnold, V. I. 1978. *Mathematical Methods in Classical Mechanics*. Berlin: Springer-Verlag.
1990. *Teoriya katastrof*. Moscow: Nauka.

Aseltine, J. A., Mancini, A. R., and Sarture, C. W. 1958. A survey of adaptive control systems. *IRE Trans. Automatic Control*, **6**(12), 102–108.

Ashwin, P., and Timme, M. 2005. When instability makes sense. *Nature*, **436**(7), 36–37.

Astolfi, A., and Ortega, R. 2003. Immension and invariance: a new tool for stabilization and adaptive control of nonlinear systems. *IEEE Trans. Automatic Control*, **48**(4), 590–605.

Astrom, K. J., and Wittenmark, B. 1961. *Adaptive Control*, 2nd edn. Reading, MA: Addison-Wesley.

Attneave, F. 1971. Multistability in perception. *Sci. Am.*, **225**(6), 63–71.

Baccus, S. A., and Meister, M. 2002. Fast and slow contrast adaptation in retinal circuitry. *Neuron*, **36**, 909–919.

Bachmayer, R., Whitcomb, L. L., and Grosenbaugh, M. A. 2000. An accurate four-quadrant nonlinear dynamical model for marine thrusters: theory and experimental validation. *IEEE J. Oceanic Engineering*, **25**(1), 146–159.

Back, A., and Chen, T. 2002. Universal approximation of multiple nonlinear operators by neural networks. *Neural Computation*, **14**, 2561–2566.

Bak, P., and Pakzusci, M. 1995. Complexity, contingency and criticality. *Proc. National Acad. Sci.*, **92**, 6689–6696.

Bak, P., and Sneppen, K. 1993. Punctuated equilibrium and criticality in a simple model of evolution. *Phys. Rev. Lett.*, **71**(24), 4083–4086.

Banham, M. R., and Katsaggelos, A. K. 1997. Digital image restoration. *IEEE Signal Processing Mag.*, **14**(2), 24–41.

Bastin, G., and Dochain, D. 1990. *On-line Estimation and Adaptive Control of Bioreactors*. Amsterdam: Elsevier.

Bastin, G., and Gevers, M. 1988. Stable adaptive observers for nonlinear time-varying systems. *IEEE Trans. Automatic Control*, **33**(7), 650–658.

Bastin, G., Bitmead, R. R., Campion, G., and Gevers, M. 1992. Identification of linearly overparametrized nonlinear systems. *IEEE Trans. Automatic Control*, **37**(7), 1073–1078.

Bellman, R. 1961. *Adaptive Control Processes – A Guided Tour*. Princeton, NJ: Princeton University Press.

1970. *Introduction to Matrix Analysis*. New York: McGraw-Hill Book Co., Inc.

Bellman, R. and Kalaba, R. 1960. Dynamic programming and adaptive control processes: mathematical foundations. *IRE Trans. Automatic Control*, **5**, 5–10.

1965. *Dynamic Programming and Modern Control Theory*. New York: Academic Press.

Bernstein, D. S. 2005. *Matrix Mathematics*. Princeton, NJ: Princeton University Press.

Bhatia, N. P., and Szego, G. P. 1970. *Stability Theory of Dynamical Systems*. Berlin: Springer-Verlag.

Birkhoff, G. D. 1927. *Dynamical Systems*. Providence, RI: American Mathematical Society Colloquium Publications.

Bischi, G.-I., Stefanini, L., and Gardini, L. 1998. Synchronization, intermittency and critical curves in a duopoly game. *Math. and Computers in Simulation*, **44**, 559–585.

Blake, A., Curwen, R., and Zisserman, A. 1994. A framework for spatiotemporal control in the tracking of visual contours. *Int. J. Computer Vision*, **11**(2), 127–145.

Boskovic, J. D. 1995. Stable adaptive control of a class of first-order nonlinearly parameterized plants. *IEEE Trans. Automatic Control*, **40**(2), 347–350.

Brambilla, M., Lugiato, L. A., Penna, V. *et al.* 1991. Transverse laser patterns. II. Variational principle for pattern selection, spatial multistability, and laser hydrodynamics. *Phys. Rev. A*, **43**(9), 5114–5120.

Brewer, D., Barenco, M., Callard, R., Hubank, M., and Stark, J. 2008. Fitting ordinary differential equations to short time course data. *Phil. Trans. Roy. Soc. A*, **366**(1865), 519–544.

Bueso, M. C., Angulo, M. C., Quian, G., and Alonso, F. J. 1999. Spatial sampling design based on stochastic complexity. *J. Multivariate Analysis*, **71**(1), 94–110.

Byrnes, C. I., and Isidori, A. 2003. Limit sets, zero dynamics, and internal models in the problem of nonlinear output regulation. *IEEE Trans. Automatic. Control*, **48**(10), 1712–1723.

Camalet, S., Duke, T., Julicher, F., and Prost, J. 2000. Auditory sensitivity provided by self-tuned critical oscillations of hair cells. *Proc. National Acad. Sci.*, **97**(7), 3183–3188.

Canudas de Wit, C., and Tsiotras, P. 1999. Dynamic tire models for vehicle traction control, in *Proceedings of the 38th IEEE Control and Decision Conference*.

Cao, C., Annaswamy, A. M., and Kojic, A. 2003. Parameter convergence in nonlinearly parametrized systems. *IEEE Trans. Automatic Control*, **48**(3), 397–411.

Carr, J. 1981. *Applications of the Center Manifold Theory*. Berlin: Springer-Verlag.

Chen, T., and Amari, S. 2001. Stability of asymmetric Hopfield networks. *IEEE Trans. Neural Networks*, **12**(1), 159–163.

Chen, T., and Chen, H. 1995. Universal approximation to nonlinear operators by neural networks with arbitrary activation functions and its application to dynamical systems. *IEEE Trans. Neural Networks*, **6**(4), 911–917.

Chizhevsky, V. N. 2000. Coexisting attractors in a CO_2 laser with modulated losses. *J. Opt. B: Quantum Semiclass. Opt.*, **2**, 711–717.

2001. Multistability in dynamical systems induced by weak periodic perturbations. *Phys. Rev. E*, **64**(3), 036223–036226.

Chizhevsky, V. N., and Corbalan, R. 2002. Phase scaling properties of perturbation-induced multistability in a driven nonlinear system. *Phys. Rev. E*, **66**(1), 016201–016205.

Cohen, M. A., and Grossberg, S. 1983. Absolute stability and global pattern formation and parallel memory storage by competitive neural networks. *IEEE Trans. Systems, Man and Cybernetics*, **13**, 815–826.

Conant, R. C., and Ashby, W. R. 1970. Every good regulator of a system must be a model of that system. *Int. J. Systems Sci.*, **1**, 89–98.

Costic, B. T., de Queiroz, M. S., and Dawson, D. M. 2000. *A new learning control approach to the active magnetic bearing benchmark system*, in *Proceedings of the 2000 American Control Conference*, vol. 4, pp. 2639–2643.

Cybenko, G. 1989. Approximation by superpositions of a sigmoidal function. *Math. Control, Signals and Systems*, **2**, 303–314.

Demontis, G. C., and Cervetto, L. 2002. Vision: how to catch fast signal with slow detectors. *News Physiol. Sci.*, **17**, 110–114.

Ding, Z. 2001. Adaptive control of triangular systems with nonlinear parameterization. *IEEE Trans. Automatic Control*, **46**(12), 1963–1968.

Ditchburn, R. W., and Ginzburg B. L. 1952. Vision with a stabilized retinal image. *Nature*, **170**(4314), 36–37.

Drakunov, S., Ozguner, U., Dix, P., and Ashrafi, B. 1995. ABS control using optimum search via sliding modes. *IEEE Trans. Control Systems Technol.*, **3**(1), 79–85.

Draper, C. S., and Lee, I. T. 1951. *Principles of Optimalizing Control Systems and an Application to the Internal Combustion Engine*. New York: The American Society of Mechanical Engineers.

1960. *Automatic Optimization of Controlled Systems* (Russian translation of Draper and Lee (1951)). Moscow: IL.

Emelyanov, S. V. 1967. *Variable Structure Control Systems*. Moscow: Nauka.

Enroth-Cugell, C., Robson, J. G., Schweitzer-Tong, D. E., and Watson, A. B. 1983. Spatio-temporal interactions in cat retinal ganglion cells showing linear spatial summation. *J. Physiol. (Lond.)*, **341**, 279–307.

Evleigh, V. W. 1967. *Adaptive Control and Optimization Techniques*. New York: McGraw-Hill Book Company.

Eykhoff, P. 1975. *System Identification. Parameter and State Estimation*. Eindhoven: University of Technology, Eindhoven.

Fairhurst, D., Tyukin, I., Nijmeijer, H., and van Leeuwen, C. 2010. Observers for canonic models of neural oscillators. *Math. Modelling of Natural Phenomena*, **5**(2), 146–184.

Farza, M., M'Saad, M., Maatoung, T., and Kamoun, M. 2009. Adaptive observers for nonlinearly parameterized class of nonlinear systems. *Automatica*, **45**, 2292–2299.

Feldbaum, A. A. 1959. *Vychislitel'nye ustroistva v avtomaticheskikh systemakh*. Moscow: Fizmatgiz.

1965. Problems of self-tuning (adaptive) systems, in *Proceedings of the 1st Conference on Self-tuning Systems*. Moscow: Nauka.

Feldkamp, L., and Puskorius, G. 1997. Fixed-weight controller for multiple systems. *Proceedings of the IEEE International Joint Conference on Neural Networks*, pp. 2268–2272.

Feldkamp, L., Puskorius, G., and Moore, P. 1996. Adaptation from fixed weight dynamic networks, in *Proceedings of the IEEE International Conference on Neural Networks*.

FitzHugh, R. 1961. Impulses and physiological states in theoretical models of nerve membrane. *Biophys. J.*, **1**, 445–466.

Florentin, J. J. 1962. Optimal, probing, adaptive control of a simple Bayesian system. *Int. J. Electronics*, **13**(2), 165–177.

Fomin, V. N., Fradkov, A. L., and Yakubovich, V. A. 1981. *Adaptivnoe upravlenie dinamicheskimi ob'ektami*. Moscow: Nauka.

Fradkov, A. L. 1979. Speed-gradient scheme and its applications in adaptive control. *Automation and Remote Control*, **40**(9), 1333–1342.

1986. Integro-differentiating velocity gradient algorithms. *Sov. Phys. Dokl.*, **31**(2), 97–98.

1990. *Adaptivnoe upravlenie v slozhnykh systemakh: bespoiskovye metody*. Moscow: Nauka.

2005. O primenenii kiberneticheskikh metodov v fizsike. *Usp. Fiz. Nauk*, **175**(2), 113–138.

Fradkov, A. L., Miroshnik, I. V., and Nikiforov, V. O. 1999. *Nonlinear and Adaptive Control of Complex Systems*. Dordrecht: Kluwer.

French, M. 2002. An analytical comparison between the nonsingular quadratic performance of robust and adaptive backstepping designs. *IEEE Trans. Automatic Control*, **47**(4), 670–675.

French, M., Szepesvari, Cs., and Rogers, E. 2000. Uncertainty, performance, and model dependency in approximate adaptive nonlinear control. *IEEE Trans. Automatic Control*, **45**(2), 353–358.

Fu, K. S. 1969. A class of self-tuning systems of automatic regulation based on statistical decision theory, in *Proceedings of the 2nd IFAC Symposium on Self-tuning Systems*. Moscow: Nauka, pp. 7–15.

Fuchs, A., and Haken, H. 1988. Pattern recognition and associative memory as dynamical processes in a synergetic system (I and II). *Biol. Cybernetics*, **60**(1), 17–22 and **60**(2), 107–109.

Gabor, D. 1946. Theory of communication. *J. Inst. Electrical Engineers*, **93**, 429–441.

Garner, W. R. 1962. *Uncertainty and Structure as Psychological Concepts*. New York: Wiley.

Gavel, D. T., and Siljak, D. D. 1989. Decentralized adaptive control: structural conditions for stability. *IEEE Trans. Automatic Control*, **34**(4), 413–426.

Gelfi, S., Stefanopoulou, A. G., Pukrushpan, J. T., and Peng, H. 2003. Dynamics of low-pressure and high-pressure fuel cell air supply systems, in *Proceedings of the 2003 American Control Conference*, vol. 3, pp. 2049–2054.

Ghosh, J., Mukherjee, D., Baloh, M., and Paden, B. 2000. Nonlinear control of a benchmark beam balance experiment using variable hyperbolic boas, in *Proceedings of the 2000 American Control Conference*, pp. 2149–2153.

Gibson, G. I. 1961. Self-optimizing or self-tuning systems of automatic regulation, in *Proceedings of the 1st World IFAC Congress*, vol. 2. Moscow: Akademii Nauk SSSR.

Gilden, D. L. 2001. Cognitive emissions of $1/f$ noise. *Psychol. Rev.*, **108**, 33–56.

Gorban, A. N. 2004. Singularities of transition processes in dynamical systems: qualitative theory of critical delays. http://ejde.math.txstate.edu/Monographs/05/.

2007. Selection theorem for systems with inheritance. *Math. Modelling of Natural Phenomena*, **2**(4), 1–45.

Gorban, A. N., and Cheresiz, V. M. 1981. Slow relaxations of dynamical systems and bifurcations of ω-limit sets. *Dokl. Math.* **24**, 645–649.

Gorban, A. N., Smirnova, E. V., and Tyukina, T. A. 2010. Correlations, risk and crisis: from physiology to finance. *Physica A*, **389**(16), 3193–3217.

Grip, H. F., Johansen, T. A., Imsland, L., and Kaasa, G. O. 2010. Parameter estimation and compensation in systems with nonlinearly parameterized perturbations. *Automatica*, **46**(1), 19–28.

Grune, L., Sontag, E., and Wirth, F. R. 1999. Asymptotic stability equals exponential stability, and ISS equals finite energy gain – if you twist your eyes. *Systems & Control Lett.*, **38**, 127–134.

Guay, M., Dochain, D., and Perrier, M. 2004. Adaptive extremum seeking control of continuous stirred tank bioreactors with unknown growth kinetics. *Automatica*, **40**(5), 881–888.

Guckenheimer, J., and Holmes, P. 2002. *Nonlinear Oscillations, Dynamical Systems and Bifurcations of Vector Fields*. Berlin: Springer-Verlag.

Gutig, R., and Sompolinsky, H. 2006. The tempotron: a neuron that learns spike timing-based decisions. *Nature Neurosci.*, **9**(3), 420–428.

Hansel, D., and Sompolinsky, H. 1992. Synchronization and computation in a chaotic neural network. *Phys. Rev. Lett.*, **68**, 718–721.

Hatfield, G., and Epstein, W. 1985. The status of the minimum principle in the theoretical analysis of visual perception. *Psychol. Bull.*, **97**(2), 155–186.

Haykin, S. 1999. *Neural Networks: A Comprehensive Foundation*. Englewood Cliffs, NJ: Prentice-Hall.

Herz, A., Suzler, B., Kuhn, R., and van Hemmen, J. L. 1989. Hebbian learning reconsidered: representation of static and dynamic objects in associative neural nets. *Biol. Cybernetics*, **60**, 457–467.

Hindmarsh, J. L., and Rose, R. M. 1984. A model of neuronal bursting using 3 coupled 1st order differential-equations. *Proc. Roy. Soc. Lond. B*, **221**(1222), 87–102.

Hofer, H., and Williams, D. R. 2002. The eye's mechanisms for autocallibration. *Optics and Photonics News*, **13**(1), 34–39.

Hopfield, J. J. 1982. Neural networks and physical systems with emergent collective computational abilities. *Proc. National Acad. Sci.*, **79**, 2554–2558.

Ilchman, A. 1997. Universal adaptive stabilization of nonlinear systems. *Dynamics and Control*, (7), 199–213.

Ioannou, P. A. 1986. Decentralized adaptive control of interconnected systems. *IEEE Trans. Automatic Control*, **31**(4), 291–298.

Ioannou, P. A., and Sun, J. 1996. *Robust Adaptive Control*. Englewood Cliffs, NJ: Prentice-Hall.

Isidori, A. 1989. *Nonlinear Control Systems: An Introduction*, 2nd edn. Berlin: Springer-Verlag.

Ito, J., Nikolaev, A., Luman, M. *et al.* 2003. Perceptual switching, eye-movements, and the bus-paradox. *Perception*, **32**, 681–698.

Ivakhnenko, A. G. 1962. *Tekhnicheskaya kibernetika systemy avtomaticheskogo upravleniya s prisposobleniem kharakteristik,* 2nd edn. Kiev: Gostekhizdat Ukrainy.

Izhikevich, E. 2007. *Dynamical Systems in Neuroscience: The Geometry of Excitability and Bursting*. Cambridge, MA: MIT Press.
Izhikevich, E. M. 2004. Which model to use for cortical spiking neurons? *IEEE Trans. Neural Networks*, **15**, 1063–1070.
Jain, A. K., Duin, R. P. W., and Mao, J. 2000. Statistical pattern recognition: a review. *IEEE Trans. Pattern Analysis and Machine Intelligence*, **22**(1), 4–37.
Jain, S., and Khorrami, F. 1997. Decentralized adaptive control of a class of large-scale interconnected nonlinear systems. *IEEE Trans. Automatic Control*, **42**(2), 136–154.
Jiang, Z.-P. 2000. Decentralized and adaptive nonlinear tracking of large-scale systems via output feedback. *IEEE Trans. Automatic Control*, **45**(11), 2122–2128.
Jiang, Z.-P., Teel, A. R., and Praly, L. 1994. Small-gain theorems for ISS systems and applications. *Math. Control, Signals and Systems*, **7**, 95–120.
Kaneko, K. 1990. Clustering, coding, switching hierarchical ordering and control in a network of chaotic elements. *Physica D*, **41**, 137–172.
1994. Relevance of dynamic clustering to biological networks. *Physica D*, **75**, 137–172.
Kaneko, K., and Tsuda, I. 2000. *Complex Systems: Chaos and Beyond*. Berlin: Springer-Verlag.
2003. Chaotic itinerancy. *Chaos*, **13**(3), 926–936.
Kanellakopoulos, I., Kokotović, P. V., and Morse, A. S. 1991. Systematic design of adaptive controllers for feedback linearizable systems. *IEEE Trans. Automatic Control*, **36**, 1241–1253.
Karsenti, L., Lamnabhi-Lagarrige, F., and Bastin, G. 1996. Adaptive control of nonlinear systems with nonlinear parameterization. *System and Control Letters*, **27**, 87–97.
Kazakevich, V. V. 1946. Sposob avtomaticheskogo regulirovaniya razlichnykh protsessov po maksimumu ili po minimumu. USSR Patent number 66335, November, 25 1943. *Byulluten' izobretenii*.
1958. Systemy ékstremal'nogo regulirovaniya i nekotorye sposoby uluchsheniya ikh kachestva i ustoichivosti, in *Avtomaticheskoe upravlenie i vychislitel'naya tekhnika*. Moscow: Mashgiz, pp. 66–96.
Kazakov, Y. M., and Evlanov, L. G. 1965. O teorii samonastraivayuschikhsya system s poiskom gradienta metodom vspomogatel'nogo operatora, in *Transaktsiya. II-go mezhd. congressa IFAC*, vol. 3. Moscow: Nauka.
Khalil, H. 2002. *Nonlinear Systems*, 3rd edn. Englewood Cliffs, NJ: Prentice-Hall.
Khlebtsevich, Y. S. 1965. Élektricheskii regulyator ékonomichnisti. Patent application 231496, April 4, 1940. USSR Patent number 170566. *Byulluten' izobretenii*.
Kitching, K. J., Cole, D. J., and Cebon, D. 2000. Performance of a semi-active damper for heavy vehicles. *ASME J. Dynamic Systems Measurement and Control*, **122**(3), 498–506.
Koch, C. 2002. *Biophysics of Computation. Information Processing in Single Neurons*. Oxford: Oxford University Press.
Kojic, A., and Annaswamy, A. M. 2002. Adaptive control of nonlinearly parameterized systems with a triangular structure. *Automatica*, **38**(1), 115–123.
Kolesnikov, A. A. 1994. *Synergeticheskaya teoriya upravleniya*. Moscow: Energoatomizdat.
2000. *Osnovy synergeticheskoi teorii upravleniya*. Moscow: ISPO-Service.
Kolmogorov, A. N., and Fomin, S. V. 1976. *Elementy teorii funktsii i funktsional'nogo analiza*. Moscow: Nauka.

Krasovskii, A. A. 1963. *Dynamika nepreryvnykh samonastraivayushikhsya system.* Moscow: Fizmatgiz.

Krasovsky, A. A., Bukov, V. N., and Shendrik, V. S. 1977. *Universal Algorithms of Optimal Control of Continuous processes.* Moscow: Nauka.

Kreisselmeier, G. 1977. Adaptive obsevers with exponential rate of convergence. *IEEE Trans. Automatic Control*, **22**, 2–8.

Krener, A. J., Kang, W., and Chang, D. E. 2004. Control bifurcations. *IEEE Trans. Automatic Control*, **49**(8), 1231–1246.

Krichman, M., Sontag, E., and Wang, Y. 2001. Input–output-to-state-stability. *SIAM J. Control and Optimization*, **39**, 1874–1928.

Krstić, M., and Kokotović, P. 1993. Transient-performance improvement with new class of adaptive controllers. *Systems and Control Lett.*, **21**, 451–461.

1996. Adaptive nonlinear output-feedback schemes with Marino–Tomei controller. *IEEE Trans. Automatic Control*, **41**(2), 274–280.

Krstić, M., and Wang Hsin-Hsiung. 2000. Stability of extremum seeking feedback for general nonlinear dynamic systems. *Automatica*, **36**, 595–601.

Krstić, M., Kanellakopoulos, I., and Kokotović, P. 1992. Adaptive nonlinear control without overparametrization. *Systems and Control Lett.*, **19**, 177–185.

1994. Nonlinear design of adaptive controllers for linear systems. *IEEE Trans. Automatic Control*, **39**(4), 738–752.

1995. *Nonlinear and Adaptive Control Design.* New York: Wiley and Sons Inc.

Kwakernaak, H. 1969. On admissible adaptive control, in *Proceedings of the 2nd IFAC Symposium on Self-tuning Systems.* Moscow: Nauka, pp. 17–22.

La Salle, J. P. 1976. Stability theory and invariance principles, in Cesari, L. Hale, J. K., and La Salle, J. P. (eds.), *Dynamical Systems, An International Symposium*, vol. 1, pp. 211–222.

Lachmann, T., and van Leeuwen, C. 2005. Individual pattern representations are context-independant, but their collective representation is context-dependent. *Q. J. Psychol.*, **58**(7), 1265–1294.

Lakatos, I. 1976. *Proofs and Refutations.* Cambridge: Cambridge University Press.

Lancaster, P., and Tismenetsky, M. 1985. *The Theory of Matrices.* New York: Academic Press.

LaSalle, J. P., and Lefschetz, S. 1961. *Stability by Liapunov's Direct Method with Applications.* New York: Academic Press.

Lawrence, D. A., Pao, L. Y., Dougherty, A. M. *et al.* 1998. Human perceptual thresholds of friction in haptic interfaces, in *Proceedings of the ASME Dynamic Systems and Control Division, ASME International Mechanical Engineering Congress & Expo, Anaheim, CA*, pp. 287–294.

Lee, R. C. 1964. *Optimal Estimation, Identification and Control.* Cambridge, MA: MIT Press.

Lee Tong-Heng and Narendra, K. S. 1988. Robust adaptive control of discrete-time systems using persistent excitation. *Automatica*, **24**(6), 781–788.

Lee, T. S., and Yuille, A. 2006. Efficient coding of visual scenes by grouping and segmentation: theoretical predictions and biological relevance, in Doya, K., Ishii, S., Rao, R., and Pougeti, A. (eds.), *Bayesian Brain, Probabilistic Approaches to Neural Coding.* Cambridge, MA: MIT Press, pp. 145–188.

Leeuwenberg, E. L. J., and Buffart, H. F. J. M. 1983. An outline of coding theory: summary of some related experiments, in Geissler, H. G., Buffart, H. F. J. M., and Leeuwenberg, E. L. J. (eds.), *Modern Issues in Perception.* Amsterdam: North-Holland, pp. 25–47.

Leopold, D. A., and Logothetis, N. K. 1999. Multistable phenomena: changing views in perception. *Trends in Cognitive Sci.*, **3**(7), 254–264.

Li Yao-Tsu, and van der Valde, W. I. 1961. Theory of nonliear self-tuning systems, in *Proceedings of the 1st World IFAC Congress*, vol. 2. Moscow: Akademii Nauk SSSR, pp.726–744.

Lin, W., and Qian, C. 2002a. Adaptive control of nonlinearly parameterized systems: a nonsmooth feedback framework. *IEEE Trans. Automatic Control*, **47**(5), 757–773.

2002b. Adaptive control of nonlinearly parameterized systems: the smooth feedback case. *IEEE Trans. Automatic Control*, **47**(8), 1249–1266.

Ljung, L. 1999. *System Identification: Theory for the User*. Englewood Cliffs, NJ: Prentice-Hall.

2008. Perspectives in system identification, in *Proceedings of the 17th IFAC World Congress on Automatic Control*, pp. 7172–7184.

Lo, J. 2001. Adaptive vs. accommodative neural networks for adaptive system identification, in *Proceedings of the IEEE International Joint Conference on Neural Networks*, pp. 1279–1284.

Loh Ai-Poh, Annaswamy, A. M., and Skantze, F. P. 1999. Adaptation in the presence of general nonlinear parameterization: an error model approach. *IEEE Trans. Automatic Control*, **44**(9), 1634–1652.

Loria, A., and Panteley, E. 2002. Uniform exponential stability of linear time-varying systems: revisited. *Systems and Control Lett.*, **47**(1), 13–24.

Loria, A., Paneteley, E. A., and Nijmeijer, H. 2001. A remark on passivity-based and discontinuous control of uncertain nonlinear systems. *Automatica*, **37**, 1481–1487.

Loria, A., Panteley, E., Popovic, D., and Teel, A. 2003. Persistency of excitation for uniform convergence in nonlinear control systems. http://arxiv.org/abs/math/0301335.

Lu, W., and Chen, T. 2003. New conditions on global stability of cohen–grossberg neural networks. *Neural Computation*, **15**, 1173–1189.

Lyapunov, A. M. 1892. *The General Problem of the Stability of Motion* [in Russian]. Kharkov: Kharkov Mathematical Society. Republished by the University of Toulouse, 1908 and Princeton University Press, 1949 (in French). Republished in English by *International Journal of Control*, 1992.

Makarova, I. M. (ed.) 2002. Novoe v synergetike: vzglyad v tret'e tysyacheletie, in, *Informatika: neogranichennye vozmozhnosti i vozmozhnye ogranicheniya*. Moscow: Nauka.

Malinetskii, G. G. (ed.) 2006. Novoe v synergetike. Novaya real'nost', novye problemy, novoe pokolonie, in *Fraktaly, kaos, veroyatnost'*. Moscow: Radiotekhnika.

Margolis, M., and Leondes, K. T. 1961. On the theory of self-tuning regulation: the method of self-tuning model, *Proceedings of the 1st World IFAC Congress*, vol. 2. Moscow: Akademii Nauk SSSR, pp. 683–698.

Marino, R. 1990. Adaptive observers for single output nonlinear systems. *IEEE Trans. Automatic Control*, **35**(9), 1054–1058.

Marino, R., and Tomei, P. 1992. Global adaptive observers for nonlinear systems via filtered transformations. *IEEE Trans. Automatic Control*, **37**(8), 1239–1245.

1993. Global adaptive output-feedback control of nonlinear systems, part I: linear parameterization. *IEEE Trans. Automatic Control*, **38**(1), 17–32.

1995a. Adaptive observers with arbitrary exponential rate of convergence for nonlinear systems. *IEEE Trans. Automatic Control*, **40**(7), 1300–1304.

1995b. *Nonlinear Control Design*. Englewood Cliffs; NJ: Prentice-Hall.

Martensson, B. 1985. The order of any stabilizing regulator is sufficient a priori information for adaptive stabilization. *Systems and Control Lett.*, **6**(2), 87–91.

Martensson, B., and Polderman, J. W. 1993. Correction and simplification to "The order of any stabilizing regulator is sufficient a priori information for adaptive stabilization." *Systems and Control Lett.*, **20**(6), 465–470.

Martinez-Conde, S., Macknik, S. L., and Hubel, D. H. 2004. The role of fixational eye movements in visual perception. *Nature Rev. Neurosci.*, **5**(3), 229–240.

Mather, G. 2006. *Foundations of perception*. Hove, MA: Psychology Press Ltd.

Meyer, K. R. 1966. On the existence of Lyapunov functions for the problem of Lur'e. *J. SIAM Control, Ser. A*, **3**(3), 373–383.

Michel, A.N., Farrel, J. A., and Porod, W. 1989. Qualitative analysis of neural networks. *IEEE Trans. Circuits and Systems*, **36**, 229–243.

Middleton, R. H., Goodwin, G. C., Hill, D. J., and Mayne, D. Q. 1988. Design issues in adaptive control. *IEEE Trans. Automatic Control*, **33**(1), 50–58.

Miller, M. I., and Younes, L. 2001. Group actions, homeomorphisms, and matching: a general framework. *Int. J. Computer Vision*, **41**(1/2), 61–84.

Milnor, J. 1985. On the concept of attractor. *Commun. Math. Phys.*, **99**, 177–195.

Mishkin, E., and Braun, L. B. Jr. (eds.). 1961. *Adaptive Control Systems*. New York: McGraw-Hill Book Co., Inc.

Mon-Williams, M., Tresilian, J. R., Strang, N. C., Kochhar, P., and Wann, J. 1998. Improving vision: neural compensation for optical defocus. *Proc. Roy. Soc. Lond. B*, **265**(1), 71–77.

Moreau, L., and Sontag, E. 2003. Balancing at the border of instability. *Phys. Rev. E*, **68**, 020901 (1–4).

Moreau, L., Sontag, E., and Arcak, M. 2003. Feedback tuning of bifurcations. *Systems and Control Lett.*, **50**, 229–239.

Morgan, A. P., and Narendra, K. S. 1992. On the stability of nonautonomous differential equations $\dot{\mathbf{x}} = [\mathbf{A} + \mathbf{B}(t)]\mathbf{x}$ with skew symmetric matrix $\mathbf{B}(t)$. *SIAM J. Control and Optimization*, **37**(9), 1343–1354.

Morosanov, I. S. 1957. Method of extremum seeking control. *Automation and Remote Control*, **18**, 1077–1092.

Morris, C., and Lecar, H. 1981. Voltage oscillations in the barnacle giant muscle fiber. *Biophys. J.*, **35**, 193–213.

Morse, A. S. 1995. Control using logic-based switching, in *Trends in Control*. Berlin: Springer-Verlag, pp. 69–113.

Morse, A. S., Mayne, D. Q., and Goodwin, G. C. 1988. Applications of hysteresis switching in parameter adaptive control. *IEEE Trans. Automatic Control*, **37**(9), 1343–1354.

Narendra, K. S., and Annaswamy, A. M. 1989. *Stable Adaptive Systems*. Englewood Cliffs, NJ: Prentice-Hall.

Narendra, K. S., and Balakrishnan, J. 1994. Improving transient response of adaptive control systems using multiple models and switching. *IEEE Trans. Automatic Control*, **39**(9), 1861–1866.

1997. Adaptive control using multiple models. *IEEE Trans. Automatic Control*, **42**(2), 171–187.

Nijmeijer, H., and van der Schaft, A. 1990. *Nonlinear Dynamical Control Systems*. Berlin: Springer-Verlag.

Nikiforov, V. O. 1998. Adaptive nonlinear tracking with complete compensation of unknown disturbances. *European J. Control*, **4**, 132–139.

O'Leary, M. D., Simone, C., Washio, T., Yoshinaka, K., and Okamura, A. M. 2003. Robotic needle insertion: effects of friction and needle geometry, in *Proceedings of the 2003 IEEE International Conference on Robotics and Automation.*

Ortega, R., Astolfi, A., and Barabanov, N. E. 2002. Nonlinear PI control of uncertain systems: an alternative to parameter adaptation. *Systems and Control Lett.*, **47**, 259–278.

Ostrovskii, I. I. 1957. Extremum regulation. *Automation and Remote Control*, **18**, 900–907.

Ott, E., and Sommerer, J. C. 1994. Blowout bifurcations: the occurence of riddled basins. *Phys. Lett. A*, **188**(1), 39–47.

Oud, W., and Tyukin, I.Yu. 2004. Sufficient conditions for synchronization in an ensemble of Hindmarsh and Rose neurons: passivity-based approach, in *Proceedings of the 6th IFAC Symposium on Nonlinear Control Systems.*

Pacejka, H. B., and Bakker, E. 1993. *The Magic Formula Tyre Model.* Supplement to *Vehicle System Dynamics*, vol. 21.

Panteley, E., Ortega, R., and Moya, P. 2002. Overcoming the detectability obstacle in certainty equivalence adaptive control. *Automatica*, **32**, 1125–1132.

Pervozvanskii, A. A. 1960. Continuous extremum control systems in the presence of random noise. *Automation and Remote Control*, **21**, 673–677.

Peterson, K. S., and Stefanopoulou, A. G. 2004. Extremum seeking control for soft landing of an electromechanical valve actuator. *Automatica*, **40**, 1063–1069.

Pogromsky, A. Yu. 1998. Passivity based design synchronizing systems. *Int. J. Bifurcation and Chaos*, **8**(2), 295–319.

Pogromsky, A. Yu., Santoboni, G., and Nijmeijer, H. 2003. An ultimate bound on the trajectories of the Lorenz system and its applications. *Nonlinearity*, **16**(5), 1597–1605.

Polderman, J. W., and Pait, F. M. 2003. Editorial to the Special Issue on Adaptive Control. *Systems and Control Lett.*, **48**, 1–3.

Pomet, J.-B. 1992. Remarks on sufficient information for adaptive nonlinear regulation, in *Proceedings of the 31st IEEE Control and Decision Conference*, pp. 1737–1739.

Prinz, A., Billimoria, C. P., and Marder, E. 2003. Alternative to hand-tuning conductance-based models: construction and analysis of databases of model neurons. *J. Neurophysiol.*, **90**, 3998–4015.

Prokhorov, D. V., Feldkamp, L. A., and Tyukin, I. Yu. 2002a. Adaptive behavior with fixed weights in recurrent neural networks, in *Proceedings of the IEEE International Joint Conference on Neural Networks*, vol. 3, pp. 2018–2022.

Prokhorov, D. V., Terekhov, V. A., and Tyukin, I. Yu. 2002b. On the applicability conditions for the algorithms of adaptive control in nonconvex problems. *Automation and Remote Control*, **63**(2), 262–279.

Pupkov, K. A., Kapalin, V. I., and Yushenko, A. S. 1976. *Functional Series in the Theory of Nonlinear Systems*. Moscow: Nauka.

Putov, V. V. 1993. Metody postroeniya adaptivnykh system upravleniya nelineinymi nestatsionarnymi dynamicheskimi ob'ektami s funktsional'nymo-parametricheskoi neopredelennost'yu, in *Dissertatsiya na soiskanie uchenoi stepeni doktora tekhn. nauk.* St. Petersburg: SPbGETU.

Rabinovic, M. I., Huerta, R., Varona, P., and Aframovich, V. S. 2008. Transient cognitive dynamics, metastability, and decision-making. *PLOS Computational Biol.*, **4**(5), 1–9.

Ritter, H., and Kohonen, T. 1989. Self-organizing semantic maps. *Biol. Cybernetics*, **61**, 241–254.

Rodieck, R. W. 1998. *The First Steps in Seeing*. Sunderland, MA: Sinauer Associates, Inc.

Saridis, G. 1977. *Self-organizing Control of Stochastic Systems*. New York: Marcel-Dekker.
Sastry, S. 1999. *Nonlinear Systems: Analysis, Stability and Control*. Berlin: Springer-Verlag.
Sastry, S., and Bodson, M. 1989. *Adaptive Control: Stability, Convergence, and Robustness*. Englewood Cliffs, NJ: Prentice-Hall.
Shang, Y., and Wah, B. W. 1996. Global optimization for neural network training. *Computer*, **29**(3), 45–54.
Sharpe, L. T., and Stockman, A. 1999. Rod pathways: the importance of seeing nothing. *Trends Neurosci.*, **22**, 497–504.
Shepard, R. 1981. Psychophysical complementarity, in Kubovy, M., and Pomeranz, J. R. (eds), *Perceptual Organization*. Hillsdale, NJ: Erlbaum, pp. 279–341.
Shi, L., and Singh, S. K. 1992. Decentralized adaptive controller design for large-scale systems with higher order interconnections. *IEEE Trans. Automatic Control*, **37**(2), 1106–1118.
Smirnakis, S., Berry, M., Warland, D. K., Biallek, W., and Meister, M. 1997. Adaptation of retinal processing to image contrast and spatial scale. *Nature*, **386**, 69–73.
Smith, M. A., Ghazizadeh, A., and Shadmehr, R. 2006. Interacting adaptive processes with different timescales underline short-term motor learning. *PLOS Biol.*, **4**(6), 1035–1043.
Smith, P. L., and Ratcliff, R. 2004. Psychology and neurology of simple decisions. *Trends Neurosci.*, **27**(3), 161–168.
Sneppen, K., Bak, P., Flyvbjerg, H., and Jensen, M. H. 1995. Evolution as a self-organized critical phenomenon. *Proc. National Acad. Sci.*, **92**, 5209–5213.
Solé, R. V., Manrubia, S. C., Benton, M., Kauffman, S., and Bak, P. 1999. Criticality and scaling in evolutionary ecology. *Trends Ecology & Evolution*, **14**(4), 156–160.
Solodovnikov, A. A. 1965. *Analyticheskie samonastraivayuschiesya systemy avtomatich-eskogo upravleniya*. Moscow: Mashinostroenie.
Sontag, E. 1990. Further facts about input to state stabilization. *IEEE Trans. Automatic Control*, **35**(4), 473–476.
2003. Adaptation and regulation with signal detection implies internal model. *Systems and Control Lett.*, **50**, 119–126.
2004. Some new directions in control theory inspired by systems biology. *Systems Biol.*, **1**(1), 9–18.
Sontag, E., and Wang, Y. 1996. New characterizations of input-to-state stability. *IEEE Trans. Automatic Control*, **41**(9), 1283–1294.
Spooner, J. T., and Passino, K. M. 1996. Adaptive control of a class of decentralized nonlinear systems. *IEEE Trans. Automatic Control*, **41**(2), 280–284.
Sragovich, V. G. 1981. *Teoriya adaptivnykh system*. Moscow: Nauka.
Sternby, J. 1980. Extremum control systems: an area for adaptive control?, in *Preprints of Joint American Control Conference*.
Stigter, J. D., and Keesman, K. J. 2004. Optimal parametric sensitivity control of a fed-batch reactor. *Automatica*, **40**, 1459–1464.
Stotsky, A. 1993. Lyapunov design for convergence rate improvement in adaptive control. *Int. J. Control*, **57**(2), 501–504.
Suemitsu, Y., and Nara, S. 2004. A solution for two-dimensional mazes with use of chaotic dynamics in a recurrent neural network model. *Neural Computation*, **16**, 1943–1957.
Taylor, J. H. 1974. Strictly positive real functions and the Lefschetz–Kalman–Yakubovich Lemma. *IEEE Trans. Circuits and Systems*, **21**(3), 310–311.
Timme, M., Wolf, F., and Geisel, T. 2002. Coexistence of regular and irregular dynamics in complex networks of pulse-coupled oscillators. *Phys. Rev. Lett.* **89**(25), 258701.

Timofeev, A. V. 1988. *Adaptivnye robototekhnicheskie kompleksy*. Leningrad: Mashinostroenie.

Tou, J. T. 1964. *Modern Control Theory*. New York: McGraw-Hill Book Co., Inc.

Truxhall, G. 1965. Self-tuning systems, in *Proceedings of the 2nd World IFAC Congress*. Moscow: Nauka, pp. 240–251.

Tsien, H. S. 1954. *Engineering Cybernetics*. New York: McGraw-Hill, Book Co., Inc.

Tsodyks, M. V., Pawelzik, K., and Markram, H. 1998. Neural networks with dynamic synapses. *Neural Computation*, **10**, 821–835.

Tsuda, I., and Fujii, H. 2007. Chaos reality in the brain. *J. Integrative Neurosci.*, **6**(2), 309–326.

Tsypkin, Y. Z. 1968. *Adaptatsiya i obuchenie v avtomaticheskikh systemakh*. Moscow: Nauka.

1970. *Osnovy teorii obuchayuschikhsya system*. Moscow: Nauka.

Tyukin, I., Steur, E., Nijmeijer, H., Fairfurst, D., Song, I., Semyanov, A., and van Leeuwen, C. 2010. State and parameter estimation for canonic models of neural oscillators. *Int. J. Neural Systems*, **20**(3), 193–207.

Tyukin, I., Steur, E., Nijmeijer, H., and van Leeuwen, C. 2008a. Non-uniform small-gain theorems for systems with unstable invariant sets. *SIAM J. Control and Optimization*, **47**(2), 849–882.

Tyukin, I., Tyukina, T., and van Leeuwen, C. 2007a. Invariant template matching in systems with temporal coding. http://pdl.brain.riken.jp/projects/template_ matching/.

Tyukin, I. Yu. 2003. Algorithms of adaptation in finite form for a class of nonlinear dynamical systems, [in Russian]. *Avtomatika i Telemekhanika*, **2**, 114–140.

Tyukin, I. Yu., Prokhorov, D. V., and Terekhov, V. A. 2003a. Adaptive control with nonconvex parameterization. *IEEE Trans. Automatic Control*, **48**(4), 554–567.

Tyukin, I. Yu., Prokhorov, D. V., and van Leeuwen, C. 2003b. Finite-form realization of the adaptive algorithms, in *Proceedings of 7th European Control Conference (ECC 2003)*.

2004. Adaptive algorithms in finite form for nonconvex parameterized systems with low-triangular structure, in *Proceedings of IFAC Workshop on Adaptation and Learning in Control and Signal Processing (ALCOSP 2004)*.

2007b. Adaptation and parameter estimation in systems with unstable target dynamics and nonlinear parametrization. *IEEE Trans. Automatic. Control*, **52**(9), 1543–1559.

2008b. Adaptive classification of temporal signals in fixed-weights recurrent neural networks: an existence proof. *Neural Computation*, **20**(10), 2564–2596.

Tyukin, I. Yu., and Terekhov, V. A. 2008. *Adaptatsiya v nelineinykh dynamicheskikh systemakh*. Moscow: Editorial URSS.

Tyukin, I. Yu., Tyukina, T. A., and van Leeuwen, C. 2009. Invariant template matching in systems with spatiotemporal coding: a matter for instability. *Neural Networks*, **22**, 425–449.

Tyukin, I. Yu., and van Leeuwen, C. 2005. Adaptation and nonlinear parameterization: nonlinear dynamics perspective, in *Proceedings of the 16th IFAC World Congress*.

Ullman, Sh., Vidal-Naquet, M., and Sali, E. 2002. Visual features of intermediate complexity and their use in classification. *Nature Neurosci.*, **5**(7), 682–687.

Utkin, V. I. 1992. *Sliding Modes in Control and Optimization*. Berlin: Springer-Verlag.

van Geit, W., de Schutter, E., and Achard, P. 2008. Automated neuron model optimization techniques: a review. *Biol. Cybernetics*, **99**, 241–251.

van Leeuwen, C. 1990. Perceptual-learning systems as conservative structures: Is economy an attractor?, *Psychol. Res.*, **52**, 145–152.

2008. Chaos breed autonomy: connectionist design between bias and baby-sitting. *Cognitive Processing*, **9**, 83–92.

van Leeuwen, C., and Raffone, A. 2001. Coupled nonlinear maps as models of perceptual pattern and memory trace dynamics. *Cognitive Processing*, **2**, 67–111.

van Leeuwen, C., Steyvers, M., and Nooter, M. 1997. Stability and intermittentcy in large-scale coupled oscillator models for perceptual segmentation. *J. Math. Psychol.*, **41**, 319–343.

van Leeuwen, C., Verver, S., and Brinkers, M. 2000. Visual illusions, solid/outline-invariance, and non-stationary activity patterns. *Connection Sci.*, **12**, 279–297.

Webster, M. A., Georgeson, M. A., and Webster, S. M. 2002. Neural adjustments to image blur. *Nature neurosci.*, **5**(9), 839–840.

Wexler, A. S., Ding, J., and Binder-Macleod, S. A. 1997. A mathematical model that predicts skeletal muscle force. *IEEE Trans. Biomed. Engineering*, **44**(5), 337–348.

Widrow, B., McCool, J. M., Larimore, M. G., and Jonson, C. R. 1976. Stationary and nonstationary learning characteristics of the LMS adaptive filter. *Proc. IEEE*, **64**(8), 1151–1162.

Wittenmark, B., and Urquhart, A. 1995. Adaptive extremal control, in *Proceedings of 34th IEEE Conference on Decision and Control*, pp. 1639–1644.

Wolfson, S., and Graham, N. 2000. Exploring the dynamics of light adaptation: the effects of varying the flickering background's duration in the probed-sinewave paradigm. *Vision Res.*, **40**(17), 2277–2289.

Yakubovich, V. A. 1968. On the theory of adaptive systems. *Dokl. Akad. Nauk. SSSR*, **182**(3), 518–521.

Yang, H., and Dillon, T. 1994. Exponential stability and oscillation of hopfield networks. *IEEE Trans. Neural Networks*, **5**, 719–729.

Yang, T. 2004. A survey on chaotic secure communication systems. *Int. J. Comp. Cogn.*, **2**, 81–130.

Ye, X., and Huang, J. 2003. Decentralized adaptive output regulation for a class of large-scale nonlinear systems. *IEEE Trans. Automatic Control*, **48**(2), 276–280.

Younger, S., Conwell, P., and Cotter, N. 1999. Fixed-weight on-line learning. *IEEE Trans. Neural Networks*, **10**(2), 272–283.

Zadeh, L. A. 1963. On the definition of adaptivity. *Proc. IEEE*, **51**, 469–470.

Zadeh, L.A., and Desoer, C. 1991. *Linear Systems Theory*. Berlin: Springer–Verlag.

Zames, G. 1966. On the input–output stability of time-varying nonlinear feedback systems. Part I: conditions derived using concepts of loop gain, conicity, and passivity. *IEEE Trans. Automatic Control*, **11**(2), 228–238.

Zubov, V. I. 1964. *Methods of A. M. Lyapunov and Their Applications*. Groningen: P. Noordhoff.

Index